Geochemistry and Mineralogy of Ni-Co Laterite Deposits

Geochemistry and Mineralogy of Ni-Co Laterite Deposits

Editors

Cristina Domènech
Cristina Villanova-de-Benavent

Basel • Beijing • Wuhan • Barcelona • Belgrade • Novi Sad • Cluj • Manchester

Editors

Cristina Domènech
Department of Mineralogy,
Petrology and Applied Geology
Universitat de Barcelona
Barcelona
Spain

Cristina Villanova-de-Benavent
Department of Mineralogy,
Petrology and Applied Geology
Universitat de Barcelona
Barcelona
Spain

Editorial Office
MDPI
St. Alban-Anlage 66
4052 Basel, Switzerland

This is a reprint of articles from the Special Issue published online in the open access journal *Minerals* (ISSN 2075-163X) (available at: www.mdpi.com/journal/minerals/special_issues/GMNCLD).

For citation purposes, cite each article independently as indicated on the article page online and as indicated below:

Lastname, A.A.; Lastname, B.B. Article Title. *Journal Name* **Year**, *Volume Number*, Page Range.

ISBN 978-3-0365-9631-0 (Hbk)
ISBN 978-3-0365-9630-3 (PDF)
doi.org/10.3390/books978-3-0365-9630-3

Contents

About the Editors

Cristina Domènech

Dr Cristina Domènech graduated from her stuides in Geology in 1996 at the Universitat de Barcelona (UB) and received her doctorate in 2001 in Geological Sciences from the Universitat Politècnica de Catalunya (UPC). After working for more than 10 years in a private environmental consultancy, she re-entered the academic world in 2013, and she is now a tenure-track 2 lecturer in the Departament de Mineralogia, Petrologia i Geologia Aplicada and member of the Mineralogia Aplicada i Medi Ambient Research Group (MAiMA) and of the Caribbean Lithosphere (CALOR) Research Group of the Universitat de Barcelona (UB). Her areas of expertise are geochemical thermodynamics and the kinetics of natural waters. She has broad experience in the modelling of the behaviour of the near and far field of nuclear waste repository; in the understanding of the geochemical the behaviour of heavy metals and trace elements, thermodynamics and kinetics of natural waters, and groundwater and soil pollution. Her research is mainly focused on the selection of thermodynamic data to perform quantitative interpretations or reactive transport modelling of field and laboratory geochemical, mineralogical and petrological data of ore deposits in order to test and validate the conceptual models derived from them.

Cristina Villanova-de-Benavent

Dr Cristina Villanova-de-Benavent graduated form her studies in Geology in 2009 at the Universitat de Barcelona (UB), and obtained her PhD in Earth Sciences at the same institution in 2016. After receiving her PhD, she worked as a part-time lecturer and researcher at the Universitat de Barcelona and later became a research fellow at the School of Environment and Technology of the University of Brighton (United Kingdom) from 2018 to 2019. Since 2020, she has been a lecturer in Ore Deposits Geology at the Faculty of Earth Sciences of the Universitat de Barcelona. She is a member of the Caribbean Lithosphere (CALOR) and the Mineral Resources for the Energy Transition (MinResET) research groups. Her research focuses on the geochemistry and mineralogy of various ore deposit types including, but not limited to, Ni-Co-PGE laterites, karst bauxites, ion adsorption REE deposits, and volcanogenic Li deposits. Her mineralogical studies, from the outcrop scale to the nanometre scale, are based on powder X-ray diffraction, micro-Raman spectroscopy, scanning and transmission electron microscopy, electron microprobe and X-ray absorption using synchrotron radiation, among others. She also became a FEEG Gemmologist in 2023.

Preface

Ni-Co laterites, which are regoliths formed after the chemical weathering of ultramafic rocks in tropical–subtropical regimes, have been the subject of study for many decades. This has provided a good picture of their structure, element distribution and mineralogy, especially in Ni-bearing minerals (e.g., garnierites, Ni-serpentines, Ni-smectites, asbolane–lithiophorite, Fe-oxyhydroxides). Interestingly, in the last decade, Ni-laterites experienced a comeback in the literature for many reasons.

In the past few years, Ni-Co laterites have surpassed Ni-sulphides as the main source of Ni, accounting for about 50% of the world's current Ni production and hosting close to 60% of the world's land-based resources. In addition, the improvement of the existing analytical techniques, and the appearance of new ones, have shed light on some of the unknowns, and given new perspectives. In particular, it has been revealed that, besides Ni, these laterite deposits usually contain other elements that are increasingly demanded (the so-called critical metals or high-tech elements). For example, Ni-laterite deposits are worthy targets of Co, Sc, platinum group elements (PGE) and/or rare earth elements (REE).

In this Special Issue, entitled "Geochemistry and Mineralogy of Ni-Co Laterite Deposits", we present an updated overview of Ni-Co laterites.

The Special Issue starts with two general descriptions of previously unknown laterite deposits, a clay-type Ni-laterite deposit from Cuba (Tauler et al.) and a magnesium silicate Ni-laterite deposit from the Philippines (Aquino et al.). This is followed by four papers focusing on the occurrence of Co, Sc and other critical metals in current tropical environments: the Dominican Republic and Venezuela (Domènech et al.) and Brazil (Dybowska et al.); as well as non-tropical environments: the Balkan peninsula (Economou-Eliopoulos et al.), Spain and Chile (González-Jiménez et al.). The following three papers deal with novel techniques or methodologies to improve the characterization of the Ni-laterite profiles, involving the use Al as a Sc proxy (Teitler et al.), water correction of in situ portable XRF analyses (Laperche et al.) and the study of hyperspectral data using quadrant scan (Zaitouny et al.). The next paper studies the silicification usually observed, yet understudied, in some Ni-laterite profiles using oxygen and silicon isotopes (Cathelineau et al.). And finally, the last contribution puts an eye on the presence of unusual minerals (diamond, SiC) in Ni-laterite breccias (El Mendili et al).

We, the Guest Editors, would like to thank the authors of all articles as well as the organizations that have financially supported their research. We especially acknowledge the editor-in-chief and the editorial board of *Minerals*, and all the reviewers for their time and dedication to this Special Issue.

Cristina Domènech and Cristina Villanova-de-Benavent

Editors

Article

Geochemistry and Mineralogy of the Clay-Type Ni-Laterite Deposit of San Felipe (Camagüey, Cuba)

Esperança Tauler [1,*], Salvador Galí [1], Cristina Villanova-de-Benavent [1], Alfonso Chang-Rodríguez [2], Kenya Núñez-Cambra [3], Giorgi Khazaradze [4] and Joaquín Antonio Proenza [1]

[1] Departament de Mineralogia, Petrologia i Geologia Aplicada, Facultat de Ciències de la Terra, Universitat de Barcelona (UB), Martí i Franquès s/n, 08028 Barcelona, Spain; gali@ub.edu (S.G.); cvillanovadb@ub.edu (C.V.-d.-B.); japroenza@ub.edu (J.A.P.)

[2] Departamento de Geología, Universidad de Moa, Las Coloradas, s/n, Moa 83330, Cuba; alfonso.chang@reduc.edu.cu

[3] Instituto de Geología y Paleontología de Cuba, Vía Blanca, Línea del Ferrocarril s/n, San Miguel del Padrón, Habana 10200, Cuba; kenya@igp.minem.cu

[4] Departament de Dinàmica de la Terra i de l'Oceà, Facultat de Ciències de la Terra, Universitat de Barcelona (UB), Martí i Franquès s/n, 08028 Barcelona, Spain; gkhazar@ub.edu

* Correspondence: esperancatauler@ub.edu

Abstract: The Ni-laterite deposit at the San Felipe plateau, located 30 km northwest of Camagüey, in central Cuba, is the best example of a clay-type deposit in the Caribbean region. San Felipe resulted from the weathering of mantle peridotites of the Cretaceous Camagüey ophiolites. In this study, a geochemical and mineralogical characterization of two profiles (83 and 84) from the San Felipe deposit has been performed by XRF, ICP-MS, quantitative XRPD, oriented aggregate mount XRD, SEM, FE-SEM, and EMPA. Core 83, with a length of 23 m and drilled in the central part of the plateau, presents a notable concentration of cryptocrystalline quartz fragments and a rather poor content of NiO, averaging 0.87 wt.%. Core 84, which is 12 m long and drilled at the border of the plateau, lacks silica fragments and presents a higher NiO content, averaging 1.79 wt.%. The smectite structural formulae reveal that they evolve from trioctahedral to dioctahedral towards the top of the laterite profiles. Quantitative XRD analyses indicate that smectite is a dominant Ni-bearing phase, accompanied by serpentine and minor chlorite. Serpentine, as smectite, is enriched in the less soluble elements Fe^{3+}, Al, and Ni towards the top of the profiles. Core 83 seems to have been affected by collapses and replenishments, whereas core 84 may have remained undisturbed.

Keywords: weathering; laterite; nickel; smectite; serpentine; clay-type; Cuba

Citation: Tauler, E.; Galí, S.; Villanova-de-Benavent, C.; Chang-Rodríguez, A.; Núñez-Cambra, K.; Khazaradze, G.; Proenza, J.A. Geochemistry and Mineralogy of the Clay-Type Ni-Laterite Deposit of San Felipe (Camagüey, Cuba). *Minerals* **2023**, *13*, 1281. https://doi.org/10.3390/min13101281

Academic Editor: Maria Economou-Eliopoulos

Received: 1 August 2023
Revised: 19 September 2023
Accepted: 21 September 2023
Published: 29 September 2023

1. Introduction

Nickel laterites are regolith materials derived from the chemical weathering of ultramafic rocks that contain economically exploitable reserves of Ni, as well as Co (e.g., [1] and references therein) and Sc [2–5].

Ni-laterites represent 70% of the world's Ni reserves and 60% of the world's current production [6,7]. These deposits develop in humid climates during periods of low tectonic activity and in areas of moderate relief [1]. There are many factors that may control the specific structure of each lateritic profile [8–13]. However, the general structure of the profile is determined by the mobility of the different elements that are released by the hydrolysis of primary (and secondary) minerals. This structure consists broadly of a silicate-bearing saprolite horizon near the ultramafic protolith, topped by an oxide zone, separated by the so-called magnesium discontinuity [14]. There are different classifications of nickel laterites depending on the morphology, lithology of the bedrock, degree of weathering, etc. [8,15]; although the most used one establishes three types of deposits according to the mineralogy of the dominant ore [16]. The main hosts of Ni are iron oxyhydroxides, concentrated in the oxide zone, in the oxide-type Ni-laterites; Ni-serpentine and garnierite found in the saprolite horizon of hydrous Mg silicate-type Ni-laterites; and smectites in clay-type Ni-laterites. A particularity of clay-type Ni-laterite deposits is that the Ni-bearing smectites are concentrated in a transition zone between the saprolite horizon and the oxide zone [16].

Detailed mineralogical studies have been carried out in numerous hydrous Mg silicate-type Ni-laterites [9,17–20]. Oxide-type Ni-laterites [2,21] and clay-type Ni-laterites (e.g., [10,22–24]) are far less common worldwide. Moreover, other deposits can be described as a mixture of clay-type and hydrous Mg silicate-type [25].

The San Felipe lateritic nickel deposit (Cuba) represents the best example of a clay-type Ni-laterite deposit in the Caribbean region, with current resources of 250 Mt at 1.43% Ni, 0.05% Co [26]. Unfortunately, limited data on the San Felipe deposit exist and are mostly in Spanish [27–31]. Hence, in this work, the geochemistry and mineralogy of two Ni-laterite profiles of the San Felipe deposit have been revisited, placing special emphasis on the abundance, textures, and mineral chemistry of Ni-bearing phyllosilicates.

2. Geological Setting

The San Felipe deposit is located on the San Felipe plateau (in Spanish, *Meseta de San Felipe*), an erosional remnant near Camagüey city, with a total extension of 50 km^2, a slight inclination from SSE to NNW, and with elevations between 140 and 190 masl. The San Felipe deposit is located at the north of the province of Camagüey, 30 km to the NW of the city of Camagüey, and 70 km SW of the city of Nuevitas (Figure 1a,b) [27–31]. The current nickel resource of the San Felipe deposit is 250 Mt at 1.43% Ni and 0.05% Co [27].

The San Felipe weathering profile is developed on the serpentinized peridotite of the Camagüey ophiolitic massif (COM). This massif is part of the so-called Northern Cuban Ophiolitic Belt [32] and has been dated as early Cretaceous [33]. These ultramafic bodies represent slices of oceanic lithosphere obducted onto the North American continental margin in the Latest Cretaceous to Late Eocene time during the collision between the volcanic arc of the Caribbean with Jurassic–Cretaceous passive margins (Figure 1c) [32,34].

Camagüey peridotites are essentially made up of mantle tectonites with subordinate gabbro sills and dykes. Mantle peridotites mainly consist of clinopyroxene-rich harzburgite, harzburgite, and minor dunite. Uplift exposed the peridotite bodies to laterization at the beginning of the Miocene [35].

In the field, the material outcropping at the top of the San Felipe plateau displays dark red and yellowish-ocher colors (Figure 2a–d), with discrete centimeter- to meter-sized ferricrete and silica blocks (Figure 2e,f). In general, the weathering profile in San Felipe includes the following, from the bottom to the top: (i) ophiolitic peridotite parent rock; (ii) saprolite; (iii) clay zone (smectite > serpentine); (iv) ferruginous clay zone (Fe oxyhydroxides > smectite); (v) oxide zone (Figure 2b); (vi) ferricrete (Figure 2c,d) with free silica concretions [28]. The latter is a characteristic feature of the San Felipe laterite profile (Figure 2e,f) [30,31].

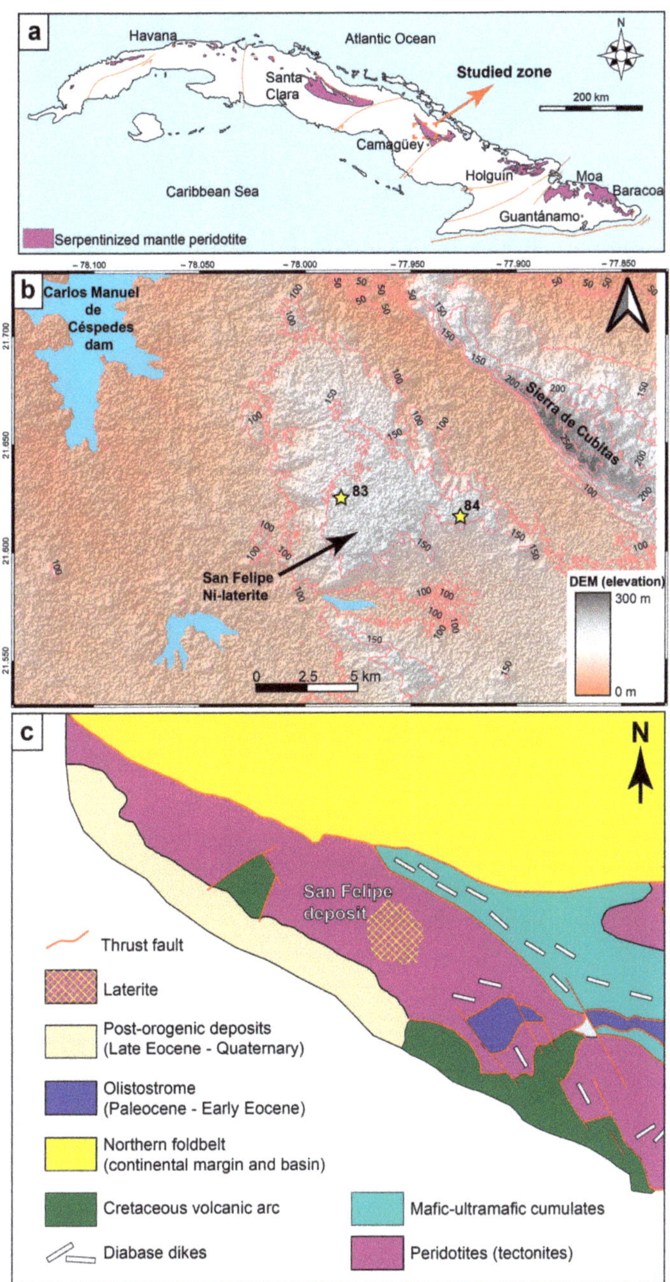

Figure 1. (**a**) Location of the Camagüey ophiolite Massif in central Cuba. (**b**) Geographic location of cores 83 and 84 (yellow stars) in the San Felipe Ni-Co deposit in Camagüey Province, Cuba. Hillshaded DEM is based on the SRTM 30 m model [36]. Contour lines represent elevations with 50 m intervals. Physiographic features are based on the naturalearthdata.com database [37]. Coordinate Reference System: EPSG:4326-WGS 84. (**c**) Simplified geological map of the study area indicating the San Felipe deposit [31].

Figure 2. Photos of the San Felipe Ni-laterite plateau. (**a**) General view at the top of the plateau, the material outcropping is of dark red color and is characterized by sparse, low vegetation. (**b**) Ocher to dark brown goethite and hematite concretion. (**c**,**d**) Centimeter- to meter-sized red blocks. (**e**) Amorphous silica or microcrystalline quartz decimeter-sized nodules, their internal translucent to opaque white appearance is shown in (**f**).

The San Felipe plateau was explored by the Asociación Económica Internacional Geominera S.A. and San Felipe Mining Ltd., in grids ranging from 1000 × 1000 m to 12.5 × 12.5 m in areas of special interest. The present study is based on two representative cores (named 83 and 84, Figure 1b). Since cores 83 and 84 represent a continuous sampling of the corresponding laterite profiles, they are going to be referred to as cores and profiles in this work equally. The textural characteristics of the hand samples are illustrated in Figure 3, and the distribution and thickness of the different units identified along the profiles are displayed in Figure 4. Core 83 was drilled in the central part of the plateau (Figure 1b), is 24 m in length, and can be divided into three horizons, from top to bottom [38]: (i) reddish oxide zone (1 m thick); (ii) reddish brown ferruginous saprolite (14 m); (iii) dark, greenish-brown saprolite (9 m). Core 84 was drilled at the border of the plateau (Figure 1b), is 12 m in length, and consists of four horizons: (i) oxide zone (1 m); (ii) ferruginous saprolite (1 m); (iii) saprolite (8 m); and (iv) green, slightly weathered peridotite (2 m). The bedrock consists of a serpentinized clinopyroxene-rich peridotite, locally intruded by gabbros, but it was not sampled in either of the cores. As observed in Figure 3a,b, samples from higher levels are more porous and crumblier and contain goethitized nodules. Downward in the profile, samples become more compact (Figure 3a–c), and fractures with a green infilling are observed (Figure 3d,e). The ferruginous saprolite in profile 83 (Figure 3f) presents fragments of cryptocrystalline quartz aggregates. In core 84, samples from the saprolite at a depth of 7 m display fractured and porous fragments of the parent rock, surrounded and filled by brown-black Fe oxyhydroxide grains (Figure 3g). At the bottom, the slight weathering on serpentinized peridotite (sample 84-12) provides a dark green to the rock (Figure 3h).

Figure 3. (**a**) Photos of the samples from core 83 (sample depth in yellow). (**b**) Photos of samples from core 84 (sample depth in yellow). (**c–h**) Photos of hand specimens embedded in epoxy resin of the two studied cores: (**c**) sample 83-01 from the oxide zone; (**d**) sample 83-04 from the ferruginous saprolite; (**e**) sample 83-10 from the ferruginous saprolite, including green infillings in fractures; (**f**) sample 83-15 with cryptocrystalline quartz, from the ferruginous saprolite; (**g**) sample 84-07 from the saprolite; (**h**) sample 84-12 from the slightly weathered serpentinized peridotite. The scale for the six images is the same as in (**a**). The sample names are composed of the code of the core (83- or 84-) and a number that indicates the depth of the sample (e.g., "83-01" refers to the sample collected at 1 m from core 83).

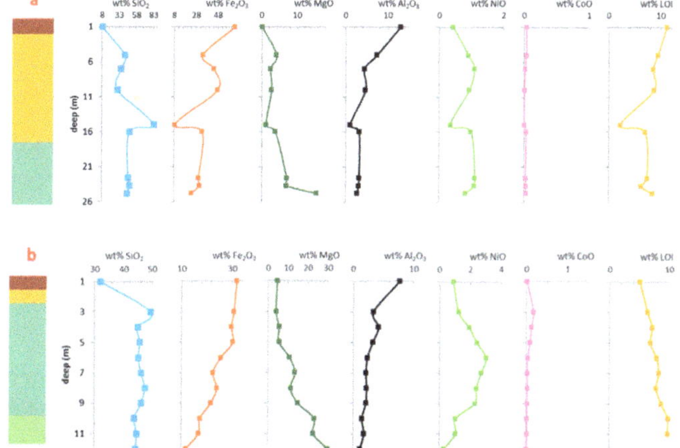

Figure 4. Simplified laterite profiles of the studied cores of San Felipe (red: oxide zone; yellow: ferruginous saprolite; light green: saprolite; olive green: serpentinized peridotite) and geochemical variation in wt.% of the bulk samples with depth: (**a**) core 83; (**b**) core 84.

3. Materials and Methods

3.1. Sampling

A total of 21 samples were selected, 9 from core 83 (top of the plateau) and 12 from core 84 (edge of the plateau). Samples 84-1 and 83-1 are from the oxide zone. Samples 83-5, 83-7, 83-10, 83-15, 83-16, and 84-3 are from ferruginous saprolite. Samples 83-23, 83-24, 83-25, 84-5, 84-6, 84-7, 84-8, 84-9, and 84-10 are from saprolite. Sample 84-11 and 84-12 are from serpentinized peridotite (Figure 3a,b)

3.2. Analytical Methods

Bulk-rock geochemical analyses on all samples (9 samples from core 83 and 12 samples from core 84) were performed at the Centro de Instrumentación Científica of the Universidad de Granada (CIC-UGR, Granada, Spain). Major element concentrations were obtained by X-ray fluorescence (XRF) spectrometry on glass beads using a Philips PV1404 spectrometer. Precision was better than $\pm 1.5\%$ for a concentration of 10 wt.%. Further details on the analytical technique can be found in [39].

Random powder X-ray diffraction (XRPD) and oriented aggregate mount X-ray diffraction (XRD, (air dry, ethylene glycol, and heated to 550 °C) analyses on all samples were obtained at Centres Científics i Tecnològics of the Universitat de Barcelona (CCiT-UB, Barcelona, Spain), in a PANalytical X'Pert PRO MPD Alpha1 powder diffractometer in Bragg–Brentano $\theta/2\theta$ geometry of 240 mm of radius, nickel filtered Cu Kα1 radiation ($\lambda = 1.5406$ Å), 45 kV, and 40 mA. The samples were scanned from 4 to 100° (2θ) with a step size of 0.017° and measuring time of 150 s per step, using an X'Celerator detector (active length = 2.122°). Mineral identification was facilitated by X'Pert Highscore search-match software using the powder diffraction database of the International Centre for Diffraction Data (ICDD). For the estimation of the relative amounts between identified mineral phases, the quantitative Rietveld analysis method [40] with the software Topas V4.2 [41] was used.

All the samples were embedded in epoxy resin, prepared as thin sections, and polished using non-aqueous fluids to prevent damage to mineral phases. The mineralogy and microtextural features of the thin sections were first examined with a Nikon Eclipse LV100 POL optical microscope (OM). The sections were later carbon coated and studied with an environmental scanning electron microscope (ESEM) Quanta 200 FEI, XTE 325/D8395, coupled with an energy dispersive X-ray spectrometer (EDS), and a JEOL JSM-7001F field emission SEM (FE-SEM), both equipped with secondary (SE) and backscattered electron (BSE) detectors, under 20 kV, at the CCiT-UB.

Quantitative electron microprobe analyses (EMPA) were also conducted on the carbon-coated polished thin sections (of 4 samples from core 83 and 5 samples from core 84) at the CCiT-UB using a five-channel JEOL JXA-8230 operating in wavelength dispersive spectroscopy (WDS) mode. The analytical conditions were 15–20 kV accelerating voltage, 10–20 nA beam current, 1–2 µm beam diameter, and 10–20 s counting time per element; XPP matrix correction was used [42]. The measurements and the calibrations were made using the following natural and synthetic standards: diopside (Si; TAP), Al_2O_3 (Al; TAP), wollastonite (Ca; PET), orthoclase (K; PET), periclase (Mg; TAP), albite (Na; TAP), Fe_2O_3 (Fe; LIF), CoO (Co; LIF), NiO (Ni; LIF), rhodonite (Mn; LIF), Cr_2O_3 (Cr; LIF), V (V; LIF), and rutile (Ti; LIF). Iron in the weathered minerals was assumed to be Fe^{3+} [10,43–45].

4. Results

4.1. Major Elements Geochemistry

The major element compositions of both profiles and loss on ignition (LOI) data are given in Supplementary Material Table S1 and in Figure 4a,b as a function of depth.

The composition in terms of SiO_2 and Al_2O_3 is similar in both profiles, with a decrease in SiO_2 and MgO, and an increase in Al_2O_3 and Fe_2O_3 to the top. The SiO_2 content in profile 83 is 45.4 wt.% at the base and 8.8 wt.% at the oxide horizon, with an anomalous abrupt increase to 84.2 wt.% at 15 m depth that is not observed in profile 84, where SiO_2 ranges from 44.5 to 32.2 wt.%. Aluminum ranges from 2.7 at the saprolite horizon to

13.0 wt.% Al_2O_3 in profile 83 and 1.2 to 7.7 wt.% in profile 84. Iron is more abundant in profile 83 and varies from 33.6 wt.% to 63.3 wt.% Fe_2O_3, reaching maximum values at the oxide zone and the ferruginous saprolite, with remarkable contents in the saprolite and an abrupt decrease at 15 m depth (8.9 wt.%). The curves for SiO_2 and Fe_2O_3 in profile 83 are almost symmetrical, indicating a strong negative correlation between them. In profile 84, Fe_2O_3 increases continuously between 11.9 wt.% and 32 wt.% from the base to the top. Magnesium is more abundant in profile 84, from 29.9 wt.% MgO at the base to 4.8 wt.% at the top, and from 15.7 to 0.2 wt.% in profile 83. In core 84 there is an important increase in NiO at 6 m deep (3.6 wt.%). Nickel varies from 0.33 wt.% to 0.91 wt.% in profile 84 and from 0.44 wt.% to 0.83 wt.% in 83. Cobalt is below 0.05 wt.% CoO in core 83 and up to 0.18 wt.% CoO in core 84. The LOI presents an opposite behavior in each core, increasing towards the top in profile 83 (8.6 to 11.3 wt.%) with an abrupt decrease at 15 m (2.3 wt.%), and decreasing towards the top in profile 84 (10 to 5.2 wt.%).

4.2. Petrography

The base of core 84 is formed by a slightly weathered, serpentinized peridotite that does not preserve most of the primary mantle mineralogy. Despite that, under the OM some of the primary textures can be observed; olivine and pyroxene have been almost completely substituted by phyllosilicates and Fe oxides during serpentinization (Figure 5). Euhedral to subhedral sections of pyroxene grains of 1 mm in length have been altered to serpentine group minerals (lizardite and chrysotile) along cleavage planes, with thin, brown goethite discontinuous intergrowths (Figure 5a). There are fractures crosscutting the grains replaced by serpentine (Figure 5a,b). Pseudomorphs or voids left by olivine grains of about 0.2 mm in diameter are surrounded by minerals of the serpentine group (lizardite and chrysotile) and Fe oxide (magnetite/maghemite) small grains. Serpentine is regularly associated with micrometric-sized grains of magnetite distributed along fractures (Figure 5c). Cores of olivine grains are replaced by microcrystalline aggregates of needle- or platelet-shaped smectite (hereafter incipient smectite), which exhibit second-order interference colors (Figure 5c,d).

The original texture of the bedrock in the saprolite and the ferruginous saprolite is lost in both profiles (from 23 to 1 m depth in core 83 and from 9 to 1 m in core 84). This part of the profile is more porous and presents foliated and bent, green smectite crystals of more than 1 mm in length that leave elongated voids due to their cleavage. These crystals are the result of an almost complete substitution of the primary minerals, mainly pyroxene, by a secondary mineral of the smectite group (hereafter type I smectite), and in some areas by irregular phyllosilicate grains that appear dark red due to the presence of iron oxyhydroxides (sample 84-08, Figure 5e). Chlorite appears as a vein bridging between smectite grains (Figure 5f). Some samples of saprolite in profile 83 (15 m) have slightly different mineralogy, with irregular brown accumulations of Fe oxyhydroxides and slightly deformed angular fragments of botryoidal silica aggregates, of approximately 2×0.2 mm (Figure 5g). This texture may suggest that these fragments come from a silicified vein that collapsed. In the ferruginous saprolite, irregular green deformed grains of phyllosilicates (lizardite and type II smectite) up to 0.2 mm in length are surrounded by irregular brown Fe oxyhydroxide aggregates (sample 83-4, Figure 5h).

SEM images of a weathered peridotite sample reveal the texture of fractured, pseudomorphed olivine grains of about 1 mm in size wrapped by layers of serpentine and smectite or kerolite-type minerals with different average Z values (Figure 6a). Dissolution voids in the olivine cores are partially filled by disordered aggregates of incipient smectite platelets and Fe oxides (magnetite/maghemite) (Figure 6b). Original orthopyroxenes hosting clinopyroxene exsolution lamellae evolve to smectite at different rates, being the clinopyroxene exsolutions more quickly substituted (to incipient smectite) (Figure 6c). In addition, abundant Cr-spinel grains are present associated with pyroxene and may occur as euhedral octahedral crystals, anhedral crystals, or as inclusions in former pyroxene (Figure 6d). In the saprolite of core 84, needles and platelets of type I smectite have been observed (Figure 6e). In the porous ferruginous saprolite, bent tabular grains of serpentine (lizardite) contain cores of type II smectite (Figure 6f). Iron oxides occur as inclusions in serpentine or within fractures crosscutting serpentine in sample 83-22 (Figure 6g). Cr-spinel can also be found as fractured, isolated, and euhedral crystals substituted by Fe oxides (Figure 6h).

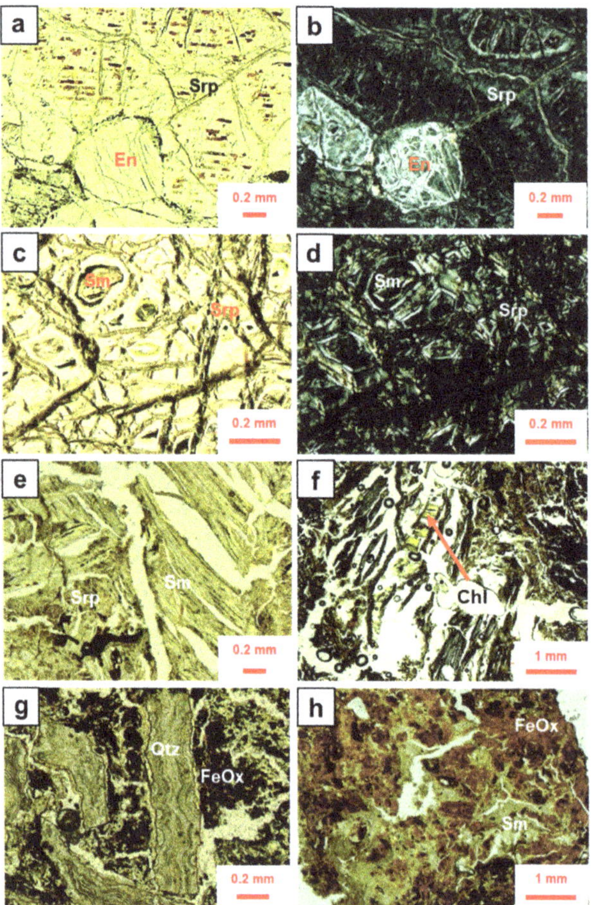

Figure 5. Optical photomicrographs taken with plane polarized light (**a**,**c**,**e**,**g**,**h**) and crossed polars (**b**,**d**,**f**) of the characteristic textures of the main mineralogy in the San Felipe Ni-laterites: (**a**,**b**) pyroxene grains with green serpentine in fractures and red-brown goethite films (sample 84-12, weathered peridotite); (**c**,**d**) olivine grain with green serpentine in fractures and black magnetite/maghemite, the nucleus of a greenish olivine grain can be observed, containing incipient smectite (sample 84-12, weathered peridotite); (**e**) fractured green serpentine grains of more than 1 mm long partially substituted by type I smectite (sample 84-08, saprolite); (**f**) chlorite as vein bridging between smectite grains; (**g**) angular fragments of botryoidal cryptocrystalline quartz aggregates surrounded by iron oxyhydroxides (sample 83-15, ferruginous saprolite); (**h**) irregular green aggregates of smectite type II, some of them dyed by iron oxyhydroxides (sample 83-04, ferruginous saprolite). Key: En = enstatite; Srp = serpentine; Sm = smectite; Chl = chlorite; Qtz = quartz; FeOx = Fe oxyhydroxides.

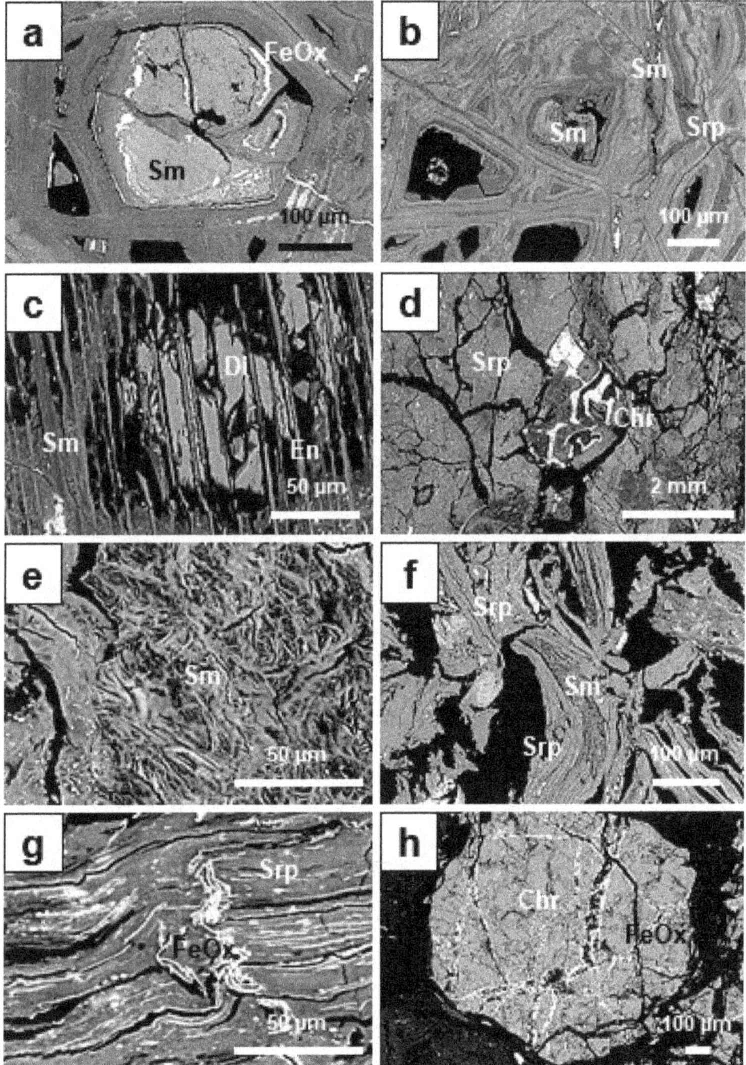

Figure 6. BSE photomicrographs of the textures observed in the San Felipe Ni-laterite: (**a**) olivine grain altered to incipient smectite surrounded by magnetite/maghemite (sample 84-12, weathered peridotite); (**b**) different generations of serpentine and smectite, note the differences in Z in serpentine, which correspond to variations in Fe and Ni contents (sample 84-12, weathered peridotite); (**c**) orthopyroxene grain with exsolutions of clinopyroxene, altered to incipient smectite (sample 84-12, weathered peridotite); (**d**) Cr-spinel as intra/inter crystals associated with pyroxene surrounded by serpentine (sample 84-09, transition between weathered peridotite and saprolite); (**e**) randomly oriented needles and platelets of type I smectite, (sample 84-09, transition between weathered peridotite and saprolite); (**f**) deformed platy serpentine crystals altered to type II smectite at the nuclei and inclusions of Fe oxyhydroxides (sample 84-02, ferruginous saprolite); (**g**) serpentine with Fe oxyhydroxides in fractures and as inclusions (sample 83-15, ferruginous saprolite); (**h**) euhedral Cr-spinel crystal with Fe oxides within fractures (sample 84-09, transition between weathered peridotite and saprolite horizon). Key: En = enstatite; Di = diopside; Srp = serpentine; Sm = smectite; Qtz = quartz; FeOx = oxides/oxyhydroxides; Chr = Cr-spinel.

4.3. Mineral Chemistry

Our results show that type I smectites (found in the saprolite of core 84, Figure 5e) and type II smectites (in both cores 83 and 84, Figure 5f) display distinct compositions (Supplementary Material Table S2, Figure 7).

The average chemical composition of type I smectite in sample 84-09 (Figure 6), in the transition between the weathered peridotite and the saprolite horizon, is 43.52 wt.% SiO_2, 2.04 wt.% Al_2O_3, 14.83 wt.% MgO, 15.46 wt.% Fe_2O_3, and 2.58 wt.% NiO, and the corresponding structural formula is: $(Mg_{1.90}Fe_{0.90}Ni_{0.19})_{\Sigma=2.99}(Si_{3.64}Al_{0.15})O_{10}(OH)_2Ca_{0.03}Na_{0.02}$. In core 84, type II smectite in the serpentine tabular grains, in the ferruginous saprolite (sample 84-02), has an average chemical composition of 46.53 wt.% SiO_2, 2.92 wt.% Al_2O_3, 3.62 wt.% MgO, 20.58 wt.% Fe_2O_3, and 2.76 wt.% NiO. The corresponding structural formula is $(Fe_{1.33}Mg_{0.51}Al_{0.21}Ni_{0.10})_{\Sigma=2.15}(Si_{3.93}Al_{0.07})O_{10}(OH)_2Ca_{0.03}Na_{0.03}$. Type II smectite from core 83 has an average of 43.06 wt.% SiO_2, 2.38 wt.% Al_2O_3, 4.57 wt.% MgO, 22.29 wt.% Fe_2O_3, and 1.06 wt.% NiO (Figure 7c), and a structural formula of $(Fe_{1.44}Mg_{0.57}Ni_{0.07})_{\Sigma=2.08}(Si_{3.82}Al_{0.18})O_{10}(OH)_2Ca_{0.02}Na_{0.03}$. The interlaminar cations are Ca and Na in both types and the occupancy is < 0.06, far from the usual values of hydrated exchangeable cations ($x \approx 0.2$-0.6) in smectites [46]. This contrasts with the incipient smectite from the slightly weathered, serpentinized peridotite in core 84, which has an average structural formula of $Mg_{2.46}Fe_{0.38}Ni_{0.02}Si_{3.69}Al_{0.09}O_{10}(OH)Ca_{0.02}Na_{0.01}$ [38] and formed as a substitution of olivine and pyroxene.

Serpentine in San Felipe can be divided, according to their Ni content, into low-Ni (so-called serpentine I) and Ni-enriched (serpentine II). The average composition of serpentine II grains in sample 84-7 (Figure 8d and Supplementary Material Table S3) is 39.92 wt.% SiO_2, 2.05 wt.% Al_2O_3, 25.33 wt.% MgO, 7.75 wt.% Fe_2O_3, and 3.23 wt.% NiO. The corresponding structural formula is $(Mg_{1.97}Fe_{0.30}Ni_{0.14}Al_{0.13})Si_{2.09}O_5(OH)_4$ (Supplementary Material Table S3, Figure 9).

The EMPA mapping of Si, Mg, Fe, Ni, and Al of a representative completely pseudomorphed olivine grain (Figure 10) gives compositions that are consistent, in the core, with a trioctahedral smectite type I rich in Fe and with a Ni content of around 3 wt.% (Supplementary Material Table S2), crosscut and surrounded by a Ni- and Fe-enriched serpentine.

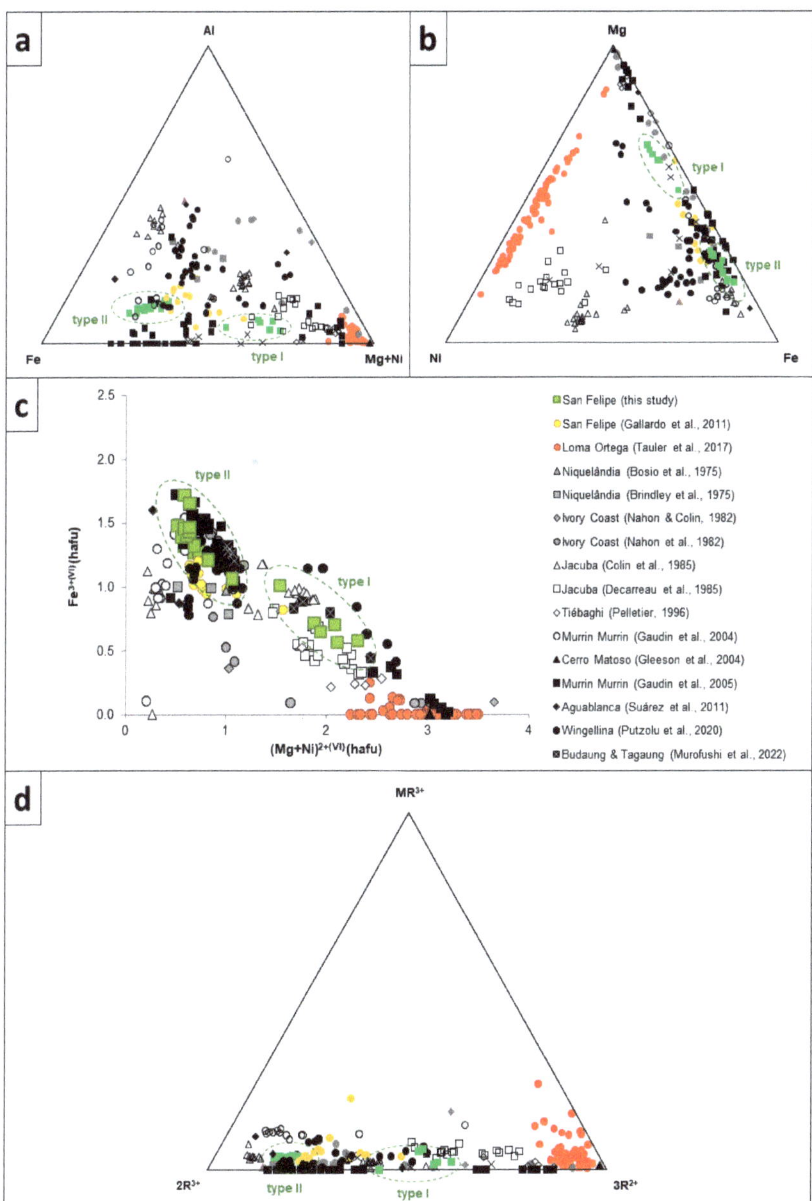

Figure 7. Mineral chemistry of Ni-smectite from San Felipe (this study and from TEM data from [38]), compared with Ni-Fe-Mg smectites worldwide, including Loma Ortega (Dominican Republic, [25]); Niquelândia, Brazil [47,48]; Ivory Coast [49,50]; Jacuba, Brazil [23,43]; Tiébaghi, New Caledonia [51]; Murrin Murrin, Australia [10,52]; Cerro Matoso, Colombia [53]; Aguablanca, Spain [54]; Wingellina, Australia [24]; Budaung and Tagaung, Nyamar [55]. (**a**) Fe^{3+}-Al-(Mg + Ni) ternary plot (based on [23]); (**b**) Ni-Mg-Fe^{3+} ternary plot; (**c**) octahedral Fe^{3+} versus (Mg + Ni) binary diagram representing the octahedral cations in smectite (modified from [23]); and (**d**) MR^{3+}-$2R^{3+}$-$3R^{2+}$- (based on [56]), in which MR^{3+} = Na + K + (Ca/2), $2R^{3+}$ = Al + Fe^{3+} − MR^{3+}, and $3R^{2+}$ = Mg + Ni + Mn. Data have been plotted as hafu (calculated on the basis of 11 oxygens), and all Fe has been considered trivalent.

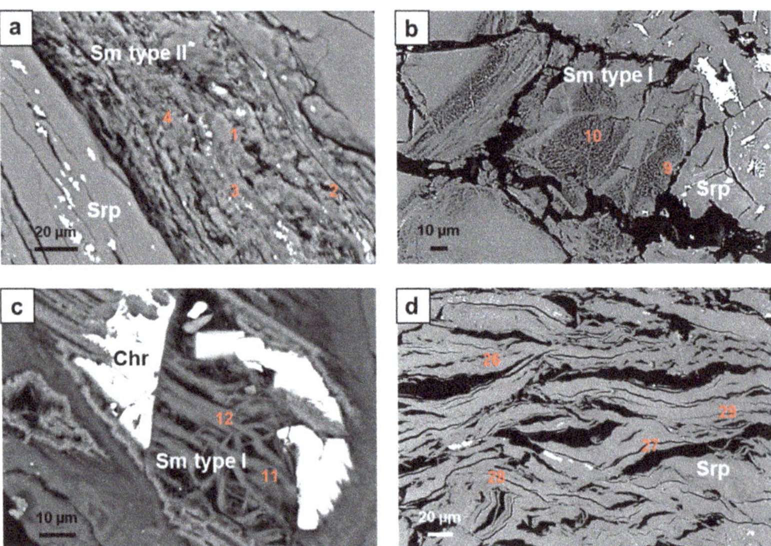

Figure 8. BSE photomicrographs of smectite and serpentine textures in San Felipe and the analyzed points displayed in Table S3: (**a**) type II smectite (sample 84-02); (**b**) type I smectite after olivine (sample 84-09); (**c**) type I smectite after pyroxene (sample 84-09); (**d**) serpentine (sample 84-07). The point analyses 84-02-1, 2, 3, and 4 correspond to smectite type II; point analyses 84-09-9 and 84-09-10 are smectite after olivine; analyses 84-09-11 and 84-09-12 are smectite after pyroxene.

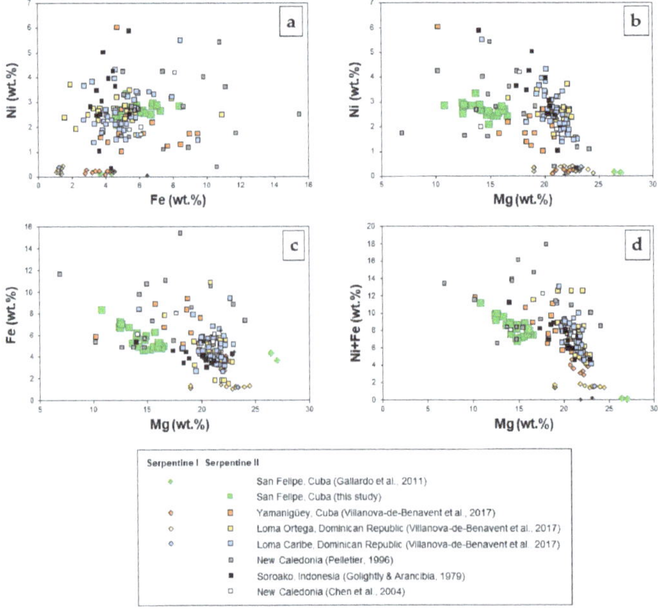

Figure 9. Composition of serpentine I and II (this study [38]) from San Felipe compared to serpentine I and II from other localities: Yamanigüey (Cuba), Loma Ortega and Loma Caribe (Dominican Republic) [45], New Caledonia [51,57], and Soroako, Indonesia [58], in terms of the major octahedral elements (Mg, Fe, Ni in wt.%). (**a**) Ni versus Fe (wt.%), (**b**) Ni versus Mg (wt.%), (**c**) Fe versus Mg (wt.%), and (**d**) Ni+Fe versus Mg (wt.%).

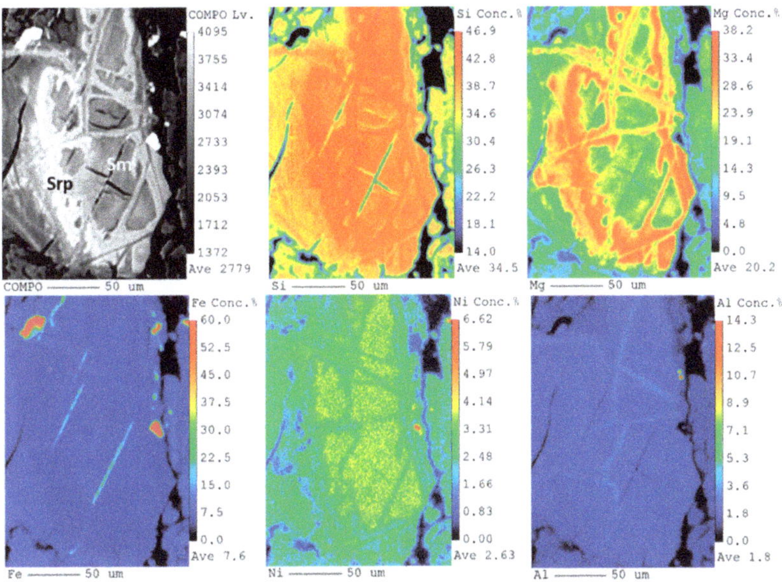

Figure 10. BSE photomicrograph (**top left**) and Si, Mg, Fe, Ni, and Al X-ray elemental maps obtained by EMPA of a type I smectite grain from sample 84-09 (transition between the serpentinized peridotite and the saprolite).

4.4. Structure Characterization of Minerals by X-ray Diffraction (XRD)

The mineral phases identified in cores 83 and 84 and their relative amounts as a function of depth are given in Supplementary Material Table S4, and Figure 11 displays the XRPD of selected samples. The minerals identified are smectite, dehydrated or collapsed smectite (with a talc-like or kerolite-like structure), serpentine (lizardite and chrysotile), chlorite, quartz, hematite, goethite, maghemite, and Cr-spinel. As discussed later, quantification of nanometric-sized, hydrated phyllosilicates and, in particular, smectite and kerolite-like structures by Rietveld profile analysis is problematic [59–61]. In sample 83-25, smectite presents an intense, broad, symmetric peak at 14.7 Å, and a kerolite faint wide reflection at 9.6 Å (Figure 8). However, at higher angles, both phases present almost coincident broad asymmetric *hk* diffraction bands, which are a typical feature of turbostratic structures [62]. Accordingly, smectite and kerolite have been quantified together due to the similarity of their XRPD patterns. Whenever the smectite broad 14.7 Å peak is important, we will refer to possible mixtures of smectite/kerolite as "smectite".

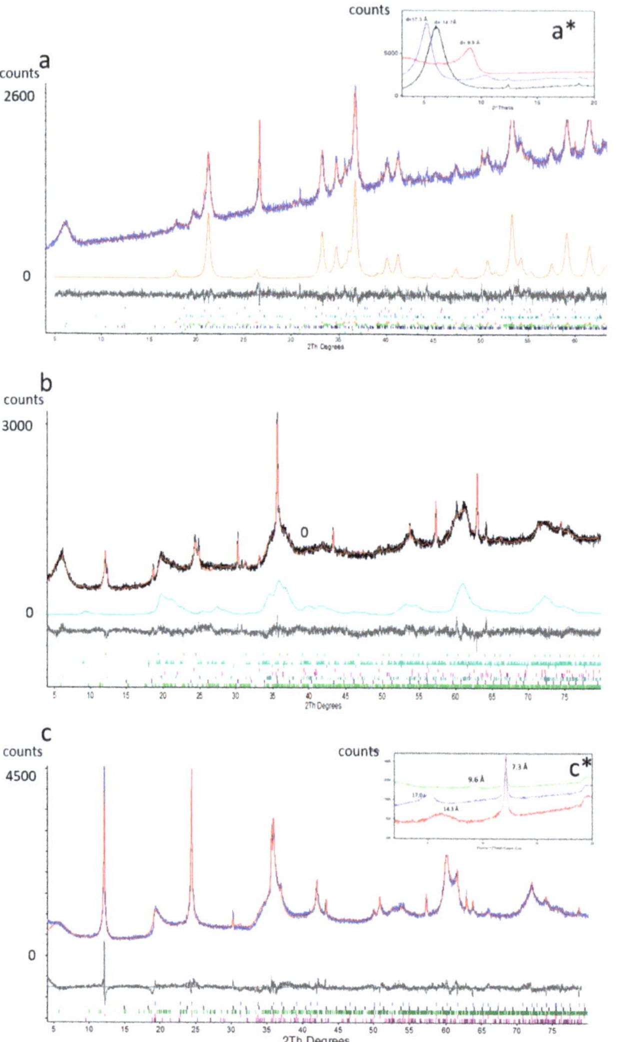

Figure 11. (**a**) XRPD refinement of sample 83-7 (ferruginous saprolite horizon): Rwp (numerical figure of merit quantifying the quality of the fit) = 3.3. Legend: blue = experimental pattern, red = calculated pattern, black = difference between experimental and calculated pattern, and orange = calculated pattern for goethite. At the bottom, the Bragg reflections for each mineral phase are shown. (**a***) XRD of oriented aggregate mounts of the same sample since 2θ degrees. Legend: black = untreated sample at room temperature, blue = treated with ETG, and red = heated to 540 °C. (**b**) XRPD refinement of sample 83-25 (saprolite horizon): Rwp = 4.5. Legend: black = experimental pattern, red = calculated pattern, gray = difference between the experimental and the calculated pattern, and green = calculated pattern for a talc-like structure (kerolite). At the bottom, the Bragg reflections for each mineral phase are shown. (**b**) XRPD refinement of sample 83-7 (ferruginous saprolite horizon): Rwp = 3.3. Legend: blue = experimental pattern, red = calculated pattern, black = difference between experimental and calculated pattern, and orange = calculated pattern for goethite. At the bottom, the Bragg reflections for each mineral phase are shown. (**c**) XRPD refinement of sample 84-12 (weathered peridotite): Rwp = 6.9, legend as in (**a**). (**c***) XRD of oriented aggregate mounts of sample 84-12. Legend: red = untreated sample at room temperature, blue = treated with ETG, and green = heated to 540 °C.

In order to get further evidence on the presence of smectite, XRD of oriented aggregate mounts have been obtained for a sample of core 83 (83-7) and another of core 84 (84-12) (Figure 11a* and Figure 11c*, respectively). In sample 83-7, the symmetrical peak at d_{001} = 14.7 Å corresponds to pure smectite with two water layers [63,64]. After heating at 540 °C, the d_{001} collapses to 9.9 Å. With ethylene glycol (EG), d_{001} = 14.7 expands to 17.3 Å as symmetrical peaks, indicating a uniform swelling of almost all the layers without mixing with the non-swelling talc/kerolite layers [59] (Figure 11b). In sample 84-12, after heating at 540 °C, the d_{001} collapses to 9.6 Å. With EG, d_{001} = 14.3 expands to 17.0 Å as a symmetrical peak (Figure 12).

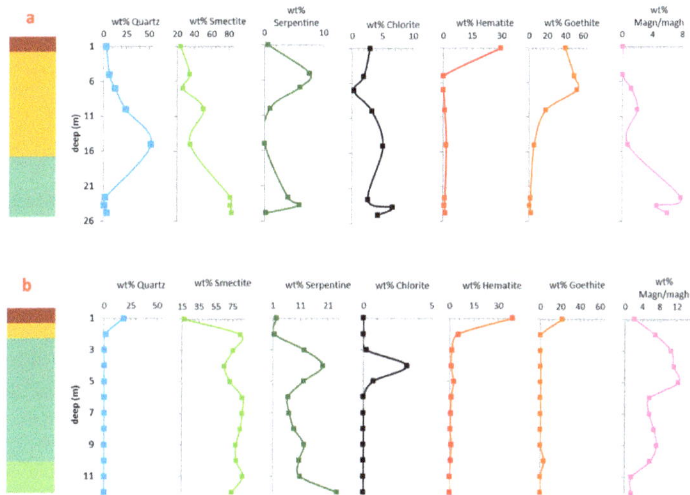

Figure 12. Variation in wt.% of mineral phases in function of depth: (**a**) core 83; (**b**) core 84. Color legend of schematic profiles as in Figure 4 (red: oxide zone; yellow: ferruginous saprolite; light green: saprolite; olive green: serpentinized peridotite).

In profile 83, smectite decreases from the base to the top (83.3 and 24.3 wt.%, respectively), and so does chlorite (4.3 at the base to 3.0 wt.% near the top). In contrast, lizardite decreases (0.3 wt.% at 25 m and 0.6 wt.% at 1 m). Profile 84 is dominated by smectite (74.0 to 17.3 wt.%) and serpentine (lizardite plus chrysotile) (24.6 wt.% at the bottom to 2.3 wt.% at the top). Near the middle of profile 83 (at 16 m depth), a highly silicified level appears, with more than 50 wt.% quartz. In core 84, quartz is less important, with only 18.2 wt.% at the top. Magnetite/maghemite decreases from the base to the top, whereas goethite and hematite increase towards the top. Near the surface, the total amount of goethite and hematite is similar in both profiles (69.3 and 60.3 wt.%), but the percentage of goethite in core 83 doubles that of core 84. In Figure 12, the variation of the major mineral phases is represented as a function of depth. It is worth noting that smectite is always dominant along both profiles (Supplementary Material Table S4).

5. Discussion

5.1. Degree of Weathering

The geochemical analyses (Figure 3) agree with the XRPD mineralogical composition (Supplementary Material Table S4, Figure 10). The SiO_2 wt.% decreases in the two profiles in the ferruginous saprolite but in profile 83 there is an abrupt increase at 15 m deep that corresponds to the occurrence of cryptocrystalline quartz millimetric to centimetric fragments (Figure 5g). This interruption in the typical regular succession of horizons explains the observed differences between profiles 83 and 84 at this depth. The increase in silica is accompanied by a reduction of Al_2O_3, Fe_2O_3, MgO, NiO, and LOI (Figure 4a,b).

The ultramafic index of alteration (UMIA) is defined as an attempt to quantify and visualize the degree of weathering of an ultramafic rock (e.g., peridotites) [2]:

$$UMIA = 100 \times (Al_2O_3 + Fe_2O_{3(T)})/(SiO_2 + MgO + Al_2O_3 + Fe_2O_{3(T)})$$

The UMIA is different in the saprolite and ferruginous saprolite of the two cores. Profile 84 presents a regular succession of horizons characteristic of this type of laterite: oxide zone (UMIA = 51), ferruginous saprolite (UMIA = 39), saprolite (UMIA = 22 to 40), and slightly weathered serpentinized bedrock (UMIA = 15). The oxide zone is underdeveloped in the two cores, and the ferruginous saprolite horizon (UMIA = 41 to 60) is more developed in core 83 with an UMIA from 30 to 40. The saprolite horizons are similar in both profiles, but the content in NiO is higher in 84. It must be noted that the total thickness of the saprolite horizon in core 83 is unknown. There are several factors that may lead to these differences, i.e., topography, position of the water table, drainage, local variations in the nature of the protolith, and degree of serpentinization.

5.2. Trioctahedral vs. Dioctahedral Smectites

Type I smectite, from the saprolite horizon, is trioctahedral in character (Supplementary Material Table S2) with a ratio between tetrahedral and octahedral cations of 1.26, near the ideal value of 1.33, and an octahedral occupancy of 2.99, where Mg is dominant. Its composition is intermediate between a saponite and nontronite and presents two types of textures: as random curved platelets filling cores of dissolved olivine (Figure 6b), and as pseudomorphs of pyroxene (Figure 6c). The octahedral occupation of Ni is <0.2.

Type II smectite is dioctahedral, close to nontronite, with a tetrahedral/octahedral ratio equal to 1.92 and an octahedral occupancy of 2.08, where Fe^{3+} is dominant (Supplementary Material Table S2). This smectite appears in the ferruginous saprolite, very often as pseudomorphs of type II serpentine (Figure 6f). The occupancy of Ni is under 0.14. It appears that, towards the top, smectites become more stable by increasing their dioctahedral character. All smectites are very deficient in interlaminar cations (Ca and Na), suggesting that they could be mixed with kerolite. Smectites in samples with less goethite (and more magnetite/maghemite) (Supplementary Material Table S4) have higher Ni contents, which attain their maximum occupancy in the structure in the central part of the saprolite in core 84 (samples 84-7,6,5) (Figure 4b). The compositional variation of the smectites of San Felipe is similar to those of Murrin Murrin, described in [65]. Similarly, a transition towards the top from a trioctahedral to a more dioctahedral coincides with the description of different weathering profiles developed in ultramafic rocks [24,65,66].

5.3. Type I Serpentine vs. Type II Serpentine

Two types of serpentine can also be distinguished. Type I, with low Ni and with excellent crystallinity, only appears surrounding pseudomorphed olivine grains in sample 84-12 (Figure 6b). The structural formula calculated by TEM in lizardite particles from this sample is $(Mg_{2.69}Fe_{0.15}Ni_{0.01}Si_{2.02}O_5(OH)_4)$ [38]. Type II is a different generation of serpentine, less crystalline and with higher Ni contents (Supplementary Material Table S3), which coexists spatially with type II smectite in both profiles. The mean structural formula for type II serpentine in sample 84-6, $(Mg_{1.97}Fe_{0.30}Ni_{0.14}Al_{0.13})Si_{2.09}O_5(OH)_4$, can be considered a solid solution of the three end-members lizardite, $Fe_2SiO_5(OH)_4$, and népouite, the latter two components being less soluble and stabilizing the serpentine as pore water becomes more acidic and oxidizing. As with smectites, Fe^{3+} in serpentines increases towards the top. Type II serpentine may be the result of topotactic reactions that incorporate Ni and Fe in type I serpentine. For a complete discussion about the solubility and stability of Mg-Fe-Ni-serpentines in laterites, see [45].

Serpentine in profiles 83 and 84 includes the crystalline form of lizardite, which is dominant, together with less chrysotile in core 84. Most lizardite in the weathered peridotite is well crystallized, showing narrow diffraction peaks (Figure 11c). The mean crystallite size calculated in the refinement of the PXRD is estimated to be around 40 nm. In addition, a small incipient less crystalline fraction of lizardite with a crystallite size of 8 nm is also observed.

5.4. XRPD Interpretation of Smectite (vs. Kerolite-Type Structures)

The quantification of mineral phases by XRPD raises several relevant issues regarding the distribution of Ni among the different phyllosilicates present in the samples. The phyllosilicates sensu lato encountered in XRPD profiles are smectite, a kerolite-type structure, serpentine (lizardite plus chrysotile), and chlorite. The identification and quantification of the latter two through XRPD is simple since both phases are well crystallized. However, identifying and estimating the amount of smectite and/or a kerolite-type structure is more problematic because part of their diffractograms would coincide. Previous works [59,63] have shown that kerolite, and its Ni equivalent pimelite, occurs generally in association with weathered ultramafic rocks as coatings or vein infillings at the bottom of weathering profiles in association with serpentine. The authors of [23] have found that smectite and kerolite were almost always intimately associated with Ni-laterites resulting from the weathering of pyroxenites. Kerolite is often a ubiquitous poorly crystalline phase present in saprolites that can be associated either with a smectite or talc. Provided the lack of reliable published crystal structures of trioctahedral and dioctahedral smectites, they can only be included with precaution in the calculation of a powder diffraction file. On the contrary, the kerolite profile can be fitted well using the crystal structure of talc, as shown in Figure 11. As a result, it cannot be ascertained by XRPD which contribution to the turbostratic profile corresponds to smectite and which to kerolite, or even to a modified/dehydrated smectite, which could be an artifact produced during the powdering of the samples. The relative amount of smectite can be estimated by the 001 reflection, which is more important in the bottom of 83 than in 84. It is worth noting that the encountered smectites have a very low content of the interlaminar cations Na^+ and Ca^{2+}, so their composition is close to a 2:1 dioctahedral or trioctahedral phyllosilicate. In particular, in the saprolite horizon of core 84, type I smectites with low Fe content approximate the composition of a talc-like structure.

5.5. Mineral Variations and Weathering Process across the Lateritic Profiles

Olivine and pyroxene relicts are absent in both profiles. The bulk composition of sample 84-12 (Supplementary Material Table S1) allows an approximate estimation of the mineral composition of the parent rock before serpentinization (and alteration of pyroxene). Assuming that only a small amount of Mg has been lost by the dissolution of a small amount of olivine, the Si/(Mg + Fe) molar ratio obtained from XRF is compatible with a peridotite-rich in orthopyroxene (~60% molar). Complete serpentinization of olivine would consume an equivalent amount of orthopyroxene (in molar units), leaving an important fraction of total pyroxene (~46% molar). Accordingly, the parent rock can be classified as an orthopyroxene-rich harzburgite. This remnant pyroxene is the base for the formation of new hydrated phases (smectite, kerolite). As weathering proceeds, part of the initial (hydrothermal, oceanic) serpentine (type I serpentine) will change its composition, incorporating Fe and Ni and becoming less soluble, while the remainder of serpentine will be dissolved incongruently, leaving some silica retained [45]. This silica is needed for the formation of smectite and possibly chlorite. In profile 84, Mg is lost progressively, from the bottom to the top, and the Mg discontinuity between the saprolite horizon and the oxide zone is less sharp compared with other lateritic profiles worldwide [20,67]. From the bottom to the top, the profile is enriched in the less soluble elements Fe, Al, and Cr. Excess silica forms quartz, while Fe forms goethite, hematite, and maghemite. Gibbsite has not been observed so it is concluded that Al has been incorporated in the iron oxides and oxyhydroxides. Indeed, goethite at the top of 83 and 84 contains a 10% mole fraction of the diaspore AlOOH component. In hematite from the same horizons, the mole fraction of corundum is only 5%.

It is worth noting that interstratified Ni clays have not been identified in San Felipe, unlike in Western Australia. According to the authors of [24], these interstratified clays formed post-laterization, during the onset of arid climate conditions in Australia. Therefore, there is no evidence of this in San Felipe.

Profile 84 follows a regular model of the formation of Ni-laterites in moderately tropical humid climates with marked dry seasons (1500 mm rainfall and 27 °C), after exposition to the surface for some million years [1,8,12]. Initial weathering of serpentinized ultramafic rocks dissolves part of type I serpentine and incipient smectite pseudomorphing pyroxene, leaching Mg and, to a lesser extent, Si, downwards. Magnetite is oxidized to maghemite or to goethite. The latter is able to retain Ni somehow so that this element is not leached immediately [68]. Goethite is the main mineral in the oxide zone. It may contain up to 2.7 wt.% NiO [31]. Serpentine and smectite incorporate Fe^{3+} and Ni that substitute Mg in the octahedral positions of both minerals. In this process, the more soluble Mg hydrous silicate component is substituted by the less soluble, which incorporates more Fe^{3+} and Ni, so that the more soluble (more Mg, less Fe^{3+} and Ni) appears below the less soluble in the profile. The rate at which the Ni retained initially by goethite is liberated, compared with the rate of its capture by the new hydrous silicates formed downwards, (incipient smectite → smectite I → smectite II; serpentine I → serpentine II) determines the content of Ni along the entire profile.

The bottom of profile 84 is poor in NiO (Supplementary Material Table S1, 0.33 wt.%, sample 84-12), and similar to the Ni content in peridotites [20]. Supergene enrichment has not yet taken place significantly, with serpentines being formed in pre-weathering conditions. In the saprolite horizon, 85% of Ni is contained in type I smectite, and only 15% in serpentine. Serpentine (Supplementary Material Table S3) is enriched in Ni with respect to the same mineral at the base of the profile; however, its amount decreases towards the top, so that its contribution to nickel content near the top is only about 5 wt.%. Sample 84-12 represents the slightly weathered, serpentinized peridotite that, nevertheless, has been deeply altered in hydrothermal (oceanic) conditions prior to weathering. Olivine and pyroxene have completely been transformed to serpentine and smectite and/or kerolite before weathering. The initial transformation of enstatite to smectite is favored by the crystallographic similarities and the chain structure of the former, along its cleavage and exsolution planes [23,65], but needs some afford of silica, which possibly comes from the incongruent dissolution of part of the olivine.

The anomalous increase in serpentine and magnetite/maghemite or goethite near the top in both profiles (samples 83-6 and 84-4) may be due to floating blocks of bedrock that have been frequently detected by GPR (ground-penetrating radar) in the San Felipe formation [31]. This is corroborated by correlation coefficients between serpentine and magnetite/maghemite or goethite for profiles 83 and 84, of 0.4 and 0.6, respectively.

It is worth noting that profile 84, from the border of the plateau, would represent an undisturbed profile that corresponds to the stationary development of regular horizons in the given lithologic and climatic conditions, and active drainage. On the contrary, profile 83, from the center of the plateau, presents fragmented microcrystalline quartz vein infillings and displays evidence of reworking along the profile. This would be a result of complex, not well-understood episodes of precipitation of silica-forming crusts, the collapse of part of the profile, and further replenishment downwards produced by active tectonics during the Pliocene [31,69].

5.6. Comparison of Smectites from Lateritic Profiles Worldwide

Supplementary Material Table S5 displays selected structural formulae of Ni-bearing smectites from Ni-laterite profiles worldwide. There are many differences between smectites based on: (i) the total layer charge and the distribution of this charge between the octahedral and the tetrahedral sheets; (ii) the Ni, Mg, and Fe content in the octahedral position; (iii) the origin, formed after olivine, or (ortho- and clino-) pyroxene or lizardite. These differences can be explained by the position of smectites in the lateritic profiles, and above all, the parent rock, topography, and climate.

The Ni-smectite from the pyroxenite of Niquelândia (Brazil) [65] is the richest in Ni (2.14 hafu and 2.71 cations in octahedral position), and the other two with remarkable Ni contents are those of Tagaung (Myanmar), and Falcondo (Dominican Republic) [25,55]: the sum of the octahedral cations being 1.8 and 1.75 hafu, and 2.71 or 3.06 cations in the octahedral position, respectively (Supplementary Material Table S5), confirming the trioctahedral character. Another smectite with dominant Ni in the octahedral position (1.03 hafu) is the one from the lower saprolite of Tagaung [55], but in this case, the sum of the octahedral cations is 2.43 (intermediate between trioctahedral and dioctahedral). Iron in the octahedral position is dominant in six smectites, two from Myanmar, three from Western Australia, and one from Cuba [38,52,55], the sum of the octahedral cations being between 2.01 and 2.27 hafu (Supplementary Material Table S5), confirming the dioctahedral character and a slight octahedral over-occupancy. Finally, Mg is the main octahedral cation in five smectites, two from the Western Ivory Coast [50], two from Western Australia [24,52], and two from Cuba [38]. Four of them yield an octahedral cation sum between 3.06 and 2.86 hafu (trioctahedral character), and the other one is 2.26 hafu (dioctahedral character) (Supplementary Material Table S5). The smectite with the highest Mg in the octahedral position (1.82 hafu, 3.06 total occupancy, trioctahedral) is from serpentinized dunite and formed after olivine.

The highest Al is 0.71 hafu in the tetrahedral position and 0.25 Al hafu in the octahedral position in the trioctahedral smectite in the transition layer in the Western Ivory Coast [50]. Iron is present in five samples in the tetrahedral position, three of them are the richest in Ni in the octahedral position (Brazil, Myanmar, Dominican Republic) [25,55,65].

Serpentine was detected by X-ray diffraction together with smectite in ten samples [25,38,50,55]. Only in Murrin Murrin (serpentinized peridotite) and Wingellina (gabbro) in Western Australia and in Niquelândia (pyroxenite) in Brazil was serpentine not reported [24,52,65].

6. Conclusions

The original mineral composition of the bedrock in the San Felipe Ni-laterite deposit is estimated to have a molar ratio between enstatite and forsterite of about 3. This mineral composition greatly influences the mineral evolution during weathering as well as the degree of supergene enrichment in Ni.

Smectite and serpentine incorporate Fe^{3+}, Al, and Ni. For smectites, this represents a transformation from trioctahedral to dioctahedral character. Even if serpentine doubles the NiO content of smectite (~3.2 and ~1.5 wt.%, respectively), smectite is much more abundant in the saprolite, so this mineral contains about 80% of the total Ni, the remainder being located in serpentine.

Interlaminar cation content in smectites is very low, suggesting that part of the X-ray turbostratic profile observed is in fact an intermediate structure between kerolite and a collapsed smectite.

Profile 84 displays a regular succession of horizons of an ideal Ni-laterite profile. This regular succession is disrupted in profile 83 by a concentration of silica, which probably reveals a complex succession of events of collapse and replenishing of the area where the laterite was developed.

Finally, considering the Ni content, the abundance and nature of the ore (smectite and serpentine), and the current resources, the San Felipe deposit could be targeted for Ni, similar to other large Ni-laterite deposits in the Caribbean, such as Moa Bay, Cerro Matoso, and Falcondo.

Supplementary Materials: The following supporting information can be downloaded at: https://www.mdpi.com/article/10.3390/min13101281/s1, Table S1: Major elements in wt.% of cores 83 and 84 (LOD = below the limit of detection); Table S2: Representative EMPA analyses (in weight percent) and structural formulae (in half-atoms per formula unit -hafu-, based on 11 oxygens) of smectite. Analyses 84-02-1, 84-02-2, 84-02-3, and 84-02-4 are from type II smectite in ferruginous saprolite (sample 84-02); 83-23-5, 83-23-6, 83-23-7, and 83-07-8 are from smectite type II in ferruginous saprolite (sample 83-23); 84-09-9 and 84-09-10 are from smectite type I from olivine; 84-09-11 and 84-09-12 are from smectite type I from pyroxene in weathered peridotite/saprolite (sample 84-09); Table S3: Representative EMPA analyses (in weight percent) and structural formulae (in half-atoms per formula unit, based on 7 oxygens) of serpentine II from sample 84-07 (saprolite horizon). The point analyses 20, 26, 27, 28, and 30 are in Figure 7; Table S4: Quantitative determination (in wt.%) of the mineral phases present in the selected samples, obtained by Rietveld refinement of XRPD. Legend: Mag = magnetite, Mgh = maghemite; Table S5: Structural formulae of Ni-smectite from various localities worldwide, ordered by Ni content in the octahedral position (hafu based on 11 oxygens).

Author Contributions: Conceptualization, E.T., S.G. and J.A.P.; methodology, E.T., S.G., C.V.-d.-B. and J.A.P.; software, E.T., S.G., C.V.-d.-B. and G.K.; validation, E.T., S.G., C.V.-d.-B. and J.A.P.; formal analysis, E.T., S.G. and C.V.-d.-B.; investigation, E.T., S.G., C.V.-d.-B., A.C.-R., K.N.-C., G.K. and J.A.P.; data curation, E.T., S.G. and C.V.-d.-B.; writing—original draft preparation, E.T. and S.G.; writing—review and editing, E.T., S.G., C.V.-d.-B., A.C.-R., K.N.-C., G.K. and J.A.P.; supervision, J.A.P.; project administration, J.A.P.; funding acquisition, J.A.P. All authors have read and agreed to the published version of the manuscript.

Funding: This research was funded by Agència de Gestió d'Ajuts Universitaris i de Recerca de Catalunya, grant number 2021 SGR 00239; and by the Ministerio de Ciencia e Innovación (MCIN) and the Agencia Estatal de Investigación (AEI) of Spain, grant number PID2019-105625RB-C21. This research has been financially supported by FEDER Funds, the Spanish projects CGL2009-10924 and CGL2012-36263.

Data Availability Statement: Not applicable.

Acknowledgments: The authors would like to thank Enrique Piñero for his inestimable support during fieldwork. The authors also thank the Assistant Editor, and the four anonymous reviewers for their accurate revisions and constructive suggestions that highly improved the quality of the manuscript.

Conflicts of Interest: The authors declare no conflict of interest.

References

1. Freyssinet, P.; Butt, C.R.M.; Morris, R.C. Ore-forming processes related to lateritic weathering. In *Economic Geology One Hundredth Anniversary Volume*; Society of Economic Geologists: Littleton, CO, USA, 2005; pp. 681–722. [CrossRef]
2. Aiglsperger, T.; Proenza, J.A.; Lewis, J.F.; Labrador, M.; Svojtka, M.; Rojas-Purón, A.; Longo, F.; Ďurišová, J. Critical metals (REE, Sc, PGE) in Ni laterites from Cuba and the Dominican Republic. *Ore. Geol. Rev.* **2016**, *73*, 127–147. [CrossRef]
3. Ulrich, M.; Muñoz, M.; Boulvais, P.; Cathelineau, M.; Cluzel, D.; Guillot, S.; Picard, C. Serpentinization of New Caledonia peridotites: From depth to (sub-)surface. *Contrib. Mineral. Petrol.* **2020**, *175*, 91. [CrossRef]
4. Teitler, Y.; Cathelineau, M.; Ulrich, M.; Ambrosi, J.P.; Munoz, M.; Sevin, B. Petrology and geochemistry of scandium in New Caledonian Ni-Co laterites. *J. Geochem. Explor.* **2019**, *196*, 131–155. [CrossRef]
5. Chassé, M.; Griffin, W.L.; Reilly, S.Y.O.; Calas, G. Scandium speciation in a world-class lateritic deposit. *Geochem. Perspect. Lett.* **2017**, *3*, 105–114. [CrossRef]
6. Dalvi, A.D.; Gordon Bacon, W.; Osborne, R.C. The past and the future of nickel laterites. In *PDAC 2004 International Convention, Trade Show & Investors Exchange*; The Prospectors and Developers Association of Canada Toronto: Toronto, ON, Canada, 2004; p. 27.
7. McRae, M.E. US Geological Survey, Mineral Commodity Summaries 2022—Nickel. Available online: https://pubs.usgs.gov/periodicals/mcs2022/mcs2022-nickel.pdf (accessed on 13 July 2023).
8. Golightly, J.P. Nickeliferous laterite deposits. In *Seventy-Fifth Anniversary Volume*; Economic Geology Publishing Company: Littleton, CO, USA, 1981; pp. 710–735. [CrossRef]
9. Gleeson, S.A.; Butt, C.R.; Elias, M. Nickel laterites: A review. *SEG Newsl.* **2003**, *54*, 11–18. [CrossRef]
10. Gaudin, A.; Decarreau, A.; Noack, Y.; Grauby, O. Clay mineralogy of the nickel laterite ore developed from serpentinised peridotites at Murrin Murrin, Western Australia. *Aust. J. Earth Sci.* **2005**, *52*, 231–241. [CrossRef]
11. Thorne, R.L.; Roberts, S.; Herrington, R. Climate change and the formation of nickel laterite deposits. *Geology* **2012**, *40*, 331–334. [CrossRef]

12. Butt, C.R.M.; Cluzel, D. Nickel laterite ore deposits: Weathered serpentinites. *Elements* **2013**, *9*, 123–128. [CrossRef]
13. Maurizot, P.; Sevin, B.; Iseppi, M.; Giband, T. Nickel-bearing laterite deposits in accretionary context and the case of New Caledonia: From the large-scale structure of earth to our everyday appliances. *GSA Today* **2019**, *29*, 4–10. [CrossRef]
14. Butt, C.R.M. Nickel Laterites and Bauxites: A Summary of Observations Made During an Overseas Trip in 1974. In *CSIRO Division of Mineralogy*; Minerals Research Laboratories: Asheville, NC, USA, 1975.
15. Golightly, J.P. Progress in understanding the evolution of nickel laterites. *Econ. Geol. Spec. Pub.* **2010**, *15*, 451–485. [CrossRef]
16. Brand, N.W.; Butt, C.R.M.; Elias, M. Nickel laterites: Classification and features. *AGSO J. Aust. Geol. Geoph.* **1998**, *17*, 81–88.
17. Wells, M.A.; Ramanaidou, E.R.; Verrall, M.; Tessarolo, C. Mineralogy and crystal chemistry of garnierites in the Goro lateritic nickel deposit, New Caledonia. *Eur. J. Miner.* **2009**, *21*, 467–483. [CrossRef]
18. Villanova-de-Benavent, C.; Proenza, J.A.; Galí, S.; García-Casco, A.; Tauler, E.; Lewis, J.F.; Longo, F. Garnierites and garnierites: Textures, mineralogy and geochemistry of garnierites in the Falcondo Ni-laterite deposit, Dominican Republic. *Ore. Geol. Rev.* **2014**, *58*, 91–109. [CrossRef]
19. Cathelineau, M.; Quesnel, B.; Gautier, P.; Boulvais, P.; Couteau, C.; Drouillet, M. Nickel dispersion and enrichment at the bottom of the regolith: Formation of pimelite target-like ores in rock block joints (Koniambo Ni deposit, New Caledonia). *Miner. Depos.* **2016**, *51*, 271–282. [CrossRef]
20. Tupaz, C.A.J.; Watanabe, Y.; Sanematsu, K.; Echigo, T. Mineralogy and Geochemistry of the Berong Ni-Co laterite Deposit, Palawan, Philippines. *Ore. Geol. Rev.* **2020**, *125*, 103686. [CrossRef]
21. Lambiv Dzemua, G.; Gleeson, S.A.; Schofield, P.F. Mineralogical characterization of the Nkamouna Co–Mn laterite ore, southeast Cameroon. *Miner. Depos.* **2013**, *48*, 155–171. [CrossRef]
22. Nahon, D.B.; Paquet, H.; Delvigne, J. Lateritic weathering of ultramafic rocks and the concentration of nickel in the Western Ivory Coast. *Econ. Geol.* **1982**, *77*, 1159–1175. [CrossRef]
23. Colin, F.; Noack, Y.; Trescases, J.J.; Nahon, D. L'alteration latéritique débutante des pyroxenites de Jacuba, Niquelândia, Brésil. *Clay Miner.* **1985**, *20*, 93–113. [CrossRef]
24. Putzolu, F.; Abad, I.; Balassone, G.; Boni, M.; Mondillo, N. Ni-bearing smectites in the Wingellina laterite deposit (Western Australia) at nanoscale: TEM-HRTEM evidences of the formation mechanisms. *Appl. Clay Sci.* **2020**, *196*, 105753. [CrossRef]
25. Tauler, E.; Lewis, J.F.; Villanova-de-Benavent, C.; Aiglsperger, T.; Proenza, J.A.; Domènech, C.; Gallardo, C.; Longo, F.; Galí, S. Discovery of Ni-smectite-rich saprolite at Loma Ortega, Falcondo mining district (Dominican Republic): Geochemistry and mineralogy of an unusual case of "hybrid hydrous Mg silicate–clay silicate" type Ni-laterite. *Miner. Depos.* **2017**, *52*, 1011–1030. [CrossRef]
26. Nelson, C.E.; Proenza, J.A.; Lewis, J.F.; López-Kramer, J. The metallogenic evolution of the Greater Antilles. *Geol. Acta* **2011**, *9*, 229–264. [CrossRef]
27. Chang-Rodríguez, A.; Rojas, A. Fases minerales portadoras de níquel presentes en el horizonte saprolítico del yacimiento San Felipe. *Minería Geol.* **2015**, *31*, 1–18.
28. Chang-Rodríguez, A.; Tauler, E.; Lavaut, W.; Rojás-Purón, A.L.; Proenza, J.A. Caracterización geoquímica del perfil litológico del yacimiento laterítico de níquel "San Felipe", Camagüey, Cuba". *Rev. Cienc. Tierra Espac.* **2015**, *16*, 9–22.
29. Chang-Rodríguez, A.; Tauler, E.; Proenza, J.A.; Rojas-Purón, A.L. Mineralogía del yacimiento laterítico de níquel San Felipe. *Minería Geol.* **2016**, *32*, 28–47.
30. Cobas-Botey, R.M. Caracterización geológica de las lateritas en diferentes regiones metalogénicas: Comparación de los yacimientos San Felipe y Piloto. *Minería Geol.* **2016**, *32*, 48–59.
31. Cobas-Botey, R.M.; Formell-Cortina, F.; Leyva-Rodríguez, C. Modelo geológico descriptivo del yacimiento laterítico San Felipe, Camagüey, Cuba. *Minería Geol.* **2017**, *33*, 251–264.
32. Iturralde-Vinent, M.A.; Garcia-Casco, A.; Rojas-Agramonte, Y.; Proenza, J.A.; Murphy, J.B.; Stern, R.G. The geology of Cuba: A brief overview and syn thesis. *GSA Today* **2016**, *26*, 4–10. [CrossRef]
33. Rojas-Agramonte, Y.; Garcia-Casco, A.; Kemp, A.; Kröner, A.; Proenza, J.A.; Lázaro, C.; Liu, D. Recycling and transport of continental material through the mantle wedge above subduction zones: A Caribbean example. *Earth Planet. Sci. Lett.* **2016**, *436*, 93–107. [CrossRef]
34. Iturralde-Vinent, M. Introduction to Cuban geology and geophysics. Ofiolitas y Arcos Volcánicos de Cuba. Miami, Florida. *Int. Geol. Correl. Programme* **1996**, *364*, 3–35.
35. Lewis, J.F.; Draper, G.; Proenza, J.A.; Espaillat, J.; Jiménez, J. Ophiolite-related ultramafic rocks (serpentinites) in the Caribbean region: A review of their occurrence, composition, origin, emplacement and nickel laterite soils. *Geol. Acta* **2006**, *4*, 237–263.
36. Farr, T.G.; Rosen, P.A.; Caro, E.; Crippen, R.; Duren, R.; Hensley, S.; Kobrick, M.; Paller, M.; Rodriguez, E.; Roth, L.; et al. The Shuttle Radar Topography Mission. *Rev. Geophys.* **2007**, *45*, RG2004. [CrossRef]
37. Kelso, N.V.; Patterson, T. Introducing natural earth data—Naturalearthdata.com. *Geogr. Tech.* **2010**, *5*, 25.
38. Gallardo, T.; Tauler, E.; Garcia-Romero, E.; Proenza, J.A.; Suarez-Barrrrios, M.; Chang, A. Caracterización Mineralógica de las Esmetitas Niquelíferas del Yacimiento de San Felipe (Camagüey, Cuba). *Macla* **2011**, *15*, 89–90.
39. Lázaro, C.; García-Casco, A.; Blanco-Quintero, I.F.; Rojas-Agramonte, Y.; Corsini, M.; Proenza, J.A. Did the Turonian–Coniacian plume pulse trigger subduction initiation in the Northern Caribbean? Constraints from 40Ar/39Ar dating of the Moa-Baracoa metamorphic sole (eastern Cuba). *Int. Geol. Rev.* **2015**, *57*, 919–942. [CrossRef]

40. Young, R.A. The Rietveld Method. In *International Union of Crystallography*; Oxford University Press: Oxford, UK, 1993; p. 298. ISBN 0198555776.
41. TOPAS. *General Profile and Structure Analysis Software for Powder Diffraction Data*, version 4.2; Bruker AXS Gmbh: Karlsruhe, Germany, 2009.
42. Pouchou, J.L.; Pichoir, F. Electron probe x-ray microanalysis applied to thin surface films and stratified specimens. *Scanning Microsc. Suppl.* **1993**, *7*, 167–189.
43. Decarreau, A.; Colin, F.; Herbillon, A.; Manaceau, A.; Nahon, D.; Paquet, H.; Trauth-Badaud, D.; Trescases, J.J. Domain segregation in Ni-Fe-Mg smectites. *Clays Clay Miner.* **1987**, *35*, 1–10. [CrossRef]
44. Roque-Rosell, J.; Villanova-de-Benavent, C.; Proenza, J.A. The accumulation of Ni in serpentines and garnierites from the Falcondo Ni-laterite deposit (Dominican Republic) elucidated by means of μXAS. *Geochim. Cosmochim. Acta* **2017**, *198*, 48–69. [CrossRef]
45. Villanova-de-Benavent, C.; Domènech, C.; Tauler, E.; Galí, S.; Tassara, S.; Proenza, J.A. Fe–Ni-bearing serpentines from the saprolite horizon of Caribbean Ni-laterite deposits: New insights from thermodynamic calculations. *Miner. Depos.* **2017**, *52*, 979–992. [CrossRef]
46. Guggenheim, S.; Adams, J.M.; Bain, D.C.; Bergaya, F.; Brigatti, M.F.; Drits, V.A.; Formoso, M.L.L.; Galán, E.; Kogure, T.; Stanjek, H. Summary of recommendations of nomenclature committees relevant to clay mineralogy: Report of the Association Internationale pour l'etude des Argiles, nomenclature committee for 2006. *Clay Miner.* **2006**, *41*, 863–877. [CrossRef]
47. Bosio, N.J.; Hurst, V.J.; Smith, R.L. Nickeliferous nontronite, a 15 Å garnierite, at Niquelândia, Goiás, Brazil. *Clays Clay Miner.* **1975**, *23*, 400–403. [CrossRef]
48. Brindley, G.W.; de Souza, J.V. Nickel containing montmorillonites and chlorites from Brazil, with remarks on schuchardtite. *Miner. Mag.* **1975**, *40*, 141–152. [CrossRef]
49. Nahon, D.B.; Colin, F. Chemical weathering of orthopyroxenes under lateritic conditions. *Am. J. Sci.* **1982**, *282*, 1232–1243. [CrossRef]
50. Nahon, D.B.; Colin, F.; Tardy, Y. Formation and distribution of Mg, Fe, Mn-smectites in the first stages of the lateritic weathering of forsterite and tephroite. *Clay Miner.* **1982**, *17*, 339–348. [CrossRef]
51. Pelletier, B. Serpentines in nickel silicate ore from New Caledonia. Australasian Institute of Mining and Metallurgy publication series—Nickel conference Bmineral to market, Kalgoorlie. *West. Aust.* **1996**, *6*, 197–205.
52. Gaudin, A.; Grauby, O.; Noack, Y.; Decarreau, A.; Petit, S. Accurate crystal chemistry of ferric smectites from the lateritic nickel ore of Murrin Murrin (Western Australia). I XRD and multi-scale chemical approaches. *Clay Miner.* **2004**, *39*, 301–315. [CrossRef]
53. Gleeson, S.A.; Herrington, R.J.; Durango, J.; Velásquez, C.A.; Koll, G. The mineralogy and geochemistry of the Cerro Matoso S.A. Nilaterite deposit, Montelíbano, Colombia. *Econ. Geol.* **2004**, *99*, 1197–1213. [CrossRef]
54. Suárez, S.; Nieto, F.; Velasco, F.; Martín, F.J. Serpentine and chlorite as effective Ni-Cu sinks during weathering of the Aguablanca sulphide deposit (SW Spain). TEM evidence for metal-retention mechanisms in sheet silicates. *Eur. J. Miner.* **2011**, *23*, 179–196. [CrossRef]
55. Murofushi, A.; Otake, T.; Sanematsu, K.; Zay Ya, K.; Ito, A.; Kikuchi, R.; Sato, T. Mineralogical evolution of a weathering profile in the Tagaung Taung Ni laterite deposit: Significance of smectite in the formation of high-grade Ni ore in Myanmar. *Min. Depos.* **2022**, *57*, 1107–1122. [CrossRef]
56. Velde, B. *Clay Minerals: A Physico-Chemical Explanation of Their Occurrence*; Elsevier: Amsterdam, The Netherlands, 1985; p. 427. ISBN 0444424237.
57. Chen, T.T.; Dutrizac, J.E.; Krause, E.; Osborne, R. Mineralogical characterization of nickel laterites from New Caledonia and Indonesia. *Int. Laterite Nickel Symp.* **2004**, 79–99.
58. Golightly, J.P.; Arancibia, O.N. The chemical composition and infrared spectrum of nickel- and iron-substituted serpentine form a nickeliferous laterite profile, Soroako, Indonesia. *Can. Mineral.* **1979**, *17*, 719–728.
59. Brindley, G.W.; Bish, D.L.; Wan, H.M. The nature of kerolite, its relation to talc and stevensite. *Min. Mag.* **1977**, *41*, 443–452. [CrossRef]
60. Hillier, S. Accurate quantitative analysis of clay and other minerals in sandstones by XRD: Comparison of a Rietveld and a reference intensity ratio (RIR) method and the importance of sample preparation. *Clay Miner.* **2000**, *35*, 291–302. [CrossRef]
61. Srodon, J. Identification and Quantitative Analysis of Clay Minerals. *Dev. Clay Sci.* **2013**, *5*, 25–49. [CrossRef]
62. Reynolds, R.C. Principles and techniques of quantitative analysis of clay minerals by X-ray powder diffraction. In *Quantitative Mineral Analysis of Clays*; Pevear, D.R., Mumpton, F.A., Eds.; CMS: Bonn, Germany, 1989; p. 437.
63. Brindley, G.W.; Brown, G. *Crystal Structures of Clay Minerals and Their X-ray Identification 1980*; Mineralogical Society: Washington, DC, USA; p. 495. ISBN 0-903056-08-9.
64. Suquet, H.; Malard, C.; Copin, E.; Pezerat, H. Variation du parametre b et de la distance basale d 001 dans une serie de saponites à charge roissante: I. États hydrates. II États "zero couche". *Clay Miner.* **1981**, *16*, 181–193. [CrossRef]
65. Colin, F.; Nahon, D.; Trescases, J.J.; Melfi, A.J. Lateritic weathering ofpyroxenites at Niquelandia, Goias, Brazil: The supergene behaviour of nickel. *Econ. Geol.* **1990**, *85*, 1010–1023. [CrossRef]
66. Paquet, H.; Duplay, J.; Nahon, D.; Tardy, Y.; Millot, G. Analyses chimiques de particules isole'es dans les populations de mine´raux argileux. Passage des smectites magnésiennes trioctaédriques aux smectites ferrifères dioctaédriques au cours de l'altération des roches ultrabasiques. *Comptes Rendus L'académie Sci. II* **1983**, *296*, 699–704.
67. Aquino, K.A.; Arcilla, C.A.; Schardt, C.; Tupaz, C.A.J. Mineralogical and geochemical characterization of the Sta. Cruz Nickel laterite deposit, Zambales, Philippines. *Minerals* **2022**, *12*, 305. [CrossRef]

68. Domènech, C.; Galí, S.; Villanova-de-Benavent, C.; Soler, J.M.; Proenza, J.A. Reactive transport model of the formation of oxide-type Ni-laterite profiles (Punta Gorda, Moa Bay, Cuba). *Miner. Dep.* **2017**, *52*, 993–1010. [CrossRef]
69. Cathelineau, M.; Boiron, M.C.; Grimaud, J.L.; Favier, S.; Teitler, Y.; Golfier, F. Pseudo-Karst Silicification Related to Late Ni Reworking in New Caledonia. *Minerals* **2023**, *13*, 518. [CrossRef]

Article

Mineralogical and Geochemical Characterization of the Sta. Cruz Nickel Laterite Deposit, Zambales, Philippines

Karmina A. Aquino [1,2,*], Carlo A. Arcilla [2,3], Christian Schardt [4] and Carmela Alen J. Tupaz [5]

1 Department of Earth Sciences, Eidgenössische Technische Hochschule Zürich (ETH Zürich), Sonneggstrasse 5, 8092 Zürich, Switzerland
2 National Institute of Geological Sciences, University of the Philippines, Velasquez St., Diliman, Quezon City 1101, Philippines; caloy.arcilla@gmail.com
3 Department of Science and Technology, Philippine Nuclear Research Institute, Diliman, Quezon City 1101, Philippines
4 Artec Umweltpraxis GmbH, Fabrikgasse 2, 08294 Lößnitz, Germany; schardt07@gmail.com
5 Faculty of the International Resource Sciences, Akita University, Akita 010-8502, Japan; catupaz@gipc.akita-u.ac.jp
* Correspondence: karmina.aquino@erdw.ethz.ch

Abstract: In this study, we present mineralogical and geochemical characterization of samples systematically collected from a nickel laterite profile at the Sta. Cruz nickel laterite deposit, Zambales, Philippines. Wavelength-dispersive X-ray fluorescence spectroscopy (WDSXRF), mass-balance element mobility calculations, transmitted and reflected light microscopy, and previously reported results from coupled X-ray diffraction (XRD) and Rietveld refinement analyses reveal that the laterite profile investigated is composed of two main horizons—the limonite and saprolite zones—separated by a thin transitional zone. Based primarily on the mineral assemblage and major element chemistry, the main zones are further subdivided into subzones: upper limonite, lower limonite, transitional zone, upper saprolite, and lower saprolite. Garnierite veins were observed cutting the upper and lower saprolite subzones. Investigation of the structure of goethite within the limonite zone via Rietveld refinement shows that the crystallinity of goethite decreases with increasing Ni content and increasing crystallite size. This suggests that upwards through the limonite zone, as goethite ages, its crystallinity increases, which possibly results in the removal of Ni from its crystal structure and eventual remobilization to the lower laterite zones.

Keywords: nickel laterites; lateritization; serpentinization; weathering; goethite ageing

Citation: Aquino, K.A.; Arcilla, C.A.; Schardt, C.; Tupaz, C.A.J. Mineralogical and Geochemical Characterization of the Sta. Cruz Nickel Laterite Deposit, Zambales, Philippines. *Minerals* **2022**, *12*, 305. https://doi.org/10.3390/min12030305

Academic Editors: Cristina Domènech and Cristina Villanova-de-Benavent

Received: 7 December 2021
Accepted: 25 February 2022
Published: 27 February 2022

Publisher's Note: MDPI stays neutral with regard to jurisdictional claims in published maps and institutional affiliations.

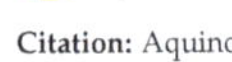

1. Introduction

Tectonic emplacement of variably serpentinized ultramafic host rock on land and subsequent chemical weathering under humid tropical to subtropical conditions results in the formation of nickel laterite deposits [1–7]. Nickel laterites typically occur as weathering mantle over ophiolite complexes, as well as komatiites and layered complexes in Archean to Phanerozoic stable cratonic platforms [2,5]. Nickel laterite deposits have a characteristic profile (from the bottom to the top): (1) bedrock consisting of partially altered ultramafics; (2) a silicate or saprolite zone characterized by Mg-silicates, such as serpentine and garnierite; and (3) an oxide or limonite zone, predominantly composed of iron oxyhydroxides, principally goethite, hematite, and maghemite [1–6]. Garnierite is a general term for a distinctively green, neoformed, fined-grained, and poorly crystalline mixture of one or more Mg-Ni phyllosilicates including serpentine, talc, chlorite, smectite, and/or sepiolite [8–13]. The term "garnierite" has been used as a field term to describe this mineral assemblage in the absence of a more detailed mineral identification. Garnierites have been classified into two groups: (a) 1:1 phyllosilicate or serpentine group and (b) 2:1 phyllosilicate group (e.g., talc, kerolite, chlorite, and sepiolite) [10,13]. They are Ni-rich and typically occur as veins along joints and shear zones, as matrix within breccias, and as coatings on saprolite blocks [4,5,14–17].

An earlier study [18] emphasizes the primary role of "goethite ageing" with the downward decrease in the bulk Ni content within the limonite of lateritic ores from New Caledonia. Goethite ageing refers to the upward increase in crystallinity of goethite within the limonite zone resulting from the expulsion of Ni from the crystal structure of goethite. Nickel is then either leached from the limonite or is sorbed on the surface of goethite crystals. This is supported by previous works [19,20] that observed a decrease in the crystallinity of goethite with increasing depth. These results suggest that as the nickel laterite profile evolves, goethite crystallinity increases, and nickel content decreases upward through the profile. Goethite ageing is therefore an important aspect of the evolution of the nickel laterite profile.

Here, we investigate samples collected from a nickel laterite profile from the Zambales Ophiolite, Philippines. We describe the mineralogy and geochemistry of the outcrop and confirm previous work on goethite ageing [18–20]. We show that small changes in the structure of goethite within the limonite zone is related to its nickel content and to the overall evolution of the laterite profile.

1.1. Zambales Ophiolite Complex

The Zambales ophiolite complex (ZOC), located in Zambales, Philippines (Figure 1a), is a generally north–south trending, east dipping complete ophiolite suite comprised of a succession of volcanic rocks, dike-sill complexes, ultramafic and mafic cumulates, residual harzburgite, and lherzolites [21–24]. The ZOC is subdivided into three massifs from north to south: the Masinloc, Cabangan, and San Antonio massifs, with each massif separated by west-northwest fault boundaries [25,26]. The Masinloc massif is made up of two blocks: the Acoje block in the north and the Coto block in the south. The Acoje block and the San Antonio massif are compositionally similar, having an island arc tholeiite (IAT) affinity, whereas the Coto block and the Cabangan massif have a signature transitional from a mid-ocean ridge basalt to island arc (MORB-IA) [22,26]. The ZOC has been dated Eocene based on the fossil assemblage of the overlying Aksitero formation [27,28]. Direct radiometric dating of various units from the ZOC concurs with the Late Eocene age of the Aksitero formation and yielded a Middle Eocene age for the ophiolite [29,30]. Schweller et al. [31] suggest an early Miocene eastward tilting and erosion of the ZOC, likely related to the subduction initiation along the ancestral Manila Trench. The Eocene age of the ophiolite implies that it is 10–15 Ma years older than the South China Sea (SCS) crust [32], suggesting that it cannot originate from the SCS [29].

1.2. Sta. Cruz Nickel Laterite Deposit

The Sta. Cruz nickel laterite deposit formed from the weathering of the ultramafic massif of the ZOC Acoje block (Figure 1). The selected nickel laterite profile is located in the municipality of Santa Cruz, northern Zambales. The deposit exhibits a typical laterite zonation, consisting of an upper limonite layer, which is underlain by the saprolite layer and bedrock. Nickel mineralization in the deposit is associated with the low-Ni, high-Fe limonite zone and the more extensive low-Fe, high-Ni saprolite zone [33]. The underlying bedrock is harzburgite with sporadic dunite lenses and minor chromitite. The harzburgite is partly serpentinized, with olivine replaced by serpentine pseudomorphs, but the grain shape has been retained. Orthopyroxene, on the other hand, is partly replaced by talc and chlorite [22].

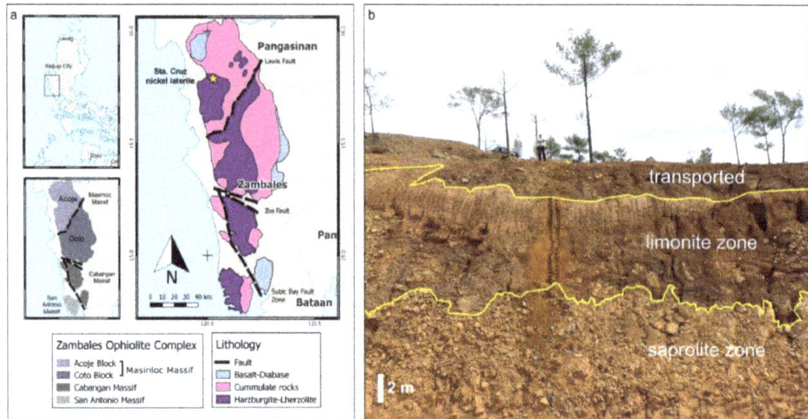

Figure 1. (**a**) Location of the study area. The Zambales Ophiolite Complex is in west Central Luzon, Philippines. It is subdivided into three massifs—the Masinloc, Cabangan, and San Antonio Massifs—separated by west–northwest fault boundaries. The Masinloc massif is subdivided into the Acoje and Coto blocks. Yellow star indicates the location of the Sta. Cruz nickel laterite deposit. (**b**) Photo of the nickel laterite profile investigated. The saprolite zone is overlain by the limonite zone. The topmost unit is a mechanically transported layer and is therefore not sampled for this study. Modified from Aquino et al. [34].

The nickel laterite profile investigated (Figure 1b) is approximately 12 m high and composed of two main units—an upper limonite zone (~7 m) and a lower saprolite zone (~5 m)—as well as a thin transitional zone (~10 cm), each characterized by a distinct set of physical, mineralogical, and geochemical properties. The detailed mineralogy of this outcrop can be found elsewhere [34,35] and is summarized in Figure 2.

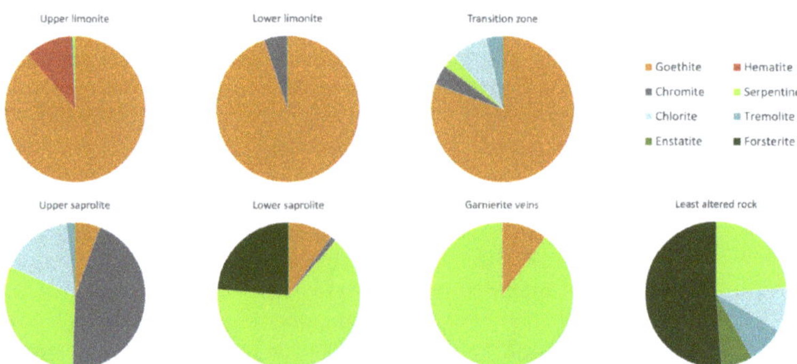

Figure 2. The laterite profile is subdivided into five subzones based on the mineralogy. The topmost zone is the upper limonite subzone, dominated by goethite with minor hematite, and is underlain by the lower limonite subzone composed of goethite and minor chromite. The transition to the saprolite is marked by the first appearance of silicate minerals and a decrease in goethite abundance. The saprolite zone is marked by the dominance of Mg-silicates and a decrease in the abundance of goethite. The upper saprolite subzone is comprised mostly of serpentine, with minor chlorite and tremolite. The lower saprolite subzone is distinguished by the presence of relict olivine. The sample taken at a depth of 10.5 m represents the least altered rock in the profile. This sample is moderately serpentinized (~40%) and has significant primary minerals. Lastly, garnierite veins are composed of about 90 wt% serpentine and 10 wt% goethite. Note that abundant chromite of primary origin was observed in the upper saprolite zone. Data are obtained from Aquino et al. [34].

2. Materials and Methods

Soil samples from the limonite and transitional zones and rock samples from the saprolite zone were collected from the nickel laterite outcrop described above (Figure 1b). The topmost 2 m of the outcrop are composed of mechanically transported materials and were not sampled for this study. Depths presented in Tables 1 and 2 and discussed throughout this paper are measured with respect to the top of the upper limonite zone. A total of 33 soil samples were collected at an interval of 20 cm, while 13 rock samples were collected at an interval of 50 cm. Samples collected near the transition zone between the limonite and saprolite zones (i.e., from 6.4 to 6.5 m) were collected at an interval of 10 cm.

2.1. Petrographic Analysis

Eight polished thin sections of the saprolite samples and the least altered rock were prepared at the Energy Research and Testing Laboratory at the Philippine Department of Energy (Taguig City, Philippines). Chromite-rich samples (>70 wt%) from the upper saprolite zone were not selected for microscopy. The samples were observed under transmitted and reflected light using a TrueVision petrographic microscope and an Olympus BX53P polarizing microscope equipped with DP74 camera at the University of the Philippines, National Institute of Geological Sciences (Quezon City, Philippines).

2.2. X-ray Diffraction Analysis

A total of 46 soil and rock samples were prepared for X-ray diffraction analyses. The samples were pulverized to ~200 mesh (<0.075 mm) using an agate mortar and pestle and were subsequently oven-dried at 105 °C for 24 h. The powdered samples were then packed on a cylindrical top-filled sample holder. The diffractograms of the samples were determined using a Bragg–Bentano Shimadzu XRD-7000 X-ray Diffractometer with CuKα radiation at the University of the Philippines, National Institute of Geological Sciences. The samples were analyzed using a step size of $1°$ per min at a $3°$ to $90°$ scan range and voltage of 30 kV. The mineral phases were then identified using the PDF4+ Minerals Database by the International Center for Diffraction Data (ICDD), as well as the Materials Data, Inc. MINERAL database [36]. Mineralogical phases were quantified via Rietveld refinement of the diffractograms using the program Siroquant, version 3.0 [34]. Furthermore, the crystallinity of the goethite phase was evaluated by observing the change in the full width at half maximum (FWHM) of the goethite (110) peak. The FWHM was calculated via the Debye–Scherrer equation [37]:

$$D = \frac{k\lambda}{\beta \cos \theta} \tag{1}$$

where k is a dimensionless shape factor, λ is the wavelength of the X-ray source (1.5418 Å for CuKα), D is the crystallite size of goethite phase as obtained from Rietveld refinement, θ is the peak position of the (110) peak in radians (0.1844), and β is the FWHM.

2.3. Whole Rock Analyses and Loss on Ignition

The samples were pulverized and dried at 105 °C for a minimum of 6 h and then cooled in a desiccator. Fused beads for each sample were then prepared using an AFM—ModuTemp automated fusion machine. The major and minor element concentrations of the samples were analyzed using a PANalytical Axios PW4400 X-ray fluorescence wavelength dispersive spectrometer. Preparation of fused beads, XRF, and loss on ignition (LOI) analyses were performed at Intertek Testing Services Philippines, Inc. (Muntinlupa City, Philippines).

2.4. Mass Balance Calculations

Mass balance calculations were performed following the isocon method [38,39]. Briefly, this method involves the evaluation of the relative changes in the concentration of an altered rock with respect to the parent rock. This is done by plotting the chemical composition of the altered rock (C^A) against the composition of the parent rock (C^O). The isocon is a line defined by one or more immobile species and the origin and has the following equation:

$$C^A = \left(\frac{M^O}{M^A}\right)C^O. \tag{2}$$

Isocons were calculated using Fe and Ti as the immobile components, as has been done in similar deposits [40,41]. The relative mass changes of a component can be evaluated as follows:

- Relative mass gain—species plotting above the isocon
- Relative mass loss—species plotting below the isocon.

Additionally, the slope of the isocon gives information on the overall change in the mass relative to the protolith (M^O/M^A).

2.5. Ultramafic Index of Alteration

For each sample, an ultramafic index of alteration (UMIA) was calculated [41]. This chemical alteration index, which is a modified version of the mafic index of alteration (MIA) [42], as well as other previously used chemical alteration indices [43,44], quantifies the chemical changes that occurred during the chemical weathering process. The UMIA is defined as:

$$\text{UMIA} = 100 \times \left[\left(\text{Al}_2\text{O}_3 + \text{Fe}_2\text{O}_{3(T)}\right) / \left(\text{SiO}_2 + \text{MgO} + \text{Al}_2\text{O}_3 + \text{Fe}_2\text{O}_{3(T)}\right)\right] \tag{3}$$

where molar ratios of the respective major elements are used. The following UMIA values are expected for each of the laterite zones: unweathered peridotite bedrock, ~3: saprolite, 4–8; limonite, 60–90 [41]. No UMIA values were previously reported for samples taken from the transition zone.

3. Results

3.1. Petrography

Samples from the limonite and transition zones are extremely weathered, with no primary minerals or textures observed in hand specimen (Table 1). The saprolite samples are also heavily weathered and altered in hand specimen and thin section (Figures 3–5). Primary minerals, such as olivine and orthopyroxene, are sparse in most samples except for the least weathered sample, N-1050 (Figure 6). These primary minerals are altered mostly to serpentine (~60–85%), although poorly crystalline Fe oxides (~5–15%) and magnetite (trace) are also present. All samples contain small amounts (<5%) of chromite, except for samples N-690 and N-1000. N-690 is adjacent to a chromitite layer located at depths between 6.8 m and 7.0 m and thus contains a significant amount of chromite (~20%, Table 2). N-1000 contains visible chromite bands (Figure 3f) in hand specimen.

Table 1. Macroscopic description of samples from the limonite and transition zones.

Depth (m)	Zone [1]	Munsell Color	Descriptive	Short Description
0	UL	7.5YR 5/8	dark brown	mostly very fine soil (<1 mm) to coarse fragments (>10 mm) of Fe oxides
0.8	UL	5YR 4/8	reddish brown	mostly very fine to fine soil (<1 mm) to medium fragments (<5 mm) of Fe oxides; slightly darker red color probably due to presence of small amounts of hematite
1.6	UL	5YR 4/8	reddish brown	fine to medium (1–5 mm) grained fragments of Fe oxides; slightly darker red color probably due to presence of small amounts of hematite

Table 1. *Cont.*

Depth (m)	Zone [1]	Munsell Color	Descriptive	Short Description
2.4	UL	5YR 4/8	reddish brown	mostly very fine soil (<1 mm) to coarse fragments (>10 mm) of Fe oxides; slightly darker red color probably due to presence of small amounts of hematite
3.2	LL	5YR 4/8	reddish brown	mostly medium to coarse grained (5–10 mm) fragments of Fe oxides; slightly darker red color probably due to presence of small amounts of hematite
4.0	LL	7.5YR 5/8	dark brown	fine soil (~1 mm) to medium fragments (<5 mm) of Fe oxides
4.8	LL	7.5YR 5/8	dark brown	fine soil (~1 mm) to coarse fragments (>10 mm) of Fe oxides
5.6	LL	7.5YR 5/8	dark brown	mostly medium to coarse grained (5–10 mm) fragments of Fe oxides
6.4	T	7.5YR 5/8	dark brown	poorly sorted mixture of mostly very coarse grained (>30 mm) and medium grained (1–5 mm) Fe oxide-rich rock fragments

[1] UL = upper limonite, LL = lower limonite, T = transition zone.

Four types of serpentine, distinguished by their occurrence and association with secondary magnetite, were observed in the saprolite of the Sta. Cruz nickel laterite deposit. Type 1 serpentine (Figures 4–6) occurs immediately adjacent to partially or completely dissolved relict olivine grains. It is characteristically magnetite-free, or, when present, magnetite occurs only in minor amounts. Moreover, type 1 serpentine is pale green and slightly pleochroic under plane-polarized light and exhibits up to first-order yellow interference colors in crossed-polarized light. Type 1 serpentine is further subdivided into types 1a and 1b serpentine. Type 1a has a thickness of not more than 30 μm, and it occurs in the least altered rock along fractures and grain boundaries in olivine and pyroxene. Type 1b occurs in the saprolite, exhibits a mesh texture around olivine fragment, and is approximately 50 to 150 μm thick. Type 2 serpentine, on the other hand, occurs as roughly parallel serpentine veins cross-cutting type 1 serpentine and is characteristically magnetite and Fe-oxide stain free. They typically occur in the lower saprolite (e.g., N-900; Figure 5). They are colorless in plane-polarized light and exhibit almost black interference colors in crossed-polarized light. Type 3 serpentine, like type 2, occurs as veins and is typically observed in the upper saprolite. It is often associated with moderate to abundant secondary magnetite, which is usually within the core of the serpentine vein (e.g., N-650; Figure 4). In plane-polarized light, type 3 serpentine is colorless to pale greenish yellow and has low relief in crossed-polarized light. It exhibits very low interference colors of up to first-order gray to white. Lastly, garnierite serpentine veins (type 4) are light green to pale green, vein-type serpentine observed to crosscut the saprolite zone (Figure 3). In thin section, type 4 serpentine is light yellowish brown in plane-polarized transmitted light and has a bright yellowish green interference color in crossed-polarized light (Figure 5).

Figure 3. Saprolite samples in hand specimen. (**a**) Sample N-650R showing black serpentine–magnetite veins (type 3) being altered into light-green serpentine (type 4, N-650V). (**b**) Sample N-1100, heavily altered saprolite sample with a garnierite vein (type 4). Note that the serpentine vein emanates from what appears to be boundaries of relict grains. (**c**) Sample N-1050, least altered saprolite sample showing pyroxene crystals about 2–3 mm in size. (**d**) Sample N-690, containing unaltered chromite disseminated in a matrix of strongly weathered Fe (hydr)oxides. (**e**) Sample N-850, heavily weathered sample with a less altered core. Small dissolution vugs later filled by Fe hydr(oxides) are also present. (**f**) Sample N-1000, showing unaltered disseminated chromite in an Fe hydr(oxide) weathered matrix. Dissolution vugs are also present.

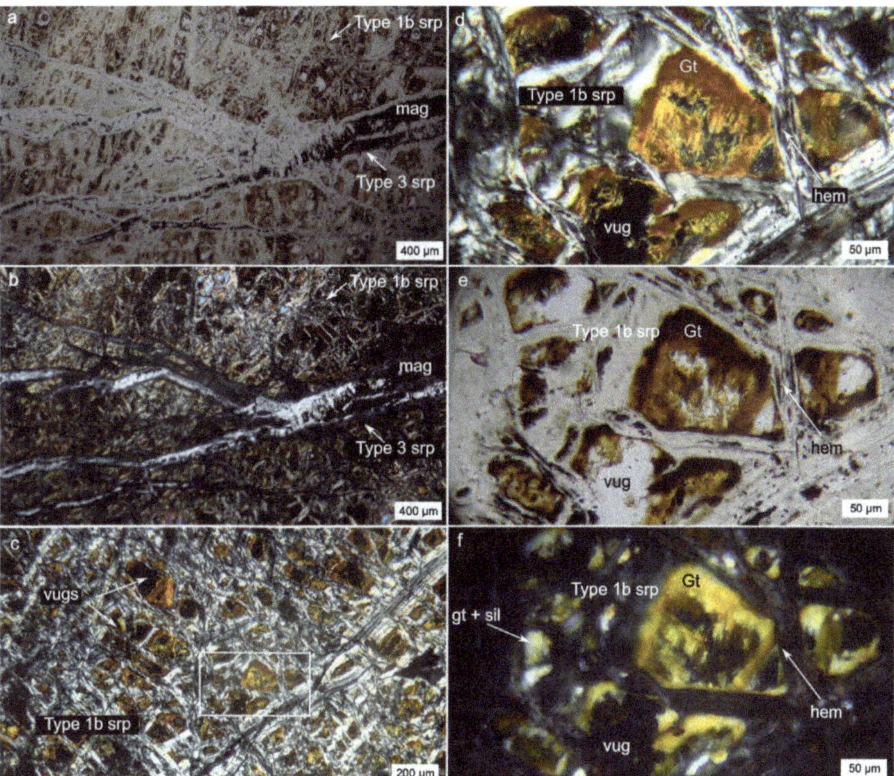

Figure 4. Upper saprolite sample N-650R in thin section. (**a**) Sample N-650R is a heavily serpentinized rock with almost no primary olivine remaining. Type 1b (mesh-type) serpentine (srp) surrounding partially or completely dissolved olivine, crosscut by type 3 (vein-type, magnetite-rich) serpentine in plane-polarized transmitted light. (**b**) Same as (**a**) in crossed-polarized transmitted light. (**c**) Occurrence of poorly crystalline Fe-oxides (gt), as well as minor amorphous silica (sil), as precipitates within dissolution vugs after olivine, in crossed-polarized transmitted light. (**d**) Detail of white box in (**c**), showing a dissolution vug filled with Fe-oxides. Surrounding mesh serpentine (type 1b) contains minor hematite (hem), likely from oxidation of magnetite. (**e**) Same as (**c**) in plane-polarized transmitted light. (**f**) Same as (**c**) in crossed-polarized reflected light, showing bright orange internal reflections of poorly crystalline Fe-oxides, red internal reflections of hematite, and white internal reflections of amorphous silica.

3.1.1. Petrography of Upper Saprolite Samples

Samples from the upper saprolite (e.g., N-650R; Figure 4) are strongly weathered and altered to serpentine (up to 85%). Most of the primary minerals are altered to serpentine, 15% poorly crystalline goethite (poorly crystalline iron oxide), and about 3% magnetite. Primary minerals, including about 5% olivine, 2% tremolite, and trace amounts of orthopyroxene, are sparse. Olivine crystals, where present, are about 40 to 200 μm in size with average size of about 100 μm, whereas orthopyroxene minerals are about 200 to 1400 μm in size. Dissolution cavities occur within sites of relict primary minerals (Figure 4). In sample N-660, the size of the cavities is approximately 50 to 300 μm, which is similar in size to the cavities and relict olivine minerals observed in sample N-650R. These cavities are often filled partially or completely with secondary, poorly crystalline goethite ± amorphous silica (Figure 4c–f). Serpentine occurs both as vein-type serpentine (type 3), as well as mesh-type serpentine (type 1b) around completely or partially dissolved olivine crystals (Figure 4c).

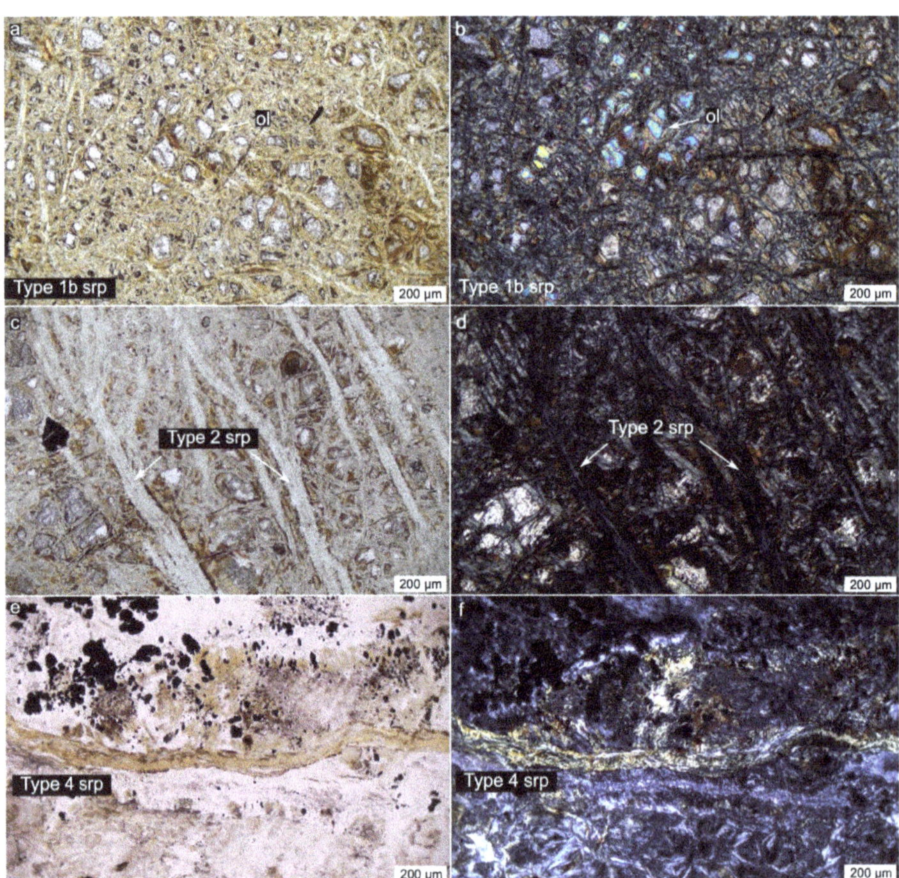

Figure 5. Thin section of samples from the lower saprolite. (**a**) Sample N-750 is strongly serpentinized and contains a significant amount of primary olivine fragments in optical continuity. Note the brown discoloration on some of the type 1b serpentines, likely due to the presence of Fe-oxide. (**b**) Same as (**a**) in crossed-polarized transmitted light. (**c**) N-900 showing roughly parallel serpentine veins (type 2 serpentine) free from Fe-oxide discoloration. Further shown are partially dissolved relict olivine grains. (**d**) Same as (**c**) in crossed-polarized transmitted light. (**e**) Photomicrograph of type 4 serpentine crosscutting the lower saprolite in plane-polarized transmitted light. (**f**) Same as (**e**) in crossed-polarized transmitted light.

3.1.2. Petrography of Lower Saprolite Samples

Samples from the lower saprolite zone (Figure 5) are less altered to serpentine (~60%) than the upper saprolite samples. Primary minerals, mostly olivine, are more abundant (up to 30%) and occur as 2–4 mm-sized groups of optically continuous crystals, with each individual fragment about 30 to 300 μm in size. Minor minerals include chromite (5%), as well as amorphous Fe oxides and trace amounts of magnetite. In sample N-750, serpentine occurs both as vein-type and mesh-type (type 1b), with the vein-type serpentine associated with fillings of magnetite (type 3). Samples taken from the deeper sections (e.g., N-900) are observed to contain both mesh-type (type 1b) and magnetite-free, subparallel serpentine veins (type 2) (Figure 5c,d). Throughout the lower saprolite, olivine crystals show dissolution features and, in places, are observed to be directly altered to brown Fe-oxides (Figure 5). Poorly crystalline goethite imparts a generally brown tinge in the lower saprolite samples as a result of either direct replacement of partially dissolved olivine (i.e., as discoloration in minerals) or infills within dissolution cavities.

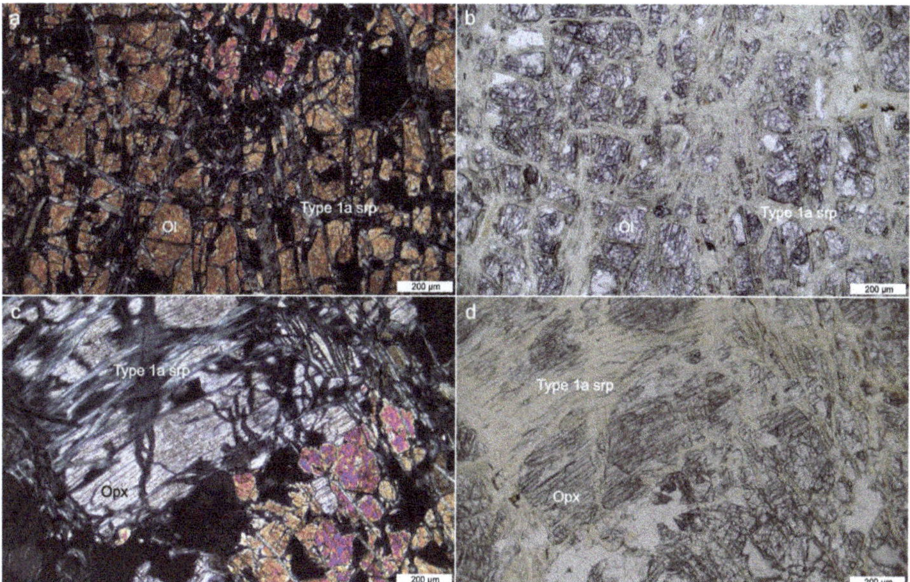

Figure 6. Photomicrographs of the least altered rock in the profile investigated. (**a**) Olivine (Ol) cut by crosscutting serpentine (type 1a) in transmitted plane-polarized light and (**b**) in crossed-polarized light. (**c**) Pyroxene (Opx) is also partially serpentinized parallel to its cleavage planes, in transmitted plane-polarized light. (**d**) Same as (**c**) in crossed-polarized transmitted light.

3.1.3. Petrography of the Least Altered Rock

Sample N-1050 represents the least altered rock in the profile (Figure 6). The sample is moderately serpentinized and contains about 40% serpentine but is also composed of abundant primary minerals including about 50% olivine, 10% pyroxene, and trace amounts of chromite. Olivine is fragmented and is cut by crosscutting vein-type type 1a serpentine approximately 20–30 µm thick, forming isolated fragments about 100 to 200 µm in size in optically continuous groups of about 2 to 4 mm. The olivine surface is rough and dissolved, as clearly seen in plane-polarized transmitted light. Dissolution cavities are also present locally in the sample and have not yet been filled with any secondary material. Unlike in other samples, olivine crystals are colorless with minimal alteration to Fe-oxides. The serpentine veins are also colorless and do not seem to have Fe-oxide staining. Pyroxene minerals are preserved in this sample and occur as large crystals more than 4 mm in size, often cut by serpentine (type 1a) parallel to cleavage traces (Figure 6).

3.2. Goethite Crystallinity and Crystallite Size Measurements

Crystallite size measurements of the goethite phase obtained from Rietveld refinement of the limonite layer reveal a generally decreasing trend with depth (Figure 7a, Table A1). Goethite crystallite sizes, which initially increase from 115 to 139 Å in the topmost upper limonite, show a pronounced decreasing trend from 139 to 112 Å in the upper limonite (i.e., depths of 0.8 to 3.2 m). Towards the mid-lower limonite, at depths of 3.6 to 4.8 m, the crystallite sizes slightly increase in value to 127 Å before decreasing again to 117 Å towards the bottom of the lower limonite. In the transition zone, the value increases again to 124 Å.

The calculated full width at half maximum (FWHM) values from the measured crystallite sizes are shown in Figure 7b and Table A1. As expected from the Debye–Scherrer equation (Equation (1)) [37], the crystallite size and the FWHM have an inverse relationship. Thus, a generally increasing trend with depth in the FWHM values is observed for the studied limonite samples. FWHM values first decrease slightly from a value of 1.09 degrees at the uppermost limonite layer (0 m) to 0.91 degrees at 0.8 m and then generally increase to a maximum of 1.11 degrees at a depth of 3.2 m. The FWHM values decrease slightly again to 0.99 degrees at 4.8 m and decrease again up to 1.07 towards the bottom of the lower limonite at a depth of 6.0 m.

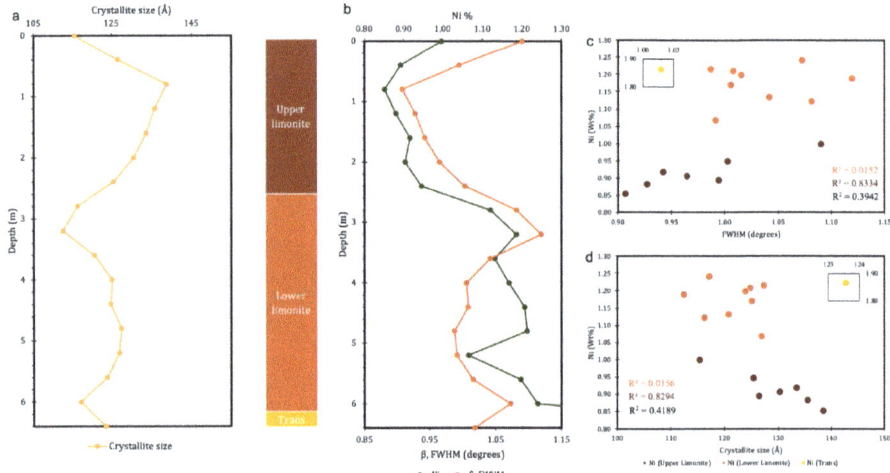

Figure 7. (**a**) Variations in crystallite size of goethite in the limonite and transition zones with depth, as calculated via Rietveld refinement. (**b**) Bulk Ni content and calculated full width at half maximum values of goethite in the limonite and transition zones with depth. (**c**) Relationship of bulk Ni content and full width at half maximum (FWHM) of goethite and (**d**) bulk Ni content and crystallite size of goethite in the limonite and transition zones. Further shown are the R^2 values of the samples from the upper limonite (red), lower limonite (orange), and whole limonite (black) zones.

3.3. Bulk Major and Minor Geochemistry

3.3.1. Limonite Zone

The limonite zone contains elevated Fe_2O_3 (up to 75.4 wt%) and Al_2O_3 (up to 14.4 wt%) and relatively low MgO (average: 1.5 wt%) and SiO_2 (average: 4.5 wt%) concentrations (Table 2, Figure 8a). Within the limonite, Fe content slightly decreases upwards. Interestingly, the Al_2O_3 content of the upper limonite zone is generally higher compared to the lower limonite zone. Within the limonite, the MgO content slightly increases downwards from an average value of 1.2 wt% in the upper limonite to 1.7 wt% in the lower limonite. Contrary to the behavior of MgO values, SiO_2 content shows a decreasing trend downwards within the limonite, with an average value of 6.6 wt% in the upper limonite to 2.9 wt% in the lower limonite. Significant concentrations of Ni, Mn, and Co are noted in the limonite zone of the investigated profile (Figure 8b). Within the limonite, NiO content ranges from 1.1 to 1.6 wt% and generally increases with depth. Average values for MnO and Co are 0.91 and 0.10 wt%, respectively. Both MnO and Co show a slightly increasing trend in the limonite zone with depth. MnO values are enriched up to 1.09 wt% at the lower limonite while Co values are enriched up to 0.16 wt% near the transition zone to the saprolite.

3.3.2. Transition Zone, Saprolite Zone, and Garnierite Veins

Fe_2O_3 shows a sharp decrease in concentration towards the transition and saprolite zones (average of 43.2 wt% and 12.5 wt%, respectively; Figure 8a). Al_2O_3 content in the transition zone is 6.2 wt%, which decreases further to an average of 3.1 wt% in the saprolite zone. From the limonite zone, MgO and SiO_2 increase to values of 12.6 wt% and 16.5 wt%, respectively, in the transition zone and to values of 31.6 wt% and 32.7 wt%, respectively, in the saprolite zone. NiO content is generally more elevated in the saprolite than in the limonite, with values of up to 3.0 wt% (Figure 8b). MnO and Co values are generally not as significant within the saprolite as in the limonite zone, with average values of 0.21 wt% for MnO and 0.02 wt% for Co.

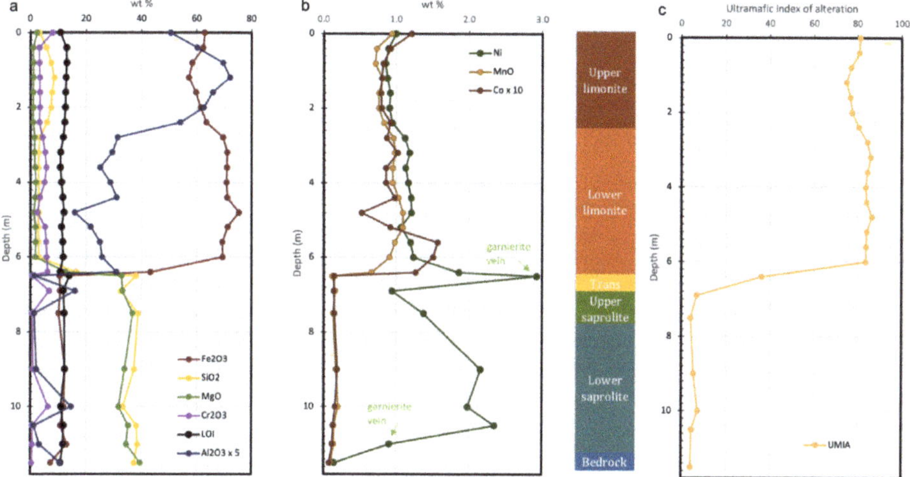

Figure 8. (**a**) Major and (**b**) minor element geochemistry of the studied profile and the corresponding laterite horizons. (**c**) Calculated ultramafic index of alteration (UMIA) [41]. Note the relatively high Ni contents of the upper garnierite vein.

Overall, the two garnierite veins have slightly higher SiO_2, MgO, and NiO and lower Fe_2O_3, Al_2O_3, and MnO than the average saprolite (Table 2). Interestingly, the serpentine vein sampled at a higher depth has a higher NiO content than the vein sampled below the lower saprolite. LOI values do not show any significant difference between each laterite horizon, with values ranging from 10.8 to 14.0 wt% within the profile.

3.4. Ultramafic Index of Alteration

The ultramafic index of alteration (UMIA) values calculated for the Sta. Cruz nickel laterite profile (Table 2, Figure 8c) are generally consistent with expected values [41]. UMIA values in the limonite zone range from 74.7 to 86.5. The upper limonite has slightly lower UMIA values than the lower limonite. Towards the transition zone, the UMIA values decrease to 36.1, before decreasing significantly to values less than 8 in the saprolite. There is no significant trend in the UMIA values between the upper and lower saprolite or the garnierite veins.

Figure 9 shows plots of the Sta. Cruz nickel laterite samples in molar ternary AF-S-M (Al_2O_3 + Fe_2O_3-SiO_2-MgO) and A-SM-F (Al_2O_3-SiO_2 + MgO-Fe_2O_3) diagrams, including their corresponding UMIA values. Both diagrams show weathering of an initially Mg- and Si-rich peridotite bedrock towards an Al- and Fe-rich limonite. There is a clear separation of the upper and lower limonite samples in both diagrams, indicating that the upper limonite contains relatively more Al and Si compared to the lower limonite.

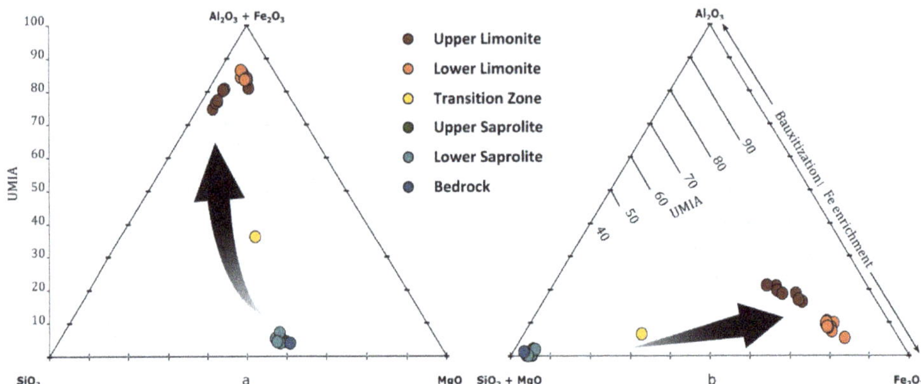

Figure 9. Chemical evolution of the laterite profile during weathering as shown by (**a**) molar ternary $Al_2O_3+Fe_2O_3$-SiO_2-MgO (AF-S-M) and (**b**) molar ternary Al_2O_3-SiO_2+MgO-Fe_2O_3 (A-SM-F) diagrams. Both diagrams show weathering of an initially Mg- and Si-rich peridotite bedrock towards an Al- and Fe-rich limonite while (**b**) highlights that Fe-enrichment is predominant over bauxitization in the Sta. Cruz nickel laterite. Further shown is the ultramafic index of alteration (UMIA) [41].

3.5. Relative Mass Changes

Relative mass changes (Figure 10, Table A2) calculated for the studied outcrop are similar to those observed in other laterite deposits [40,45]. The limonite zone is characterized by an almost complete removal of SiO_2 (i.e., up to 99% relative mass loss) and MgO (up to 100% mass loss). Ni is also leached by up to about 50% in the limonite, with the amount of depletion decreasing with depth. Co and MnO, on the other hand, are associated with mass gains in the limonite. Co mass gains increase in the lower limonite towards the transition zone. A similar behavior is observed for Mn but to a lesser degree.

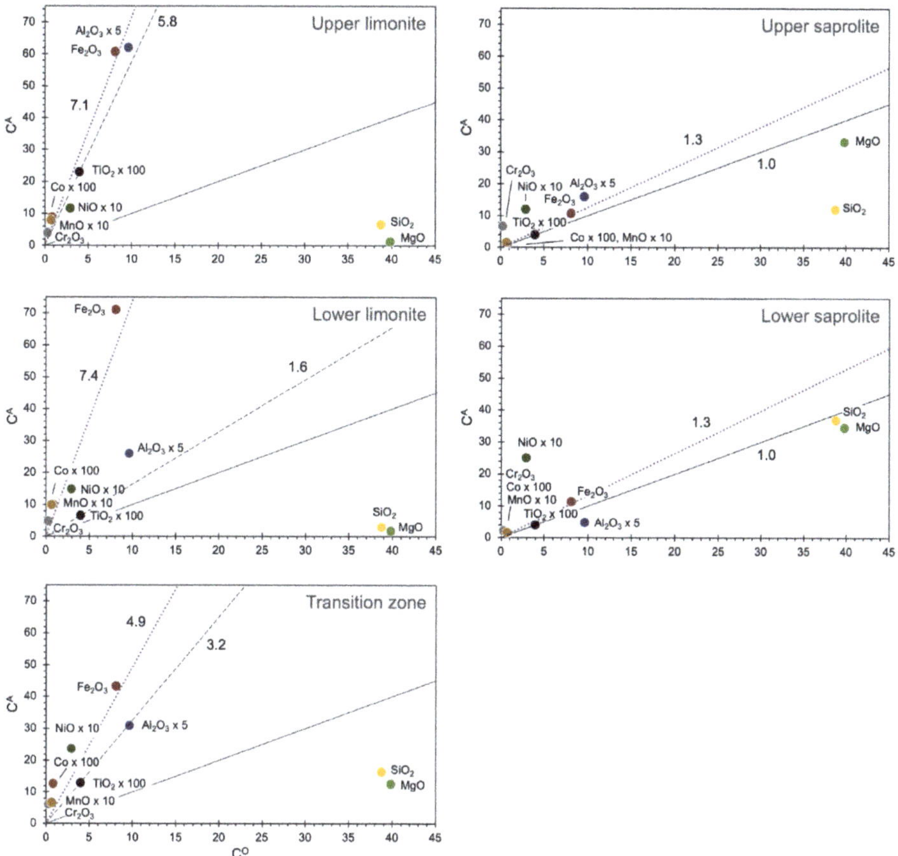

Figure 10. Isocon diagrams for each laterite zone in the outcrop investigated. Altered compositions (C^A) are the average composition of the samples from each zone. The composition of the parent rock (C^O) is the average composition of two peridotite samples collected near the study area (Table 2). Dashed line is the isocon defined by the immobile species Ti and the origin, while dotted line is the best-fit isocon defined by Fe, Ti, and the origin. Solid line represents $M^O/M^A = 1$. The numbers are the slope of the isocon or M^O/M^A. Species plotting along the isocon did not change their mass during transformation, species plotting above the isocon experienced mass gains, whereas species plotting below the isocon experienced mass loss during transformation.

The transition zone is characterized by a slight decrease in the associated SiO_2 and MgO mass loss. Ni starts to experience a slight mass gain (+64%), while Co mass gains are the highest around this zone. Towards the saprolite zone, MgO and SiO_2 mass losses decrease to an average of −34% and −33%, respectively. The saprolite zone is also characterized by mass gains of NiO, MnO, and Co, with the NiO mass gains reaching more than +600%. Interestingly, the least altered rock (N-1050) is associated with significant Ni mass gain of +668%.

Table 2. Bulk geochemistry and ultramafic index of alteration (UMIA) of samples from the Sta. Cruz nickel laterite. All compositions are in wt%.

Sample	Horizon [1]	Depth (m)	SiO_2	TiO_2	Al_2O_3	Fe_2O_3	Cr_2O_3	MnO	NiO	Co	MgO	CaO	BaO	Na_2O	K_2O	P_2O5	SO_3	LOI	Total	UMIA
N-000	UL	0.0	3.28	0.12	10.08	62.91	7.91	0.94	1.27	0.12	2.42	<0.01	<0.005	<0.01	<0.01	0.01	0.29	10.78	100.0	81.1
N-040	UL	0.4	5.75	0.21	12.03	62.34	3.14	0.73	1.14	0.09	1.02	<0.01	<0.005	0.02	<0.01	0.02	0.39	12.96	99.7	80.8
N-080	UL	0.8	7.43	0.29	13.94	58.27	3.23	0.71	1.08	0.08	1.09	<0.01	<0.005	0.02	0.01	0.02	0.34	13.03	99.5	76.9
N-120	UL	1.2	8.53	0.31	14.42	57.04	3.33	0.81	1.12	0.08	1.07	<0.01	<0.005	<0.01	0.02	0.02	0.30	12.84	99.8	74.7
N-160	UL	1.6	7.74	0.26	13.17	59.67	3.33	0.76	1.17	0.08	1.01	<0.01	<0.005	0.01	0.01	0.02	0.30	12.65	100.1	76.6
N-200	UL	2.0	7.38	0.24	12.51	61.43	3.38	0.76	1.15	0.08	1.04	<0.01	<0.005	<0.01	<0.01	0.01	0.30	12.57	100.8	77.3
N-240	UL	2.4	5.92	0.18	10.78	63.52	3.16	0.83	1.20	0.09	0.95	<0.01	<0.005	0.02	<0.01	0.01	0.24	12.34	99.2	80.5
N-280	LL	2.8	3.27	0.08	6.27	69.67	4.38	0.96	1.43	0.09	1.51	<0.01	0.02	0.02	<0.01	0.01	0.18	11.86	99.7	84.4
N-320	LL	3.2	2.85	0.06	5.88	71.17	5.26	0.98	1.51	0.10	1.42	<0.01	0.01	0.01	<0.01	0.01	0.17	10.84	100.2	85.9
N-360	LL	3.6	2.84	0.06	5.02	71.05	5.76	0.95	1.44	0.09	1.75	<0.01	0.01	0.02	<0.01	0.01	0.15	10.81	99.9	84.5
N-400	LL	4.0	2.90	0.06	5.74	70.90	5.06	0.95	1.49	0.09	1.99	<0.01	<0.005	0.01	<0.01	0.01	0.16	11.19	100.5	83.7
N-440	LL	4.4	3.09	0.07	6.20	71.19	3.49	1.03	1.54	0.10	1.80	<0.01	<0.005	0.02	<0.01	0.01	0.17	11.66	100.3	84.1
N-480	LL	4.8	2.89	0.04	3.18	75.38	2.47	1.09	1.54	0.05	1.24	<0.01	0.06	0.03	<0.01	0.01	0.14	12.07	100.1	86.5
N-520	LL	5.2	2.76	0.06	4.33	71.31	5.24	1.09	1.36	0.09	1.86	<0.01	0.03	0.02	<0.01	0.01	0.12	11.25	99.4	84.2
N-560	LL	5.6	2.86	0.08	5.00	69.43	5.70	0.98	1.52	0.16	1.86	<0.01	<0.005	0.02	<0.01	0.01	0.13	11.65	99.2	83.8
N-600	LL	6.0	3.05	0.08	5.15	69.53	5.85	0.91	1.58	0.15	1.79	0.01	<0.005	<0.01	<0.01	0.01	0.14	11.82	99.9	83.6
N-640	T	6.4	16.48	0.13	6.18	43.23	6.16	0.66	2.37	0.13	12.57	0.65	<0.005	0.04	<0.01	0.01	0.06	10.77	99.3	36.1
N-650	V	6.5	38.04	<0.01	0.29	10.47	0.63	0.12	3.72	0.02	32.87	<0.01	<0.005	<0.01	<0.01	<0.001	0.02	13.98	100.1	
N-690	US	6.9	32.52	<0.01	3.19	10.68	6.73	0.16	1.20	0.01	33.16	<0.01	0.01	0.19	<0.01	<0.001	<0.001	11.87	99.7	6.7
N-750	LS	7.5	38.74	<0.01	0.28	10.16	0.57	0.13	1.75	0.02	36.85	0.02	<0.005	0.01	<0.01	<0.001	0.01	12.20	100.7	4.1
N-900	LS	9.0	37.34	<0.01	0.43	12.51	0.90	0.17	2.75	0.02	33.94	0.01	<0.005	<0.01	<0.01	<0.001	0.01	12.36	100.4	5.3
N-1000	LS	10.0	33.31	0.04	2.93	11.96	6.33	0.20	2.52	0.02	31.83	<0.01	<0.005	<0.01	<0.01	<0.001	0.01	11.01	100.1	7.2
N-1050	LS	10.5	38.27	<0.01	0.25	10.88	0.54	0.14	2.99	0.01	35.29	0.01	<0.005	<0.01	<0.01	<0.001	0.01	11.98	100.4	4.5
N-1100	V	11.0	38.55	0.01	0.64	13.08	0.65	0.14	1.15	0.01	34.60	<0.01	<0.005	<0.01	<0.01	<0.001	0.01	11.82	100.7	
PH-S-15A	BR		37.42	0.04	2.16	7.29	0.19	0.08	0.19	0.01	39.61	1.99	0.01	<0.01	0.01	0.01		10.89	99.9	4.0
PH-S-15B	BR		40.13	0.04	1.70	8.93	0.36	0.06	0.40	0.01	40.05	1.49	<0.01	<0.01	0.01	0.01		7.73	100.5	4.2

[1] UL = upper limonite, LL = lower limonite, T = transition zone, V = vein, US = upper saprolite, LS = lower saprolite, BR = bedrock.

4. Discussion

4.1. Laterite Zonation

Based on mineralogy, geochemistry, and calculated UMIA, the Sta. Cruz nickel laterite deposit can be divided into two main zones—an upper limonite layer and a lower saprolite layer—separated by a transition zone. The transition zone is characterized by mineralogy and geochemistry intermediate between the limonite and saprolite layer. The limonite is further subdivided into two layers: upper and lower limonite. Similarly, the saprolite can be divided into upper and lower saprolite zones. Lastly, garnierite veins were observed cutting the upper and lower saprolite layers.

4.1.1. Limonite Zone

The limonite zone represents the most evolved layer of the laterite profile investigated, with a UMIA value of 78. It is an Fe- and Al-rich layer, completely devoid of primary minerals and made up of the mineral assemblage goethite + hematite for the upper limonite and goethite + chromite for the lower limonite. Mg and Si are almost completely leached out of this layer, with relative mass losses of almost 100%. The upper limonite contains a slightly higher amount of SiO_2 than the lower limonite, resulting in a lower UMIA for the upper limonite. This is probably due to the presence of amorphous silica, a common reaction product of weathering, along with amorphous Fe-oxides. Amorphous silica could not be detected by X-ray diffraction due to the absence of a crystalline structure. Within the limonite, MnO, NiO, and Co generally increase with depth and are enriched in the lower limonite. Bulk chemistry of the samples reveals a significant amount of Al_2O_3 of more than 10 wt%, although Al-bearing minerals, such as gibbsite, were not observed in the limonite zone. Goethite with 18% Al substitution best fits the X-ray diffractogram of samples from the upper limonite [35]. Therefore, in the absence of mineral chemistry data, it is inferred here that Al is hosted in the limonite by goethite. We hypothesize that Ni within the limonite is hosted primarily by goethite, with the nickel contents observed to be directly proportional to the FWHM of the goethite (110) peak and inversely proportional to the goethite crystallite size (Figure 7). A stronger correlation (R^2 = 0.83) is observed in the upper limonite than in the lower limonite (R^2 = 0.016) for both FWHM and crystallite size. One possibility is that Ni is not exclusively hosted by goethite in the lower limonite zone. For example, Mn oxyhydroxides, which are often reported in lower oxide zones of Ni laterites [5,46], including two other deposits from the Philippines [47,48], can host Ni. In particular, significant amounts of Ni (up to 15.6 wt%) have been reported to be associated with Mn oxyhydroxides from the Intex deposit in Mindoro, Philippines [48]. Mn oxyhydroxides are difficult to detect with the methods performed in this paper, but the slight increase in Mn and Co concentrations towards the lower limonite may hint at their presence in the Sta. Cruz laterite. Because phases other than goethite may host Ni, a weaker correlation between the bulk Ni contents and both goethite crystallite size and FWHM of the goethite (110) peak is observed. The stronger correlation between bulk Ni and goethite crystallinity in the upper limonite support our hypothesis that goethite is the main host of Ni.

4.1.2. Transition Zone

The transition zone (~6.5 m) is a thin layer between the saprolite and limonite zone and is characterized by an abrupt mineralogical, textural, and geochemical transition from the limonite into the saprolite layer. This layer is comprised of the mineral assemblage goethite + lizardite + chlorite + tremolite + chromite. Goethite is the dominant mineral in the limonite zone, whereas lizardite, together with minor amounts of chlorite and tremolite, is characteristic of the saprolite zone. The change in the mineral assemblage in the transition zone is reflected in the geochemistry: abrupt increase in Mg and SiO_2 corresponds to the first appearance of Mg-silicates and a decrease in the Fe_2O_3 associated with a decrease in abundance of goethite. The transition zone has a UMIA of 36, a value in between the expected value of at least 60 for the limonite zone and the expected value of 4–8 for the saprolite zone.

Ni in the transition zone is hosted predominantly by goethite, but other phases, such as serpentine and chlorite, may also host Ni. Thus, bulk Ni content cannot be used to determine the effect of Ni content on the structure of goethite in the sample from the transition zone. For both FWHM and crystallite size, the sample from the transition zone is a clear outlier in the generally observed trends.

4.1.3. Saprolite Zone

The saprolite zone is comprised mostly of serpentine and has a geochemistry dominated by MgO and SiO_2. The upper saprolite (6.6 to 6.9 m) is composed of the assemblage lizardite + chlorite + chromite + goethite $\pm$ tremolite. The lower saprolite (7.5 to 11 m) contains a significant amount of primary minerals, mostly olivine, and is made up of the mineral assemblage lizardite + olivine + goethite. Except for the topmost sample from the upper saprolite, samples from the saprolite zone have UMIA values between 4 and 7. Fe is hosted in the saprolite within serpentine, chromite, and poorly crystalline Fe-oxides, occurring as alteration of olivine and serpentine. Minor amounts of Al_2O_3 in the saprolite suggest the absence of Al-bearing minerals, although small amounts of Al may substitute for Si in the serpentine crystal structure [49,50]. Significant concentrations of Al were observed in saprolite samples containing chlorite.

Four types of serpentine have been recognized in the saprolite zone of the deposit. Type 1a serpentine is interpreted to have formed earlier during the initial hydrothermal serpentinization of the ZOC [51–54]:

$$(Mg,Fe,Ni)_2SiO_4 + 1.5\ H_2O \rightarrow 0.5\ (Mg,Fe,Ni)(OH)_2 + 0.5\ (Mg,Fe,Ni)_3Si_2O_5(OH)_4. \tag{4}$$
$$\text{Olivine} \qquad\qquad\qquad \text{Fe-Brucite} \qquad\qquad \text{Serpentine}$$

This is supported by the pseudomorphic nature of type 1a serpentine after primary minerals. Although both olivine and orthopyroxene can be altered to serpentine, olivine is more abundant and has more Ni content (0.24 to 0.36 wt%) than orthopyroxene (0.07 to 0.08 wt%) in the ZOC peridotite [22]. It follows that olivine is the main source of Ni in the deposit. Over time and with further reaction with water, serpentinization may continue, resulting in the continued formation of serpentine along olivine grain boundaries and thickening of type 1a serpentine to form type 1b serpentine, which were observed in the upper levels of the saprolite.

One major consequence of the hydration of peridotite is the associated volume increase that occurs as a rock composed of dense minerals is altered to less-dense secondary minerals [55–57]. Volume expansion associated with serpentinization has been discussed earlier [56]. More recently, a volume increase of 44 $\pm$ 8% [55] has been experimentally measured associated with serpentinization for a duration of 10 to 18 months. The associated volume expansion can both seal existing fractures in the system and induce new fractures at the same time. These tension cracks or cross-fractures [56] are formed during earlier, moderately high-temperature (>200 °C) serpentinization, although we recognize that this process may continue at lower temperatures (<100 °C) [58–64]. Relatively high Ni content in the least. altered rock relative to unaltered protolith (1.15% vs. 0.29 wt% NiO), which contains only type 1 serpentine, may hint at the lower temperature formation of the serpentine, although further analyses will be needed to confirm this. Roughly parallel serpentine veins (type 2 serpentine) may later form in these cracks, likely from solutions leached [4,5,12] from the upper horizons of the profile. Type 2 serpentine have been observed to cut earlier formed type 1 serpentine in the saprolite samples (Figure 5).

Type 3 serpentine are characteristically magnetite rich. As suggested earlier [54], magnetite may form from the oxidation of pre-existing Fe-rich serpentine via:

$$Fe_3Si_2O_5(OH)_4 + 0.5\ O_{2(aq)} \rightarrow Fe_3O_4 + 2\ SiO_{2(aq)} + 2\ H_2O. \tag{5}$$
$$\text{Serpentine} \qquad\qquad\qquad \text{Magnetite}$$

Thus, we interpret type 3 serpentine to have formed from oxidation of earlier-formed type 1 or type 2 serpentine. Lastly, type 4 serpentine (garnierite) were observed to either directly crosscut oxidized saprolite samples or alter pre-existing serpentine, suggesting that type 4 serpentine is formed during weathering after the formation of the saprolite zone. In addition, the upper garnierite vein contains significant amounts of Ni (up to 3.72 wt% NiO) and is similar to garnierite serpentine described in previous studies [12,48].

4.2. Goethite Ageing and Hematite Formation

Structural investigation of the mineral goethite allows us to highlight the importance of this mineral in the development of the oxide zone of the laterite deposit. Reaction of oxidized rainwater and dissolution of primary minerals in the upper horizon result in the leaching of mobile elements such as Mg and Si downwards the profile while immobile elements, such as Fe and Al, are enriched in the limonite zone. Residually enriched Fe can be oxidized to form poorly crystalline goethite [3,5,65,66]:

$$4\,Fe^{2+} + O_2 + H_2O \rightarrow 4FeO(OH) + 8\,H^+$$
$$\text{Goethite} \tag{6}$$

Goethite initially has low crystallinity [3,18,65], which is attributed to the incorporation of elements, including Ni, that introduce structural defects in the goethite crystal structure [18]. As this stage, lateritization proceeds and goethite ages, and Ni and other impurities are expelled from the structure of goethite and subsequently leach away towards the lower horizons [7,18,65]. The expulsion of Ni from goethite results in goethite ageing or the upward increase in the crystallinity of goethite within the limonite zone as lateritization progresses [18–20]. In the Sta. Cruz deposit, goethite ageing is supported by the observed correlation between the bulk Ni content in the limonite and FWHM of the goethite (110) peak. Ni expelled during goethite ageing may contribute to the supergene Ni enrichment in the saprolite zone and in the garnierite veins. Higher up in the profile, hematite may form from the dehydroxylation of goethite according to [67]:

$$FeO(OH) \rightarrow 0.5\,H_2O + 0.5\,Fe_2O_3.$$
$$\text{Goethite} \qquad\qquad \text{Hematite} \tag{7}$$

Petrographic investigation of saprolite samples revealed that secondary magnetite is altered to hematite, especially at higher depths (e.g., N-650, N-660; Figure 4). In the presence of oxygen, magnetite can be altered to maghemite and then to hematite [67], according to the following reaction:

$$Fe_3O_4 + 0.25\,O_2 \rightarrow 1.5\,Fe_2O_3.$$
$$\text{Magnetite} \qquad\qquad \text{Maghemite/Hematite} \tag{8}$$

Over time, the limonite zone expands as oxide-forming reactions continue to progress, deepening the limonite-saprolite zone boundary and increasing the thickness of the limonite zone. The expansion of the limonite zone is affected by a number of factors, including time, the rate of erosion, relief, and climate. A moderate to low-lying relief in a tropical climate provides the ideal conditions for laterite formation. Low to moderate relief results in a limited erosion rate allowing the preservation of the newly formed limonite zone and at the same time providing a low-lying water table. Since leaching occurs above the water table, a deep water table allows leaching of a thicker layer of rock and, consequently, an enhanced supergene Ni enrichment below the water table [1,6].

5. Conclusions

We report the mineralogy and geochemistry of a laterite profile formed from the alteration of the ultramafic rocks from the Zambales Ophiolite Complex. Combined mineralogical–geochemical analysis reveal that the Sta. Cruz nickel laterite deposit is composed of two main zones—the limonite and saprolite zones—separated by a thin transition zone. Late-stage, Ni-rich serpentine garnierite veins are observed to crosscut the saprolite zone. The limonite zone is characterized by Mg and Si mass loss and residual enrichment of Fe, Al, and Ni. The saprolite zone, on the other hand, is associated with Mg and Si mass gains. Similar to other deposits, the limonite zone is dominated by goethite with minor hematite, while the saprolite zone is dominated by the mineral serpentine. At least four types of serpentine are characterized and distinguished based on occurrence and association with magnetite. Structural investigation of goethite within the limonite zone resulted in a negative correlation between the bulk nickel content and crystallinity of goethite and between bulk nickel content and crystallite size, emphasizing the important role that goethite plays during the formation of the limonite zone.

Author Contributions: K.A.A. conceptualized the study, collected the samples, performed the analyses, and wrote the first draft of the manuscript. C.A.A. and C.S. supervised the study. C.A.A. provided resources to conduct the study. C.A.A., C.S. and C.A.J.T. reviewed, revised, and edited the manuscript. All authors have read and agreed to the published version of the manuscript.

Funding: This research received funding from the University of the Philippines—Office of the Vice Chancellor for Research and Development—Thesis and Dissertation Grant Project No.: 181801 TNSE and the University of the Philippines—National Institute of Geological Sciences Research Grant to K.A.

Data Availability Statement: Data are contained within the article or in Appendix A.

Acknowledgments: Arnulfo Santiago is thanked for welcoming and allowing us to collect the samples used in this study. We also thank Gerald Quiña and Jumar Valdez for their assistance during the fieldwork. James Cesar Refran and James Jimenez are thanked for helping with the photography of the hand samples and thin sections. We also thank the student assistants of the Earth Materials Science laboratories for their assistance in the sample preparation. We thank the academic editors and three anonymous reviewers for constructive comments that greatly improved the manuscript.

Conflicts of Interest: The authors declare no conflict of interest.

Appendix A

Table A1. Goethite crystallite size obtained from Rietveld refinement and calculated full width at half maximum values (FWHM).

Sample	Crystallite size	FWHM
N-000	115	1.09
N-040	127	0.99
N-080	139	0.91
N-120	136	0.93
N-160	134	0.94
N-200	130	0.96
N-240	125	1.00
N-280	116	1.08
N-320	112	1.12
N-360	121	1.04
N-400	125	1.01
N-440	125	1.01
N-480	127	0.99
N-520	127	0.99
N-560	124	1.02
N-600	117	1.07
N-640	124	1.02

Table A2. Calculated relative mass changes of the samples from the Sta. Cruz laterite.

Sample	Horizon [1]	Depth (m)	SiO$_2$	Al$_2$O$_3$	Fe$_2$O$_3$	Cr$_2$O$_3$	MnO	NiO	MgO	Co	LOI
N-000	UL	0	−99	−27	8	307	90	−40	−99	101	16
N-040	UL	0.4	−98	−13	8	62	47	−46	−100	53	39
N-080	UL	0.8	−97	1	0	66	43	−48	−100	36	40
N-120	UL	1.2	−97	5	−2	71	63	−47	−100	33	38
N-160	UL	1.6	−97	−4	3	72	53	−45	−100	31	36
N-200	UL	2	−97	−9	6	74	53	−45	−100	33	35
N-240	UL	2.4	−98	−22	10	63	67	−43	−100	53	33
N-280	LL	2.8	−99	−56	17	119	88	−34	−99	40	27
N-320	LL	3.2	−99	−59	19	163	92	−30	−100	64	16
N-360	LL	3.6	−99	−65	19	188	86	−34	−99	39	16
N-400	LL	4	−99	−60	19	153	86	−31	−99	39	20
N-440	LL	4.4	−99	−56	19	74	102	−29	−99	58	25
N-480	LL	4.8	−99	−78	26	24	113	−29	−100	−15	30
N-520	LL	5.2	−99	−70	19	162	113	−37	−99	48	21
N-560	LL	5.6	−99	−65	16	185	92	−30	−99	153	25
N-600	LL	6	−99	−64	16	192	78	−27	−99	143	27
N-640	T	6.4	−91	−35	8	361	93	64	−94	206	16
N-690	US	6.9	−33	32	5	1876	84	225	−34	33	27
N-750	LS	7.5	−25	−89	−5	60	42	350	−30	35	31
N-900	LS	9	−27	−83	16	151	85	605	−36	70	33
N-1050	LS	10.5	−35	15	1	49	53	668	−33	17	29

[1] UL = upper limonite, LL = lower limonite, T = transition zone, US = upper saprolite, LS = lower saprolite. Calculated for each component using the equation $\Delta C/CO = (MA/MO)(CA/CO) - 1$ [38,39] using the Fe and Ti isocon.

References

1. Golightly, J. Nickeliferous laterite deposits. *Econ. Geol.* **1981**, *75*, 710–735.
2. Brand, N.W.; Butt, C.R.M.; Elias, M. Nickel Laterites: Classification and Features. *J. Aust. Geol. Geophys.* **1998**, *17*, 81–88.
3. Elias, M. Nickel laterite deposits—Geological overview, resources and exploitation. *Cent. Ore Depos. Res. Univ. Tasmania* **2002**, *4*, 205–220.
4. Gleeson, S.A.; Butt, C.R.M.; Elias, M. Nickel Laterites: A Review. *SEG Newsl.* **2003**, *54*, 9–16. [CrossRef]
5. Butt, C.R.M.; Cluzel, D. Nickel laterite ore deposits: Weathered serpentinites. *Elements* **2013**, *9*, 123–128. [CrossRef]
6. Golightly, J.P. Progress in Understanding the Evolution of Nickel Laterites. In *The Challenge of Finding New Mineral Resources: Global Metallogeny, Innovative Exploration, and New Discoveries*; Goldfarb, R.J., Marsh, E.E., Monecke, T., Eds.; Society of Economic Geologists: Littleton, CO, USA, 2010; Volume 15, pp. 451–485.
7. Trescases, J.-J. Weathering and geochemical behaviour of the elements of ultramafic rocks in New Caledonia. *Bur. Miner. Resour. Geol. Geophys. Canberra* **1973**, *141*, 149–161.
8. Pecora, W.T. Nickel-silicate and associated nickel-cobalt-manganese-oxide deposits near Sao Jose do Tocantins, Goiaz, Brazil. *US Geol. Surv. Bull.* **1944**, *935*, 247–305.
9. Faust, G. The hydrous nickel-magnesium silicates—The garnierite group. *Am. Mineral.* **1966**, *51*, 279.
10. Brindley, G.W.; Hang, P.T.H.I. The Nature of Garnierites—I Structures, Chemical Compositions and Color Characteristics. *Clays Clay Miner.* **1973**, *21*, 27–40. [CrossRef]
11. Galí, S.; Soler, J.M.; Proenza, J.A.; Lewis, J.F.; Cama, J.; Tauler, E. Ni enrichment and stability of Al-free garnierite solid-solutions: A thermodynamic approach. *Clays Clay Miner.* **2012**, *60*, 121–135. [CrossRef]
12. Villanova-de-Benavent, C.; Proenza, J.A.; Galí, S.; García-Casco, A.; Tauler, E.; Lewis, J.F.; Longo, F. Garnierites and garnierites: Textures, mineralogy and geochemistry of garnierites in the Falcondo Ni-laterite deposit, Dominican Republic. *Ore Geol. Rev.* **2014**, *58*, 91–109. [CrossRef]
13. Brindley, G.W. The Nature and Nomenclature of Hydrous Nickel-Containing Silicates. *Clay Miner.* **1974**, *10*, 271–277. [CrossRef]
14. Wells, M.A.; Ramanaidou, E.R.; Verrall, M.; Tessarolo, C. Mineralogy and crystal chemistry of "garnierites" in the Goro lateritic nickel deposit, New Caledonia. *Eur. J. Mineral.* **2009**, *21*, 467–483. [CrossRef]
15. Villanova-De-Benavent, C.; Nieto, F.; Viti, C.; Proenza, J.A.; Galí, S.; Roqué-Rosell, J. Ni-phyllosilicates (garnierites) from the Falcondo Ni-laterite deposit (Dominican Republic): Mineralogy, nanotextures, and formation mechanisms by HRTEM and AEM. *Am. Mineral.* **2016**, *101*, 1460–1473. [CrossRef]
16. Roqué-Rosell, J.; Villanova-de-Benavent, C.; Proenza, J.A. The accumulation of Ni in serpentines and garnierites from the Falcondo Ni-laterite deposit (Dominican Republic) elucidated by means of μXAS. *Geochim. Cosmochim. Acta* **2017**, *198*, 48–69. [CrossRef]
17. Putzolu, F.; Abad, I.; Balassone, G.; Boni, M.; Mondillo, N. Ni-bearing smectites in the Wingellina laterite deposit (Western Australia) at nanoscale: TEM-HRTEM evidences of the formation mechanisms. *Appl. Clay Sci.* **2020**, *196*, 105753. [CrossRef]

18. Dublet, G.; Juillot, F.; Morin, G.; Fritsch, E.; Fandeur, D.; Brown, G.E. Goethite aging explains Ni depletion in upper units of ultramafic lateritic ores from New Caledonia. *Geochim. Cosmochim. Acta* **2015**, *160*, 1–15. [CrossRef]
19. Kuhnel, R.A.; Roorda, H.J.; Steensma, J.J.; Kühnel, R.A.; Roorda, H.J.; Steensma, J.J.; Kuhnel, R.A.; Roorda, H.J.; Steensma, J.J.; Kühnel, R.A.; et al. The crystallinity of minerals-A new variable in pedogenetic processes: A study of goethite and associated silicates in laterites. *Clays Clay Miner.* **1975**, *23*, 349–354. [CrossRef]
20. Trescases, J.-J. *L'évolution Géochimique Supergène des Roches Ultrabasiques en Zone Tropicale et la Formation des Gisements Nickélifères de Nouvelle-Calédonie*; Université Louis Pasteur: Strasbourg, France, 1975; Volume 78, ISBN 2709903628.
21. Abrajano, T.A.; Pasteris, J.D.; Bacuta, G.C. Zambales ophiolite, Philippines I. Geology and petrology of the critical zone of the Acoje massif. *Tectonophysics* **1989**, *168*, 65–100. [CrossRef]
22. Hawkins, J.W.; Evans, C. A Geology of the Zambales Range, Luzon, Philippine Islands—Ophiolite derived from an island arc-backarc basin pair. In *The Tectonic and Geologic Evolution of Southeast Asian Seas and Islands, Part 2*; Geophysical Monographs Series; Hayes, D.E., Ed.; American Geophysical Union: Washington, DC, USA, 1983; Volume 27, pp. 95–123.
23. Rossman, D.L.; Castañada, G.C.; Bacuta, G.C. Geology of the Zambales ophiolite, Luzon, Philippines. *Tectonophysics* **1989**, *168*, 1–22. [CrossRef]
24. Yumul, G.P.; Dimalanta, C.B. Geology of the Southern Zambales Ophiolite Complex, (Philippines): Juxtaposed terranes of diverse origin. *J. Asian Earth Sci.* **1997**, *15*, 413–421. [CrossRef]
25. Bacuta, G.C. Geology of some Alpine-type chromite deposits in the Philippines.pdf. *J. Geol. Soc. Philipp.* **1979**, *33*, 44–80.
26. Yumul, G.P.; Dimalanta, C.B.; Faustino, D.V.; De Jesus, J.V. Translation and docking of an arc terrane: Geological and geochemical evidence from the southern Zambales ophiolite complex, Philippines. *Tectonophysics* **1998**, *293*, 255–272. [CrossRef]
27. Garrison, R.E.; Espiritu, E.; Horan, L.J.; Mack, L.E. Petrology, sedimentology, and diagenesis of hemipelagic limestone and tuffaceous turbidites in the Aksitero Formation, central Luzon, Philippines. *US Geol. Surv. Prof. Pap.* **1979**, *1112*, 15–16.
28. Amato, F.L. Stratigraphic paleontology in the Philippines. *Philipp. Geol.* **1965**, *19*, 1–24.
29. Fuller, M.; Haston, R.; Almasco, J. Paleomagnetism of the Zambales ophiolite, Luzon, northern Philippines. *Tectonophysics* **1989**, *168*, 171–203. [CrossRef]
30. Encarnación, J.P.; Mukasa, S.B.; Obille, E.C. Zircon U-Pb geochronology of the Zambales and Angat Ophiolites, Luzon, Philippines: Evidence for an Eocene arc-back arc pair. *J. Geophys. Res.* **1993**, *98*, 19991. [CrossRef]
31. Schweller, W.J.; Karig, D.E.; Bachman, S.B. Original Setting and Emplacement History of the Zambales Ophiolite, Luzon, Phillipines, from Stratigraphic Evidence. In *The Tectonic and Geologic Evolution of Southeast Asian Seas and Islands: Part 2*; Hayes, D.E., Ed.; American Geophysical Union: Washington, DC, USA, 1983; Volume 27, pp. 124–138, ISBN 978-1-118-66409-4.
32. Taylor, B.; Hayes, D.E. The tectonic evolution of the South China Basin. In *The Tectonic and Geologic Evolution of Southeast Asian Seas and Islands*; American Geophysical Union: Washington, DC, USA, 1980; Volume 23, pp. 89–104, ISBN 1118663799.
33. de Santiago, A.P.B. *Evaluation of Ni-Bearing Saprolite Resources Contained in Filipinas Mining Corporation MPSA No. 268-2008-III Located Barangay Guinabon, Sta. Cruz, Zambales, Philippines*; Muntinlupa City, Philippines, 2015.
34. Aquino, K.A.; Arcilla, C.A.; Schardt, C.S. Mineralogical Zonation of the Sta. Cruz Nickel Laterite Deposit, Zambales, Philippines Obtained from Detailed X-ray Diffraction Coupled with Rietveld Refinement. *J. Geol. Soc. Philipp.* **2019**, *73*, 1–14.
35. Aquino, K.A. *Spatio-Temporal Evolution of Laterization: Insights from Detailed Mineralogical Characterization and Reactive Transport Modelling of Sta. Cruz Nickel Laterite Deposit, Zambales, Philippines*; University of the Philippines: Quezon City, Philippines, 2018.
36. Nickel, E.H.; Nichols, M.C. *Mineral Database*; MDI Minerals Data: Livermore, CA, USA, 2003.
37. Cullity, B.D.; Stock, S.R. *Elements of X-ray Diffraction*; Prentice Hall: Englewood Cliffs, NJ, USA, 2001; p. 664.
38. Grant, J.A. Isocon analysis: A brief review of the method and applications. *Phys. Chem. Earth* **2005**, *30*, 997–1004. [CrossRef]
39. Grant, J.A. The isocon diagram-a simple solution to Gresens' equation for metasomatic alteration. *Econ. Geol.* **1986**, *81*, 1976–1982. [CrossRef]
40. Quesnel, B.; de Veslud, C.L.C.; Boulvais, P.; Gautier, P.; Cathelineau, M.; Drouillet, M. 3D modeling of the laterites on top of the Koniambo Massif, New Caledonia: Refinement of the per descensum lateritic model for nickel mineralization. *Miner. Depos.* **2017**, *52*, 961–978. [CrossRef]
41. Aiglsperger, T.; Proenza, J.A.; Lewis, J.F.; Labrador, M.; Svojtka, M.; Rojas-Purón, A.; Longo, F.; Ďurišová, J. Critical metals (REE, Sc, PGE) in Ni laterites from Cuba and the Dominican Republic. *Ore Geol. Rev.* **2016**, *73*, 127–147. [CrossRef]
42. Babechuk, M.G.; Widdowson, M.; Kamber, B.S. Quantifying chemical weathering intensity and trace element release from two contrasting basalt profiles, Deccan Traps, India. *Chem. Geol.* **2014**, *363*, 56–75. [CrossRef]
43. Parker, A. An Index of Weathering for Silicate Rocks. *Geol. Mag.* **1970**, *107*, 501–504. [CrossRef]
44. Ohta, T.; Arai, H. Statistical empirical index of chemical weathering in igneous rocks: A new tool for evaluating the degree of weathering. *Chem. Geol.* **2007**, *240*, 280–297. [CrossRef]
45. Muñoz, M.; Ulrich, M.; Cathelineau, M.; Mathon, O. Weathering processes and crystal chemistry of Ni-bearing minerals in saprock horizons of New Caledonia ophiolite. *J. Geochem. Explor.* **2019**, *198*, 82–99. [CrossRef]
46. Roqué-Rosell, J.; Mosselmans, J.F.W.; Proenza, J.A.; Labrador, M.; Galí, S.; Atkinson, K.D.; Quinn, P.D. Sorption of Ni by "lithiophorite-asbolane" intermediates in Moa Bay lateritic deposits, eastern Cuba. *Chem. Geol.* **2010**, *275*, 9–18. [CrossRef]
47. Tupaz, C.A.J.; Watanabe, Y.; Sanematsu, K.; Echigo, T. Spectral and chemical studies of iron and manganese oxyhydroxides in laterite developed on ultramafic rocks. *Resour. Geol.* **2021**, *71*, 377–391. [CrossRef]

48. Tupaz, C.A.J.; Watanabe, Y.; Sanematsu, K.; Echigo, T.; Arcilla, C.; Ferrer, C. Ni-co mineralization in the intex laterite deposit, Mindoro, Philippines. *Minerals* **2020**, *10*, 579. [CrossRef]
49. Evans, B.W.; Hattori, K.; Baronnet, A. Serpentinite: What, Why, Where? *Elements* **2013**, *9*, 99–106. [CrossRef]
50. Rinaudo, C.; Gastaldi, D.; Belluso, E. Characterization of Chrysotile, Antigorite, and Lizardite by FT-Raman Spectroscopy. *Can. Mineral.* **2003**, *41*, 883–890. [CrossRef]
51. Beard, J.S.; Frost, B.R.; Fryer, P.; McCaig, A.; Searle, R.; Ildefonse, B.; Zinin, P.; Sharma, S.K. Onset and progression of serpentinization and magnetite formation in Olivine-rich troctolite from IODP hole U1309D. *J. Petrol.* **2009**, *50*, 387–403. [CrossRef]
52. Frost, B.R.; Evans, K.A.; Swapp, S.M.; Beard, J.S.; Mothersole, F.E. The process of serpentinization in dunite from New Caledonia. *LITHOS* **2013**, *178*, 24–39. [CrossRef]
53. Klein, F.; Bach, W.; Jöns, N.; McCollom, T.; Moskowitz, B.; Berquó, T. Iron partitioning and hydrogen generation during serpentinization of abyssal peridotites from 15° N on the Mid-Atlantic Ridge. *Geochim. Cosmochim. Acta* **2009**, *73*, 6868–6893. [CrossRef]
54. Frost, B.R.; Beard, J.S.; Frost, R.B.; Beard, J.S.; Frost, B.R.; Beard, J.S.; Frost, R.B.; Beard, J.S. On silica activity and serpentinization. *J. Petrol.* **2007**, *48*, 1351–1368. [CrossRef]
55. Klein, F.; Le Roux, V. Quantifying the volume increase and chemical exchange during serpentinization. *Geology* **2020**, *48*, 552–556. [CrossRef]
56. O'Hanley, D.S. Solution to the volume problem in serpentinization. *Geology* **1992**, *20*, 705–708. [CrossRef]
57. Evans, B.W. The Serpentinite Multisystem Revisited: Chrysotile Is Metastable. *Int. Geol. Rev.* **2004**, *46*, 479–506. [CrossRef]
58. Leong, J.A.M.; Howells, A.E.; Robinson, K.J.; Cox, A.; Debes, R.V.; Fecteau, K.; Prapaipong, P.; Shock, E.L. Theoretical Predictions Versus Environmental Observations on Serpentinization Fluids: Lessons From the Samail Ophiolite in Oman. *J. Geophys. Res. Solid Earth* **2021**, *126*, 1–28. [CrossRef]
59. de Obeso, J.C.; Kelemen, P.B. Major element mobility during serpentinization, oxidation and weathering of mantle peridotite at low temperatures. *Philos. Trans. A Math. Phys. Eng. Sci.* **2020**, *378*, 20180433. [CrossRef]
60. Templeton, A.S.; Ellison, E.T.; Glombitza, C.; Morono, Y.; Rempfert, K.R.; Hoehler, T.M.; Zeigler, S.D.; Kraus, E.A.; Spear, J.R.; Nothaft, D.B.; et al. Accessing the Subsurface Biosphere Within Rocks Undergoing Active Low-Temperature Serpentinization in the Samail Ophiolite (Oman Drilling Project). *J. Geophys. Res. Biogeosci.* **2021**, *126*, 1–30. [CrossRef]
61. Cardace, D.; Meyer-dombard, D.R.A.R.; Woycheese, K.M.; Arcilla, C.A.; Brazelton, W.; Carolina, E. Feasible metabolisms in high pH springs of the Philippines. *Front. Microbiol.* **2015**, *6*, 1–17. [CrossRef]
62. Streit, E.; Kelemen, P.; Eiler, J. Coexisting serpentine and quartz from carbonate-bearing serpentinized peridotite in the Samail Ophiolite, Oman. *Contrib. Mineral. Petrol.* **2012**, *164*, 821–837. [CrossRef]
63. Kelemen, P.B.; Matter, J.; Streit, E.E.; Rudge, J.F.; Curry, W.B.; Blusztajn, J. Rates and mechanisms of mineral carbonation in peridotite: Natural processes and recipes for enhanced, in situ CO_2 capture and storage. *Annu. Rev. Earth Planet. Sci.* **2011**, *39*, 545–576. [CrossRef]
64. Ternieten, L.; Früh-Green, G.L.; Bernasconi, S.M. Carbon Geochemistry of the Active Serpentinization Site at the Wadi Tayin Massif: Insights From the ICDP Oman Drilling Project: Phase II. *J. Geophys. Res. Solid Earth* **2021**, *126*, e2021JB022712. [CrossRef]
65. Freyssinet, P.; Butt, C.R.M.; Morris, R.C.; Piantone, P. Ore-forming processes related to lateritic weathering. In *Economic Geology 100th Anniversary Volume*; Society of Economic Geologists: Littleton, CO, USA, 2005; Volume 1, pp. 681–722.
66. Leong, J.A.M.; Shock, E.L. Thermodynamic constraints on the geochemistry of low-temperature, continental, serpentinization-generated fluids. *Am. J. Sci.* **2020**, *320*, 185–235. [CrossRef]
67. Cornell, R.M.; Schwertmann, U. *The Iron Oxides: Structure: Properties, reactions, Occurences and Uses*; Wiley-VCH: Hoboken, NJ, USA, 2003; ISBN 3527302743.

Article

Co–Mn Mineralisations in the Ni Laterite Deposits of Loma Caribe (Dominican Republic) and Loma de Hierro (Venezuela)

Cristina Domènech [1,*], Cristina Villanova-de-Benavent [2,*], Joaquín A. Proenza [2], Esperança Tauler [2], Laura Lara [2], Salvador Galí [2], Josep M. Soler [3], Marc Campeny [4] and Jordi Ibañez-Insa [5]

[1] Grup MAiMA, Mineralogia Aplicada, Geoquímica i Geomicrobiologia, Departament de Mineralogia, Petrologia i Geologia Aplicada, Facultat de Ciències de la Terra, Universitat de Barcelona (UB), Martí i Franquès s/n, 08028 Barcelona, Spain

[2] Departament de Mineralogia, Petrologia i Geologia Aplicada, Facultat de Ciències de la Terra, Universitat de Barcelona (UB), Martí i Franquès s/n, 08028 Barcelona, Spain; japroenza@ub.edu (J.A.P.); esperancatauler@ub.edu (E.T.); laauraalara17@gmail.com (L.L.); ga-li@ub.edu (S.G.)

[3] Institute of Environmental Assessment and Water Research, IDAEA-CSIC, 08034 Barcelona, Spain; josep.soler@idaea.csic.es

[4] Departament de Mineralogia, Museu de Ciències Naturals de Barcelona, Passeig Picasso s/n, 08003 Barcelona, Spain; mcampenyc@bcn.cat

[5] Geosciences Barcelona (GEO3BCN-CSIC), Lluís Solé i Sabarís s/n, 08028 Barcelona, Spain; jibanez@geo3bcn.csic.es

* Correspondence: cristina.domenech@ub.edu (C.D.); cvillanovadb@ub.edu (C.V.-d.-B.)

Citation: Domènech, C.; Villanova-de-Benavent, C.; Proenza, J.A.; Tauler, E.; Lara, L.; Galí, S.; Soler, J.M.; Campeny, M.; Ibañez-Insa, J. Co–Mn Mineralisations in the Ni Laterite Deposits of Loma Caribe (Dominican Republic) and Loma de Hierro (Venezuela). *Minerals* **2022**, *12*, 927. https://doi.org/10.3390/min12080927

Academic Editor: Maria Boni

Received: 29 June 2022
Accepted: 20 July 2022
Published: 22 July 2022

Publisher's Note: MDPI stays neutral with regard to jurisdictional claims in published maps and institutional affiliations.

Abstract: Cobalt demand is increasing due to its key role in the transition to clean energies. Although the main Co ores are the sediment-hosted stratiform copper deposits of the Democratic Republic of the Congo, Co is also a by-product of Ni–Co laterite deposits, where Co extraction efficiency depends, among other factors, on the correct identification of Co-bearing minerals. In this paper, we reported a detailed study of the Co mineralisation in the Ni–Co laterite profiles of Loma Caribe (Dominican Republic) and Loma de Hierro (Venezuela). Cobalt is mainly associated with Mn-oxyhydroxide minerals, with a composition between Ni asbolane and lithiophorite, although a Co association with phyllosilicates has also been recorded in a Loma de Hierro deposit. In Loma Caribe, Co-bearing Mn-oxyhydroxide minerals mainly developed colloform aggregates, and globular to spherulitic grains, while in Loma de Hierro, they displayed banded colloform, fibrous or tabular textures. Most of the compositional analyses of Mn-oxyhydroxides yielded 20 and 40 wt.% Mn, with Ni and Co up to 16 and 10 wt.%, respectively. In both profiles, Mn-bearing minerals were mainly found in the transition from the oxide horizon to the saprolite, as observed in other laterite profiles in the world, where the precipitation of Mn-bearing minerals is enhanced because of the pore solution saturation and pH increase.

Keywords: nickel; cobalt; manganese; lithiophorite; asbolane; critical metals; laterite

1. Introduction

In 2020, the European Commission published a new list of critical raw materials considered essential for the economy of Europe, according to their potential supply risk and their economic and industrial importance [1]. This list contains 30 raw materials, compared with the 14, 20 and 27 that were indicated in the previous lists of 2011, 2014 and 2017, respectively. One of these critical materials is Co, which is a transition metal that has been in the list since 2011. The main reason to consider Co as a critical raw material is that worldwide production comes primarily from the sediment-hosted stratiform copper deposits of the Democratic Republic of the Congo (accounting for 70% of the world's production) [1–3], which is also the main supplier to Europe. Cobalt is essential in many industries such aerospace, textile, electronics, automotive, renewable energies, and energy-intensive industries, where it is mainly used in batteries, super alloys, catalysts and magnets [2]. Cobalt, together with Ni, rare earth elements (REE) and Li, is of strategic importance in the transition to cleaner energies [4,5]. Recent forecasts indicate that Europe would need up to five times more Co in 2030, and 15 times more Co in 2050 for electric vehicle batteries and energy storage, compared with the 2019 supply, this increase also being linked to the increase in demand of other metals such as Ni or Mn [1]. The strategy of EU is based on the diversification of global supply chains and the enhancement of circularity and resource efficiency.

In addition to the Congo deposits, Co is also a by-product of the mining of magmatic Ni–Cu sulphide deposits hosted in mafic and ultramafic rocks, such as Sudbury (Canada) or Norilsk (Russia), and of Ni–Co laterite deposits in Australia, New Caledonia, Indonesia, the Philippines, Brazil, Cameroon, Greece and the Caribbean region ([2,3,6] and references therein). Cobalt content in laterite deposits is estimated to comprise 36% of terrestrial Co deposits [7]. Cobalt has also been identified in manganese nodules and crusts on ocean floors [3].

Ni–Co laterite deposits result from the weathering of ultramafic rocks (generally serpentinised peridotites) at tropical latitudes [8–11]. In a typical Ni–Co laterite profile, the unaltered or serpentinised peridotite is found at the bottom, covered by a saprolite horizon that contains relicts of the parent rock with secondary serpentine, goethite and Mg–Ni phyllosilicates. An oxide horizon is formed over the saprolite zone, consisting of Fe-(goethite, hematite, maghemite), Al-(gibbsite) and occasionally, Mn-oxides and hydroxides. In some cases, an iron cap or ferricrete has developed over the oxide horizon. According to the main Ni mineralogy, Ni–Co laterites are classified into oxide type (where Ni is mainly associated with Fe-oxyhydroxides), clay silicate type (where Ni is concentrated in smectite group minerals), and hydrous silicate type (rich in Mg–Ni phyllosilicates, including Ni-serpentines and garnierites) ([8,12] and references therein). Cobalt is usually accumulated within the ferruginous unit or at the transition from the saprolite to the oxide horizon, and is mostly concentrated into poorly crystalline Fe- and Mn-oxyhydroxides.

Although in some deposits Co is found in primary minerals (e.g., sulphides, carbonates), in their alteration products (e.g., heterogenite (CoOOH)), or in laterite deposits, Co is usually associated with Mn-oxyhydroxides, in particular, asbolane, lithiophorite, or the so-called asbolane–lithiophorite intermediates [13].

Due to the low grade, its irregular distribution, and the complex mineralogy of the ores, the Co-production from laterite ores is generally low [14,15]. Moreover, the Co beneficiation mainly depends on its oxidation state and the relative proportion of Co to Cu or Ni [13]. Therefore, a proper identification and characterisation of Co-bearing minerals in the context of a laterite deposit is relevant, in order to improve the extraction efficiency. Recently, the number of studies dealing with how Co (and also Ni, Mn and Sc) are distributed in these deposits, have increased [6].

There are large Ni–Co laterite deposits with economic grades of Co (e.g., Murrin Murrin or Kalgoorlie in Australia, Goro in New Caledonia, or Nkamouna in Cameroon) [2]. However, recent studies have pointed out that Ni–Co laterites from the Caribbean area also contain significant amounts of Co that could be exploited in the near future [16–18].

Therefore, in this study we present a detailed characterisation of Mn-bearing minerals of two profiles from two Caribbean Ni–Co laterite deposits, with the goal of determining their mineralogy and composition and ultimately understanding their formation processes. More precisely, the study is centered in the Loma Caribe and the Loma de Hierro profiles, for which no previous work on manganese mineralisations is available. This study also aims to compare the Co-Mn-bearing phases identified in the Caribbean area with those reported in other deposits worldwide.

2. Geological Setting

The Loma Caribe Ni laterite deposit belongs to the Falcondo mining area, located in the central part of the Dominican Republic, in the northern Caribbean (Figure 1a). This deposit, together with other six other satellite deposits ([16] and references therein) developed during the Miocene after weathering of the Loma Caribe peridotite, which occurs as a serpentinised belt of ultramafic rocks, approximately 4–5 km wide and 95 km long, NW of Santo Domingo (Figure 1b). The Loma Caribe peridotite represents a harzburgitic oceanic mantle, as part of a dismembered Cretaceous ophiolitic complex [19]. The Loma Caribe deposit is classified as hydrous Mg silicate type, and, based on its geochemical footprint, textures and mineralogy, has been divided into several zones, which are, from bottom to top, the serpentinised peridotite, the saprolite horizon, the oxide horizon and a duricrust zone [16]. The protolith is a serpentinised peridotite rich in olivine and enstatite, crosscut by serpentine minerals (i.e., lizardite). The saprolite horizon consists of Ni-rich lizardite with relics of olivine and enstatite, and with minor clinochlore, maghemite and goethite. In the oxide horizon, ore samples contain goethite, hematite, gibbsite and chromian spinel [16] (Figure 2).

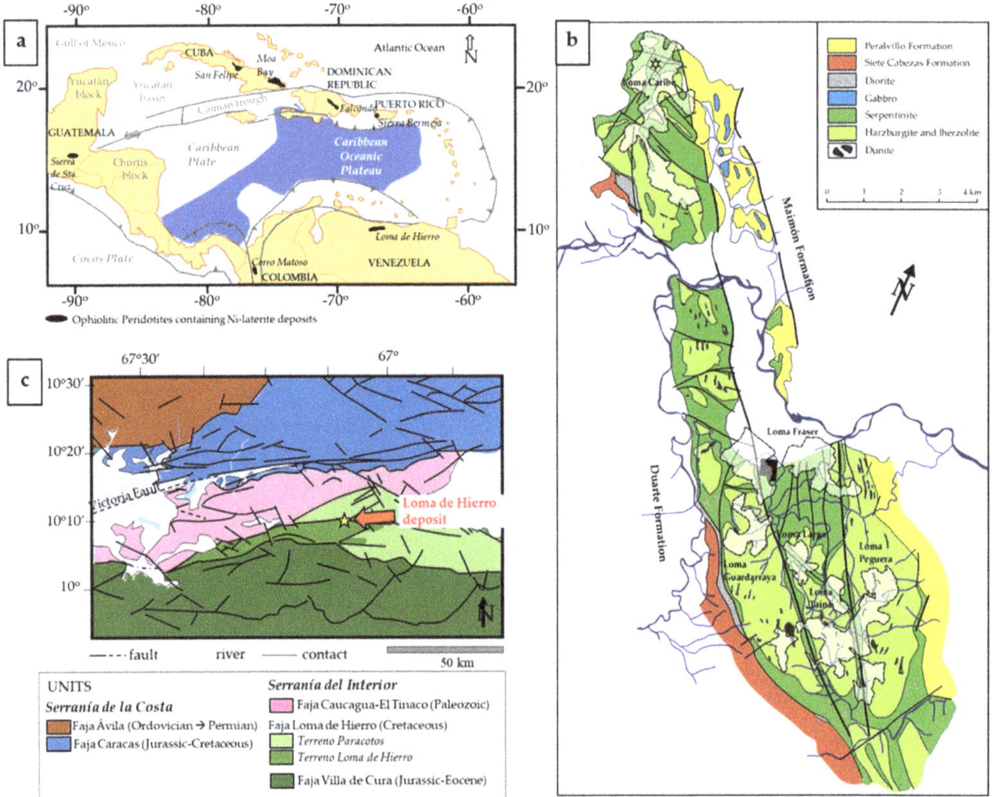

Figure 1. Loma Caribe (Dominican Republic) and Loma de Hierro (Venezuela) Ni laterite deposits: (**a**) location in the Caribbean region; (**b**) geological map of Loma Caribe (redrawn from [18]); and (**c**) geological map of Loma de Hierro (simplified from [20]).

Ore minerals are mainly secondary Ni serpentine and garnierites (fine grained, poorly crystalline, Ni-bearing serpentine, talc, chlorite, smectite and sepiolite) [16,18]. The Falcondo mining district has 67.8 Mt of indicated Ni resources at an average grade of 1.5 wt.% Ni [16].

The Loma de Hierro laterite deposit is located in the northcentral part of Venezuela. This deposit is placed on the Faja Loma de Hierro, which belongs to an elongated, deformed belt in the Caribbean Plate southern margin (Figure 1a), represented by the Cretaceous Dutch and Venezuelan Islands and the Cordilleras of Venezuela [21] (Figure 1c).

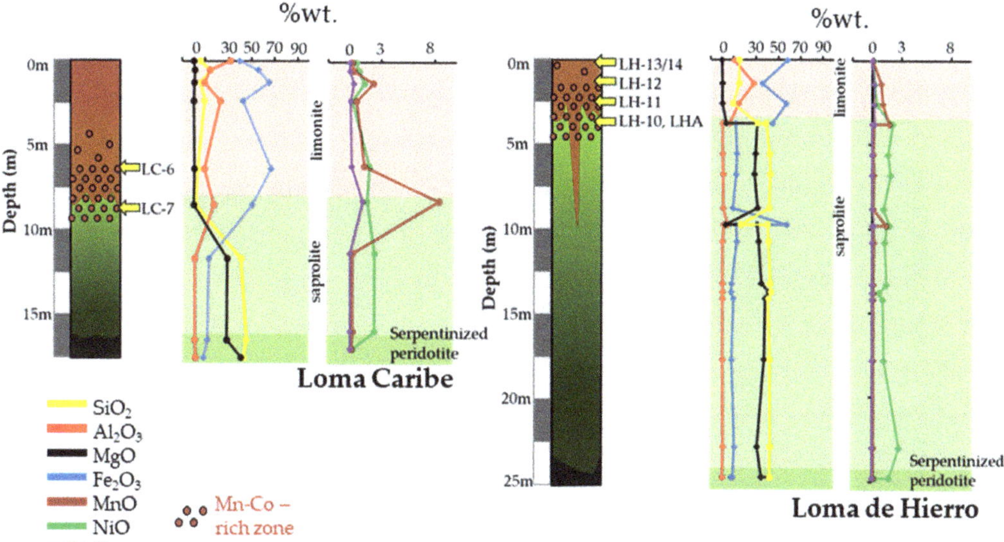

Figure 2. Ni–Co laterite profile from Loma Caribe (Falcondo, Dominican Republic) and Loma de Hierro (Venezuela) ore deposits, showing the location of the samples studied in this paper with major and relevant minor element contents.

The exploited area has a length of ~15 km and a width of 1 to 7 km [17]. This deposit formed over the partially serpentinised peridotites of the Loma de Hierro ophiolite, a W–E aligned unit (~100 km^2) within the Serranía de la Costa in the south Caribbean Plate margin (Figure 1) [21,22]. The Loma de Hierro ophiolite is made up of harzburgite and dunite (Loma de Níquel mantle peridotite), gabbros (Gabro de Mesia; 127 + 1.9/−4.3 Ma) and basaltic rocks with MORB affinity (Basalto de Tiara) [23–25]. Several studies [24,25] have described the Loma de Hierro ophiolite as being a fragment of proto-Caribbean lithosphere.

The Loma de Hierro deposit is also considered as a hydrous Mg silicate type Ni deposit, developed from a partially serpentinised harzburgite peridotite. Over this peridotite, and as in the case of Loma Caribe, several horizons have been defined: lower saprolite, upper saprolite and a poorly developed oxide zone [17]. The protolith is comprised of olivine and orthopyroxene, with minor amounts of lizardite, chrysotile, kerolite and chlorite. The lower saprolite contains large amounts of olivine, orthopyroxene, lizardite and chrysotile. In some samples, the presence of garnierites is observed. The upper saprolite is poorer in olivine and orthopyroxene, and richer in lizardite, chrysotile, kerolite and chlorite. The saprolite zone is enriched in Ni (1.1–2.7 wt.% NiO). The oxide zone is characterised by high contents of goethite, hematite and gibbsite. Ore minerals are secondary Ni serpentine and Ni-rich kerolite–pimelite garnierite mixtures (~22 wt.% NiO).

Economic concentrations of Mn and Co are observed in both profiles. Co–Mn mineralisations mainly occur as fine-grained, black, semi-metallic or dull coatings and open space infillings near the transition between the oxide and the saprolite horizons (Figure 3).

In Loma Caribe, two Mn–Co-rich levels have been described: (i) at the transition zone between the saprolite and the oxide horizon (8.4 wt.% MnO; 1.23 wt.% Co); (ii) at the lower part of the upper oxide zone (2.2 wt.% MnO; 0.4 wt.% Co) [16]. In Loma de Hierro deposit, the MnO and Co reach grades of up to 1.40 wt.% and 0.14 wt.%, respectively [17].

Figure 3. Field occurrence of Co–Mn mineralisations in Loma Caribe (**a–c**) and Loma de Hierro (**d–f**). Panorama view of the mining pits (**a,d,e**) and close-up views of Co–Mn mineralisations (**b,c,f**).

3. Materials and Methods

3.1. Sampling and Sample Preparation

A total of seven samples were selected for the present study, two from Loma Caribe, (LC-6 and LC-7) and five from Loma de Hierro (LH-10, LH-11, LH-12, LH-13/14, and LHA) (Figure 2). All samples except LHA have been studied previously [16,17] but they have been revisited here with a focus on Mn–Co-bearing minerals. Sample LHA from Loma de Hierro was collected from a Mn–Co rich zone, and was divided into two subsamples according to their colour, which were named LHA-1 (whitish sample) and LHA-2 (blackish sample).

3.2. Analythical Methods

Samples LC6, LC7, a subsample of LC7 with higher concentration of blackish material (rich in Mn-oxyhydroxides) named LC7_black, and samples LHA-1 and LHA-2 were analysed using X-ray powder diffraction (XRPD). Samples were powdered using an agate mortar and pestle. XRPD analysis was carried out at the Centres Científics i Tecnològics of the Universitat of Barcelona (CCiT-UB) with a PANalytical X'Pert PRO MPD Alpha1 powder diffractometer in Bragg–Brentano $\theta/2\theta$ geometry of 240 mm of radius, nickel filtered Cu Kα1 radiation (λ = 1.5406 Å), 45 keV and 40 mA. The samples were scanned from 4 to 100° (2θ) with a step size of 0.017° and a measuring time of 150 s per step, using an X'Celerator detector (active length = 2.122°) and a variable divergence slit. Mineral identification was performed by X'Pert Highscore search–match software using the Powder Diffraction File (PDF-2) of the International Centre for Diffraction Data (ICDD), and quantitative mineral phase analyses were obtained by full profile Rietveld refinement using XRPD data and TOPAS V4.2 software [26,27]. Sample LH-10 was also analysed by powder angle-dispersive micro-XRD in the microdiffraction/high-pressure station of the BL04-MSPD beamline at ALBA Synchrotron, Barcelona. This beamline is equipped with a Kirkpatrick–Baez mirror that allows a focus of the monochromatic beam to around 15 μm × 15 μm (full width at half maximum) with a Rayonix SX165 CCD detector.

For the above measurements, a wavelength of 0.4246 Å was selected, as determined from the absorption K-edge of Sn (29.2 keV). The data were acquired by using the synchrotron through-the-substrate X-ray microdiffraction technique [28]. Small amounts of powdered material were placed on top of a Kapton® tape attached to a glass substrate. The sample-to-detector distance and the beam centre position was calibrated from LaB$_6$ diffraction scans obtained under exactly the same conditions as the sample. The calibration and subsequent integration of the CCD images were performed with Dioptas software [29]. Phase identification on the integrated powder scans was carried out by using the package Diffrac.EVA from Bruker, together with the Powder Diffraction File (PDF-2) and the Crystallography Open Database (COD).

Samples LC6, LC7, LC7_black, LHA-1 and LHA-2 were analysed by differential thermal analysis coupled with thermogravimetry (DTA-TG) using a Netzsch STA 409C/CD instrument. Approximately 150 mg of powdered sample was put in an alumina crucible under a N_2(g) atmosphere with a flow rate of 80 mL min^{-1}, in a range between 25 to 800 °C at 10 °C min^{-1} using ~80 mg of an alumina standard (Perkin-Elmer 0419-0197). The same powder samples were analysed by Fourier transform infrared spectroscopy (FTIR) using a 2000 FTIR spectrometer (Perkin–Elmer System, Waltham, MA, USA) at a range for vibrational spectra of 400–4000 cm^{-1}, at the Departament de Mineralogia, Petrologia i Geologia Aplicada de la Facultat de Ciències de la Terra (Universitat de Barcelona).

The major, minor and trace element compositions discussed in this paper were drawn from references [16,17] except data for sample LHA that were obtained by X-ray fluorescence (XRF) and inductively coupled mass spectrometry (ICP-MS; after acid dissolution) at the Actlabs Laboratories (Ontario, Canada) as described in the aforementioned papers.

Polished sections and polished thin sections of the selected samples were examined under a transmitted and reflected light optical microscope at the Departament de Mineralogia, Petrologia i Geologia Aplicada de la Facultat de Ciències de la Terra (Universitat de Barcelona), while samples containing Co and Mn mineralisations were carbon coated and studied in a scanning electron microscope (SEM) Quanta 200 FEI, XTE 325/D8395, with an INCA energy dispersive spectrometer (EDS) microanalysis system, and a field emission SEM (FE-SEM) Jeol JSM-7100 equipped with an INCA Energy 250 EDS, under 20 kV and 5 nA, at the CCiT-UB. Co–Mn-bearing minerals were analysed by electron microprobe analyser (EMPA) at the CCiT-UB by using a JEOL JXA-8230 electron microprobe, equipped with five wavelength-dispersive spectrometers and an energy-dispersive spectrometer. The operating conditions were 20 kV accelerating voltage, 15 nA beam current, 2 μm beam diameter and a counting time of 20 s per element. The calibration standards used were: wollastonite (Si, Ca), corundum (Al), orthoclase (K), hematite (Fe), periclase (Mg), rhodonite

(Mn), NiO (Ni), metallic Co (Co), rutile (Ti), albite (Na), Cr_2O_3 (Cr), V (V), Sc (Sc), and baryte (Ba).

Thermodynamic calculations were carried out with GibbsStudio software [30] based on the PHREEQC code [31] and the thermodynamic data llnl.dat database supplied with PHREEQC code and modified, when specified, to complete Co and Mn aqueous species and solid-phases thermodynamic data.

4. Results

4.1. Mineralogy and Petrography

In Loma Caribe deposits, and according to XRPD (samples LC-6, LC-7 and LC-7_black), lithiophorite is ubiquitous with clear peaks at 9.3 and 4.7 Å (Figure 4a,b). Those d_{hkl} are comparable to those obtained by Burlet and Vanbrabant [32] (9.45 and 4.69 Å in lithiophorite, 4.76 Å in asbolane, and 9.59 and 4.76 Å in lithiophorite–asbolane intermediates), and by Putzolu and co-authors [15] (4.84 Å in asbolane), confirming the observations by Aiglsperger and co-authors [16]. Gibbsite, goethite and chromite were also identified in those samples. High concentrations (12%) of lithiophorite were quantified by Rietveld in sample LC-7_black (Figure 4b).

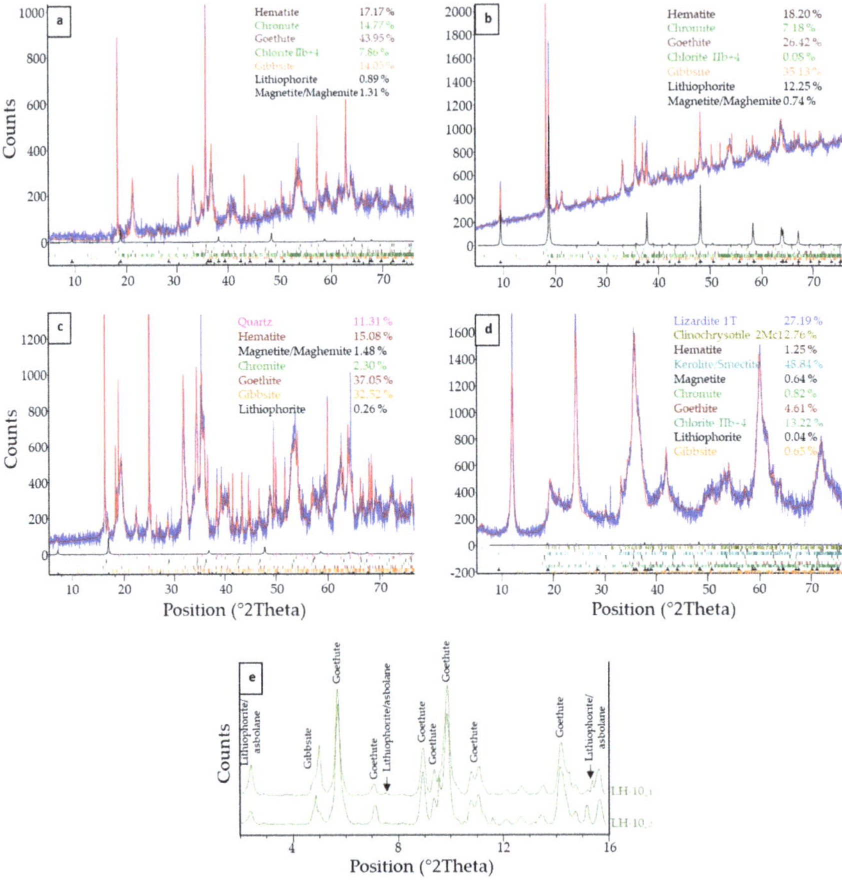

Figure 4. (**a–d**) X-ray diffraction patterns obtained for samples LC-6 (**a**), LC-7-black (**b**), LH-11 (**c**), and LH-10 (**d**). In blue, experimental data; in red, Rietveld fit, and in black, lithiophorite spectra. Mineral percentages obtained after Rietveld interpretation (λ = 1.5406 Å). (**e**) Two micro-XRD measurements of sample LH-10 obtained at ALBA synchrotron (λ = 0.4246 Å).

Although in minor amounts (up to 0.3%), lithiophorite was also identified in the Loma de Hierro samples (Figure 4d,e). The main mineral phases were goethite, gibbsite, quartz, lizardite and smectite. Micro-XRD scans of sample LH-10 (Figure 4e) were dominated by typical XRD peaks of goethite. However, whereas the LH-10$_{-2}$ curve exhibited the XRD signature of gibbsite, the LH-10$_{-1}$ curve showed several features that can be attributed to lithiophorite and/or asbolane (Figure 4e). Although it is difficult to distinguish these two minerals solely by XRD due to their similar structure and lattice spacings, it should be noted that the PDF-2 pattern for lithiophorite (00-041-1378) provides a better match to the experimental results.

Thermogravimetry showed that samples from Loma Caribe experience a mass loss of about 4% before 260–280 °C, probably due to the loss of adsorbed water for temperatures below 110–160 °C [33,34] and to the transformation of gibbsite to boehmite, which is reported to occur at temperatures between 150 and 260 °C [34] (Figure 5a). Between 260 and 320 °C, the mass loss is between 5 and 8%. The latter process can be ascribed to the loss of the OH^{-} groups in the reaction from goethite to hematite (Figure 5a). This reaction (i.e., $2FeOOH \rightarrow Fe_2O_3 + H_2O$) occurs at temperatures between 300 and 344 °C [35] or 270 and 330 °C [36], and it is commonly associated with a mass loss of 10% or less if there are secondary phases. This mass loss coincides with an endothermic peak around 325 °C in the DTA curve (Figure 5b) that is considered as the result of the breakdown of goethite [34–36].

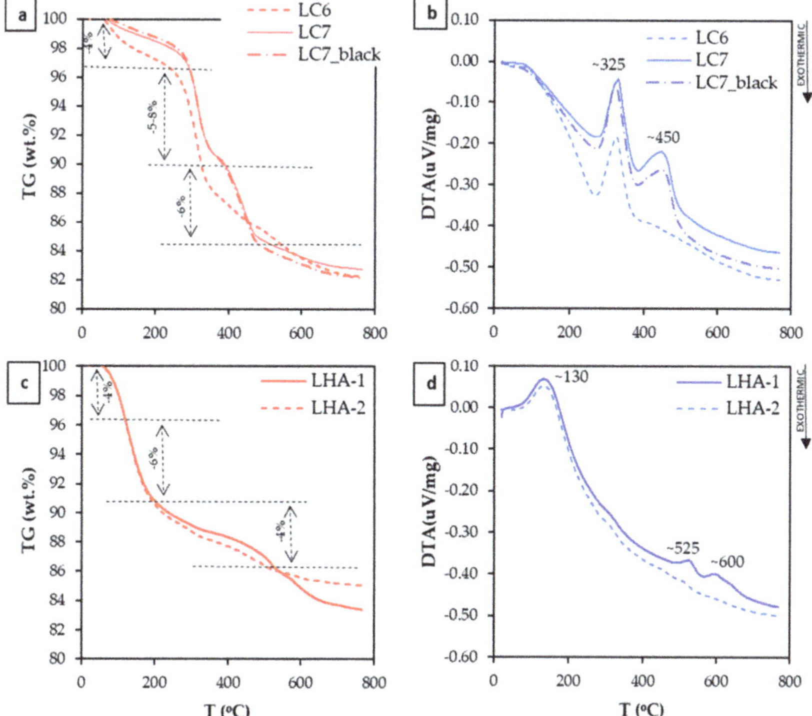

Figure 5. TG and DTA curves of samples LC6, LC7 and LC7_black from Loma Caribe deposit (**a,b**), and of samples LHA-1 and LHA-2 from Loma de Hierro (**c,d**).

Differential thermal analyses of samples LC7 and LC7_black revealed an additional endothermic peak at T~450 °C, associated with a mass loss of ~6%. This peak can be attributed to the dehydroxylation of AlOOH ($AlOOH \rightarrow Al_2O_3 + H_2O$) (even though it was not detected by XRD) or of an amorphous Al-hydroxide, that may occur between 450–600 °C and 400–500 °C, respectively, or to the loss of OH in lithiophorite (430–500 °C) [37]. Llorca and co-authors [38] also measured an endothermic peak around 420–460 °C in samples of asbolane, lithiophorite and asbolane–lithiophorite intermediates of New Caledonia that was attributed to the loss of hydroxyl groups.

Samples LHA-1 and LHA-2 from Loma de Hierro showed distinctive TG and DTA analyses, compared with those of Loma Caribe (Figure 5c,d). Specifically, an endothermic peak at ~130 °C, associated with a mass loss around 10% (sample LHA-1), as well as two minor peaks around 525 and 600 °C could be distinguished. The peak at 130 °C was due to the loss of adsorbed water from clays such as smectite [37], although [39] indicated that it could also be attributed to one of the endothermic reactions shown by asbolane. Peaks at 525 and 600°C could be associated with asbolane (570–625 °C) [38]. However, research by Medeiros Ribeiro and co-authors [39] suggests that the peak around 600 °C refers to the dehydroxylation of the interlayer hydroxide of chlorite.

Depending on the crystallinity of Mn-oxyhydroxides, the characteristic IR frequencies appeared as a single broad band, or in distinct bands in the range of 425–530 cm^{-1} (associated with Mn–O bonds of the MnO$_6$ octahedra) and in the range of 3100–3500 cm^{-1} (stretching modes of OH groups) [40,41]. The band around 1600 cm^{-1} was the water bending mode, that may suggest that at least some of the OH is H$_2$O [41,42]. According to IR spectra (Figure 6), samples were mostly poorly crystalline, although peaks around 3400, 1600 and 1000 cm^{-1} could be assigned to lithiophorite, while those around 3600–3500 cm^{-1} were related to gibbsite [43,44]. The shadowed area (800–600 cm^{-1}) in samples LHA-1 and LHA-2 included some peaks related to Mn-oxides and clays [45], and the peak around 1200 cm^{-1} could be related to kaolinite. It must be noted that mineral contamination can also affect IR spectra.

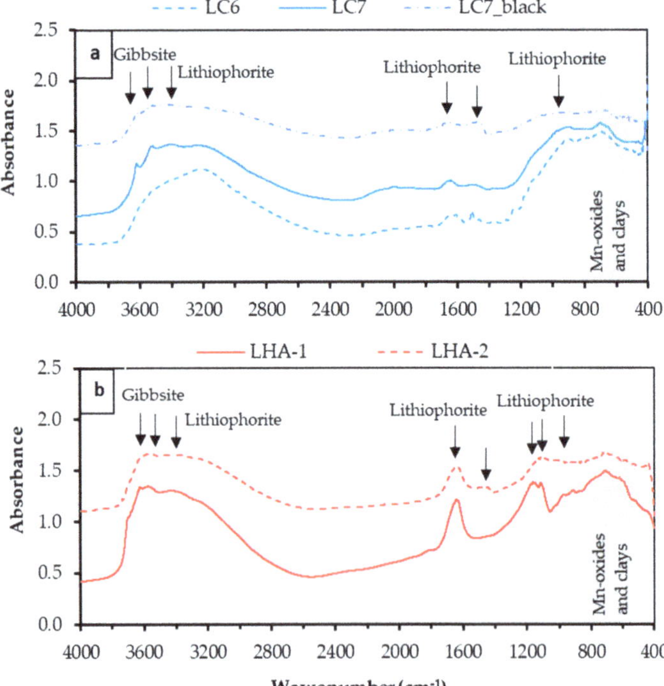

Figure 6. IR spectra of (**a**) samples LC6, LC7 and LC7_black from Loma Caribe (Dominican Republic), and (**b**) samples LHA-1 and LHA-2 from Loma de Hierro (Venezuela). Arrows indicate the peak positions. The grey shadowed band represents the peaks associated with Mn-oxides.

4.2. Co-Mn Mineralisations

4.2.1. Mn-Oxyhydroxides

In the Loma Caribe samples, Co-rich Mn-oxyhydroxides were in close association with Fe–Al-oxyhydroxides. Three major textural types were identified.

The first type was characterised by banded and colloform aggregates (Figure 7), which is the most common texture observed in Mn-oxyhydroxides. Some bands within the colloform aggregates appeared to be formed by parallel and/or radial fibrous aggregates (Figure 7a–c). Figure 8 shows the spatial distribution of some elements within these colloform aggregates and fibrous bands. In general, a close spatial association between Mn, Ni, Co and Al was observed, while, conversely, Si had a mirrored distribution, and no significant variations of Fe or Mg were observed. Colloform aggregates locally displayed distinct external rims (Figure 7a) and in rare occasions, these particles had a uniform core (Figure 7c).

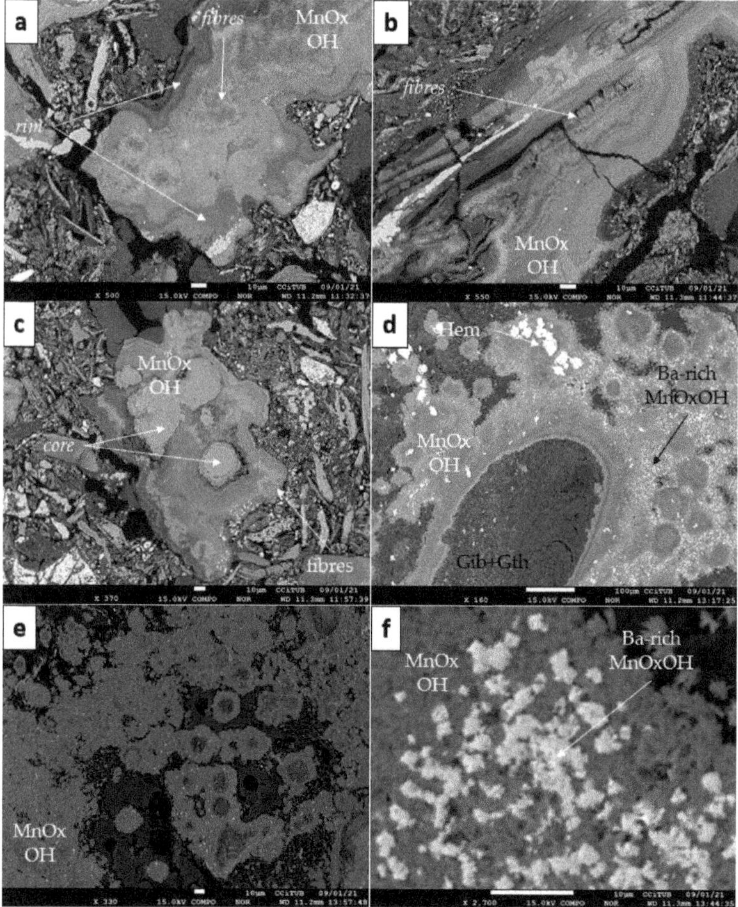

Figure 7. Back-scattered electron (BSE) photomicrographs of the Mn-oxyhydroxide textures identified in Loma Caribe (Dominican Republic) samples: (**a–c**) colloform aggregates including distinct features (rims, core and fibres; see text for explanation); (**d,e**) globular spherulitic aggregates with a gibbsite and goethite (Gib + Gth) core (**d**), and as small, isolated spherulitic aggregate particles (**e**); (**f**) Ba-rich angular grains within the globular spherulitic aggregates. Note the flaky appearance of the Mn-oxyhydroxides in the background in (**f**).

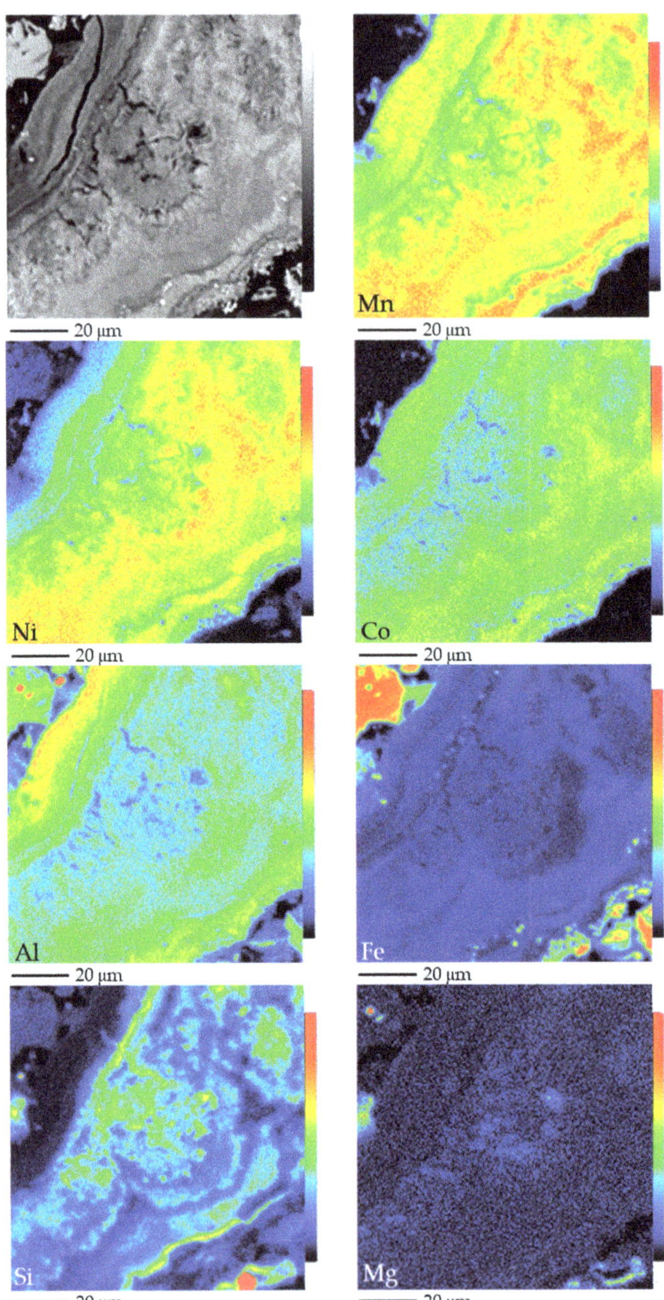

Figure 8. BSE photomicrograph (top left) and Mn, Ni, Co, Al, Fe, Si and Mg X-ray elemental maps obtained by EMPA of banded colloform aggregates from Loma Caribe. Colour scale from black to red indicates increasing element content.

The second Mn-oxyhydroxide type was slightly less abundant, and consisted of globular to spherulitic aggregates, with sizes from some tens to several hundreds of micrometres;

locally, larger aggregates appeared to be hollow-cored, filled with goethite and/or gibbsite (Figure 7d,e). These globular to spherulitic aggregates were composed of short, flaky crystals (Figure 7f). The least abundant occurrences were discrete, micrometre-sized, Ba-rich angular grains (with high mean atomic number Z in BSE images, Figure 7d,f) disseminated within some globular-spherulitic aggregates (of lower mean atomic number Z). These last two types were only observed in the saprolite horizon sample.

In Loma de Hierro (Venezuela), Mn-oxyhydroxide grains are a minor component of the mineralogical assemblage. According to their textural features, three types were identified (Figure 9). Firstly, as in Loma Caribe, they typically displayed banded colloform features, as kidney-shaped (Figure 9a), elongated (Figure 9b), and sub-rounded (Figure 9c,d) grains, measuring a few hundreds of micrometres in diameter or length, surrounded by goethite and other oxides. Certain rounded grains displayed one or more concentric rims having an homogeneous compositional footprint (Figure 9c,d). Locally, Ni and Co were concentrated in the outer rims, while the Mn grade was higher in the core (Figure 10).

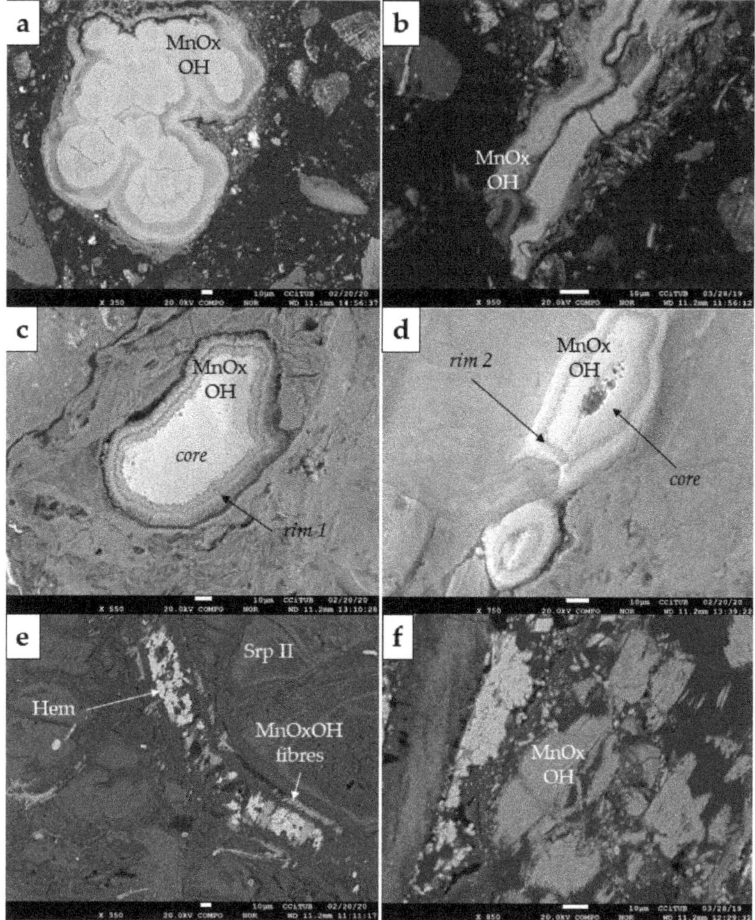

Figure 9. BSE photomicrographs of the textural features displayed by Mn-oxyhydroxides in Loma de Hierro Ni laterites (Venezuela): (**a–d**) colloform aggregates developing kidney-shaped grains (**a**), elongated grains (**b**), and sub-rounded individual grains with distinct multi-phase rims (**c,d**); (**e**) fibres within veinlets, coexisting with hematite (Hem) and secondary serpentine (Serp-II); (**f**) fragmented tabular aggregates.

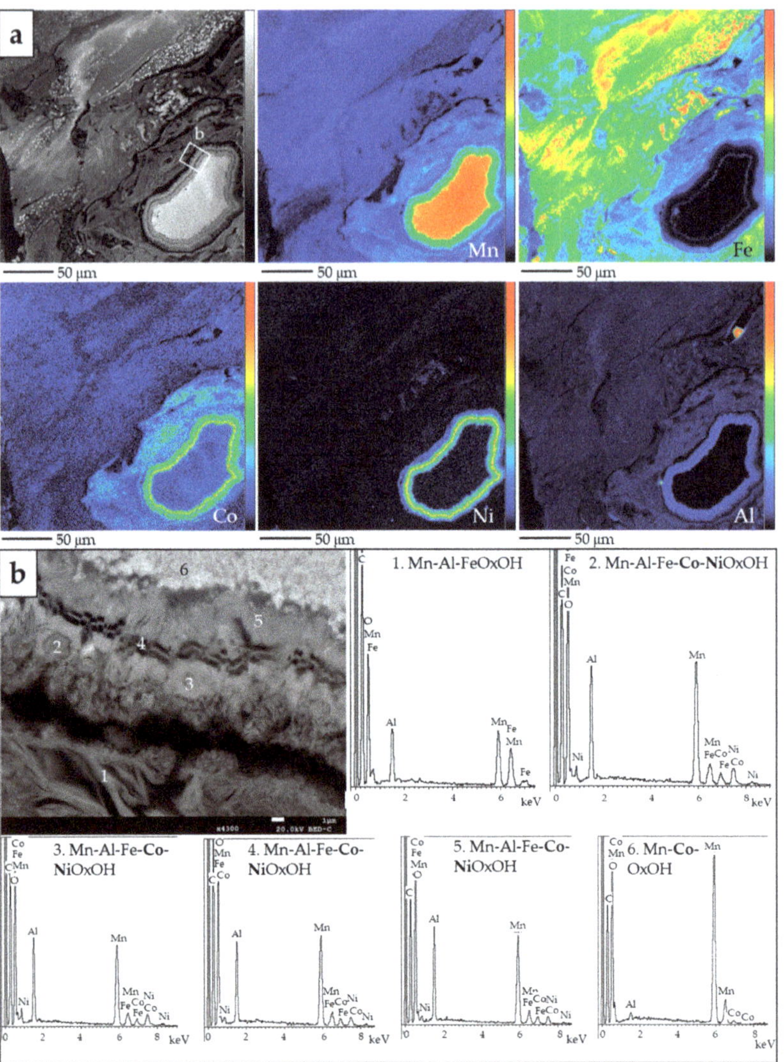

Figure 10. (**a**) BSE photomicrograph (top left) and Mn, Fe, Co, Ni and Al X-ray elemental maps obtained by EMPA, of the same sub-rounded multi-phase Mn-oxyhydroxide grain in Figure 9c (Loma de Hierro), in a matrix of Mn- and Fe-oxyhydroxides (colour scale from black to red indicates increasing element content); (**b**) magnified image of the outer rim of the same grain, displaying variations in composition in terms of Mn, Al, Fe, Ni and Co.

The second type, less common, were short fibres of a few tens of micrometres in length, in close association with Fe-oxide and within serpentine veinlets (Figure 9e).

Finally, the least common Mn-oxyhydroxide in Loma de Hierro occurred as fragmented tabular aggregates, dispersed in the goethite matrix (Figure 9f).

With regard to the chemical composition of Loma Caribe Mn-oxyhydroxides, colloform aggregates and rims, fibrous bands and globular spherulitic aggregates showed high variations in MnO (14–34 wt.%), NiO (4–21 wt.%), CoO (4–12 wt.%), SiO$_2$ (0.4–11 wt.%), Al$_2$O$_3$ (7–22 wt.%) and FeO (< 9 wt.%). However, the uniform cores within the colloform aggregates had a considerably higher amount of MnO (average of 54 wt.%) while CoO,

FeO, Al$_2$O$_3$, and SiO$_2$ contents were below 1 wt.%, and Ni content was close to 4 wt.% NiO. The angular, high-Z grains were characterised by high Ba and Mn contents (10–11 wt.% BaO and 54–59 wt.% MnO) and low amounts of Ni and Co (0.7 wt.% NiO, 1 wt.% CoO) (Table 1).

In Loma de Hierro, the MnO content in the colloform and tabular aggregates varied from 20 to 41 wt.%. They contained 2–12 wt.% Al$_2$O$_3$, 2 -28 wt.% FeO, 8–20 wt.% NiO, 3–7 wt.% CoO, <2 wt.% SiO$_2$, and <5 wt.% MgO (Table 2). However, grain cores were richer in Mn (76–80 wt.% MnO) but were Ni and Co poor (<0.3 wt.% NiO and <2 wt.% CoO). These cores were overgrown by a rim with high Al (12–15 wt.%) and relatively low Mn (29–37 wt.%) (Table 2; Figure 9c), or another with high Mn (71–73 wt.%) and low Al (2–3 wt.% Al$_2$O$_3$) (Table 2; Figure 9d). The short fibres had Mn, Fe, Mg, Co and Ni values similar to those of colloform and tabular aggregates, but were richer in Si (up to 14 wt.% SiO$_2$, partly probably due to contamination of the analyses) (Table 2).

Most of the compositional analyses of Mn-oxyhydroxides had a Mn content within 20 and 40 wt.% Mn, with contents of Ni and Co up to 16 and 10 wt.%, respectively. Some analyses from the Mn–Co rich zone (sample LHA) were the richest in Mn content (~60 wt.% of Mn) although they showed very low Ni and Co contents (Figure 11).

In general, samples from Loma de Hierro were richer in Mn than those of Loma Caribe (Figure 11a,b). Both sets of samples showed similar Ni contents but those of Loma Caribe were found to be Co richer. A common feature observed in all samples was a positive correlation between Co and Al contents (Figure 11c).

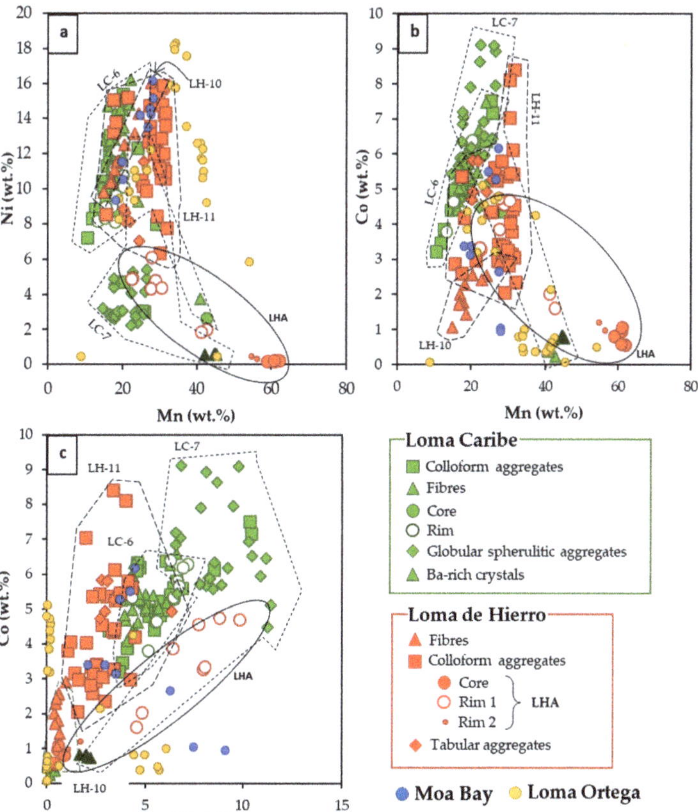

Figure 11. Binary diagrams of (**a**) Ni vs. Mn, (**b**) Co vs. Mn, and (**c**) Co vs. Al (in wt.%), of the Mn–Co minerals analysed in this study, and from Loma Ortega [46] and Moa Bay [47].

Table 1. Representative EMP analyses (in weight percent) of the studied Co–Mn-oxyhydroxide types in Loma Caribe (Dominican Republic). <d.l.: below detection limit; n.a: not analysed; F: fibres; CA: colloform aggregates; C: core; R: rim; GS: globular spherulitic aggregates; Ba: Ba-rich minerals.

Label	Texture	SiO_2	Al_2O_3	CaO	K_2O	Na_2O	MgO	NiO	CoO	FeO	MnO	Cr_2O_3	BaO	TiO_2	V_2O_3	Sc_2O_3	Total
LC-6_01	F	6.24	8.37	0.04	<0.01	0.01	0.37	17.22	6.73	6.84	26.01	<0.01	0.01	n.a	n.a	n.a	71.84
LC-6_02	F	3.01	8.18	0.05	0.04	0.09	0.37	18.82	6.96	7.82	27.90	<d.l.	0.11	n.a	n.a	n.a	73.34
LC-6_03	F	3.99	8.71	0.03	0.03	0.01	0.12	20.60	8.52	8.72	28.63	0.03	<d.l.	n.a	n.a	n.a	79.38
LC-6_04	F	4.14	9.47	0.02	0.02	0.03	0.24	14.97	6.89	9.32	23.57	0.05	0.04	n.a	n.a	n.a	68.75
LC-6_05	F	2.09	9.50	0.07	<0.01	0.02	0.13	19.23	6.38	6.21	23.09	0.03	0.02	n.a	n.a	n.a	66.76
LC-7_01	F	0.49	12.47	0.02	0.17	0.61	0.64	11.82	9.51	4.02	31.08	<d.l.	<d.l.	n.a	n.a	n.a	70.82
LC-6_06	CA	4.36	8.91	0.06	<0.01	0.01	0.11	19.11	7.88	8.72	26.86	<0.01	<d.l.	n.a	n.a	n.a	76.02
LC-6_07	CA	6.65	10.47	0.04	<0.01	0.06	0.08	16.32	6.12	7.47	20.68	<d.l.	0.09	n.a	n.a	n.a	67.98
LC-6_08	CA	1.40	10.73	0.03	0.02	0.06	0.07	16.63	6.30	6.45	21.62	0.03	0.09	n.a	n.a	n.a	63.43
LC-6_09	CA	7.70	10.91	0.05	0.02	0.18	0.15	13.95	6.64	5.79	22.35	0.02	0.08	n.a	n.a	n.a	67.86
LC-7_02	CA	0.39	5.95	0.11	0.43	1.13	2.02	10.20	5.56	6.24	37.47	0.02	0.42	n.a	n.a	n.a	69.94
LC-7_03	CA	1.09	11.39	0.04	0.16	0.46	0.35	15.70	8.11	3.72	30.91	<0.01	<d.l.	n.a	n.a	n.a	71.93
LC-7_04	CA	1.91	16.03	0.05	0.14	0.37	0.05	6.01	7.79	12.62	25.95	0.05	<d.l.	n.a	n.a	n.a	70.97
LC-7_05	CA	0.50	19.67	0.04	0.05	0.18	0.02	3.85	9.12	4.08	33.27	0.10	<d.l.	n.a	n.a	n.a	70.87
LC-6_10	C	0.39	0.24	0.03	0.03	0.04	7.78	3.36	0.30	0.63	55.20	<d.l.	0.66	n.a	n.a	n.a	68.66
LC-6_11	C	0.61	0.76	0.04	0.05	0.04	6.65	4.76	0.46	0.91	52.85	<d.l.	0.57	n.a	n.a	n.a	67.71
LC-6_12	R	11.27	10.38	0.09	0.03	0.16	0.11	11.43	5.88	5.13	19.96	0.02	<d.l.	n.a	n.a	n.a	64.46
LC-6_13	R	1.52	9.69	<d.l.	0.03	0.06	0.07	10.41	4.83	7.46	17.20	0.04	0.06	n.a	n.a	n.a	51.38
LC-6_14	R	2.11	8.16	0.10	<0.01	<d.l.	0.21	15.53	6.81	8.13	23.74	<d.l.	<d.l.	n.a	n.a	n.a	64.79
LC-6_15	R	1.56	13.53	0.04	0.03	0.08	0.11	13.74	7.99	4.58	25.72	0.02	<d.l.	n.a	n.a	n.a	67.40
LC-6_16	R	7.92	12.20	0.18	0.04	0.19	0.11	14.23	6.80	6.49	22.84	<d.l.	<d.l.	n.a	n.a	n.a	71.00
LC-6_17	R	1.34	11.85	0.09	0.04	0.04	0.25	16.23	8.13	4.91	26.05	0.02	<d.l.	n.a	n.a	n.a	68.95
LC-7_06	GS	0.26	15.33	0.07	0.11	0.24	0.16	6.85	11.36	1.90	34.17	0.02	0.33	n.a	n.a	n.a	70.81
LC-7_07	GS	3.40	12.39	0.31	0.05	0.49	0.24	5.60	8.71	11.08	22.74	0.65	0.06	n.a	n.a	n.a	65.71
LC-7_08	GS	5.46	16.01	0.14	0.12	0.37	0.11	3.48	7.10	13.70	21.66	0.86	0.02	n.a	n.a	n.a	69.03
LC-7_09	GS	2.63	14.82	0.13	0.07	0.18	0.08	3.90	10.04	8.30	25.64	0.26	<d.l.	n.a	n.a	n.a	66.05
LC-7_10	GS	7.29	15.57	0.17	0.10	0.32	0.10	3.43	7.49	9.97	22.02	0.39	<d.l.	n.a	n.a	n.a	66.84
LC-7_11	GS	1.73	19.73	0.06	0.05	0.13	0.05	3.58	8.29	6.69	30.43	0.19	<d.l.	n.a	n.a	n.a	70.92
LC-7_12	GS	0.98	21.11	0.02	<0.01	0.09	0.02	3.69	8.20	2.89	33.13	0.05	0.09	n.a	n.a	n.a	70.26
LC-7_13	GS	1.69	21.26	0.02	0.01	0.09	0.04	2.85	5.70	10.58	28.71	0.18	0.10	n.a	n.a	n.a	71.23
LC-7_14	Ba	0.30	3.91	0.04	0.51	0.36	0.05	0.70	0.96	2.75	58.51	0.08	9.87	n.a	n.a	n.a	78.04
LC-7_15	Ba	0.25	3.20	0.05	0.51	0.29	0.02	0.74	1.04	3.16	57.82	0.10	10.51	n.a	n.a	n.a	77.70
LC-7_16	Ba	0.27	3.81	<0.01	0.41	0.27	0.03	0.71	1.04	2.79	57.63	0.07	10.56	n.a	n.a	n.a	77.59
LC-7_17	Ba	0.47	4.21	<d.l.	0.43	0.26	0.02	0.71	0.91	6.33	54.30	0.17	10.51	n.a	n.a	n.a	78.32

Table 2. Representative EMP analyses (in weight percent) of the studied Co–Mn-oxyhydroxide types in Loma de Hierro (Venezuela). <d.l.: below detection limit; F-fibres; CA–colloform aggregates; C-core; R-rim; T-tabular.

Label	Texture	SiO_2	Al_2O_3	CaO	K_2O	Na_2O	MgO	NiO	CoO	FeO	MnO	Cr_2O_3	BaO	TiO_2	V_2O_3	Sc_2O_3	Total
LH-10_01	F	3.47	1.06	0.03	0.04	0.01	4.60	18.19	3.23	12.48	31.30	0.03	0.05	n.a.	n.a.	n.a.	74.50
LH-10_02	F	3.93	1.81	0.08	0.02	0.04	6.29	20.40	3.71	3.87	35.48	<d.l.	0.24	n.a.	n.a.	n.a.	75.86
LH-10_03	F	5.56	0.82	0.06	0.06	<d.l.	6.16	13.20	2.19	24.20	21.70	<d.l.	n.a.	<0.01	<d.l.	n.a.	73.94
LH-10_04	F	4.75	0.98	0.07	0.04	0.06	5.80	16.76	3.08	12.47	29.73	<d.l.	n.a.	0.02	<0.01	n.a.	73.76
LH-11_01	CA	0.12	2.01	0.08	n.a.	n.a.	3.12	18.77	3.78	9.96	40.03	<d.l.	n.a.	0.01	0.02	<d.l.	77.90
LH-11_02	CA	0.20	2.26	0.05	n.a.	n.a.	3.48	18.15	4.83	10.12	39.41	<d.l.	n.a.	<d.l.	<d.l.	<d.l.	78.50
LH-11_03	CA	0.13	2.68	0.06	n.a.	n.a.	2.07	20.02	5.17	8.90	38.13	<d.l.	n.a.	<d.l.	<d.l.	<d.l.	77.16
LH-11_04	CA	0.16	2.98	0.04	n.a.	n.a.	4.26	13.52	8.95	8.96	37.55	<d.l.	n.a.	<d.l.	0.01	<d.l.	76.44
LH-11_05	CA	0.37	6.35	0.04	n.a.	n.a.	3.76	16.23	7.37	4.21	40.69	<d.l.	n.a.	<d.l.	n.a.	<d.l.	79.02
LH-11_06	CA	0.30	7.53	0.03	n.a.	n.a.	4.51	13.98	10.70	2.08	39.03	<d.l.	n.a.	<d.l.	<0.01	<d.l.	78.16
LHA_1	C	0.04	1.02	0.02	0.12	<d.l.	<d.l.	0.27	1.23	0.45	78.56	<d.l.	0.12	n.a.	n.a.	n.a.	81.82
LHA_2	C	0.13	1.72	<0.01	0.15	0.08	0.05	0.25	1.05	1.62	75.97	0.01	0.26	n.a.	n.a.	n.a.	81.30
LHA_3	C	0.05	1.06	<0.01	0.08	<d.l.	<0.01	0.30	1.38	0.45	78.74	0.02	0.07	n.a.	n.a.	n.a.	82.15
LHA_4	R	0.08	8.58	<0.01	0.16	0.07	0.03	2.54	2.06	3.05	54.91	0.03	0.50	n.a.	n.a.	n.a.	72.02
LHA_5	R	0.13	9.13	0.03	0.22	<d.l.	0.10	2.37	2.59	3.55	52.98	<d.l.	0.72	n.a.	n.a.	n.a.	71.81
LHA_6	R	0.08	18.56	<d.l.	0.08	0.10	0.04	5.61	5.97	1.47	39.21	<d.l.	0.12	n.a.	n.a.	n.a.	71.24
LHA_7	R	0.09	2.10	0.02	0.59	0.09	0.03	0.37	1.25	0.49	72.74	<d.l.	1.30	n.a.	n.a.	n.a.	79.07
LHA_8	R	0.09	3.26	0.04	0.90	0.05	0.02	0.57	1.55	0.56	70.65	0.01	1.31	n.a.	n.a.	n.a.	79.01
LH-11_07	T	1.41	5.25	0.04	n.a.	n.a.	2.50	10.43	7.42	24.66	26.30	0.80	n.a.	0.04	0.03	<d.l.	78.88
LH-11_08	T	1.19	5.79	0.02	n.a.	n.a.	2.45	10.36	7.38	22.38	28.17	0.66	n.a.	0.02	0.03	<d.l.	78.44
LH-11_09	T	0.92	6.43	0.02	n.a.	n.a.	2.72	14.72	6.72	17.50	33.08	0.46	n.a.	0.02	0.02	<d.l.	82.61
LH-11_10	T	1.35	5.15	0.03	n.a.	n.a.	2.74	11.27	6.00	28.23	26.08	0.69	n.a.	0.04	0.04	<d.l.	81.62

4.2.2. Mn-Bearing Phyllosilicates

In addition, Co was identified in the phyllosilicates of Loma de Hierro. According to XRPD, the clay component was dominated by chlorite and smectite. These clay species occurred as micrometre-sized platy individual crystals (Figure 12). In some cases, an inner core with lower Z, and an outer rim with higher Z was distinguished (in BSE photomicrographs, Figure 12). The core contained 27–34 wt.% SiO_2, 25–29 wt.% Al_2O_3, <2 wt.% FeO, 6–12 wt.% MnO, <0.5 wt.% NiO, and 2–4 wt.% CoO (Table 3). The rim yielded 19–20 wt.% SiO_2, 23–24 wt.% Al_2O_3, 3–4 wt.% FeO, 18–19 wt.% MnO, <0.9 wt.% NiO, and 6–7 wt.% CoO. Interestingly, both contained <1 wt.% MgO (Figure 12).

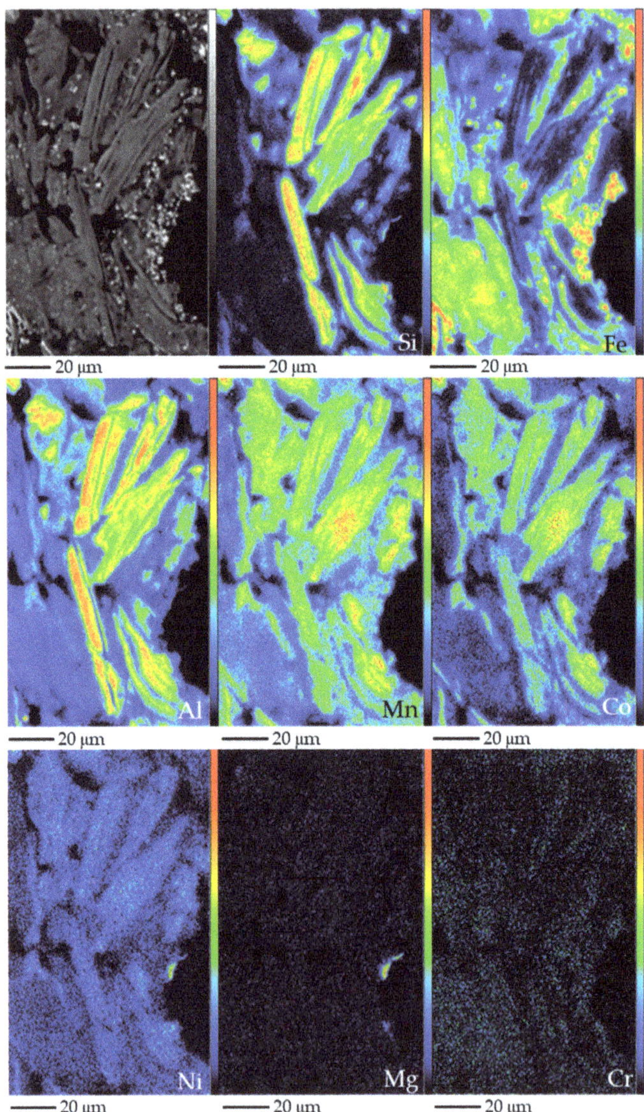

Figure 12. BSE photomicrograph of sample LHA (top left), and Si, Fe, Al, Mn, Co, Ni, Mg and Cr X-ray elemental maps obtained by EMPA, of an aggregate of Mn–Ni–Co-bearing chlorite. Colour scale from black to red indicates increasing element content.

Table 3. Representative EMP analyses (in weight percent) of the studied Co-bearing phyllosilicates from Loma de Hierro (Venezuela). <d.l.: below detection limit; C-core; R-rim.

Label	Texture	SiO_2	Al_2O_3	CaO	K_2O	Na_2O	MgO	NiO	CoO	FeO	MnO	Cr_2O_3	BaO	Total
LHA_10	C	34.35	28.60	0.06	0.06	0.00	0.08	0.30	1.54	1.13	5.78	0.04	<d.l.	71.95
LHA_11	C	26.50	24.96	0.03	0.04	0.06	0.04	0.54	3.95	1.49	12.08	<d.l.	0.04	69.73
LHA_12	R	19.91	24.12	0.04	0.13	0.02	0.01	0.86	5.93	3.29	18.22	0.02	0.02	72.57
LHA_13	R	18.76	23.36	0.03	0.06	<d.l.	0.02	0.79	6.53	3.41	18.95	<d.l.	0.05	71.96

5. Discussion

5.1. Co-Mn Bearing Minerals in Loma Caribe and Loma de Hierro

Most of the Loma Caribe and Loma de Hierro Mn-bearing minerals from the oxide horizon had a composition between Ni asbolane (($Ni,Co)_xMn(O,OH)_4 \cdot nH_2O$) and lithiophorite (($Al,Li)MnO_2(OH)_2$) (Figure 13). These minerals mainly formed colloform to tabular textures in the case of Loma de Hierro, while in Loma Caribe they were also present as fibrous aggregates. The purest Ni asbolane was found in sample LH-10 of Loma de Hierro. However, in the saprolite horizon of Loma Caribe, Mn-bearing minerals were mainly globular or spherulitic lithiophorites. These observations were consistent with the results of DRX, DTA-TG and FTIR, which clearly indicated the occurrence of lithiophorite in sample LC-7.

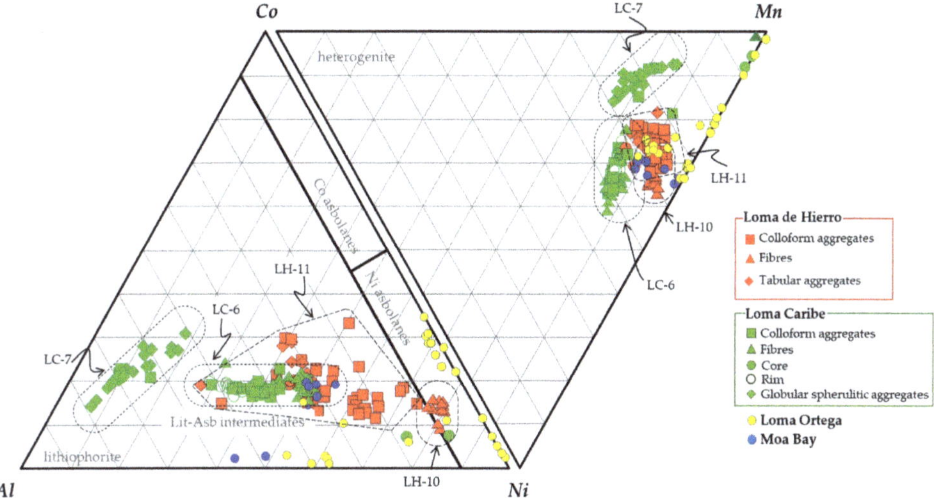

Figure 13. Ternary plots representing Mn, Ni, Co, Al contents (cation proportions calculated on the basis of 100 oxygens) obtained from EMPA analyses on Mn-oxyhydroxides (lithiophorites, lithiophorite–asbolane intermediates and asbolane) from Loma Caribe and Loma de Hierro (this study), Loma Ortega [46] and Moa Bay [47].

The highest Mn contents were found in sample LHA, from Loma de Hierro. They were explained by the abundance of grains with a core rich in Mn (up to 80 wt.% MnO, Table 2), but poor in Co and Ni. Hence, these were not considered lithiophorite nor lithiophorite–asbolane intermediates. These grains were surrounded by rims with alternate Mn-rich or Co/Ni-rich compositions.

Putzolu and and co-authors [15] described two different generations of romanèchite, which despite their close paragenetic association, showed different chemical compositions. Romanèchite I had an average BaO content of 7.64 wt.%, with low amounts of NiO and CoO (average values of 1.21 and 0.72 wt.%, respectively), while romanèchite II had lower BaO (average 2.68 wt.%) and higher NiO and CoO (average values of 2.90 and 1.48 wt.%, respectively). Ba-rich Mn-oxyhydroxides identified in Loma Caribe (Figure 7f) presented slightly higher BaO contents (up to 10.5 wt.%, Table 1, Figure 7f) than those of romanèchite I of [15], although NiO and CoO contents were lower (less than 0.7 and 1.4 wt.%, respectively). Pure romanèchite has the structural formulae $Ba_{0.66}Mn_5O_{10} \cdot 34H_2O$, in which the content of Ba is around 20 wt.% BaO [48]. In our case, as in [15], Ba content was far below the stoichiometric value.

Furthermore, the association of Mn (and Co) with phyllosilicates (Table 3) is not uncommon, as it has also been observed in serpentine minerals [49] and smectite clays [50].

5.2. Comparison with Other Co-Mn Mineralisations in Lateritic Systems

The Ni and Co of Mn-oxyhydroxides from the Loma Caribe and Loma de Hierro were similar to those of other Ni laterites of the Caribbean region. Their Ni grade was similar to that of Moa Bay, and a bit lower than that measured in Loma Ortega [46,47], however, the Co contents were considerably higher in Loma Caribe and Loma de Hierro when compared with those of Loma Ortega and Moa Bay (Figure 11a,b).

The Al content in Mn-oxyhydroxides object of this study was higher in Loma Caribe and Loma de Hierro than in Loma Ortega and Moa Bay (Figure 11c). In the hybrid hydrous Mg silicate–clay silicate type Ni laterite deposit of Loma Ortega (Falcondo mining district, Dominican Republic), Co-bearing Mn phases were observed as coatings or vein fillings, close to quartz, goethite and serpentine II, replacing pyroxene, olivine and serpentine along grain boundaries and fractures [46]. As shown in Figure 13, they were mostly Ni asbolanes with Mn and Ni contents between 11 and 70 wt.% MnO, and up to 23 wt.% NiO, respectively. In most cases, Co content was below 1 wt.% CoO, although in some cases it graded up to 6.5 wt.% CoO. Al and Fe were highly variable, ranging from 0.1 to 11.5 wt.% Al_2O_3 and from 0.1 to 21 wt.% FeO. In general, in these samples, the Fe and Al contents were anti-correlated. In Moa Bay, an oxide type laterite deposit in Cuba, Mn-bearing minerals, mostly lithiophorite-asbolane intermediates (Figure 13), are found as veins and coatings along fractures at the lowest part of the oxide horizon [47]. In these minerals, Mn content ranges from 23 to 37 wt.% MnO, and present high amounts of Ni (12–21 wt.% NiO) and Co (1–8 wt.% CoO). In the analysed samples, Al and Fe contents ranged from 4 to 17 wt.% Al_2O_3 and from 3 to 11.4 wt.% Fe_2O_3, again inversely correlated.

The chemical footprint of Mn minerals reported in this study was also similar to other lithiophorite, asbolane–lithiophorite intermediates, and cryptomelane recorded in other localities (Table 4). Most data reported in the literature come from studies from New Caledonia, where Co-bearing minerals occur as colloform or fibrous cryptocrystalline to microcrystalline aggregates, intimately intergrown with Mn–Fe-oxides and other poorly crystallised minerals, which appear as spots or coatings, onion peel structures, pseudomorphs of silicates or (rhizo)concretions [51]. The identified Co-bearing minerals in New Caledonia were phyllomanganates (asbolane, lithiophorite and intermediates) with a variable range of compositions between Ni and Co; being lithiophorite, more abundant in those weathering profiles with plagioclase lherzolite as a protolith ([51] and references therein), although other Co and Mn minerals were also identified (heterogenite, cryptomelane, ramsdellite, todorokite). Several authors [38,51–55] have investigated the crystal chemistry and composition of several samples of Ni and Co-bearing minerals of Poro, Tiébaghi and Koniambo.

In the Nkamouna deposit (Cameroon), lithiophorite is the main Mn-phase, and is commonly endowed in Co. It was identified as bluish to black cryptocrystalline to fine-grained concretions, commonly associated with gibbsite and magnetite [56]. Other phases were pyrolusite, cryptomelane (associated with pyrolusite) and lithiophorite–asbolane intermediates.

In the Wingellina oxide-type laterite deposits (Western Australia), Mn-oxyhydroxides were studied by [15], who identified lithiophorite–asbolane intermediate dominant phases and some other minor phases such as romanèchite, ernienickelite–jianshuiite, manganite, and birnessite. Mn-oxyhydroxides of the Palawan deposit (Philippines) were studied by [57], who identified lithiophorite–asbolane intermediates as the main host of Ni and Co in the oxide horizon, while [58,59] studied the asbolane from the products of weathering of the Lipovsk serpentine deposit (Middle Urals, Russia), which occur as platy fragments associated with goethite.

Table 4. Manganese, Co and Ni contents of Mn-bearing minerals from different laterite deposits (number of samples within parentheses).

Mn-Bearing Minerals	Laterite Deposit	MnO$_2$ (wt.%)	CoO (wt.%)	NiO (wt.%)	Reference
Asbolane	Lipovsk (Middle Urals, Russia) (1)	94.2	–	6.3	[59]
	Lipovsk (Middle Urals, Russia) (2)	46.9–49.1	6.9	11.4	[58]
	Urals (Russia) (3)	55.1–66.7	<6.9	11.4–21.6	[38]
	New Caledonia (1)	48.3	21.1	10.7	[53]
	New Caledonia (2)	48.3–66.7	<21.1	10.7–17.9	[38]
	New Caledonia (2)	58.9–65.2	1.2–1.7	10.1–12.6	[54]
	New Caledonia (3)	40.1–66.7	0.9–5.1	17.9–20.1	[52]
	Moa Bay (Cuba) (3)	31.3–43.3	1.2–7.8	19.3–20.5	[47]
Birnessite	New Caledonia (3)	85.1–91.9	0.8–1.5	0.1–1.0	[54]
Cryptomelane	Nkamouna (Cameroon) (5)	34.0–80.9	2.7–16.9	1.1–6.9	[56]
Heterogenite	New Caledonia (1)	0.2	69.3	4.7	[38,52,53]
	Shaba (Democratic Republic of the Congo) (1)	0.5	76.4	1.9	[38]
Lithiophorite	New Caledonia (1)	46.0	7.1	1.5	[38,52]
	Postmasburg (South Africa) (1)	47.0	–	–	[38]
	Nkamouna (Cameroon) (12)	29.3–50.5	3.1–10.8	1.7–7.9	[56]
	New Caledonia (2)	44.1–46.0	7.1–8.9	1.3–1.4	[53]
	Loma Ortega (Dominican Republic) (27)	14.1–85.6	0.1–6.5	0.6–23.2	[46]
Lithiophorite–asbolane intermediate	Palawan (Philippines) (4)	34.3–44.7	1.3–8.1	3.7–19.5	[57]
	New Caledonia (1)	45.4	12.6	2.7	[52]
	Wingellina (Australia) (8)	33.7–55.1	0.1–2.5	10.5–14.1	[15]
	Nkamouna (Cameroon) (5)	20.2–43.8	2.3–17.0	0.7–7.5	[56]
Romanechite	Wingellina (Australia) (4)	51.9–69.3	0.7–1.6	1.1–2.9	[15]
Phyllomanganate	New Caledonia (4)	74.6–88.7	1.2–1.5	1.4–6.2	[53]
Pyrolusite	Nkamouna (Cameroon) (2)	70.0–96.4	0.7–1.2	0.9–1.0	[56]
Ramsdellite	New Caledonia (1)	95.3	0.8	–	[52]
Todorokite	New Caledonia (1)	61.9	0.9	0.4	[52]
Co–Ni-bearing Mn-hydroxides	Wingellina (Australia) (6)	54.1–90.8	0.2–3.0	2.3–14.1	[15]
Mn-oxide	New Caledonia (1)	38.7	11.9	5.8	[53]
Unspecified	New Caledonia (9)	13.6–38.8	4.2–10.1	15.9–22.6	[55]

5.3. Formation of Co-Mn Bearing Minerals in Laterite Deposits

Mn-bearing minerals in Loma Caribe and Loma de Hierro were mainly found in the transition from the oxide horizon to the saprolite (Figure 2). Their formation at this level within the laterite profile is not uncommon, as it has also been observed in other deposits worldwide [55,60].

The shift of soil solution from acidic to slightly alkaline is the main trigger for the formation of Mn oxides at this interface in the laterite profile ([55] and references therein). In the oxide zone, Mn is commonly sunk by goethite, hematite or other Fe-oxyhydroxides. Due to weathering, temperature, and humidity, Mn can be leached from Fe-bearing minerals and transported downwards as colloidal complexes or aqueous complexes [55,61,62]. In the transition to the saprolite horizon, due to the increase in pH and solution saturation, Mn-oxides precipitation may occur [61]. Although being thermodynamically favoured, oxidation of Mn(II) is very slow in natural waters, but Fe(III) and especially bacterial and fungal activity can accelerate this process by several orders of magnitude [61,63,64]. The presence of lithiophorite, however, suggests that the oxidation of Mn(II) is a slow process [58], as the structure of this mineral needs time to form.

Figure 14 shows predominance diagrams (Eh–pH) of the Mn–Co–H$_2$O system calculated with the Gibbs Studio code [30] based on Phreeqc [31] and the llnl.dat database from [31]. The thermodynamic data for Co aqueous species $Co(OH)_3^-$ and $Co(OH)^+$ were taken from the ThermoChimie database [65] and [66]. Due to the lack of thermodynamic data for lithiophorite, asbolane or lithiophorite–asbolane intermediate compounds, birnessite was considered instead, in the construction of predominance diagram. Birnessite, being part of the same phyllomanganate group, can be thus considered representative also for lithiophorite and lithiophorite–asbolane [48]. Furthermore, it has to be emphasised that lithiophorite is considered to form after the alteration of birnessite (and other less mature Mn phases as romanechite) [55]. Calculations were performed allowing only the formation of birnessite, $Mn(OH)_3(s)$ and $Mn(OH)_2(s)$. Other Mn solid phases such as pyrolusite, todorokite, hausmannite and manganite were not allowed to precipitate since their higher thermodynamic stability inhibits the formation of birnessite. The outcome of the thermodynamic calculation showed that an increase in pH in the oxidised zone allows the formation of Mn-oxyhydroxides.

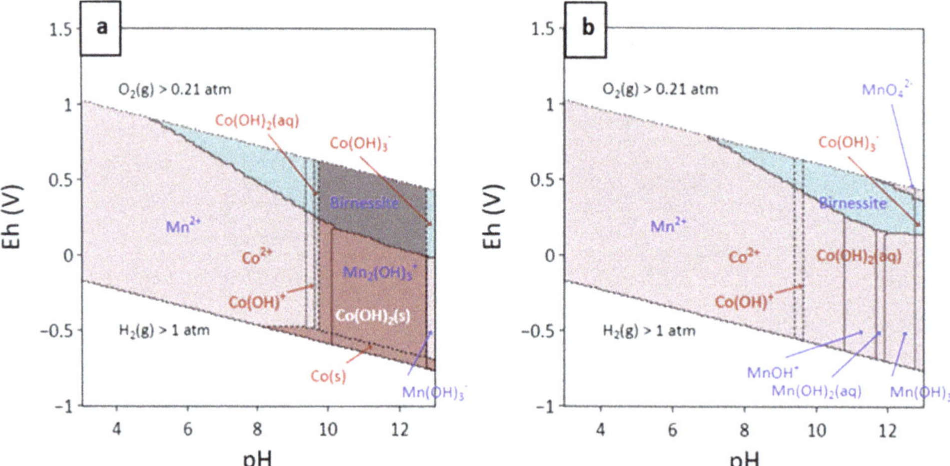

Figure 14. Predominance diagram (Eh–pH) of the Mn–Co–H_2O system calculated with Gibbs Studio [30] and the llnl.dat database from [31] with some modifications (see text for explanation), at 25 °C, with a total aqueous concentration of Mn and Co of 10^{-6} M (**a**) and 10^{-10} M (**b**).

The type of mineral that precipitates depends on the chemical conditions, water composition, Eh and pH. The authors of [56] also observed rim/core textures in the Nkamouna laterite (Cameroon), with pyrolusite at the centre of cryptomelane-lined cavities. The authors considered that as the presence of other cations inhibits the formation of pure Mn-solid phases, pyrolusite should have formed after the precipitation of cryptomelane, which would have removed other cations from the solution. In Loma Caribe and Loma de Hierro, core grains were richer in Mn than rims. This observation agrees with those of [15,56]. However, textural features suggest that rims overgrew the core instead of being formed first (Figure 9c).

In Loma Caribe and Loma de Hierro, lithiophorite and lithiophorite–asbolane intermediates were the most common minerals observed. Most researchers consider that these minerals are a weathering product of previous Mn-oxide minerals [15,56,67,68], but in other cases there is no direct evidence of a primary mineral [57]. In Loma Caribe and Loma de Hierro, Mn-bearing mineral textures did not indicate alteration from previous bearing minerals, so lithiophorite precipitation from soil solution cannot be discarded [56].

Due to its higher content in Al, it is considered that the stability of lithiophorite is higher than that of other Mn minerals, as Al inhibits electron transfer hindering Mn(III) reduction, and, therefore, mineral dissolution under supergene regime ([52] and references therein).

Formation of lithiophorite implies the incorporation of Al in the system. In either Loma Caribe or Loma de Hierro, Al-bearing minerals such as gibbsite and Mn-bearing phyllosilicates have been identified, confirming the availability of Al in the system, that ultimately may come from clinopyroxene and plagioclase minerals from the protolith (lherzolite, clinopyroxene-rich harzburgite, gabbro and diorite) [23–25,69,70].

Mn-oxyhydroxides, and in particular those of biogenic origin, present a high sorption capacity because of their structure, variable (but low) crystallinity with a large surface area, and a very low point of zero charge that promotes negative charged surface areas under common soil pH values [64,71,72]. Phyllomanganates also present a high cation exchange capacity [73]. Cobalt is easily adsorbed onto Mn-bearing minerals' surfaces, and despite Co(II) being the most thermodynamically stable oxidation state in environmental conditions, it is easily oxidised to Co(III) by reduction of Mn(IV) to Mn(III) under supergene conditions [7,71]. With ageing, Co enrichment increases, and the initially adsorbed Co becomes permanently bound [56,74]. Repetition of this process would have led to an enrichment of Co in relation to Mn. This is consistent with the highest content of Co in the saprolite samples from Loma Caribe (Figure 11b). At higher pH, precipitation of Co hydroxide minerals is enhanced (Figure 14a), which might represent an explanation of the Co enrichment in saprolite [51]. Cobalt behaviour is also affected by Eh conditions. Under reducing conditions, Co is released into solution because Co(III) is directly reduced or by reductive dissolution of Mn and/Fe oxides (Figure 14) [75]. All these oxidation–reduction processes take place under supergene conditions, implying good drainage and oxygenated conditions, such as those near a fluctuating water table [51,56]. Therefore, Co–Mn-bearing minerals can be considered a reliable indicator of the position of the paleo-water table within a laterite profile.

Similar to Ni, the primary sources of Mn and Co in Loma Caribe and Loma de Hierro can be found in the protolith [16,17]. Continuous weathering under tropical conditions leaches most of the soluble elements of the protolith, and the least soluble ones, such as Fe and Al, accumulate, forming their own minerals [6,51]. Nickel, Mn and Co are first retained (e.g., sorption processes and/or replacement of Fe) in Fe and Al-bearing minerals [6], and ultimately leached and concentrated in the saprolite-to-oxide zone interface.

6. Conclusions

The Co mineralisations in the Ni–Co laterite profiles of Loma Caribe and Loma de Hierro were mainly associated with Mn-oxyhydroxide minerals, with a composition between Ni asbolane and lithiophorite, found in the transition between the saprolite and the oxide zone. In Loma Caribe, Co-bearing Mn-oxyhydroxide minerals showed colloform aggregate, and globular to spherulitic granular textures, while in Loma de Hierro, they displayed banded colloform, fibrous or tabular textures. In Loma de Hierro, Mn–Co phyllosilicates were also identified. Most of the compositional analyses of Mn-oxyhydroxides yielded 20 and 40 wt.% Mn, with Ni and Co up to 16 and 10 wt.%, respectively, and only in a few cases were Mn contents of up to 54 wt.% MnO on average, measured.

These type of Co mineralisations have also been identified in other laterite deposits worldwide. Formation of Mn-oxyhydroxides in the transition zone is mainly due to the change of pH and Eh occurring in this area and to the saturation of pore solution that may lead to the formation of these minerals by precipitation. Co is considered to be firstly sorbed onto Mn-bearing minerals, after being oxidised because of Mn reduction under supergene conditions, becoming permanently bound into Mn-minerals. Therefore, these minerals are considered to be an indicator of the water level during laterite profile formation.

Author Contributions: Conceptualisation, C.D., C.V.-d.-B. and J.A.P.; methodology, C.D., C.V.-d.-B., E.T. and L.L.; software, C.D.; investigation, C.D., C.V.-d.-B., L.L., S.G., J.M.S., J.I.-I. and M.C., resources, J.A.P.; writing—original draft preparation and editing, C.D., C.V.-d.-B. and J.A.P.; writing—review, E.T., J.M.S., S.G., J.I.-I. and M.C.; supervision, J.A.P.; project administration, J.A.P.; funding acquisition, J.A.P. All authors have read and agreed to the published version of the manuscript.

Funding: This research was supported by grant PID2019-105625RB-C21 funded by MCIN/AEI/ 10.13039/501100011033 and by the Caribbean Lithosphere Reserve Group (http://caribbeanlithos. com/). IDAEA-CSIC is a Severo Ochoa Center of Research Excellence (Spanish Ministry of Science and Innovation, Project CEX2018-000794-S).

Data Availability Statement: Not applicable.

Acknowledgments: We are grateful to Maite Garcia Vallès and to the staff of the Centres Científics i Tecnològics of the Universitat de Barcelona, Barcelona (Spain) (CCiT-UB) for their assistance in measurements. Micro-XRD measurements were performed at BL04-MPSD beamline of ALBA Synchrotron, Cerdanyola del Vallès, Barcelona (experiment AV-2019023550). We would like to thank Oriol Vallcorba and Catalin Popescu for their support during the XRD experiments at ALBA Synchrotron. IDAEA-CSIC is a Severo Ochoa Center of Research Excellence (Spanish Ministry of Science and Innovation, Project CEX2018-000794-S). We are also grateful to the reviewers for improving the quality of this manuscript.

Conflicts of Interest: The authors declare no conflict of interest.

References

1. European Commission. Report from the Commission to the European Parliament, the Council, the European Economic and Social Committee and the Committee of the Regions: Critical Raw Materials Resilience: Charting a Path towards greater Security and Sustainability. Available online: https://eur-lex.europa.eu/legal-content/EN/TXT/?uri=CELEX:52020DC0474 (accessed on 15 September 2021).
2. Slack, J.F.; Kimball, B.E.; Shedd, K.B. Cobalt. In *Critical Mineral Resources of the United States—Economic and Environmental Geology and Prospects for Future Supply*; Schulz, K.J., DeYoung, J.H., Jr., Seal, R.R., II, Bradley, D.C., Eds.; U.S. Geological Survey Professional Paper: Reston, VA, USA, 2017; Chapter F; pp. F1–F40. [CrossRef]
3. US Geological Survey. *Mineral Commodity Summaries 2020: US Geological Survey*; US Geological Survey: Reston, VA, USA, 2020; 200p. [CrossRef]
4. Alves Dias, P.; Blagoeva, D.; Pavel, C.; Arvanitidis, N. *Cobalt: Demand-Supply Balances in the Transition to Electric Mobility*; EUR 29381 EN; Publications Office of the European Union: Luxembourg, 2018; ISBN 978-92-79-94311-9. [CrossRef]
5. Seck, G.S.; Hache, E.; Barnet, C. Potential bottleneck in the energy transition: The case of cobalt in an accelerating electro-mobility world. *Resour. Policy* **2021**, *75*, 102516. [CrossRef]
6. Santoro, L.; Putzolu, F.; Mondillo, N.; Herrington, R.; Najorka, J.; Boni, M.; Dosbaba, M.; Maczurad, M.; Balassone, G. Quan-titative mineralogical evaluation of Ni-Co laterite ores through XRPD-QPA- and automated SEM-based approaches: The Wingellina (Western Australia) case study. *J. Geochem. Explor.* **2021**, *223*, 106695. [CrossRef]
7. Newsome, L.; Arguedas, A.F.S.; Coker, V.; Boothman, C.; Lloyd, J. Manganese and cobalt redox cycling in laterites; Biogeochemical and bioprocessing implications. *Chem. Geol.* **2019**, *531*, 119330. [CrossRef]
8. Butt, C.R.M.; Cluzel, D. Nickel Laterite Ore Deposits: Weathered Serpentinites. *Elements* **2013**, *9*, 123–128. [CrossRef]
9. Freyssinet, P.; Butt, C.R.M.; Morris, R.C.; Piantone, P. Ore-Forming Processes Related to Lateritic Weathering. In *One Hundredth Anniversary Volume*; Society of Economic Geologists: Littleton, CO, USA, 2005; pp. 681–722. [CrossRef]
10. Golightly, J.P. *Nickeliferous Laterite Deposits*; Society of Economic Geologists: Littleton, CO, USA, 1981; pp. 710–735. [CrossRef]
11. Golightly, J.P.; Goldfarb, R.J.; Marsh, E.E.; Monecke, T. Progress in Understanding the Evolution of Nickel Laterites. In *The Challenge of Finding New Mineral Resources: Global Metallogeny, Innovative Exploration, and New Discoveries*; Society of Economic Geologists: Littleton, CO, USA, 2010; Volume 15, pp. 451–485. [CrossRef]
12. Villanova-De-Benavent, C.; Domènech, C.; Tauler, E.; Galí, S.; Tassara, S.; Proenza, J.A. Fe–Ni-bearing serpentines from the saprolite horizon of Caribbean Ni-laterite deposits: New insights from thermodynamic calculations. *Miner. Depos.* **2016**, *52*, 979–992. [CrossRef]
13. Dehaine, Q.; Tijsseling, L.T.; Glass, H.J.; Törmänen, T.; Butcher, A.R. Geometallurgy of cobalt ores: A review. *Miner. Eng.* **2020**, *160*, 106656. [CrossRef]
14. Elias, M.; Donaldson, M.J.; Giorgetta, N.E. Geology, mineralogy, and chemistry of lateritic nickel-cobalt deposits near Kalgoorie, Western Australia. *Econ. Geol.* **1981**, *76*, 1775–1783. [CrossRef]
15. Putzolu, F.; Balassone, G.; Boni, M.; Maczurad, M.; Mondillo, N.; Najorka, J.; Pirajno, F. Mineralogical association and Ni-Co deportment in the Wingellina oxide-type laterite deposit (Western Australia). *Ore Geol. Rev.* **2018**, *97*, 21–34. [CrossRef]
16. Aiglsperger, T.; Proenza, J.A.; Lewis, J.F.; Labrador, M.; Svojtka, M.; Rojas-Purón, A.; Longo, F.; Ďurišová, J. Critical metals (REE, Sc, PGE) in Ni laterites from Cuba and the Dominican Republic. *Ore Geol. Rev.* **2016**, *73*, 127–147. [CrossRef]
17. Domènech, C.; Galí, S.; Soler, J.M.; Ancco, M.P.A.; Meléndez, W.; Rondón, J.; Villanova-De-Benavent, C.; Proenza, J.A. The Loma de Hierro Ni-laterite deposit (Venezuela): Mineralogical and chemical composition. *Boletín Soc. Geológica Mex.* **2020**, *72*, A050620. [CrossRef]

18. Villanova-de-Benavent, C.; Proenza, J.A.; Galí, S.; García-Casco, A.; Tauler, E.; Lewis, J.F.; Longo, F. Garnierites and garnierites: Textures, mineralogy and geochemistry of garnierites in the Falcondo Ni laterite deposit, Dominican Republic. *Ore Geol. Rev.* **2014**, *58*, 91–109. [CrossRef]
19. Lewis, J.F.; Jiménez, J.G. Duarte complex in the La Vega–Jarabacoa–Janico Area, Central Hispaniola: Geological and geochemical features of the sea floor during the early stages of arc evolution. In *Geologic and Tectonic Development of the North America–Caribbean Plate Boundary in Hispaniola*; Mann, P., Draper, G., Lewis, J.F., Eds.; Geological Society of America Special Papers: Boulder, CO, USA, 1991; Volume 262, pp. 115–142.
20. Hackley, P.C.; Urbani, F.; Karlsen, A.W.; Garrity, C.P. *Mapa Geologico de Venezuela a Escala 1:750,000*; USGS Publications Open-File Report 2006-1109 Warehouse: Virgina, VA, USA, 2006. [CrossRef]
21. Giunta, G.; Beccaluva, L.; Coltorti, M.; Mortellaro, D.; Siena, F.; Cutrupia, D. The peri-Caribbean ophiolites: Structure, tectono-magmatic significance and geodynamic implications. *Caribb. J. Earth Sci.* **2000**, *36*, 1–20.
22. Urbani, F. Una revisión de los terrenos geológicos del sistema montañoso del Caribe, Norte de Venezuela. *Boletín Geología* **2018**, *23*, 118–216.
23. Marvin, B.; Valencia, V.; Grande, S.; Urbani, F.; Hurtado, R. Geocronología U–Pb en cristales de zircón de la metadiorita de la Guacamaya, gabro de la ofiolita de Loma de Hierro y gabro de El Chacao, estados Aragua y Guárico. In Proceedings of the V Simposio Venezolano de Geociencias de las Rocas Ígneas y Metamórficas, Caracas, Venezuela, 28–29 November 2013.
24. Neill, I.; Kerr, A.C.; Chamberlain, K.R.; Schmitt, A.K.; Urbani, F.; Hastie, A.R.; Pindell, J.L.; Barry, T.L.; Millar, I.L. Vestiges of the proto-Caribbean seaway: Origin of the San Souci Volcanic Group, Trinidad. *Tectonophysics* **2014**, *626*, 170–185. [CrossRef]
25. Urbani, F.; Rodríguez, J.A. *Atlas Geológico de la Cordillera de la Costa*; Ediciones Fundación Geos y Funvisis: Caracas, Venezuela, 2004; 146p.
26. Cheary, R.W.; Coelho, A.A. A fundamental parameters approach to X-ray line-profile fitting. *J. Appl. Crystallogr.* **1992**, *25*, 109–121. [CrossRef]
27. Coelho, A.A. *TOPAS-Academic, version 4.2*; TOPAS-Academic: Brisbaine, Australia, 2007.
28. Rius, J.; Vallcorba, O.; Frontera, C.; Peral, I.; Crespi, A.; Miravitlles, C. Application of synchrotron through-the-substrate microdiffraction to crystals in polished thin sections. *IUCrJ* **2015**, *2*, 452–463. [CrossRef]
29. Prescher, C.; Prakapenka, V.B. DIOPTAS: A program for reduction of two-dimensional X-ray diffraction data and data ex-ploration. *High Press. Res.* **2015**, *35*, 223–230. [CrossRef]
30. Nardi, A.; de Vries, L.M. *GibbsStudio, version 3.1.1*; Barcelona Science Technologies SL: Barcelona, Spain, 2017. Available online: https://gibbsstudio.io/ (accessed on 31 October 2020).
31. Parkhurst, D.L.; Appelo, C.A.J. Description of input and examples for PHREEQC version 3—A computer program for specia-tion, batch-reaction, one-dimensional transport, and inverse geochemical calculations. *US Geol. Surv. Tech. Methods* **2013**, *6*, 497. Available online: https://pubs.usgs.gov/tm/06/a43/pdf/tm6-A43.pdf (accessed on 1 July 2022).
32. Burlet, C.; Vanbrabant, Y. Study of the spectro-chemical signatures of cobalt-manganese layered oxides (asbolane-lithiophorite and their intermediates) by Raman spectroscopy. *J. Raman Spectrosc.* **2015**, *46*, 941–952. [CrossRef]
33. Katsiapi, A.; Tsakiridis, P.; Oustadakis, P.; Agatzini-Leonardou, S. Cobalt recovery from mixed Co–Mn hydroxide precipitates by ammonia–ammonium carbonate leaching. *Miner. Eng.* **2010**, *23*, 643–651. [CrossRef]
34. Wang, Y.; Xing, S.; Zhang, Y.; Li, Z.; Ma, Y.; Zhang, Z. Mineralogical and thermal characteristics of low-grade Jinlong bauxite sourced from Guangxi Province, China. *J. Therm. Anal.* **2015**, *122*, 917–927. [CrossRef]
35. Gialanella, S.; Girardi, F.; Ischia, G.; Lonardelli, I.; Mattarelli, M.; Montagna, M. On the goethite to hematite phase transformation. *J. Therm. Anal. Calorim.* **2010**, *102*, 867–873. [CrossRef]
36. De Aquino, T.F.; Riella, H.; Bernardin, A.M. Mineralogical and Physical–Chemical Characterization of a Bauxite Ore from Lages, Santa Catarina, Brazil, for Refractory Production. *Miner. Process. Extr. Met. Rev.* **2011**, *32*, 137–149. [CrossRef]
37. Földvári, M. *Handbook of Thermogravimetric System of Minerals and Its Use in Geological Practice*; Occasional Papers of the Geological Institute of Hungary; Geological Institute of Hungary: Budapest, Hungary, 2011; ISBN 978-963-671-288-4.
38. Llorca, S.; Monchoux, P. Supergene cobalt minerals from New Caledonia. *Can. Mineral.* **1991**, *29*, 149–161.
39. Ribeiro, P.P.M.; de Souza, L.C.M.; Neumann, R.; dos Santos, I.D.; Dutra, A.J.B. Nickel and cobalt losses from laterite ore after the sulfation-roasting-leaching processing. *J. Mater. Res. Technol.* **2020**, *9*, 12404–12415. [CrossRef]
40. Palchik, N.A.; Moroz, T.N.; Grigorieva, T.N.; Miroshnichenko, L.V. Manganese minerals from the Miassovo freshwater lake: Composition and structure. *Russ. J. Inorg. Chem.* **2014**, *59*, 511–518. [CrossRef]
41. White, W.N.; Vito, C.; Scheetz, B.E. The mineralogy and trace element chemistry of black manganese oxide deposits from caves. *J. Cave Karst Stud.* **2009**, *271*, 136–143.
42. Ling, F.T.; Post, J.E.; Heaney, P.J.; Kubicki, J.D.; Santelli, C.M. Fourier-transform infrared spectroscopy (FTIR) analysis of triclinic and hexagonal birnessites. *Spectrochim. Acta Part A Mol. Biomol. Spectrosc.* **2017**, *178*, 32–46. [CrossRef]
43. Chukanov, N.V. *Infrared Spectra of Mineral Species: Extended Library*; Springer: Dordrecht, The Netherlands, 2014. [CrossRef]
44. Potter, R.M.; Rossman, G.R. The tetravalent manganese oxides: Identification, hydration, and structural relationships by infrared spectroscopy. *Am. Mineral.* **1979**, *64*, 1199–1218.
45. Carmichael, S.K.; Doctor, D.H.; Wilson, C.G.; Feierstein, J.; McAleer, R. New insight into the origin of manganese oxide ore deposits in the Appalachian Valley and Ridge of northeastern Tennessee and northern Virginia, USA. *GSA Bull.* **2017**, *129*, 1158–1180. [CrossRef]

46. Tauler, E.; Lewis, J.F.; Villanova-De-Benavent, C.; Aiglsperger, T.; Proenza, J.A.; Domènech, C.; Gallardo, T.; Longo, F.; Galí, S. Discovery of Ni-smectite-rich saprolite at Loma Ortega, Falcondo mining district (Dominican Republic): Geochemistry and mineralogy of an unusual case of "hybrid hydrous Mg silicate—Clay silicate" type Ni-laterite. *Miner. Depos.* **2017**, *52*, 1011–1030. [CrossRef]

47. Roqué-Rosell, J.; Mosselmans, F.; Proenza, J.A.; Labrador, M.; Galí, S.; Atkinson, K.D.; Quinn, P.D. Sorption of Ni by "lithiophorite-asbolane" intermediates in Moa Bay lateritic deposits, eastern Cuba. *Chem. Geol.* **2010**, *275*, 9–18. [CrossRef]

48. Post, J.E. Manganese oxide minerals: Crystal structures and economic and environmental significance. *Proc. Natl. Acad. Sci. USA* **1999**, *96*, 3447–3454. [CrossRef]

49. Dublet, G.; Juillot, F.; Brest, J.; Noël, V.; Fritsch, E.; Proux, O.; Olivi, L.; Ploquin, F.; Morin, G. Vertical changes of the Co and Mn speciation along a lateritic regolith developed on peridotites (New Caledonia). *Geochim. Cosmochim. Acta* **2017**, *217*, 1–15. [CrossRef]

50. Putzolu, F.; Abad, I.; Balassone, G.; Boni, M.; Mondillo, N. Ni-bearing smectites in the Wingellina laterite deposit (Western Australia) at nanoscale: TEM-HRTEM evidences of the formation mechanisms. *Appl. Clay Sci.* **2020**, *196*, 105753. [CrossRef]

51. Maurizot, P.; Sevin, B.; Lesimple, S.; Bailly, L.; Iseppi, M.; Robineau, B. Chapter 10 Mineral resources and prospectivity of the ultramafic rocks of New Caledonia. In *New Caledonia: Geology, Geodynamic Evolution and Mineral Resource*; Maurizot, P., Mortimer, N., Eds.; Geological Society: London, UK, 2020; Volume 51, pp. 247–277. [CrossRef]

52. Llorca, S.M. Metallogeny of supergene cobalt mineralization, New Caledonia. *Aust. J. Earth Sci.* **1993**, *40*, 377–385. [CrossRef]

53. Manceau, A.; Llorca, S.; Calas, G. Crystal chemistry of cobalt and nickel in lithiphorite and asbolane from New Caledonia. *Geoch. Cosmoch. Acta* **1987**, *51*, 105–113. [CrossRef]

54. Ploquin, F.; Fritsch, E.; Guigner, J.M.; Esteve, I.; Delbes, L.; Dublet, G.; Juillot, F. Phyllomanganate vein-infillings in faulted and Al-poor regoliths of the New Caledonian ophiolite: Periodic and sequential crystallization of Ni–asbolane, Alk–birnessite and H–birnessite. *Eur. J. Mineral.* **2019**, *31*, 335–352. [CrossRef]

55. Ulrich, M.; Cathelineau, M.; Muñoz, M.; Boiron, M.-C.; Teitler, Y.; Karpoff, A.M. The relative distribution of critical (Sc, REE) and transition metals (Ni, Co, Cr, Mn, V) in some Ni-laterite deposits of New Caledonia. *J. Geochem. Explor.* **2018**, *197*, 93–113. [CrossRef]

56. Dzemua, G.L.; Gleeson, S.A.; Schofield, P.F. Mineralogical characterization of the Nkamouna Co–Mn laterite ore, southeast Cameroon. *Miner. Depos.* **2012**, *48*, 155–171. [CrossRef]

57. Tupaz, C.A.J.; Watanabe, Y.; Sanematsu, K.; Echigo, T. Mineralogy and geochemistry of the Berong Ni-Co laterite deposit, Palawan, Philippines. *Ore Geol. Rev.* **2020**, *125*, 103686. [CrossRef]

58. Chukhrov, F.V.; Gorshkov, A.I.; Vitovskaya, I.V.; Drits, V.A.; Sivtsov, A.V. On the Nature of Co-Ni Asbolane; a Component of Some Supergene Ores. In *Ore Genesis: The State of the Art*; Amstutz, G.C., El Goresy, A., Frenzel, G., Kluth, C., Moh, G., Wauschkuhn, A., Zimmermann, R.A., Eds.; Springer: Berlin/Heidelberg, Germany, 1982; pp. 230–239. [CrossRef]

59. Gorshkov, A.I.; Bogdanov, Y.A.; Sivtsov, A.V.; Mokhov, S.V. A new Mg-AI-Ni asbolane. *Doklady Akad. Nauk* **1995**, *342*, 781–784. (In Russian)

60. Marsh, E.; Anderson, E.; Gray, F. Chapter H of Mineral deposit models for resource assessment. In *Nickel-Cobalt Laterites—A Deposit Model*; 2010–5070–H; Paper for U.S. Geological Survey Scientific Investigations: Denver, CO, USA, 2013; 38p. Available online: https://pubs.usgs.gov/sir/2010/5070/h/ (accessed on 28 June 2022).

61. Burns, R.G.; Burns, V.M. Manganese oxides. In *Marine Minerals*; Ribbe, P.H., Ed.; Mineralogical Society of America: Washington, DC, USA, 1979.

62. Quantin, C.; Becquer, T.; Berthelin, J. Mn-oxide: A major source of easily mobilizable Co and Ni under reducing conditions in New Caledonia Ferralsols. *Comptes Rendus Geosci.* **2002**, *334*, 273–278. [CrossRef]

63. Aoshima, M.; Tani, Y.; Fujita, R.; Tanaka, K.; Miyata, N.; Umezawa, K. Simultaneous Sequestration of Co^{2+} and Mn^{2+} by Fungal Manganese Oxide through Asbolane Formation. *Minerals* **2022**, *12*, 358. [CrossRef]

64. Villalobos, M.; Bargar, J.; Sposito, G. Trace Metal Retention on Biogenic Manganese Oxide Nanoparticles. *Elements* **2005**, *1*, 223–226. [CrossRef]

65. Giffaut, E.; Grivé, M.; Blanc, P.; Vieillard, P.; Colàs, E.; Gailhanou, H.; Gaboreau, S.; Marty, N.; Madé, B.; Duro, L. Andra thermodynamic database for performance assessment: ThermoChimie. *Appl. Geochem.* **2014**, *49*, 225–236. [CrossRef]

66. Plyasunova, N.V.; Zhang, Y.; Muhammed, M. Critical evaluation of thermodynamic of complex formation of metals ions in aqueous solution: IV hydrolysis and hydroxo-complex of Ni^{2+} at 298.15K. *Hydrometallurgy* **1998**, *48*, 153–169. [CrossRef]

67. Cui, H.; You, L.; Feng, X.; Tan, W.; Qiu, G.; Liu, F. Factors Governing the Formation of Lithiophorite at Atmospheric Pressure. *Clays Clay Miner.* **2009**, *57*, 353–360. [CrossRef]

68. Rao, D.; Nayak, B.; Acharya, B. Cobalt-rich lithiophorite from the Precambrian Eastern Ghats manganese ore deposit of Nishikhal, south Orissa, India. *Mineralogia* **2010**, *41*, 11–21. [CrossRef]

69. Farré-De-Pablo, J.; Proenza, J.A.; González-Jiménez, J.M.; Aiglsperger, T.; Garcia-Casco, A.; Escuder-Viruete, J.; Colás, V.; Longo, F. Ophiolite hosted chromitite formed by supra-subduction zone peridotite–plume interaction. *Geosci. Front.* **2020**, *11*, 2083–2102. [CrossRef]

70. Marchesi, C.; Garrido, C.J.; Proenza, J.A.; Hidas, K.; Varas-Reus, M.I.; Butjosa, L.; Lewis, J.F. Geochemical record of subduction initiation in the sub-arc mantle: Insights from the Loma Caribe peridotite (Dominican Republic). *Lithos* **2016**, *252-253*, 1–15. [CrossRef]

71. Decrée, S.; Pourret, O.; Baele, J.-M. Rare earth element fractionation in heterogenite (CoOOH): Implication for cobalt oxidized ore in the Katanga Copperbelt (Democratic Republic of Congo). *J. Geochem. Explor.* **2015**, *159*, 290–301. [CrossRef]
72. Nicholson, K. Contrasting mineralogical-geochemical signatures of manganese oxides; guides to metallogenesis. *Econ. Geol.* **1992**, *87*, 1253–1264. [CrossRef]
73. Tebo, B.M.; Bargar, J.R.; Clement, B.G.; Dick, G.J.; Murray, K.J.; Parker, D.; Verity, R.; Webb, S.M. Biogenic Manganese Oxides: Properties and Mechanisms of Formation. *Annu. Rev. Earth Planet. Sci.* **2004**, *32*, 287–328. [CrossRef]
74. Wu, Z.; Lanson, B.; Feng, X.; Yin, H.; Tan, W.; He, F.; Liu, F. Transformation of the phyllomanganate vernadite to tectomanganates with small tunnel sizes: Favorable geochemical conditions and fate of associated Co. *Geochim. Cosmochim. Acta* **2021**, *295*, 224–236. [CrossRef]
75. Beak, D.G.; Kirby, J.K.; Hettiarachchi, G.M.; Wendling, L.; McLaughlin, M.J.; Khatiwada, R. Cobalt Distribution and Speciation: Effect of Aging, Intermittent Submergence, In Situ Rice Roots. *J. Environ. Qual.* **2011**, *40*, 679–695. [CrossRef]

Article

Evolution of the Piauí Laterite, Brazil: Mineralogical, Geochemical and Geomicrobiological Mechanisms for Cobalt and Nickel Enrichment

Agnieszka Dybowska [1], Paul F. Schofield [1,*], Laura Newsome [2,3], Richard J. Herrington [1], Julian F. W. Mosselmans [4], Burkhard Kaulich [4], Majid Kazemian [4], Tohru Araki [4], Thomas J. Skiggs [5], Jens Kruger [5], Anne Oxley [1,6], Rachel L. Norman [1] and Jonathan R. Lloyd [2]

[1] Department of Earth Sciences, Natural History Museum, London SW7 5BD, UK
[2] Williamson Research Centre, School of Earth and Environmental Sciences, University of Manchester, Manchester M13 9PL, UK
[3] Camborne School of Mines, University of Exeter, Penryn TR10 9FE, UK
[4] Diamond Light Source Ltd., Harwell Science and Innovation Campus, Chilton OX11 0DE, UK
[5] School of Ocean and Earth Science, University of Southampton, Southampton SO14 3ZH, UK
[6] Brazilian Nickel PLC, 10 Lonsdale Gardens, Tunbridge Wells TN1 1NU, UK
* Correspondence: p.schofield@nhm.ac.uk

Citation: Dybowska, A.; Schofield, P.F.; Newsome, L.; Herrington, R.J.; Mosselmans, J.F.W.; Kaulich, B.; Kazemian, M.; Araki, T.; Skiggs, T.J.; Kruger, J.; et al. Evolution of the Piauí Laterite, Brazil: Mineralogical, Geochemical and Geomicrobiological Mechanisms for Cobalt and Nickel Enrichment. *Minerals* **2022**, *12*, 1298. https://doi.org/10.3390/min12101298

Academic Editors: Cristina Villanova-de-Benavent and Cristina Domènech

Received: 6 September 2022
Accepted: 3 October 2022
Published: 14 October 2022

Publisher's Note: MDPI stays neutral with regard to jurisdictional claims in published maps and institutional affiliations.

Abstract: The Piauí laterite (NE Brazil) was initially evaluated for Ni but also contains economic concentrations of Co. Our investigations aimed to characterise the Co enrichment within the deposit; by understanding the mineralogy we can better design mineral processing to target Co recovery. The laterite is heterogeneous on the mineralogical and lithological scale differing from the classic schematic profiles of nickel laterites, and while there is a clear transition from saprolite to more ferruginous units, the deposit also contains lateral and vertical variations that are associated with both the original intrusive complex and also the nature of fluid flow, redox cycling and fluctuating groundwater tables. The deposit is well described by the following six mineralogical and geochemical units: SAPFE, a clay bearing ferruginous saprolite; SAPSILFE, a silica dominated ferruginous saprolite; SAPMG, a green magnesium rich chlorite dominated saprolite; SAPAL, a white-green high aluminium, low magnesium saprolite; saprock, a serpentine and chlorite dominated saprolite and the serpentinite protolith. Not all of these units are 'ore bearing'. Ni is concentrated in a range of nickeliferous phyllosilicates (0.1–25 wt%) including serpentines, talc and pimelite, goethite (up to 9 wt%), magnetite (2.8–14 wt%) and Mn oxy-hydroxides (0.35–19 wt%). Lower levels of Ni are present in ilmenites, chromites, chlorite and distinct small horizons of nickeliferous silica (up to 3 wt% Ni). With respect to Co, the only significant chemical correlation is with Mn, and Mn oxy-hydroxides contain up to 14 wt% Co. Cobalt is only present in goethite when Mn is also present, and these goethite grains contain an average of 0.19 wt% Co (up to a maximum of 0.65 wt%). The other main Co bearing minerals are magnetite (0.41–1.89 wt%), chlorite (up to 0.45 wt%) and ilmenite (up to 0.35 wt%). Chemically there are three types of Mn oxy-hydroxide, asbolane, asbolane-lithiophorite intermediates and romanechite. Spatially resolved X-ray absorption spectroscopy analysis suggests that the Co is present primarily as octahedrally bound Co^{3+} substituted directly into the MnO_6 layers of the asbolane-lithiophorite intermediates. However significant levels of Co^{2+} are evident within the asbolane-lithiophorite intermediates, structurally bound along with Ni in the interlayer between successive MnO_6 layers. The laterite microbial community contains prokaryotes and few fungi, with the highest abundance and diversity closest to ground level. Microorganisms capable of metal redox cycling were identified to be present, but microcosm experiments of different horizons within the deposit demonstrated that stimulated biogeochemical cycling did not contribute to Co mobilisation. Correlations between Co and Mn are likely to be a relic of parent rock weathering rather than due to biogeochemical processes; a conclusion that agrees well with the mineralogical associations.

Keywords: Ni-Co laterite; critical metals; Piauí; metal cycling; cobalt speciation; biogeochemistry; redox cycling

---

1. Introduction

Nickeliferous laterite deposits are an increasingly important resource for a range of critical metals such as Ni and Co, e.g., [1–5], but also including rare earth elements (REE), platinum group elements (PGE) and green technology elements such as Sc e.g., [6–9]. Nickel-cobalt lateritic deposits form in tropical and sub-tropical environments through extensive weathering of ultramafic rocks. Variations in climate, weathering and erosional environment have a significant impact on the nature of laterite deposits and, based largely upon their nickeliferous mineralogy, they are often classified as one of three distinct deposit types; oxide dominated, clay rich or hydrous Mg-silicate e.g., [10–14].

Many of the globally most significant Ni-Co laterite deposits occur in the humid tropical regions within New Caledonia, Colombia, Cuba, Indonesia, Philippines, Dominican Republic and Venezuela e.g., [11,15]. In addition to these deposits are those within more temperate, semi-arid and arid regions such as Australia, USA, Central America, Brazil, Madagascar, Kazakhstan, Russia, Turkey, Greece, and the Balkan region of Europe e.g., [11,14,16], and these deposits are often referred to as palaeolaterites or palaeodeposits e.g., [11,12,15]. Distinct laterite deposits are distributed throughout Brazil though they are less common in the south and northeastern regions [17,18] and are mostly correlated to the Sul Americano and Velhas (Lower and Upper Tertiary, respectively) levelling cycles [18]. These Brazilian lateritic deposits are broadly consistent with the general description for laterites from around the world and there is a numerical dominance of hydrous Mg-silicate deposit types over oxidised types [18]. One of these Brazilian nickeliferous laterites is that of São João do Piauí, northeastern Brazil e.g., [19,20], and while being generally consistent with the standard laterite description, the profile maintains an additional complexity due to a series of mafic intrusions within the original dunite source.

The nickel-cobalt laterite deposit at São João do Piauí, located in Piauí State of northeastern Brazil (Figure 1), was acquired by Brazilian Nickel Ltd. (BRN) in 2014 with a view to applying heap leach technology e.g., [21], for ore processing. The deposit resource is 72 Mt with 1.00% Ni and 0.05% Co with 73% of the resource measured [22]. By the end of 2017 a demonstration plant had successfully established a low-impact heap leaching, purification and recovery process for the hydrometallurgical recovery of nickel and cobalt as mixed-metal (Ni and Co) hydroxide products. This demonstration technology produced heap extractions in excess of 80% Ni. By the end of 2022 it is hoped that BRN will be continuously producing Ni and Co, aiming towards 1400 t of Ni and 35 t of Co per year [23].

In this manuscript we investigate the factors that contribute to the mobilisation and enrichment of cobalt in a nickel-laterite deposit. We provide a description of the mineralogy, geochemistry, geomicrobiology and evolution of the São João do Piauí nickel-cobalt laterite deposit with an emphasis on the mobility and potential cycling of cobalt.

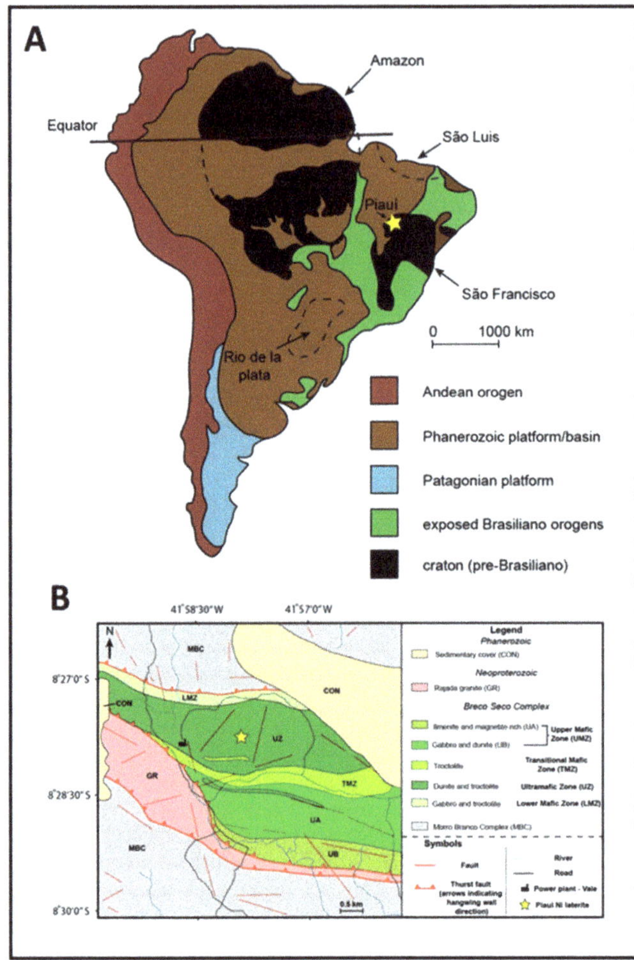

Figure 1. (**A**) The Amazonian, São Luis, São Francisco and Rio de la Plata continental cratons of South America with the poorly exposed areas indicated with dashed lines and the location of the Piauí laterite shown by a star. Modified after [24]. (**B**) A geological map of the Brejo Seco Complex with the location of the Piauí deposit shown by a star. Modified after [25].

2. Geological Context

The São João do Piauí deposit is situated in the igneous mafic-ultramafic Brejo Seco–Piauí Complex in the São Francisco craton [20,24,25]. The complex is bound between two contrasting tectonic settings with the metasedimentary Morro Branco Complex to the north and the syn-collisional granite of the Rajada Suite to the south (Figure 1). The deposit is set within a terrain of intensely deformed volcanic, volcano–sedimentary and sedimentary rocks of Precambrian age that have been intruded by mafic-ultramafic bodies. These mafic-ultramafic bodies are differentiated and comprise serpentinised ultramafics at their base with the upward progression towards troctolites, metagabbros and anorthosites [22,25]. The entire sequence has been intruded by granitic and granodioritic bodies and is unconformably overlain by sediments of the Sierra Grande Formation, Pimenteiras Formation and Cabeças Formation [25,26].

The Brejo Seco–Piauí Complex is dominantly gabbroic in nature and has been divided into four zones [25,27] with a north to south progression starting with the lower mafic zone followed by the ultramafic zone, transitional mafic zone and the upper mafic zone. The lower mafic zone is an olivine-rich troctolite cumulate, while the ultramafic zone is a serpentinised dunite which grades through a complex of diorites, gabbros and troctolites forming the transitional mafic zone and upper mafic zone [27]. The ultramafic body at São João do Piauí lies within the ultramafic zone, is capped by a silcrete crust and has undergone intense and extensive laterisation resulting in the formation of the nickel-cobalt laterite deposit [18,19,25].

The ultramafic unit of the ultramafic zone outcrops in a zone some 3.5 km in strike running east–west and 1.5 km wide at the widest point north–south. A pronounced silica cap forms a 100 m high plateau that is related to the palaeosurface of the Sul Americano and Velhas levelling cycle [18]. This silica cap overlies and protects the underlying mineralised weathered ultramafic unit itself, and locally the cap forms a topographic plateau that measures some 2 km × 0.2–0.7 km in extent. The mineralised profile is 30 to 50 m thick below this cap and is partly exposed on the slopes of the dissected plateau, although colluvium derived from the silica cap obscures much of the geology and the bulk of the geology of the deposit has only been revealed by drilling [28].

3. Profile Description and Sample Collection

The laterite profile at the Piauí deposit site was exposed in a small trial open pit excavated to remove 10,000 tonnes of ore for mineral test work, completed in early 2016 shortly before the authors visited the site. The maximum depth of the exposed profile was approximately 30 m. Figure 2 shows the profile of the Piauí laterite as viewed from within the test pit with the major lithological units indicated.

Figure 2. Profile of the Piauí deposit showing the rock units used during fieldwork ((**A**) looking west; (**B**) looking southwest). Unit indicated with # is a silicified ferruginous saprolite (scale bar = 3 m).

The profile as sampled is rather atypical if compared to the classic published schematic profiles [4]. At the base of the exposed profile is a yellow-white (occasionally yellow-green), coarse to medium grained, strongly leached blocky saprolite that preserves its source rock adcumulate texture and contains occasional clusters of magnetite/chromite. This unit is labelled saprock in Figure 2B and shown in more detail in Figure 3A. Overlying the saprock is an iron oxide rich silicified ferruginous saprolite that varies in colour from deep red-brown to orange-ochre and occasionally sandy yellow close to the highly irregular contact with the saprock (Figure 3A). The extent of the silicification is highly variable, present either as a siliceous stockwork (Figure 4C,D) or as an extensive boxwork (Figure 4A). The silicified ferruginous saprolite is volumetrically the most significant lithological unit exposed in the profile.

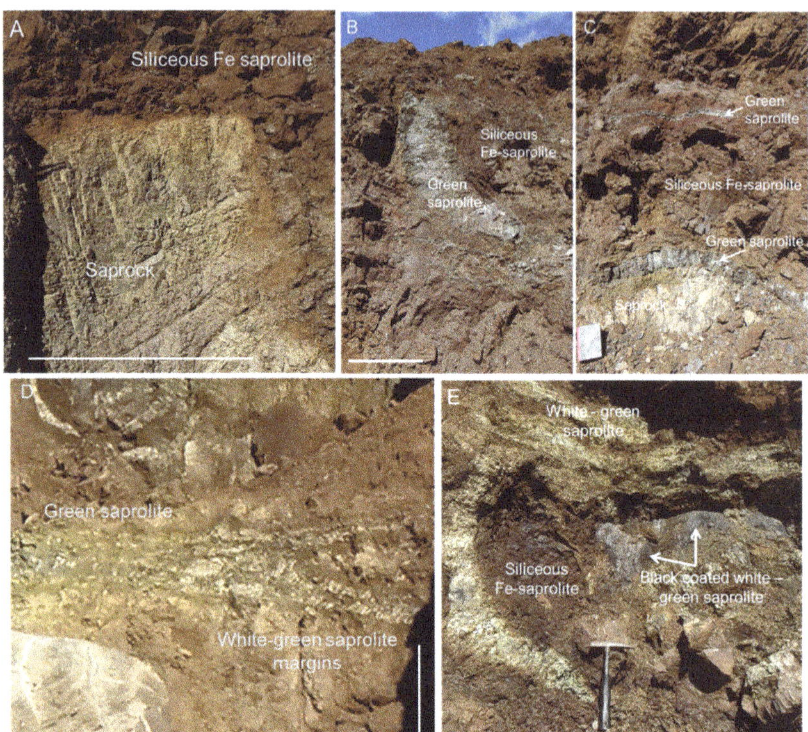

Figure 3. Details from the rock units observed in the profile of the Piauí deposit (**A**) Contact between the saprock and the overlying siliceous saprolite. Scale bar is 1.5 m. (**B,C**) Examples of the green saprolite components within the siliceous Fe-saprolite. Scale bar on B is 1 m. (**D**) Green saprolite layer with white-green margins. Scale bar is 1 m. (**E**) Thick desiccating black coating on kaolinite rich white-green saprolite.

Within the silicified ferruginous saprolite several layers of green saprolite are developed varying in thickness from a few cm up to ~2 m. These green layers generally have white-green outer zones or margins, and, occasionally, white-green layers up to ~30 cm are present within the green saprolite where multiple layers merge (Figure 3D). The layers are generally horizontal in nature, but are occasionally steeply dipping (Figures 3B,C and 4B). The interface of the white-green saprolite layers with the silicified ferruginous saprolite are often found in association with black staining or concentrations of black Mn-oxide rich saprolite (Figures 3E and 4E–G). The green or white-green layers are generally friable though yield preserved igneous textures in places and, while the green parts of these layers contain disseminated magnetite/chromite and are weakly magnetic, the white-green layers contain ilmenite rather than magnetite and are therefore not magnetic. Rarely, elongate nodules or veinlets (10s of cm in length and 2–5 cm thick) of massive bright green serpentine or garnierite are present along the interface between these green or white-green saprolite layers and the silicified ferruginous saprolite.

Overlying the silicified ferruginous saprolite is a red-brown or orange-ochre coloured friable granular ferruginous saprolite dominated by iron oxides and clay minerals (e.g., Figure 2A). Within both the ferruginous saprolite and silicified ferruginous saprolites are magnesite and kaolinite veins, either horizontal or very steeply dipping, that vary in thickness from ~1 cm up to ~15 cm. At the very top of the profile an iron cap has developed at the surface, partly formed of magnetic iron oxides, often containing colluvium of residual and transported material. The serpentinised dunite protolith did not outcrop within the test pit.

Figure 4. Veining and boxwork textures from the saprolite units observed in the profile of the Piauí deposit (**A**) Finely meshed silica boxwork within the silicified Fe-saprolite. (**B**) Flat and steeply dipping, clay-rich, green saprolite veins commonly observed throughout the ferruginous saprolite units. Scale bar is 30 cm. (**C**) Discrete fine-scale silica veining within the silicified ferruginous saprolite. Scale bar is 15 cm. (**D**) Silicification within the silicified ferruginous saprolite facilitated by the discrete fine-scale, late-stage silica veins. (**E–G**) Examples of black mineral aggregates and coatings associated with the white-green saprolitic layers. Scale bar is 50 cm in (**E**) and (**F**).

In the test pit, two distinct sampling strategies were employed to sample the range of lithologies in the profile. Firstly, a suite of samples was collected from across the profile in order to define the variation of the major lithological units and also characterise specific features including veins, nodules, clasts and unusual textural components. It should be noted that the serpentinised dunite protolith was not sampled from within the test pit but from a locality exposing unweathered bedrock, approximately 1.5 km from the pit towards the processing plant.

Secondly, a depth profile of fresh samples was collected from the newly exposed laterite section for geomicrobiological investigation. These samples were collected from depths of 1 m to approximately 5 m below the April 2016 ground level. Samples were collected from the exposed rock face after removing approximately 15 cm of surface material with a clean stainless steel trowel sterilised with 70% ethanol. For microbial community profiling, a similarly sterilised hand auger was used to recover undisturbed sediment cores directly into sterilised plastic core liners, which were immediately sealed to minimise air ingress. Bulk samples were collected for mineralogical characterisation and microcosm experiments and placed into sterilised containers with ambient air head space. The 10 samples were stored in cool boxes and shipped back to the UK for analysis.

The Supplementary Materials contains a schematic of the Piauí test pit (Figure S1) and details the spatial distribution of the samples (Figures S1–S3).

4. Experimental Methods

4.1. Whole Rock Geochemistry

Thirty-nine samples were analysed for a suite of 53 elements by inductively coupled plasma-atomic emission spectroscopy (ICP-AES) and inductively coupled plasma atomic mass spectrometry (ICP-MS) at the OMAC Laboratories Ltd., Loughrea, Ireland. For the following suite of elements samples were prepared by lithium borate fusion and analysed by ICP-MS: Ba, Ce, Cr, Cs, Dy, Er, Eu, Ga, Gd, Hf, Ho, La, Lu, Nb, Nd, Pr, Rb, Sm, Sn, Sr, Ta, Tb, Th, Tm, U, V, W, Y, Yb and Zr. Samples were prepared by 4 acid digestion and analysed with ICP-AES for the remaining suite of elements: Ag, As, Cd, Co, Cu, Li, Mo, Ni, Pb, Sc, Tl, Zn and majors Si, Al, Fe, Ca, Mg, Na, K, Cr, Ti, Mn, P and Ba. For quality control purposes, a set of standards, blanks and sample duplicates were analysed as independent laboratory checks.

4.2. Bulk Mineralogy Using X-ray Diffraction

X-ray powder diffraction (XRD) patterns were collected with a Panalytical XPert Pro MPD diffractometer (PANalytical, Almelo, the Netherlands) equipped with an Xcelerator real-time strip with an active detector length of 2.122° and using either Cu Kα_1 or Co Kα. Data were recorded in continuous mode over the 3–70 2Theta range with 0.017° steps, 175 s per step and a scan rate of 0.01° s^{-1}. The total scan time was less than 2 h. Each sample was powdered first using a mortar and pestle and then processed further by wet milling using a Retsch XRD Mill McCrone to obtain a fine powder with narrow particle size distribution (down to 3 μm). The powders were than packed into aluminium sample wells using a back loading method with portions of loose powder pressed firmly to achieve good packing density and a random orientation. Mineral identification from the XRD patterns was performed using search-match routines with reference patterns from the ICDD database. The following reference patterns were used for mineral identification: goethite (#00-029-0713), hematite (#00-033-0664), maghemite (#00-039-1346), talc (#00-013-0558), quartz (#00-046-1045), chlorite (#00-029-701), kaolinite (#00-083-0971), lizardite (#00-050-1625), antigorite (#00-002-0095), ilmenite (#01-073-1255), nontronite (#00-29-1497), vermiculite (#01-76-0847).

4.3. Spatially Resolved Mineralogy Using Electron Microscopy

Electron microprobe analysis (EMPA) was performed using wavelength-dispersive spectrometry on a Cameca SX100 electron microprobe. Operating conditions were: 20 kV accelerating voltage, 20 nA current and a 1 μm spot size. Standards used were a combination of natural minerals for Na (jadeite), Mg (olivine), Al (corundum), Ca (wollastonite), Si (fayalite) and Ti (rutile); synthetic compounds for Ti and Mn ($MnTiO_3$), P ($ScPO_4$), K (KBr), Fe (FeO), Ni (NiO_2) and pure metals (Co). All data were matrix-corrected using the Cameca version of the PAP PhiRhoZ programme after [29,30].

Quantitative X-ray microanalysis was performed using the Oxford Instruments INCA XMax Energy Dispersive Spectrometer (EDS, Oxford Instruments, Santa Barbara, CA, USA) on the Zeiss EVO 15LS scanning electron microscope (SEM, Zeiss, Oberkochen, Germany).

Objective lens to specimen working distance was kept constant at 8.5 mm (fixed focus). The electron beam accelerating voltage was 20 kV, and electron beam current 3 nA. X-ray acquisition live-time was 20 s. Quant optimisation was performed on cobalt metal, typically every 3 h. Beam current was monitored regularly during the analysis session. Large area energy-dispersive X-ray (EDX) maps were collected on the polished block samples. Rectangular areas were selected over a sample and split into individual tiles using a minimum magnification of 250x per tile. Acquisition conditions for EDX smartmap on individual tiles were set as follows: dwell time of 100 µs, process time of 2, minimum 30 frames, overlap of 20 pixels. Large area EDX maps were then montaged from individual tiles using the Automate function in the Oxford Instruments INCA software.

FEI Quanta 650 FEG SEM (Thermo Fisher Scientific, Brno, Czech Republic) was used to collect smartmaps in samples where high spatial resolution was required. EDX maps were collected using the Bruker Nano XFlash® energy dispersive detector operating at 12 kV with a sample working distance fixed at approximately 25–26 mm. Maps were processed using ESPIRIT version 1.9 software. Pixel resolution for the maps collected varied from 50 nm to 0.4 µm depending on the area mapped and acquisition time for each map was at least 45 min.

4.4. Synchrotron Analysis

Micro-focus X-ray absorption spectroscopy (µXAS) measurements were carried out on the microfocus beamline I18 at Diamond Light Source Ltd. (DLS), Chilton, UK [31,32]. Data were collected in fluorescence mode with polished resin blocks mounted vertically in reflection geometry and the Si-drift fluorescence detector with XSPRESS3 read-out electronics set at 90° to the incident beam. X-ray energies were selected using a Si(111) double crystal monochromator. Tests for self-absorption from the samples were carried out by performing an XAS measurement at each edge with the sample set at varying degrees to the incident beam. No changes to the spectral structure were evident as a function of incident angle, suggesting that self-absorption effects, if present at all, were minimal. µXAS spectra were acquired with the sample at 45° using a beam focussed to either 3.4 µm vertically and 2.8 µm horizontally or 2.2 µm (vertically) and 2.2 µm (horizontally). X-ray absorption near edge structure (XANES) and X-ray absorption fine structure (XAFS) spectra were acquired at the Co K-edge (~7709 eV) and Ni K-edge (~8333 eV) to 12 Å^{-1}. A minimum of 2 scans per point were collected and no beam related degradation of the sample was observed within the asbolane-lithiophorite minerals analysed.

XAS spectra were calibrated, background subtracted and normalised usi ng Athena [33], and Artemis [33] was used for XAFS analysis. Shell-by-shell fits of the XAFS spectra were developed with path lengths, Debye–Waller (σ^2) factors, coordination number (CN) and the relative energy shift (δE) allowed to vary independently while the amplitude function ($S_0{}^2$) remained fixed. Further shells were only added to the model if they were shown to significantly improve the fit.

Scanning transmission X-ray microscopy (STXM) and L-edge XANES measurements were performed at the I08 beamline at DLS, UK. Focused ion beam sections 100 nm thick were extracted from Mn oxy-hydroxide grains in sample F030. Spatially correlated energy dependent image stacks (~7 × 8 µm, 70 × 80 pixels) were acquired at the Co (770–810 eV), Mn (630–668 eV) and Fe (700–735 eV) L-edges with a nominal beam size of ~100 nm and a dwell time per pixel of 7 ms (Co, Mn) or 10 ms (Fe). For all measurements the energy step size was 0.1 eV over the main L$_3$ and L$_2$ edges. The MANTiS program [34] was used to align the image stacks, normalize to the background X-ray intensity (I0) and extract XANES spectra from any pixel or specific regions of interest (ROIs) in the image stacks.

4.5. Identification of Microorganisms in Field Samples

The composition and abundance of microbial communities in different laterite horizons were characterised by Illumina MiSeq amplicon sequencing and quantitative polymerase chain reaction (qPCR). Samples were selected to represent the range of lithologies present and different depths beneath the 2016 ground level. DNA was extracted from six sediment core or bulk samples using a PowerSoil kit (Qiagen); two samples yielded insufficient DNA, therefore, another extraction was performed using a FastDNA spin kit for soils (MP Biomedicals), which yielded sufficient DNA from one sample (Table S5, Supplementary Materials). To confirm the lithological unit classification and geochemistry of the microbiology samples, the chemical composition was analysed by X-ray fluorescence (XRF), mineralogy by XRD (Bruker D8 Advance), and total organic carbon (TOC) was measured using a Shimadzu SSM5000A. The pH was measured using a calibrated electrode 1 h after adding 1 g of laterite (wet weight) to 10 mL deionised water. Full details of the DNA amplification, primers, sequencing, and qPCR are provided in the Supporting Material. Briefly, DNA was sequenced using the Illumina MiSeq platform: for prokaryotes, the 16S rRNA gene was amplified using the primers 515F and 806R; for fungi, the ITS2 region was amplified using the primers ITS4F and 5.8SR. For qPCR, the primers 8F and 1492R were used to quantify prokaryotes and the primers EUKF and EUKR were used to quantify eukaryotes.

Microcosm experiments were performed on seven samples of different laterite horizons to investigate whether the microbial community present in different horizons could mobilise metals by stimulating the development of reducing conditions. Laterite sediment microcosms comprised 3 g laterite, 30 mL of sterile artificial groundwater and either 10 mM glucose or a mixture of 5 mM acetate and 5 mM lactate as a source of organic matter and as an electron donor [35]. The headspace was degassed with an 80:20 N_2:CO_2 mix and the bottles were incubated in the dark at 30 °C. The sediment microcosms were periodically monitored for geochemical changes by removing an aliquot of sediment slurry using a N_2 degassed needle and syringe. The Ferrozine assay was used to measure Fe(II) and Fe(III); Fe(II) was assessed after digesting 0.1 mL of sediment slurry in 4.9 mL 0.5 N HCl for 1 h, and then total Fe after an additional 1 h digest in 0.5 N hydroxylamine-HCl [36,37]. Supernatant was separated from the sediment slurry by centrifugation (16,200× g, 5 min). Major anions and volatile fatty acids were measured by ion chromatography (Dionex ICS 5000), and pH and Eh by calibrated electrodes. To measure aqueous metal concentrations the supernatant was diluted into 2% nitric acid and analysed by ICP-AES.

5. Results

5.1. Whole Rock Geochemistry

The bulk chemistries of the samples (Tables 1 and 2, with additional elements in Tables S1 and S2 of the Supplementary Materials, with some minor and trace elements presented in Figure 5) describe the geochemical variation across the whole profile. Combining these data with hand specimen and field observations (e.g., Figures 2–4), and adapting the terminology adopted in the report of [28] and the previous work cited therein [38] we have subdivided the deposit into seven fundamental lithological units: Serpentinite, Saprock, SAPSILFE, SAPMG, SAPAL, SAPSFE and COB (Table 3).

The average bulk chemistries for these lithological units (Table 4) shows that across the whole deposit, the composition of the samples (excluding the silica nodules and Mn oxide rich sample) is dominated by SiO_2 (average 36 wt%), with average concentrations for Fe_2O_3, MgO and Al_2O_3 of 19 wt%, 17 wt% and 8.6 wt%, respectively. The Co content varies widely from 47 ppm to 2960 ppm, with an average of 543 ppm, but is strongly enriched in the SAPAL unit. The Ni content is heterogeneous (0.28–9.2 wt%) with an average of 2.70 wt%.

Table 1. Bulk chemical analyses of the Piauí laterite samples Serpentinite, Saprock, SAPFE, SAPSILFE, Green Silica Nodules and Mn-Oxide (Additional elements are presented in (Table S1 of the Supplementary Materials).

Analyte	Unit	Serpentinite			Saprock		SAPFE				SAPSILFE				Green Silica Nodules					Mn Oxide
		F046	F047	F049	F017	F039	F018	F019	F022	F028	F030	F032	F043	F044	F008	F009	F011	F012	F013	F050
SiO$_2$	wt%	38.7	40.1	39.1	37.4	38.8	12.6	24.7	22.2	20.5	48.2	28.6	45.9	64.3	88.4	94.1	92.8	90.5	85	6.31
Fe$_2$O$_3$	wt%	11.65	9.74	10.8	11.5	18.45	56.9	37	45.5	49.6	32.7	40.6	40	23.3	0.37	3.48	0.71	0.72	1.26	10.1
MnO	wt%	0.22	0.21	0.24	0.14	0.17	0.69	0.44	0.61	0.43	0.29	0.44	0.34	0.16	0.01	0.04	0.01	0.01	0.02	40.3
Al$_2$O$_3$	wt%	2.3	2.31	2.11	1.19	2.33	5.79	3.97	4.67	5.06	3.58	4.69	2.45	2.52	0.1	0.04	0.01	0.12	0.14	3.39
CaO	wt%	0.04	0.06	0.05	0.02	0.04	0.03	0.03	0.04	0.04	0.04	0.08	0.02	0.02	0.04	0.02	0.01	0.02	0.01	0.05
MgO	wt%	34.8	34.6	34.9	35.7	26.5	6.17	14.95	9.25	10.4	3.4	7.99	0.51	2.48	4.19	0.65	1.4	2.16	5.29	2.01
Cr$_2$O$_3$	wt%	1.54	1.655	1.83	0.89	1.58	4.2	2.94	3.09	3.5	2.68	2.62	1.74	1.7	0.02	0.008	0.004	0.009	0.019	0.167
TiO$_2$	wt%	0.03	0.03	0.03	0.02	0.04	0.1	0.07	0.12	0.11	0.07	0.06	0.08	0.08	0.06	0.05	<0.01	0.01	0.02	0.01
Ni	wt%	0.329	0.345	0.343	0.732	0.824	2.18	3.11	1.43	1.78	0.911	1.33	0.394	0.279	3.14	0.683	1.125	1.415	1.975	3.55
LOI	wt%	12.1	12.3	12.1	13.7	13.65	8.75	11.8	11.9	8.73	7.59	12.15	7.46	5.08	2.66	0.73	1.18	1.61	2.85	13.35
Cd	ppm	<0.5	<0.5	1	<0.5	0.6	<0.5	1.7	1.4	1.8	0.6	0.8	<0.5	0.5	1.9	0.7	1.1	1.2	1.5	3.8
Co	ppm wt%	138	102	130	158	224	812	628	610	616	318	632	194	136	25	3	49	47	56	4.36
Cu	ppm wt%	11	15	10	299	579	1160	1410	723	782	671	303	713	418	157	39	172	233	286	1.02
Li	ppm	<10	<10	<10	<10	<10	<10	<10	<10	<10	<10	<10	<10	<10	<10	<10	<10	<10	<10	<10
Mo	ppm	<1	<1	1	<1	2	3	2	2	1	<1	1	3	2	1	<1	<1	<1	1	<1
Pb	ppm	10	3	9	2	<2	4	3	17	<2	<2	3	14	3	6	4	4	40	<2	<20
Sc	ppm	5	5	5	6	6	26	20	21	20	16	21	19	11	1	<1	1	2	5	8
Zn	ppm	138	145	148	71	114	379	781	233	342	228	292	141	133	27	12	37	154	69	686
Ba BaO	ppm wt%	8.4	3.2	5.1	61.7	64.5	157	198	129.5	161	165.5	262	29.1	38	7	3.8	2.6	22	10.4	9.93
Ce	ppm	0.8	0.4	0.6	<0.5	1.9	<0.5	<0.5	1.4	6.1	2.4	16.6	14.8	8.9	<0.5	0.3	0.1	0.4	0.2	1150
Dy	ppm	0.13	0.26	0.21	<0.05	0.33	0.14	0.4	0.36	1.37	0.51	2.31	1.36	1.77	0.12	<0.05	<0.05	0.06	0.05	36.8
Er	ppm	0.08	0.13	0.13	<0.03	0.29	0.18	0.44	0.31	0.73	0.27	1.01	0.72	1.32	0.12	0.06	<0.03	0.05	<0.03	18.5
Eu	ppm	0.09	0.12	0.1	<0.03	0.19	0.05	0.13	0.1	0.5	0.19	0.77	0.32	0.46	<0.03	<0.03	<0.03	<0.03	<0.03	12.85
Gd	ppm	0.21	0.24	0.35	0.05	0.56	0.15	0.29	0.29	1.86	0.78	2.66	1.17	1.99	0.1	0.1	<0.05	0.06	<0.05	40.9
Ho	ppm	0.04	0.05	0.05	0.02	0.06	0.06	0.08	0.11	0.26	0.12	0.43	0.26	0.39	0.04	0.02	0.01	0.01	0.02	6.62

Table 1. *Cont.*

Analyte	Unit	Serpentinite			Saprock		SAPFE				SAPSILFE				Green Silica Nodules					Mn Oxide
		F046	F047	F049	F017	F039	F018	F019	F022	F028	F030	F032	F043	F044	F008	F009	F011	F012	F013	F050
La	ppm	0.8	0.5	0.7	<0.5	2.6	<0.5	1.3	1.4	11.1	3	12.8	5.5	10.2	0.6	0.3	0.1	0.3	0.3	255
Lu	ppm	0.02	0.03	0.03	0.02	0.03	0.07	0.05	0.07	0.11	0.05	0.13	0.15	0.18	0.03	0.02	0.01	<0.01	0.02	2.45
Nd	ppm	0.8	0.9	0.9	<0.1	2.1	0.3	0.7	1.2	9.9	2.7	15.4	7.4	9.2	<0.1	0.2	<0.1	0.3	0.2	265
Pr	ppm	0.2	0.16	0.18	<0.03	0.43	<0.03	0.06	0.2	1.98	0.79	3.02	1.42	1.95	<0.03	0.04	<0.03	0.06	0.03	73.9
Sm	ppm	0.17	0.35	0.22	<0.03	0.24	0.09	0.08	0.27	1.41	0.65	2.84	0.93	1.61	0.17	0.05	<0.03	0.06	<0.03	56.9
Tb	ppm	0.02	0.04	0.04	0.01	0.05	0.02	0.05	0.06	0.25	0.1	0.39	0.21	0.29	0.01	0.01	<0.01	0.01	0.02	6.83
Tm	ppm	0.03	0.03	0.02	<0.01	0.03	0.05	0.03	0.03	0.09	0.05	0.11	0.11	0.19	0.01	0.01	0.01	0.02	0.01	2.89
Yb	ppm	0.08	0.15	0.12	0.07	0.21	0.28	0.39	0.44	0.55	0.29	0.85	0.69	1.25	0.06	0.09	0.03	0.04	0.1	18.7
ΣREE	ppm	3.47	3.36	3.65	0.17	9.02	1.39	4	6.24	36.21	11.9	59.32	35.04	39.7	1.26	1.2	0.26	1.37	0.95	1947.34

Table 2. Bulk chemical analyses of the Piauí laterite samples SAPAL, SAPMG, magnesite and kaolinite (Additional elements are presented in Table S2 of the Supplementary Materials).

Analyte	Unit	SAPAL			SAPMG															Magnesite	Kaolinite
		F014	F015	F033	F002	F005	F006	F007	F023	F024	F025	F026	F029	F031	F034	F038	F040	F041	F003	F035
SiO_2	wt%	44.4	46.8	42	31	37.4	32.6	36.6	42.5	38.5	38.1	31.5	31.2	33.9	34.3	30	37.3	36.8	1.4	42.4
Fe_2O_3	wt%	7.32	3.36	7.54	14	9.47	11.9	7.36	11.65	14.6	11.15	14.1	27.4	14.45	16.2	3.28	9.63	7.82	0.2	10.3
MnO	wt%	0.34	0.4	2.21	0.31	0.06	0.34	0.4	0.11	0.25	0.13	0.43	0.25	0.79	0.89	0.23	0.1	0.39	0.01	0.3
Al_2O_3	wt%	24.9	27.9	22	11.7	8.36	13.75	12.4	5.38	3.38	4.24	14.25	3.33	11.4	13.3	19.3	7.13	11.6	0.08	31
CaO	wt%	0.09	0.08	0.1	0.05	0.09	0.13	0.12	0.04	0.04	0.06	0.14	0.04	0.19	0.16	0.11	0.06	0.11	0.25	0.04
MgO	wt%	3.26	2.69	5.09	19.5	18.6	16.9	15.55	21.2	24.8	27.9	23.7	23.1	21	13.05	30.7	17	13.05	47.9	0.7
Cr_2O_3	wt%	0.12	0.02	0.05	0.24	0.79	0.12	0.13	0.51	2.04	1.23	0.05	1.53	0.06	0.24	0.07	0.87	0.15	0.01	0.43
TiO_2	wt%	0.11	0.12	0.67	2.2	0.04	1.38	1.81	0.03	0.06	0.03	1.77	0.12	1.89	1.31	0.65	0.05	2.09	<0.01	0.22
Ni	wt%	1.685	1.4	2.31	6.45	8.65	6.36	8.48	1.105	2.11	2.67	1.28	1.855	2.2	2.2	1.17	9.22	8	0.0355	0.498
LOI	wt%	17.65	16.85	16.2	11.85	14.1	14.1	15.05	16.45	14.45	14.5	13.65	11.75	14.3	15.05	14.95	13.7	15.35	51.1	14
Cd	ppm	<0.5	<0.5	0.7	2.1	0.9	2.3	3.4	0.6	1.6	2.2	1.3	1.4	1.9	1.4	1.4	1.2	2.6	<0.5	<0.5
Co	ppm	835	996	2960	1240	295	546	715	108	289	185	47	346	156	1510	236	394	728	13	516
Cu	ppm	881	848	1170	507	659	593	625	353	324	294	409	1140	594	1510	211	822	632	66	652

Table 2. *Cont.*

Analyte	Unit	SAPAL			SAPMG														Magnesite	Kaolinite
		F014	F015	F033	F002	F005	F006	F007	F023	F024	F025	F026	F029	F031	F034	F038	F040	F041	F003	F035
Li	ppm	10	10	20	20	20	100	70	30	<10	10	80	10	70	60	110	30	60	<10	<10
Mo	ppm	2	<1	2	2	3	3	2	1	<1	<1	1	<1	1	1	2	1	1	<1	<1
Pb	ppm	4	<2	<2	10	11	8	10	4	3	7	2	7	7	14	6	<2	<2	<2	5
Sc	ppm	13	13	29	26	5	29	24	16	9	9	32	11	60	19	10	6	24	<1	16
Zn	ppm	317	345	394	1100	502	609	724	268	161	126	172	378	290	429	347	573	732	12	169
Ba	ppm	317	251	3600	44	30.8	77.6	90.6	46.8	67.9	84.1	209	49.8	537	1520	78.2	32.9	132	17.6	72.9
Ce	ppm	64.7	19.7	689	7.8	2.5	17.3	28.9	1.9	2.8	4.6	48.2	1.9	78.8	91.2	32.9	12.3	36.2	<0.5	8.5
Dy	ppm	32.8	2.79	22.3	1.83	2.14	14	20.7	1.25	0.92	1.16	4.03	0.44	8.96	4.24	7.81	2.97	26.8	0.14	0.89
Er	ppm	14.65	1.72	10	1.09	1.27	7.22	11.3	0.66	0.54	0.77	1.96	0.2	4.67	2.47	4.69	1.7	14.15	0.08	0.42
Eu	ppm	14.6	1.75	9.3	0.51	0.44	4.41	6.05	0.6	0.27	0.28	1.14	0.08	1.65	1.39	3.28	0.67	7.98	0.06	0.2
Gd	ppm	50.8	3.94	30.6	2.49	2.1	20.4	30.3	1.86	0.96	1.27	4.09	0.41	9.07	5.49	15.25	2.73	40.4	0.14	0.83
Ho	ppm	6.05	0.6	3.93	0.41	0.45	2.85	4.33	0.25	0.24	0.25	0.77	0.08	1.76	0.9	1.65	0.65	5.38	0.02	0.15
La	ppm	292	24.3	207	6.9	9	147	211	3.2	2.8	2.9	11.1	2.3	26.7	23.7	129.5	11.8	260	<0.5	2.9
Lu	ppm	1.8	0.3	1.18	0.13	0.22	0.67	1.31	0.07	0.11	0.08	0.23	0.04	0.69	0.33	0.44	0.29	1.39	0.02	0.08
Nd	ppm	312	21.7	220	10.1	7.5	95.8	111	7.4	2.2	2.9	17.9	1.6	36.1	24.6	92.1	12.9	154.5	0.3	3.5
Pr	ppm	67	5.54	48.6	1.82	1.88	20	26.7	1.49	0.46	0.52	3.59	0.33	7.48	5.47	19.25	2.49	36.2	0.04	0.67
Sm	ppm	50.2	3.84	35.6	2.21	1.44	15.1	19.2	2.17	0.68	0.67	4.15	0.31	8.05	4.92	11.5	2.36	27.4	<0.03	0.62
Tb	ppm	6	0.56	4.06	0.34	0.37	2.65	3.8	0.27	0.13	0.19	0.65	0.07	1.43	0.75	1.4	0.44	4.9	0.02	0.14
Tm	ppm	2.1	0.27	1.34	0.17	0.24	0.9	1.54	0.09	0.06	0.08	0.25	0.03	0.71	0.32	0.5	0.27	1.74	<0.01	0.06
Yb	ppm	11.7	1.6	8.61	0.73	1.45	5.22	8.51	0.45	0.43	0.43	1.89	0.21	4.21	2.11	3.13	1.66	9.09	0.03	0.44
ΣREE	ppm	926.4	88.61	1292	36.53	31	353.5	484.6	21.66	12.6	16.1	99.95	8	190.3	167.9	323.4	53.23	626.1	0.85	19.4

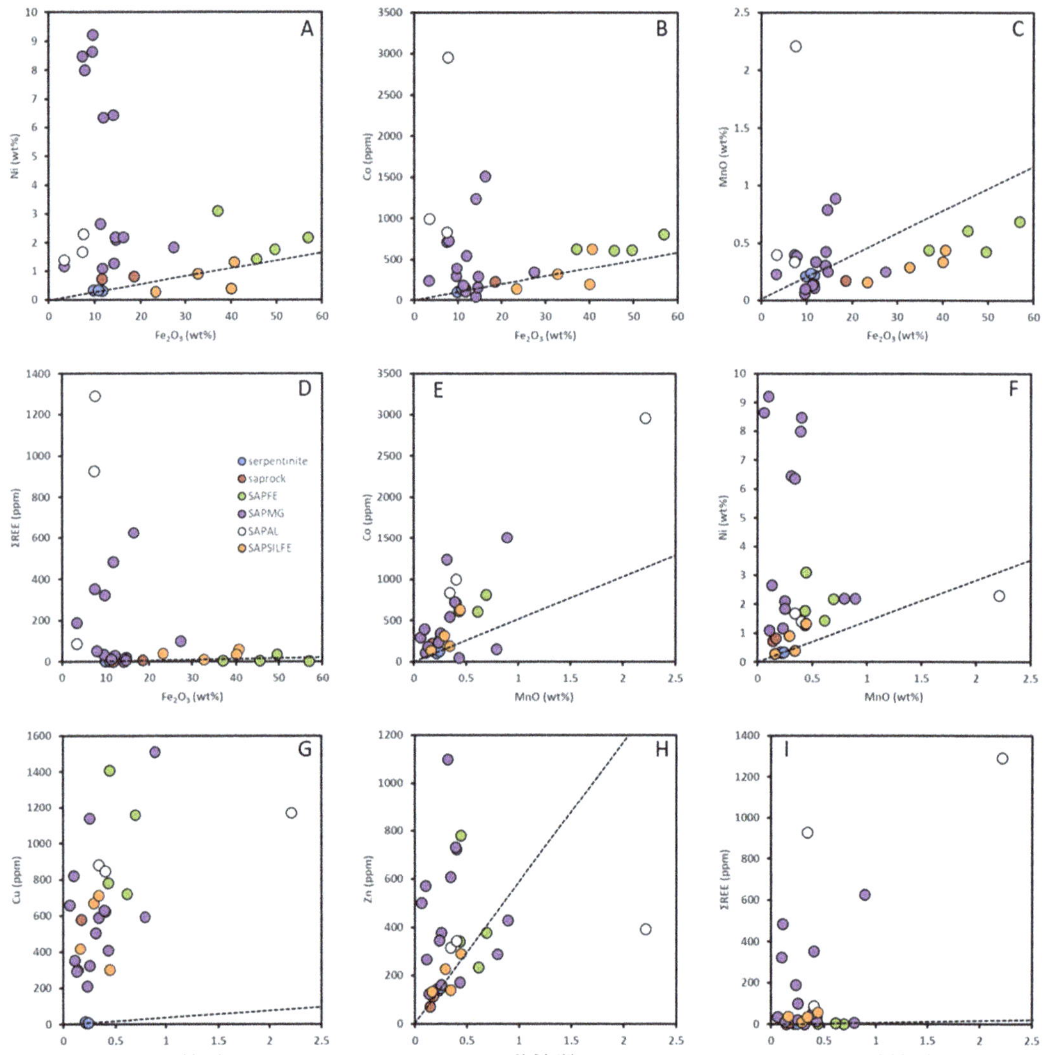

Figure 5. (**A–D**) Concentrations of Ni, Co, MnO and ΣREE plotted against wt% Fe$_2$O$_3$. (**E–I**) Concentrations of Co, Ni, Cu, Zn and ΣREE as a function of wt% MnO. The dotted lines represent the equal residual enrichment trends e.g., [15] qualitatively showing relative enrichment with respect to the serpentinite.

The serpentinite is primarily MgO and SiO$_2$ with about 10% Fe$_2$O$_3$. The saprock has depleted MgO and elevated Fe$_2$O$_3$ compared to the serpentinite though sample F017 is very similar geochemically to the serpentinite, with depleted Al$_2$O$_3$, MnO and Cr$_2$O$_3$ the only clear differences in terms of major elements. Sample F017 was from the base of the test pit and was potentially close to the serpentinite protolith (which was not actually exposed). The ferruginous saprolite SAPFE has the lowest SiO$_2$ (20 wt%) and highest Fe$_2$O$_3$ (47 wt%) contents across the profile and its MgO content is depleted to the range 6.2–15 wt%. With respect to the green and white-green saprolitic lithologies, SAPMG has higher MgO and Fe$_2$O$_3$, and lower SiO$_2$ and Al$_2$O$_3$ than SAPAL, with the MgO and Fe$_2$O$_3$ in the ranges 13–30 wt% and 3.3–16 wt% in SAPMG, and SAPAL containing 42–44 wt% and 22–28 wt%

SiO$_2$ and Al$_2$O$_3$, respectively. SAPMG and SAPAL have the highest levels of TiO$_2$ from the whole profile. The silicified ferruginous saprolite SAPSILFE has the highest SiO$_2$ values (average 46.8 wt%) of the lithological units, though this highly variable (29–64 wt%) with the Fe$_2$O$_3$ content lowest for SAPSILFE samples with the highest SiO$_2$ and vice versa. Indeed, as the SiO$_2$ content of the SAPSILFE decreases, the chemistry becomes increasingly similar to that of the ferruginous saprolite SAPFE. The Cr$_2$O$_3$ values are highest in the two ferruginous saprolites SAPFE (3.4 wt%) and SAPSILFE (2.2 wt%), compared to 1.7 wt% in the serpentinite protolith. The saprock, SAPMG and SAPAL are increasingly depleted in Cr$_2$O$_3$ with respect to the serpentinite protolith. TiO$_2$ is significantly more concentrated in the green and white-green saprolites SAPMG (0.93 wt%) and SAPAL (0.55 wt%) than in any other unit (0.03–0.10 wt%).

Table 3. Definitions of the lithological units used in this work.

Lithological Unit	Description
Serpentinite	Serpentinised dunite source rock that maintains its broadly igneous though pseudomorphic texture and is geochemically described by MgO > ~35 wt%, SiO$_2$ >~40 wt% and Al$_2$O$_3$ <~5 wt%.
Saprock	Highly leached yellow to white adcumulate with MgO between 25 and 35 wt%, high SiO$_2$ and low Al$_2$O$_3$.
SAPSILFE	Silicified ferruginous saprolite (silicious Fe-saprolite c/f Figures 2–4) and is characterised by high Fe$_2$O$_3$ > ~25 wt% and high silica SiO$_2$ >~ 25 wt%.
SAPMG	Predominantly green saprolitic unit with MgO >~13 wt% and Al$_2$O$_3$ in the range 3–20 wt%.
SAPAL	White-green (occasionally pink) saprolitic unit with high Al$_2$O$_3$ >~20 wt% and low MgO <~5 wt%.
SAPFE	Red brown ferruginous saprolite with high iron oxide Fe$_2$O$_3$ > ~40 wt% and low silica SiO$_2$ <~ 25 wt%.
COB	Undifferentiated cover comprising an iron oxide sedimented breccia of silicified laterite.

Table 4. Average bulk chemistries in wt% for the Piauí laterite lithological units as described in this study and some green silica nodules from the SAPSILFE unit. F050 is an aggregate of black mineral grains and coatings within the SAPSILFE unit directly associated with a white-green SAPAL layers (see Figure 4F). * Co is listed in ppm except for [¥] where the value for F050 is in wt%.

Lithological Unit	n	SiO$_2$	Fe$_2$O$_3$	MnO	Al$_2$O$_3$	MgO	CaO	Cr$_2$O$_3$	TiO$_2$	Ni	Co *
SAPFE	4	20.00	47.25	0.54	4.87	10.19	0.04	3.43	0.10	2.13	666
SAPMG	14	35.12	12.36	0.33	9.97	20.43	0.10	0.57	0.96	4.41	485
SAPAL	3	44.40	6.07	0.98	24.93	3.68	0.09	0.06	0.30	1.80	1597
SAPSILFE	4	46.75	34.15	0.31	3.31	3.60	0.04	2.19	0.07	0.73	320
Saprock	2	38.10	14.98	0.16	1.76	31.10	0.03	1.24	0.03	0.78	191
Serpentinite	3	39.30	10.73	0.22	2.24	34.77	0.05	1.68	0.03	0.34	123
F050	1	6.31	10.1	40.3	3.39	2.01	0.05	0.167	0.01	3.55	4.36 [¥]
Green silica nodules	5	90.16	1.308	0.018	0.082	2.738	0.02	0.012	0.035	1.6676	36

Geochemical analysis enables the recognition of general trends (Figure 5) for the elements Mn, Ni and Co, and the trace elements Cu, Zn and rare earth elements (REE). The green and white-green saprolite lithologies SAPMG and SAPAL have the highest concentrations of Co and Mn and, aside from one SAPFE sample, also for Ni. Both Co and Ni within the serpentinite, saprock, SAPFE and SAPSILFE are linearly correlated with

Fe_2O_3. (Figure 5A,B). Concentrations of Ni reach > 9 wt% in SAPMG and 1.4–2.3 wt% in SAPAL (Table 2, Figure 5A), and while there is some degree of correlation between the Ni and MnO contents across all lithologies (Figure 5F) the SAPMG samples with the highest Ni values (>~6 wt%) appear to have a fairly constant MnO concentration. Co is strongly enriched in the SAPAL samples and most of the SAPMG samples, but no distinct enrichment with respect to the serpentinite protolith is evident for the other lithological units (Tables 1 and 2 and Figure 5B). Similarly, MnO is enriched in SAPAL and many of the SAPMG samples, but is depleted with respect to the serpentinite for the other lithologies (Tables 1 and 2 and Figure 5C). The Co and MnO are strongly correlated (Figure 5E) throughout the profile. REEs appear to be enriched in the SAPAL and about half of the SAPMG samples (Table 2, Figure 5D), with these REE enriched samples showing a possible correlation with MnO (Figure 5I).

Appreciable concentrations of Ni (0.5–3.1 wt%) are also found in a layer of green silica nodules from within the silicified ferruginous saprolite which had the lowest concentration of Co amongst all samples tested. The highest concentration of Co (4.4 wt%) was measured in the Mn oxide rich mineral aggregates associated with the SAPAL units which also contain high concentrations of Ni (3.55 wt%). The magnesite veins have very low concentrations of Ni and Co, while the kaolinite veins have 0.5 wt% Ni and >500 ppm Co as well as elevated levels of Cu (652 ppm).

Cu is enriched in all units of the profile with respect to the serpentinite protolith (Tables 1 and 2), with concentrations consistently higher than generally reported for other Ni-Co laterite deposits e.g., [6,8,9,15]. Considering the saprock, SAPFE and SAPSILFE units, the concentration of Cu is correlated to some degree with that of Fe_2O_3 and MnO (Figure 5G). Zn is enriched in the SAPMG and SAPAL units compared to concentrations in the serpentinite, saprock, SAPFE and SAPSILFE units (Figure 5H). The concentrations of Zn in the green and white-green saprolite units are generally within the upper bounds of the range reported by [8] and [15], though are consistently higher than reported elsewhere e.g., [9,39]. Sc concentrations range up to 60 ppm, and, while slightly low, are within the range presented for most other laterite deposits e.g., [6,8,9,15,39,40], though well below those reported for the Syerston–Flemington deposit [41].

Chondrite normalised REE concentrations averaged across each profile unit are shown in Figure 6 (sample specific REE plots are given in Figure S4 of the Supplementary Materials). SAPMG and SAPAL show a negative slope across the sequence with SAPMG showing a distinct negative Ce anomaly. The serpentinite protolith has a generally flat trend, while the saprock displays a negative slope from La to Tb followed by a flat trend from Tb to Lu. The silicified ferruginous saprolite SAPSILFE shows a negative sloping trend while the SAPFE unit displays a negative slope from La to Er and a positive slope from Er to Lu.

5.2. Bulk Mineralogy

XRD analyses reveal the differences in bulk mineralogy between the lithological units within the profile (Figure 7). The serpentinite is dominated by lizardite with traces of chlorite, chrysotile, chromite and hematite evident in the XRD patterns. The overlying saprock also contains significant serpentine minerals, predominantly lizardite, but in association with chlorite minerals, Fe-oxides (hematite, maghemite and goethite), with traces of chromite and magnetite subsequently confirmed from SEM analyses.

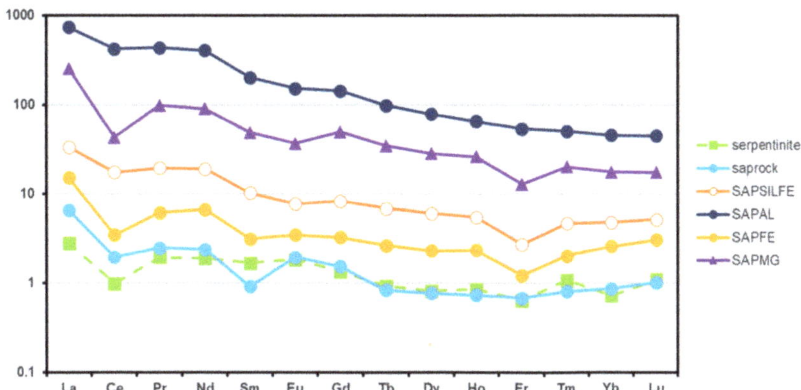

Figure 6. Chondrite normalised REE plots averaged within the profile units. REE concentrations are normalised to the CM values of [42].

The ferruginous saprolite SAPFE is dominated by Fe-oxides, with goethite generally the most abundant, but occasionally hematite was the primary Fe-oxide mineral. Traces of serpentine minerals remain in this unit, however there is a far more diverse suite of layered silicates in SAPFE than for the serpentinite and saprock with chlorite, vermiculite, saponite and nontronite all observed by XRD. Quartz was identified in about half of the SAPFE samples. The silicified ferruginous saprolite SAPSILFE is dominated by goethite, hematite, maghemite, quartz and hydrated silica, with traces of nontronite, chlorite and saponite commonly observed. The relative variation of the different Fe-oxide minerals within and between the two ferruginous units (SAPFE and SAPSILFE) can be seen in Figure S5 (Supplementary Materials), associated with the bulk Ni content, highlighting that the samples with the highest concentrations of Ni are all mineralogically dominated by chlorite. XRD of the green silica nodules reveals that appreciable levels of talc are present within the predominantly quartz samples

The green saprolitic SAPMG units are dominated by chlorite with traces of other layered silicates lizardite, vermiculite and saponite, along with quartz being commonly observed. Ilmenite is also frequently observed in SAPMG, with the presence of magnetite and chromite confirmed as trace minerals using SEM analyses. The XRD patterns of the SAPMG samples with the highest concentrations of Ni show very little mineralogical variation, being dominated by chlorite (Figure S5), with vermiculite occasionally observed. The white-green saprolite SAPAL, on the other hand, is dominated by kaolinite with vermiculite and saponite, and only traces of chlorite and occasional quartz are observed.

5.3. Petrography and Mineral Chemistry of the Laterite Profile

5.3.1. The Serpentinite Protolith

The serpentinite protolith is a highly serpentinised dunite consisting predominately of serpentine and opaque minerals (Figure 8A,B). It is characterised by the strong or complete pseudomorph replacement of primary olivine into serpentine minerals, primarily lizardite, along with fibrous chrysotile in the form of thick mm-sized veins (Figure 8C), which gives the rock a mesh-like texture on both macro- and micro-scales, and chlorite is also occasionally observed. These are type-I serpentines e.g., [43]. The prevalence of lizardite in the serpentinite samples, as suggested by our characterisation, further from the saprock boundary may suggest that the serpentinisation occurred below 300–320 °C e.g., [44]. The alteration of serpentine into serpentine-II is evident along trans-granular cracks and fractures, but also takes place as a coating along grain edges. Our characterisation suggests that lizardite is the dominant component of serpentine-II. Secondary magnetite has precipitated in fissures and the proportion of magnetite increases with increased weathering of the serpentinite. Secondary Fe-oxides are also seen to have precipitated between serpentine grains, most significantly within chrysotile veins, and silica is observed in samples considered to be close to the serpentinite–saprock boundary, where magnetite is altering to hematite (Figure 8B). The fibrous chrysotile veins are sheared in places and talc-magnesite alteration of serpentine-I minerals is retained locally within some veins (Figure 8C,D). With increased weathering, the birefringence of the chrysotile increases, possibly correlating with an increased Ni content and suggesting that some chrysotile is second generation serpentine-II type.

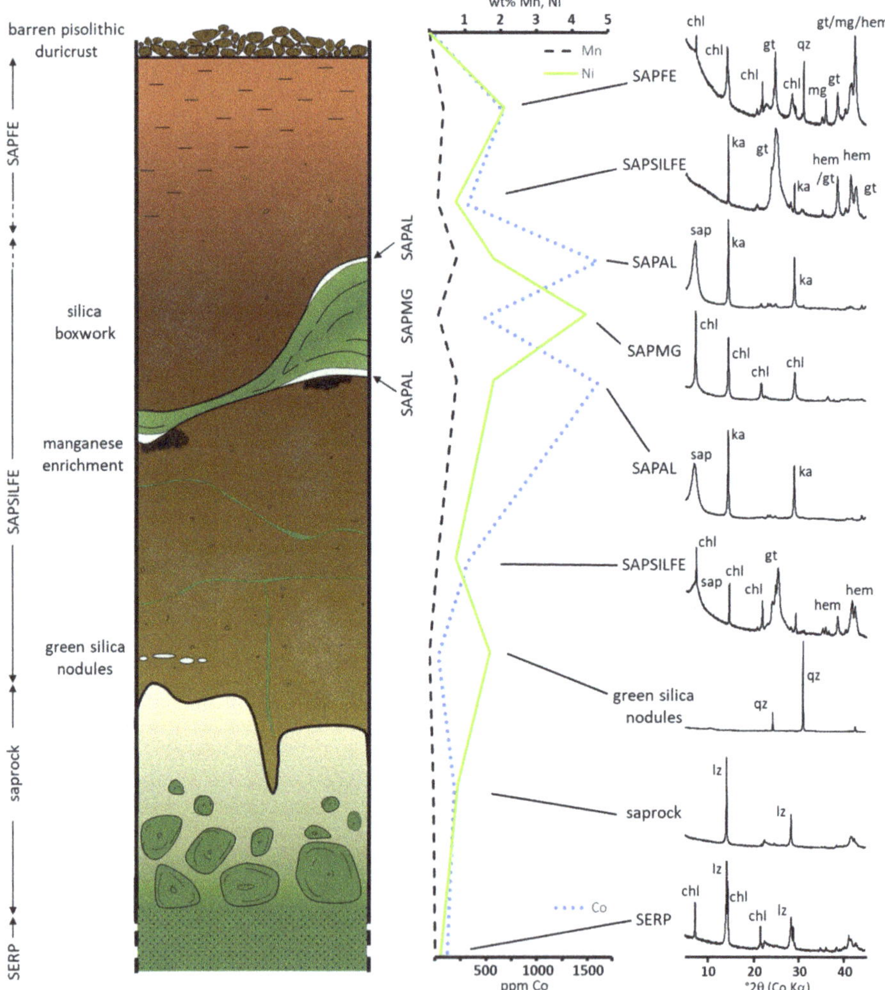

Figure 7. Representative XRD patterns from the different lithological units of the Piauí deposit (**right**). The XRD patterns are plotted with a stylized schematic profile for the Piauí laterite (**left**) and the associated average Co, Ni and Mn contents from Table 4 (**centre**). NB this is not a true representation of the profile of the deposit; the Piauí laterite is vertically and horizontally variable and the schematic shown here condenses the major features into a single representative profile. XRD patterns are from samples (top to bottom): SAPFE (F019), SAPSILFE (F043), SAPAL (F014), SAPMG (F006), SAPAL (F015), SAPSILFE (F030), green silica nodules (F008), saprock (F017), serpentinite (F046). chl = chlorite, lz = lizardite, ka = kaolinite, qz = quartz, sap = saponite, gt = goethite, hem = hematite, mg = maghemite.

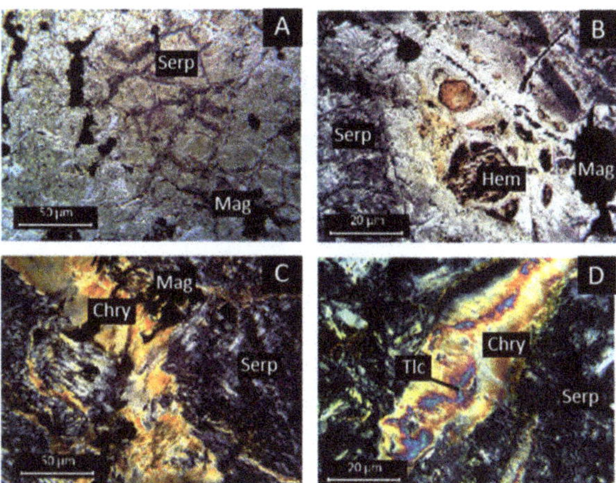

Figure 8. Photomicrographs of the serpentinite protolith. (**A**) Serpentinised olivine minerals (Serp) displaying meshwork textures. Secondary magnetite (Mag) has formed in fractures along serpentine grain boundaries. (**B**) Pseudomorph replacement of olivine by iron oxides (Hem = hematite). (**C**) Faulted fibrous chrysotile (Chry) veins and deformed secondary magnetite crystals suggesting local tectonic activity. (**D**) Talc (Tlc) replacement of serpentine minerals within chrysotile veins.

While XRD shows lizardite to be the most abundant phase with traces of hematite, chlorite and chrysotile, SEM-EDX also identifies magnetite, chromite and Mn oxy-hydroxides. EPMA analyses show that Ni is most abundant in the magnetite and Mn oxy-hydroxides, and present at <0.5 wt% levels in chromite, serpentine and hematite (Table 5). The hematite additionally contains almost 1 wt% Cr_2O_3 (Table 5). Co was detected in Mn oxy-hydroxides, magnetite and chromite (Table 5). Mn oxy-hydroxides generally show higher concentrations of Co than Ni (Table 5).

Table 5. Average wt% chemical compositions of the main minerals in the serpentinite. *n* = number of EPMA analyses; nd = not detected; na = not analysed.

Mineral	*n*	Al_2O_3	MgO	Fe_2O_3	Cr_2O_3	SiO_2	MnO	Ni	Co
Ferritchromite	81	2.03	3.98	59.58	29.69	na	2.67	0.16	0.06
Serpentine	106	0.41	38.22	2.42	nd	44.82	nd	0.38	nd
Hematite	110	nd	1.04	97.85	0.98	0.92	0.36	0.30	nd
Magnetite	48	nd	na	51.71	nd	na	0.10	8.93	1.18
Mn oxy-hydroxide	87	0.88	nd	7.48	nd	nd	28.05	6.61	7.41

SEM imaging of the serpentinite revealed the presence of small (<2 µm in diameter) faceted particles of roughly spherical morphology (Figure 9) similar (though smaller) to those described by [15]. EPMA revealed them to be Fe-rich with unusually high Co and Ni concentrations, though their small particle size precluded precise measurement of their chemistry. Based on their different morphology to other Fe oxides in the serpentinite samples, elemental composition (in particular an unusual enrichment in Co, no detectable Al and Cr) these particles were tentatively identified as magnetite and subsequently confirmed via STXM Fe L-edge XAS spectra. Highly altered chromites are identified within the serpentinite that show evidence of former martitisation and extensive weathering (Figure 9C). These chromites are similar to the ferritchromites described by [45] within the Nkamouna serpentinite, southeast Cameroon, and also display overgrowths of highly Ni enriched Fe-oxides on the grain rims (Figure 9D).

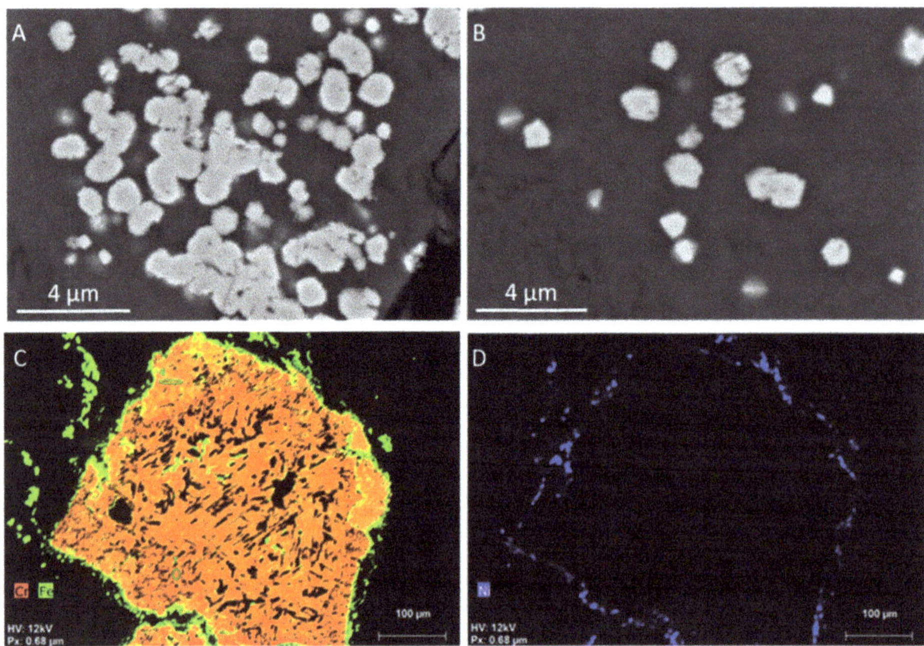

Figure 9. (**A,B**) SEM backscatter electron (BSE) images of small neoformed grains of Co enriched magnetite within a serpentine matrix within the serpentinite protolith. (**C,D**) Element maps ((CEDX) red = Cr; green = Fe; (**D**) blue = Ni) of highly altered relict chromite grains with outer rims of newly formed Ni enriched Fe-oxides.

5.3.2. Co and Ni Mineralogy and Chemistry of the Saprolite Units

Cobalt concentrations vary widely across the Piauí profile, with the lowest concentration observed in a sample of SAPMG (47 ppm) and the highest concentration in a sample SAPAL (2960 ppm), although 4.36 wt% Co was detected in a sample of Mn oxy-hydroxide (F050, Table 1). While Co is clearly enriched in the majority of the green SAPMG and white-green SAPAL saprolitic layers (Figure 5), the clearest correlations observed from bulk chemical analyses across the profile (Tables 1 and 2) were between Co and Mn with r = 0.86 (Figure 5) and between Co and Ni for all units excluding SAPMG and SAPAL (r = 0.87). This is supported by SEM-EDX analysis which revealed that Co is predominantly associated with Mn oxy-hydroxides and its distribution within the samples is much more discrete compared to that of Ni. A range of minerals are identified as Co hosts in the Piauí profile (Table 6).

The highest average concentrations of Co are associated with Mn oxy-hydroxides, thus reflecting the significant Co-Mn correlation. Four different textures and morphologies of Mn oxy-hydroxide grains frequently occur in the Piauí samples, including: aggregates of plates/layers (5–10 μm thick) with highly zoned and chemically complex grains (>~25 μm), thin (~10 μm thick) coatings on grain surfaces or in intragranular spaces, and highly fractured and veined grains > 100 μm in size (Figure 10A–C respectively). The fourth morphological type is occasionally observed within the ferruginous saprolites and potentially reflects a grain size reduction and slight deformation of pre-existing Mn-oxyhydroxide grains. Mn-oxyhydroxides infill spaces within the ferruginous saprolitic fabric, producing very small grains and mineral coatings (Figure 10D).

Table 6. Host minerals for Co in the Piauí laterite profile. n = the total number of points analysed for each mineral across all samples tested. The average levels of MnO in manganoan goethite is 1.29 wt%, in manganoan ilmenite is 24.4 wt%, in aluminian chromite is 0.20 wt% and in ferritchromite is 2.67 wt%. Data are from EMPA analysis.

Mineral	Concentration (wt%)			
	n	Min	Max	Average
Mn oxy-hydroxide	284	0.35	14.2	3.44
Manganoan goethite	89	0.03	0.65	0.19
Magnetite	52	0.41	1.89	1.14
Chlorite	552	0.03	0.45	0.05
Manganoan ilmenite	42	0.04	0.35	0.12
Aluminian chromite	133	0.03	0.06	0.04
Ferritchromite	63	0.03	0.16	0.06

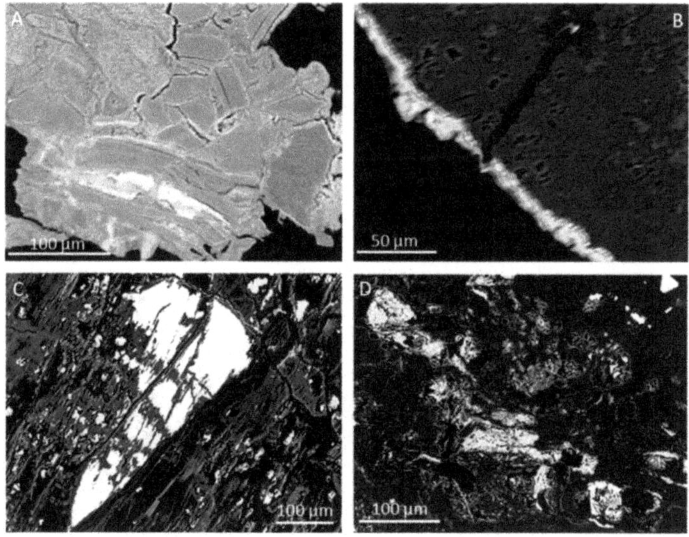

Figure 10. BSE images of the four main textures of Mn oxy-hydroxide grains within the Piauí profile, as described in the text. (**A–C**) are from SAPMG and (**D**) is form SAPFE.

The Mn oxy-hydroxide phases vary chemically as a function of their host profile unit as highlighted by a Co-Ni-Al diagram (Figure 11A). Based on the Al content, these phases can be classified as asbolane and asbolane-lithiophorite intermediates e.g., [3,46,47]. Barium was also detected in some grains and an inverse relationship was observed between the Ba content and Co enrichment (Figure 11B). The presence of Ba indicates a transition of Mn oxy-hydroxides from the phyllomanganate lithiophorite/asbolane type to the hollandite group mineral romanechite. It is possible that there are multiple types of romanechite within the Piauí deposit, similar, to some degree, to the two generations described for the Wingellina laterite by [46]. The romanechites at Piauí were most commonly observed in SAPAL with romanechite-I having higher Ba contents (>2 wt% Ba) but reduced Co and Ni (average 1.82 wt% and 6.24 wt%, respectively), while romanechite-II had reduced Ba content (<2 wt% Ba) with enrichment of Co and Ni (average 4.69 wt% and 12.15 wt%, respectively).

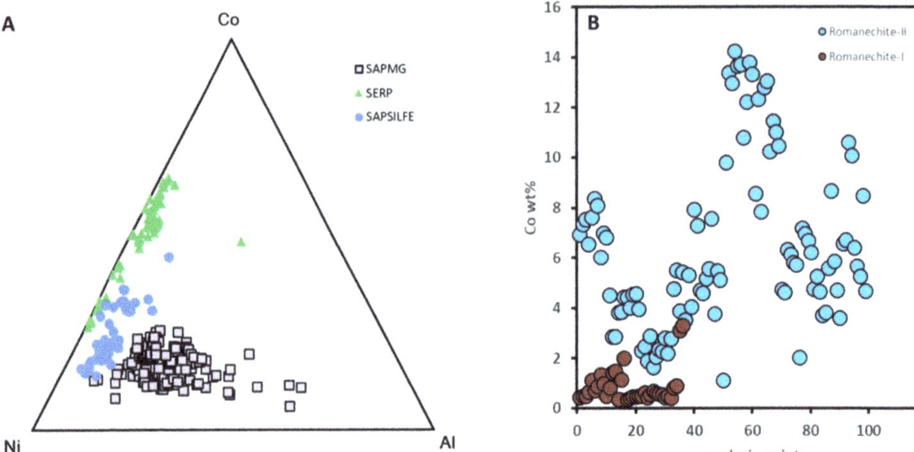

Figure 11. (**A**) Relative contents of Al, Ni and Co (wt%) measured by EMPA in Mn oxy-hydroxide grains from three units of the Piauí profile, and (**B**) the Co content (wt%) in the romanechite Mn oxy-hydroxide grains separated by Ba concentration. The Ba content cut-off between romanechite-I and romanechite-II was 2 wt%.

The highest concentration of Co from any of the bulk samples was found in F050 (4.36 wt%), an aggregate of black mineral grains and coatings both within and directly adjacent to a white-green SAPAL saprolitic layer (Figure 4F). SEM-EDX imaging and Fourier transform infrared (FTIR) spectra (Figure S6, Supplementary Materials) indicated the presence of only one type of Ba-bearing Mn oxy-hydroxide, with the chemical composition most closely matching romanechite, with concentrations of 12.5 wt% Ba, 48.0 wt% Mn, 1.8 wt% Fe, 3.0 wt% Co, 0.63 wt% Ni and 0.97 wt% Cu. Cobalt concentrations varied considerably from 1.5 to 10.4 wt% between all the grains probed. In contrast to other Mn oxy-hydroxides in the Piauí samples, identified as asbolane and asbolane-lithiophorite intermediates, this romanechite phase had Co concentrations significantly higher than Ni. Moreover, this romanechite contradicts the inverse relationship between Ba and Co observed in all other romanechite type Mn oxy-hydroxides of the Piauí deposit (e.g., Figure 11B). This suggests a third type or generation of romanechite at Piauí, with romanechite-III being relatively enriched in Ba and Co (12.5 wt% and 3.0 wt% respectively) and relatively depleted in Ni (0.63 wt%) and appearing in concentrated locations at the interface between the ferruginous saprolitic units and the white-green SAPAL saprolite layers.

In this study we employed a detection limit of 0.03 wt% for Co in our EMPA analyses of goethite in order to reliably account for the impact of the Fe induced background and provide consistency across this study and associated work e.g., [48,49]. Cobalt was detected in goethite only when Mn was also present in the goethite, with the Mn either substituted into the goethite at the atomic level or possibly present as nanoscale cobaltiferous Mn oxy-hydroxide inclusions. Over half of the goethite analyses showed no Co present, and, as such, there are two types of goethite in the Piauí deposit, Co-free and Co bearing manganoan goethite (Table 6). Although Co was only present in manganoan goethite, there is a distinct correlation between Co and Ni within the goethite (Figure 12), with correlation values of 0.65 between Co and Ni compared to 0.56 between Co and Mn.

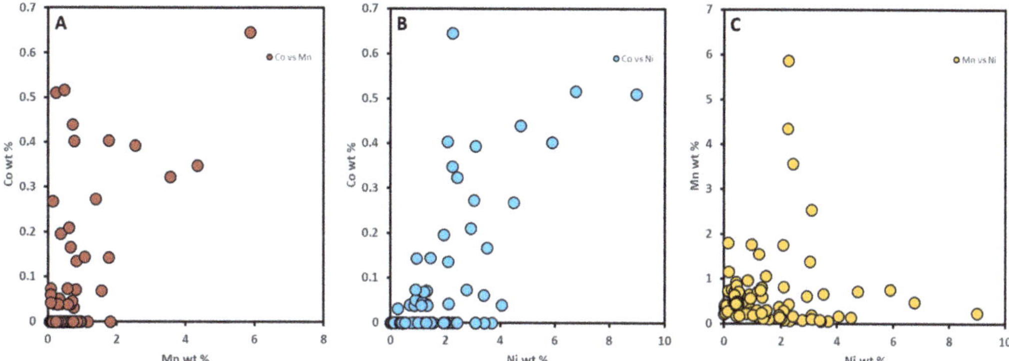

Figure 12. Relationship between (**A**) Co and Mn, (**B**) Co and Ni, and (**C**) Ni and Mn as measured by EMPA in goethite grains from the saprolitic units of the Piauí profile.

Chlorite, which is an important mineral host for Ni at Piauí, also contained small quantities of Co with an average Co concentration of 0.05 wt%, however ranging widely with a maximum concentration of 0.45 wt% (Table 6). There is a positive relationship between the Co and Ni content in the chlorite, with the grains most enriched in Ni also having the highest Co concentration.

The highest concentrations of Co in the chlorite phase were identified in SAPMG, particularly sample F002, which also had a high bulk Co concentration of 1240 ppm (Table 2). SEM-EDX mapping of this sample indicated that the primary hosts for Co were chlorite and Mn-rich ilmenite with no Mn oxy-hydroxides identified. It is of course possible that Mn oxy-hydroxides could be present as very small heterogeneously distributed grains not detected or not present in the small subsample analysed, and Mn oxy-hydroxides were detected in other Piauí samples with similar bulk Mn concentration to F002. High resolution imaging and mapping of the chlorites in sample F002 revealed distinct Ni enriched silicate inclusions within the chlorite fabric, that have an elongate laminate morphology up to ~60 μm long and <10 μm thick, that have levels of Co far higher than in the surrounding chlorite (Figure 13). Co concentrations in these inclusions range from 0.10 wt% to 0.45 wt% with an average of 0.24 wt%, consistently higher than the Co levels in the host chlorite matrix (Figure 13). Average chemical analyses of these Co-bearing Ni-silicate inclusions have compositions 19.95 wt% Ni, 4.95 wt% Mg, 3.10 wt% Al, 16.63 wt% Si and 5.92 wt% Fe, consistent with a cobalt enriched nimite.

Of all the saprolitic units, sample F033 from the white-green SAPAL lithology has the highest concentrations of Co (2960 ppm, Table 2, Figure 3E). The bulk mineralogy from XRD comprises two crystalline phases, chlorite and vermiculite, and despite containing over 2 wt% MnO, no specific Mn oxy-hydroxide minerals were identified in the XRD pattern. SEM-EDX imaging, however, revealed the presence of dispersed Mn oxy-hydroxides primarily as grain boundary coatings and as void infillings, and EMPA analysis showed that these Mn oxy-hydroxides had high concentrations of Co between 0.64 and 4.53 wt% and Ni between 2.50 and 18.8 wt%, whereas the chlorite within the sample contained 0.94 wt% Ni and no detectable Co. While the majority of the Mn oxy-hydroxides in this SAPAL sample are asbolane-lithiophorite intermediates, there are also areas of elevated Ba where the EMPA analyses reveal the presence of a romanechite-type phase similar to that described for F050 (also within or adjacent to SAPAL).

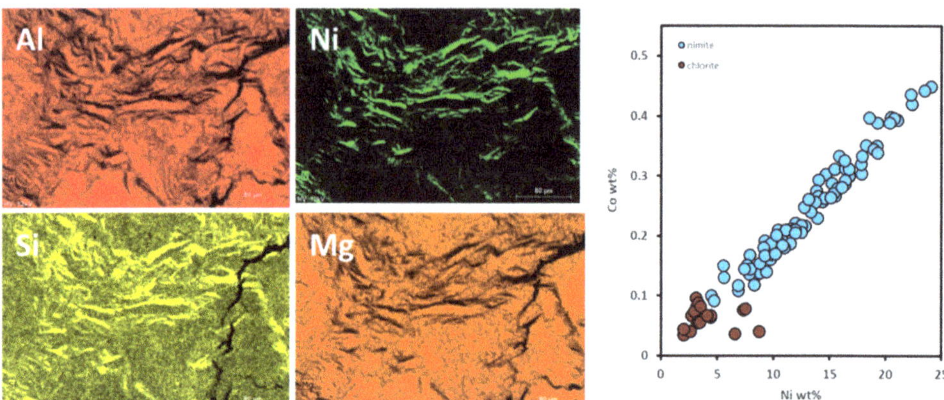

Figure 13. (**Left**) Al, Si, Ni and Mg element maps overlain the secondary electron (SE) image of Co enriched Ni silicate inclusions within the chlorite of SAPMG sample F002. (**Right**) Cobalt levels from EMPA analyses of the Co enriched Ni silicate inclusions. Blue filled circles = nimite, brown filled circles = chlorite.

Although there are limited chemical correlations with Ni across the profile (for Ni:Mn and Ni:Fe in all units excluding SAPMG and SAPAL r = 0.71 and 0.68, respectively), XRD and bulk chemistry showed that the bulk Ni contents of the five saprolitic lithologies is ultimately linked to their mineralogy. The samples with the highest Ni concentrations in SAPMG are dominated by the presence of chlorite and other layered silicates, while goethite, which dominates SAPFE and SAPSILFE, consistently shows significant Ni incorporation, albeit at lower levels compared to chlorites. SEM-EDX element maps reveal the ubiquitous presence of Ni associated with chlorite, whereas Mn and Ti are present in discrete mineral phases either as inclusions within the chlorite or filling in pores and cracks. Ni was detected by SEM-EDX and/or EMPA in 11 different minerals (Table 7), whereby a relationship between Ni and Mg within the silicate phases becomes evident (Figure 14). While the pimelite and nickeliferous talc (found exclusively within the green silica nodules, see below) has fairly well-defined Ni concentrations, there is a huge variation in the Ni abundance in chlorites from the SAPMG lithology. Nickeliferous chlorite is also reported in the Vermelho deposit of the Pará region of Brazil with 2–3 wt% NiO recorded in the chlorite of the silicate zone and up to 12 wt% NiO in chlorites from the oxidised zone [50]. By contrast, smectites In the clay bearing saprolites from Barro Alto (Goiás State, Brazil) only contain up to 3.3 wt% Ni [51]. Two distinct groupings of serpentine can be observed at Piauí (Figure 10), highlighting the persistence of low Ni serpentine-I (Table 5) in the serpentinite and the second-generation serpentine-II with elevated Ni contents e.g., [15,43] found mostly in SAPMG and the saprock, but also in the serpentinite.

The goethite within the saprolite units incorporated a range of trace elements detectable by EMPA (following the criteria of [52]) including Co, Ni, Mn, Al, Si and Mg with the maximum wt% element measured being 0.65, 0.90, 5.87, 3.05, 4.65 and 2.25 wt%, respectively. The highest total trace element substitution in goethite totalled 12.15 wt%. The Al, Si and Mg all increase linearly in concentration as Fe decreases while Co, Ni and Mn contents seem to be unrelated to Fe content. As reported above, there is a distinct correlation between Ni and Co in the goethite, but no clear relationship between Ni and Mn (Figure 12).

Table 7. Host minerals for Ni in the saprolitic units of the Piauí laterite profile. *n* = the total number of points analysed for each mineral across all samples tested. Data are from EMPA analysis.

Mineral	Concentration (wt%)			
	n	Min	Max	Average
Chlorite	554	0.1	24.0	3.2 (med)
Nickeloan talc	101	5.8	15.3	13.5
Pimelite	28	19.5	25.0	21.7
Serpentine	154	0.04	3.52	0.40 (med)
Manganoan ilmenite	66	0.03	1.36	0.35
Ilmenite	119	0.03	1.55	0.27
Aluminian chromite	158	0.08	0.18	0.13
Ferritchromite	81	0.06	1.25	0.16
Goethite	116	0.03	9.0	1.32
Magnetite	52	2.84	14.1	8.60
Mn oxy-hydroxide	284	0.35	18.8	8.49

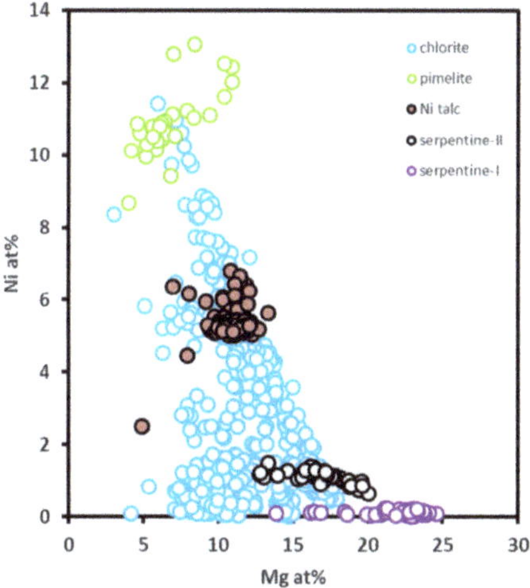

Figure 14. Partitioning of Ni within the silicate minerals across the Piauí laterite profile. The chlorite analyses all come from the SAPMG lithology. The Ni-rich serpentine analyses come from serpentinite-II, mostly in the SAPMG lithology and occasionally in the serpentinite. The low Ni analyses come from serpentine-I in the serpentinite.

5.3.3. Ni mineralogy and Chemistry of the Green Silica Nodules

Within the SAPSILFE lithology close to the boundary with the saprock, three horizontal layers of green silica plates and nodules were observed. Bulk chemical analyses of these nodules (Tables 1 and 4) show that they contain up to 3.14 wt% Ni as well as up to 5.29 wt% MgO and 3.48 wt% Fe_2O_3. XRD analyses identified the presence of talc minerals along with quartz. SEM-EDX and EMPA analyses showed the presence of two distinct talc chemistries and associated textures (Figure 15), and that the Ni was entirely partitioned into these talc minerals.

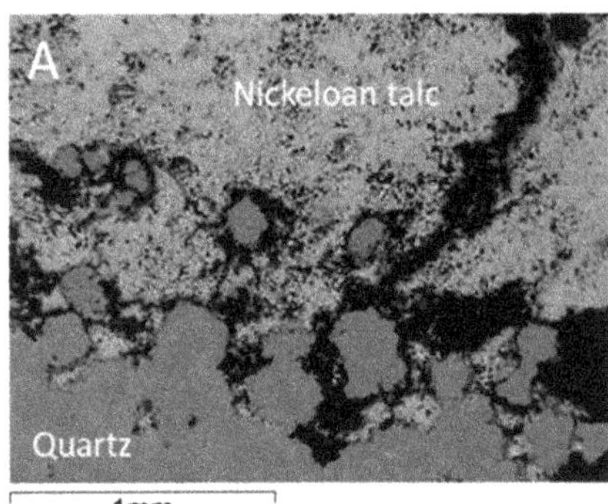

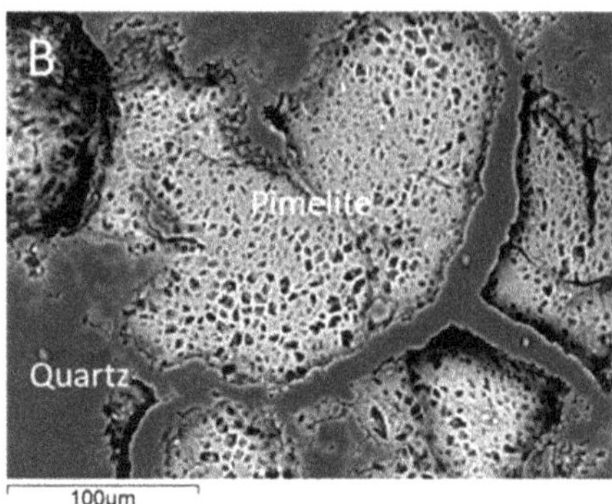

Figure 15. BSE images of silica nodules with (**A**) nickeloan talc and (**B**) pimelite.

Nickeloan talc was observed as very porous massive clusters (Figure 15A) and microcrystalline inclusions within the quartz matrix, and pimelite occurs as pockets of very porous material embedded within the quartz matrix (Figure 15B). The average Ni content in the nickeloan talc was 13.5 wt% with small variations observed (Table 7, Figure 14). Other elements detected in this phase included Al, Fe, Ca, Cu (all below 1 wt%) and very small quantities of Co, mostly at the detection limit of the electron microprobe (0.03 wt%). Average composition of the pimeleite phase was 28.9 wt% NiO, 9.5 wt% MgO and 45 wt% SiO_2, while cobalt was not detected in this phase.

5.4. Cobalt and Nickel Speciation in Asbolane-Lithiophorite Intermediates

Microfocus XAS experiments were performed on asbolane-lithiophorite intermediates from the SAPAL sample F033. This sample contains the highest abundance of Co (2960 ppm) and Mn (1.71 wt%) of any of the rock samples, it was collected from a region of the SAPAL unit containing black coatings (Figure 3E), and SEM and EMPA analysis showed asbolane-lithiophorite intermediates mostly as grain boundary coatings and as void infillings.

Co K-edge XANES spectra, extracted EXAFS and their Fourier transforms (FT) are presented in Figure 16. There is an energy shift of ~5 eV for the absorption edge and white line intensity positions in the XANES spectra of Co^{2+} bearing erythrite ($Co_3(AsO_4)_2.8H_2O$) compared to those of Co^{3+} bearing heterogenite (CoOOH) (Figure 16A). As reported by [53] we also found that the XANES spectrum for Co in Co-doped synthetic goethite had an absorption edge and white line position very similar to that of the Co^{2+} standard (see Figure 16A), indicating that any Co within goethite in the Piauí samples may be present as Co^{2+}.

While the Co XANES spectra of the asbolane-lithiophorite intermediates within F033 all appear broadly similar to each other with a general shape similar to that of Co^{3+} substituted phyllomanganates e.g., [35,53], there is a clear shoulder to the absorption edge at ~7725 eV, equivalent to the white line position of the Co^{2+} standard. This low energy shoulder is slightly more pronounced in the XANES spectra for grains 1 and 2 compared to those of the grains 3 and 4. Moreover, there is a shift to lower energy in the position of the XANES white line for grains 1 and 2 compared to those of grains 3 and 4. The XANES spectra suggested that some Co^{2+} may be present in these asbolane-lithiophorite intermediates with a Co^{2+}/Co^{3+} ratio higher in grains 1 and 2 compared to grains 3 and 4.

The EXAFS spectra for grains 1 and 2 are similar to each other, as are the spectra for grains 3 and 4 (Figure 16B). The principal difference between the spectra of grains 1 and 2 compared to those of grains 3 and 4 is in the region 4.8–6 Å^{-1} where the spectra of grains 1 and 2 show a small feature at ~5.7 Å^{-1} on top of the rising slope of the asymmetric maximum at ~6.8 Å^{-1}, whereas the spectra for grains 3 and 4 show a distinct feature at ~5.6 Å^{-1}, separated from the more symmetrical maximum at ~6.8 Å^{-1}. Additionally, the intensity of the feature at ~6.75 Å^{-1} is greater for the spectra of grains 3 and 4 than it is for grains 1 and 2. The first peak in the FT of the EXAFS spectrum of grain 1 (Figure 16C) has a clear shoulder on the high R side, whereas this is marked by a single peak for the FTs of the EXAFS spectra for the other three grains. Detailed results of the EXAFS modelling are provided in Table S3 of the Supplementary Materials.

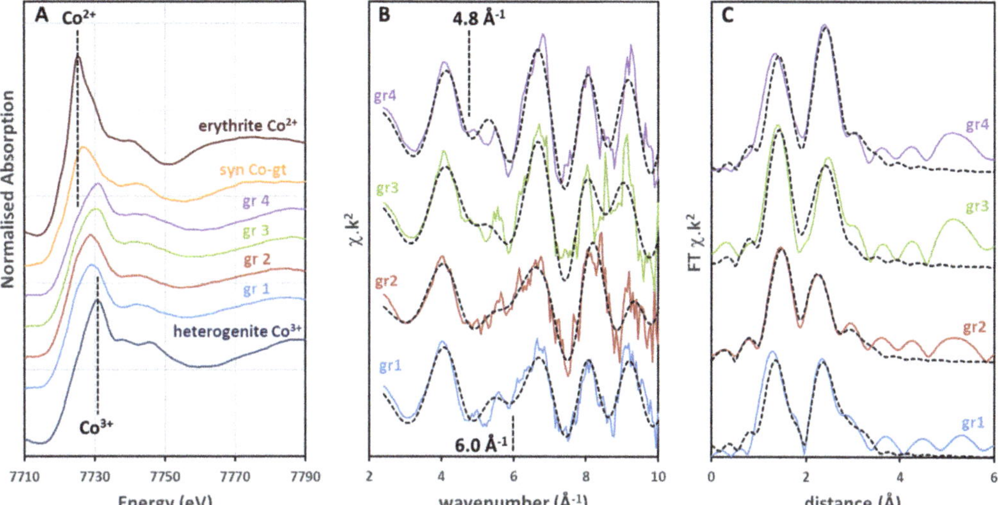

Figure 16. (**A**) Co K-edge XANES spectra of standards heterogenite, erythrite and synthetic Co-doped goethite plotted with those for 4 asbolane-lithiophorite grains. (**B**) EXAFS spectra for the same asbolane-lithiophorite intermediates points as shown in (**A**). (**C**) Fourier transforms of the Co EXAFS spectra from (**B**). Dashed lines in (**B**,**C**) are the modelled fits to the experimental data (solid lines).

The best fit model for the spectrum of grain 3 is with a two shell model, such that there is an oxygen shell at a distance of 1.91 (2) Å consistent with Co^{3+} in octahedral coordination [53–56] and a cationic shell with Co-(Co,Mn) interatomic distances of 2.83 (2) Å, typical for Mn-Mn edge sharing octahedra [54–56]. These distances are consistent with Co^{3+} substituting for Mn in the Mn layer of the asbolane-lithiophorite intermediates e.g., [56]. The best fit model for the spectra of grains 2 and 4 are similar, though they required a three shell model, with a Co-O distance of 1.96 (1) Å and 1.92 (2) Å, respectively, a second shell Co-(Co,Mn) distance of 2.80 (1) Å and 2.85 (2) Å, respectively, and a 2nd cationic shell (of a smaller intensity than the first cationic shell) with a Co-(Co,Mn) distance of 3.46 (4) Å and 3.53 (6) Å, respectively, which is typical for Mn-Mn corner sharing octahedra [56,57].

The spectrum of grain 1 however did not fit well to a 2 or 3 shell model as determined for grains 2–4, with Co-O first shell fitting particularly poorly. Instead, it became evident that a second Co species with a longer Co-O distance was required, and, as such, the best fit model for this spectrum requires two specific clusters. The first cluster represents Co^{3+} substituting for Mn in the Mn layer of asbolane-lithiophorite intermediates with a Co-O distance of 1.90 (5) Å, a Co-(Co,Mn) distance of 2.83 (3) Å and a Co-(Co,Mn) distance of 3.54 (9) Å. The second cluster comprises a single shell with a Co-O distance of 2.09 (7) Å, indicative of octahedral Co^{2+} [56], and the fit to the data is optimum with this cluster being present at the 40% level. This mixed Co^{3+} and Co^{2+} model also fits well to the additional shoulder apparent on the first peak of the FT for grain 1.

A third shell Co-(Co,Mn) distance of ~3.5 Å is indicative of both Co^{3+}, substituting for Mn in the Mn layer of the asbolane-lithiophorite intermediate, but also of octahedral Co^{2+} within the interlayer [56]. This shell is required for fitting the spectrum of grains 1 and 2, not required for fitting the spectrum of grain 3 and marginally improved the fit for grain 4. The spectra of both grains 3 and 4 are well modelled by a single Co^{3+} species.

However, the presence of this third shell is essential for the fit of the spectrum of grain 2 and, together with the lengthened Co-O distance in the model for the spectrum of grain 2, strongly suggests that a significant proportion of the Co in grain 2 is present as Co^{2+}, potentially up to 30% based upon the Co-O distance alone. While the two-cluster model does fit the spectrum of grain 2 well, it does not show a significant improvement to the fit compared to the single cluster 3 shell model. Nevertheless, supporting the qualitative assessment of the XANES, there is strong evidence from the EXAFS analysis that Co^{2+} is present at significant levels within the interlayer in some of these asbolane-lithiophorite intermediates.

To try and further confirm the presence of Co^{2+} within the asbolane-lithiophorite intermediates, as suggested by the Co K-edge XANES and EXAFS analyses, we performed a STXM study to extract L-edge XANES spectra for Co, Ni and Mn from asbolane-lithiophorite intermediate grains in sample F033. L-edge XANES spectra are excellent fingerprints for transition metal oxidation states in minerals, e.g., [58–61], and have been used recently to characterise Ni and Co in laterite materials e.g., [35,62]. The application of STXM for extracting L-edge XANES spectra provides a spatial context for the spectra.

The Co $L_{2,3}$ XANES spectra are presented in Figure 17 with the points from which they were extracted shown in the STXM images. The L_3-edge spectrum for octahedrally coordinated Co^{3+} in heterogenite is dominated by a strong absorption peak at 783.6 eV with smaller shoulders on the low (781.6 eV) and high (785.9 eV) energy sides (Figure 17). The L_3-edge spectrum for octahedral Co^{2+} in erythrite has a much broader main feature with more spectral detail than for Co^{3+} that is centred around 781.8 eV. Six spectra (A-F) extracted from the STXM energy stack for a F033 asbolane-lithiophorite intermediate are shown in Figure 17. Spectra A, B and C are very similar to that of heterogenite, indicating that the Co is trivalent and octahedrally coordinated. Spectra D, E and F, however, show a progressively increasing intensity around 781.8 eV and a distinct change in the shape of the absorption peak at 783.6 eV. This is strongly indicative of the presence of Co^{2+}. Also shown in Figure 17 are oxidation state images for Co^{3+} and Co^{2+} extracted from the STXM energy stacks, demonstrating that the Co^{2+} is located in a distinct area within the asbolane-lithiophorite grain. L-edge XANES analysis for Mn and Fe show no variation across the sample indicating that the nature of the asbolane-lithiophorite intermediate does not change and there are no goethite inclusions associated with these changes in Co oxidation state (within the 100 nm resolution of the experiment).

The Ni K-edge XANES spectra (Figure 18A) of the same grains described above are very similar, with a sharp white line and a broad two-component feature at ~8400 eV, as described for a Ni species developing a $Ni(OH)_2$ 'brucite-like' layer [47]. The EX-AFS spectra (Figure 18B) and their associated FTs (Figure 18C) are similar for each grain, with some variation evident within the EXAFS in the range 7–9 Å^{-1}, and are similar to those presented by [47] and [63]. All these spectra are well fitted to a 2 shell model (Table S4, Supplementary Materials) with a Ni-O interatomic distance of 2.03–2.06 (1) Å and a Ni-Ni distance of 2.97–3.08 (3) Å, indicative of Ni as $Ni(OH)_2$. e.g., [47,55,63]. Although we found no evidence for Ni-Al distances, as observed by [47] in their Co-poor sample, it is likely that these $Ni(OH)_2$ 'brucite-like' layers are associated with the $Al(OH)_3$ layers in these asbolane-lithiophorite intermediates.

5.5. Composition and Diversity of the Microbial Community Present across the Piauí Laterite

DNA was extracted from laterite sediment and sequenced to investigate the composition and diversity of the microbial community and to look for the presence of potential metal cycling microorganisms. The prokaryotic community was profiled in samples from five different laterite horizons by 16S rRNA gene amplicon sequencing (Figure 19A,B). Three samples were selected for fungal community characterisation, and two were successfully profiled by sequencing the ITS region (Figure 19C).

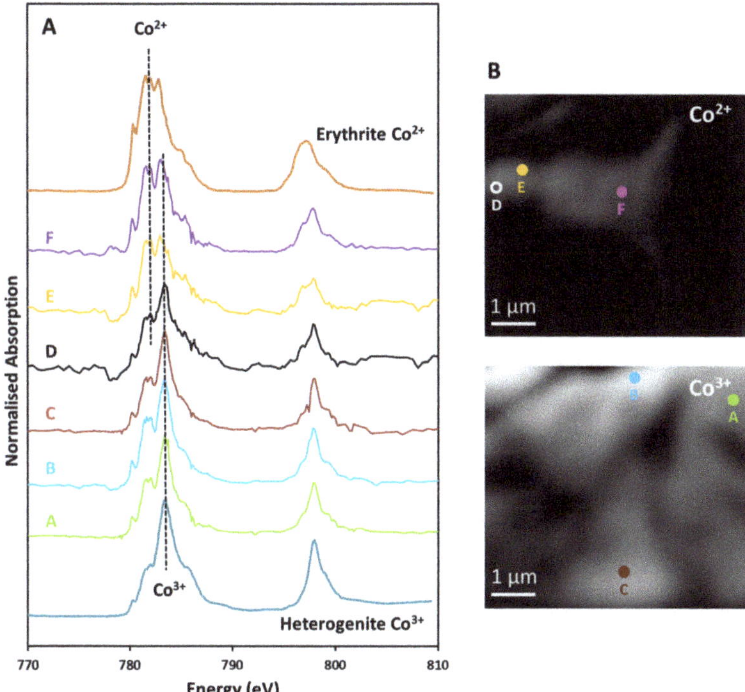

Figure 17. (**A**) $L_{2,3}$ XANES spectra for Co extracted from STXM image stacks. The positions of the absorption maxima associated with Co^{2+} and Co^{3+} are indicated on the spectra plot. (**B**) Oxidation state images showing the distribution of Co^{2+} (**top**) and Co^{3+} (**bottom**) within the Mn oxy-hydroxide grain from sample F033. The points from which the spectra have been extracted are shown on the oxidation state images.

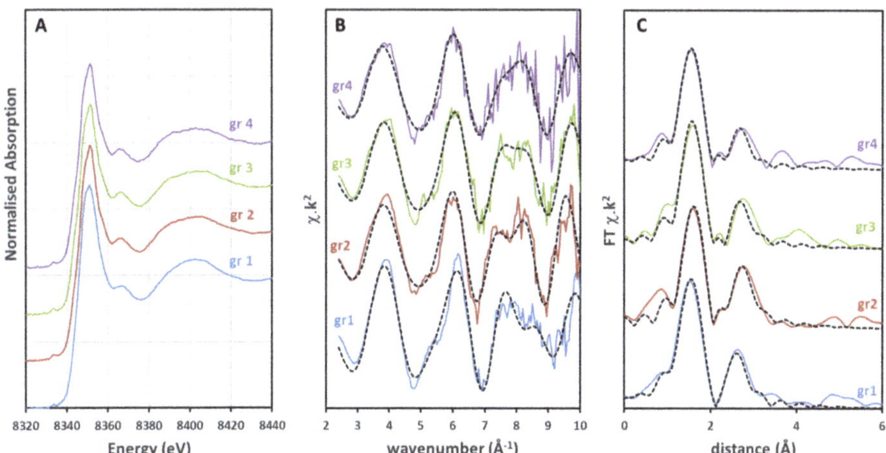

Figure 18. (**A**) Ni K-edge XANES spectra of 4 asbolane-lithiophorite intermediate grains from sample F033. (**B**) EXAFS spectra for the same asbolane-lithiophorite intermediates points as shown in (**A**). (**C**) Fourier transforms of the Ni EXAFS spectra from (**B**). Dashed lines in (**B**) and (**C**) are the modelled fits to the experimental data (solid lines).

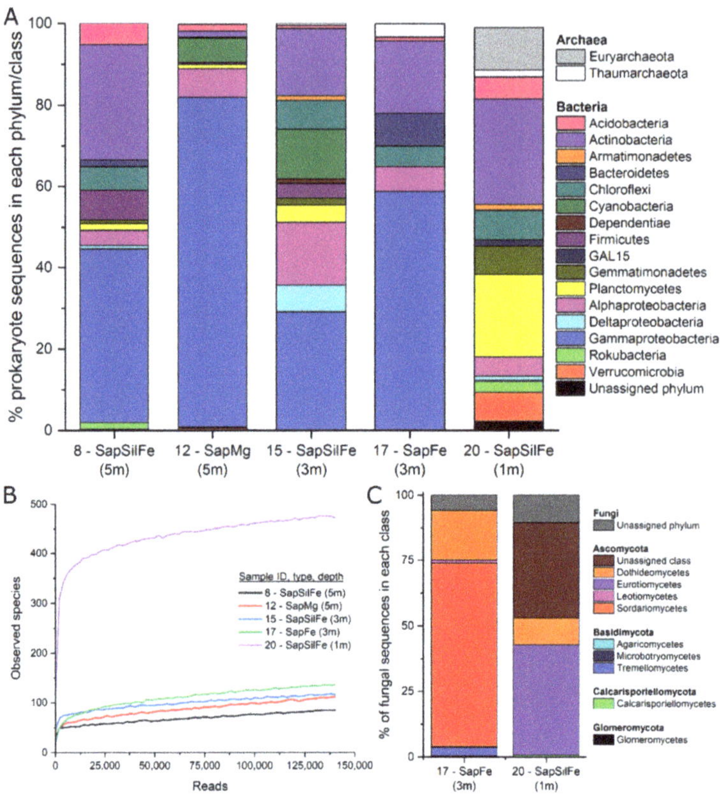

Figure 19. (**A**) Prokaryote community composition at the phylum/class level, sequences > 1% in at least one sample shown. (**B**) Rarefaction curve for prokaryotes. (**C**) Fungal community composition at the class level.

5.5.1. Microbial Abundance and Diversity

The highest number (478) of observed prokaryotic species (operational taxonomic units, OTUs) and the greatest Shannon diversity (7.38) were found in sample 20, a SAPSILFE collected approximately 1 m below ground level (mbgl) (Figure 19B, Table S5 Supplementary Materials). The number of OTUs was much lower (79–125) in the four other samples. The Shannon diversity was higher in the three SAPSILFE samples (4.3–7.4) compared to the SAPMG (3.3) and SAPFE (2.9) samples, but there is no trend with sample depth.

Correlation plots between prokaryotic diversity indicators (Shannon diversity and number of OTUs) and geochemical variables (Table S5) were made to identify whether laterite geochemistry may control microbial diversity. There is a positive relationship between the diversity indicators and concentrations of K ($R^2 > 0.68$), but few significant trends were observed, including for essential elements (e.g., Mg, P, Mn, Na, Ca) and major laterite components (Fe, Si). Positive correlations are evident for Al, Co and Ni concentrations (elements sometimes considered toxic), and total organic carbon, but these are strongly influenced by sample 20, which has the highest concentrations of these variables and the highest prokaryotic diversity; the trends were not observed in the remaining four samples.

Of the five samples where 16S rRNA gene copies were quantified using qPCR, the highest values were from sample 17 (1270 fg g^{-1} laterite), a SAPFE, followed by sample 20 (44 fg g^{-1} laterite), a SAPSILFE, while the other three samples were below detection (Table S5). The value for sample 17 is broadly comparable with a previous study of Cuban laterites, which reported data in the order of 1×10^7–1×10^8 copy number per gram dry sample [64] compared to 2.3 x 10^6 here (result converted from fg g^{-1} into copy number g^{-1}). The value for sample 20 is within the range reported for composite samples from Piauí, which contain 32–49 fg g^{-1} laterite [35].

There were also very low numbers of observed fungal species, with only 21 and 32 fungal OTUs present in the two samples sequenced (samples 17 and 20) and insufficient fungal DNA present for sequencing in sample 8, a SAPSILFE from 5 mbgl. The amount of 18S rRNA gene copies quantified by qPCR (note this will include all eukaryotes not just fungi) was 125–167 fg g^{-1} laterite.

Overall, the microbial diversity in the Piauí laterite is low, perhaps reflecting the challenges of living in a such a highly weathered, oxidising environment, with low concentrations of organic carbon and essential elements available to support growth (Table S5). There do not appear to be any significant trends in microbial abundance and diversity with depth, or by lithological unit type, or with geochemical conditions.

5.5.2. Prokaryotic Community Composition

The most common phyla/classes observed in the Piauí laterite samples are the Actinobacteria and Gammaproteobacteria (in particular, the *Burkholderiaceae* family) (Figure 19A). The closest phylogenetic relatives to the five most frequently detected OTUs in each sample were identified using Blastn (Table S6 Supplementary Materials). Commonly reported are sequences closely related to those from soil environments, e.g., from studies of the rhizosphere, endophytes, crops, ferralsol and also from human impacted environments such as from studies of the impact of wetland restoration, wastewater irrigation, mining and heavy metal contamination. Sequences closely related to those from studies of clays, lithifying microbialites, ureolysis, soils with elevated CO_2 conditions and bacteria involved in hydrocarbon biodegradation are also common.

The dataset was analysed to identify the presence of metal-cycling prokaryotes, which could potentially reduce or oxidise Mn, Fe and S minerals and therefore contribute to the mobilisation and enrichment of Co via indirect redox transformations [35]. Many sequences were assigned to the *Burkholderiaceae* (0.1% to 78%), some of which are known to oxidise Mn(II) and Fe(II) [65]. Some sequences are closely related to genera with known Mn(II)-oxidising activity, including *Sphingopyxis* (0%–1.8%), *Streptomyces* (0%–2.0%) and *Pseudomonas* (0%–3.1%) [66]. Although some sequences were assigned to the Nitrospirales (0.1%–0.9%) which contain known Fe(II) oxidisers, none were assigned to the Fe(II)-oxidising *Leptospirillum* genus [67]. Other Fe(II)-oxidising groups include *Acidithiobacillus* sp. (0%–7.6%); these are acidophilic prokaryotes, although the pH of the laterites was circumneutral (5.1–7.4). *Acidithiobacillus* and *Sulfurifustis* sp. (0%–5.0%) are acidophiles known to oxidise sulphur [68–70]. In general, relatively few sequences were assigned to genera closely related to known Fe(III)- and sulfate reducers, although some sequences were aligned with *Desulfosporosinus* sp. (0%–6.3%), which are known to reduce Mn(IV), Fe(III) and sulfate, in some cases at low pH [71,72]. Microbial siderophores are able to complex metals, especially Fe, and enhance their mobility. Some sequences in the Piauí laterite are closely related to well-known siderophore producing genera included *Acinetobacter* (0%–1.9%), *Pseudomonas* (0%–3.1%) and *Streptomyces* (0%–2.0%) [73–76].

Other prokaryotes of interest include nitrogen cycling organisms, which are fairly common in all samples, including sequences closely related to known nitrogen fixers such as *Burkholderia* (0%–61%), *Ralstonia* (0%–2.6%) and *Sphingomonas* (0%–4.0%) spp. [77,78], and anaerobic ammonium oxidisers including sequences from the Nitrospirales (0.1%–0.9%) and *Nitrosomonadaceae* (0%–2.3%), and most closely related to *Nitrososphaera viennensis* (0%–3.2%) [67,79,80]. Sequences assigned to *Massilia* spp. (0%–55%) are closely related to species known to hydrolyse urea, reduce nitrate and metabolise organic compounds, and *Massilia* spp. have previously been isolated from mining impacted soils [81,82]. Phosphorus mobilising prokaryotes are also prevalent, including sequences assigned to *Burkholderia* (0%–61%), *Pseudomonas* (0%–3.1%) and *Ralstonia* (0%–2.6%) spp., all of which have been shown to solubilise inorganic phosphorus and/or mineralise organic phosphorus compounds [83].

Overall, the composition of the prokaryote communities in the laterite sediments shows the presence of sequences closely related to organisms known to oxidise Mn, Fe and S minerals at different horizons within the deposit. These microbes can trap cobalt e.g., through the formation of Mn(IV)/Fe(III) oxides, or mobilise cobalt e.g., through the dissolution of sulphide minerals. As well as this, sequences closely related to organisms known to reduce metals (which could mobilise Co associated with Mn(IV)/Fe(III) oxides under anoxic conditions) and those that secrete siderophores (which could solubilise Co associated with Fe(III) oxides) were also identified, which could support the hypothesis that microbes contribute to the weathering of laterite and enhance the mobility of cobalt. Although the concentrations of elements essential for growth in the laterite were low, sequences closely related to organisms capable of fixing N and mobilising P were present, suggesting these may be able to provide these elements and support microbial growth.

5.5.3. Fungal Community Composition

The Piauí laterite contained very few fungal OTUs identified by ITS sequencing (Table S5). Samples 17 and 20 predominantly contained sequences assigned to Ascomycota (Figure 19C). Of the five most abundant OTUs for each sample (ten in total), only five sequences could be identified with confidence (Table S7 Supplementary Material). Four were from sample 17 and comprised common environmental yeast and mould and a pathogenic fungus that infects plant roots, the other was a versatile saprotroph from sample 20. Siderophore-producing fungi were identified to be present including *Rhodotorula* sp. (0%–0.2%) and *Aspergillus* sp. (0%–5.5%) [84].

5.5.4. Microcosm Experiments

Previous data for composite Piauí samples showed that adding an organic substrate stimulated the development of metal-reducing conditions and enhanced the recovery of cobalt [35]. Here, microcosm experiments were set up to investigate whether this would also be the case for individual horizons of the Piauí deposit collected for these experiments, stimulated using glucose or an acetate/lactate mix as a source of carbon and electron donor, and to see whether the heterogeneity in the microbial community was reflected in the biogeochemical response to redox cycling.

No significant changes in sediment texture were observed during the 105 day incubation, unlike in previous experiments comprising composite samples from the Piauí laterite [35]. Geochemical monitoring showed that nitrate reduction was stimulated in all samples with both electron donors (Figures S7 and S8 Supplementary Materials). Trace amounts of Mn and Fe were released to the aqueous phase, but there was no clear evidence of Fe(II) production or sulfate reduction with either electron donor. Volatile fatty acids (VFAs) were measured as a proxy for glucose breakdown products or to show consumption of acetate/lactate; the overall results from these experiments showed the indigenous microbial community was stimulated by the addition of organic substrates, but this did not lead to substantial microbial metal reduction and dissolution of cobalt.

The results of the sediment microcosm study show that, for the most part, the microbial community is able to metabolise carbon compounds and reduce nitrate. This is consistent with the results of the DNA sequencing analysis which identified that the microbial community is dominated by prokaryotes involved in C and N cycling. There was no conclusive evidence that microbial metal reduction had been stimulated in the seven samples of the different Piauí laterite horizons. This is in contrast to earlier work that demonstrated that microbial Mn(IV) and Fe(III) reduction was stimulated by the addition of organic electron donors in composite samples of the Piauí laterite; samples collected from heaps created from the excavation of 10,000 tonnes of laterite [35]. This suggests that mineral processing, including the physical disaggregation of Piauí sediment into composite samples during mining (and the consequent exposure to air and rainfall) may have somehow primed the microbial community to favour the capacity for metal reduction.

6. Discussion

6.1. Mineralogical Development the Piauí Lateritic Profile

The dunite of the ultramafic zone of the Brejo Seco–Piauí Complex originally comprised >90 vol% olivine with substantial chromite (<7 vol%) and a small proportion of plagioclase, clinopyroxene, orthopyroxene and phlogopite [27]. Serpentinisation replaced the olivine with serpentine-I and magnetite (e.g., Figure 8) with the ongoing replacement of chromite by ferrichromite and magnetite (e.g., Figure 9) [27].

During the laterisation process, the serpentine-I and any remaining olivine and pyroxene is altered into Ni-bearing serpentine-II, Fe-oxides including Ni-rich goethite, hematite and magnetite. Where Al is available, Ni bearing chlorite and/or other clays (mostly smectite) formed. Continued leaching, particularly of Mg and Si, results in the formation of the saprock and ferruginous saprolite that comprise the majority of the Piauí laterite and a significant Mg-discontinuity between these two units. This typical progression for laterite formation under tropical to sub-tropical climatic conditions, e.g., [11], can be visualised in a Si-Al-Fe (SAF) ternary plot (dotted line, Figure 20A) and quantified using the index of laterisation (IOL), as defined by [85] and expressed in Equation (1):

$$\text{IOL} = 100 \times [(\text{Al}_2\text{O}_3 + \text{Fe}_2\text{O}_3)/(\text{SiO}_2 + \text{Al}_2\text{O}_3 + \text{Fe}_2\text{O}_3)] \tag{1}$$

The IOL (calculated in wt%) gives average values of 24 and 30 for the serpentinite and saprock at Piauí, slightly higher than similar lithologies formed from serpentinised harzburgite and peridotite [86]. The oxide zone at Piauí represented by the SAPFE lithology, however, has IOL values between 63 and 83 within the moderately laterised division of the IOL scheme [85,86], reflecting the presence of numerous phyllosilicates and clays. Had laterisation continued under these conditions, it may be expected that the SAPFE lithology would have completely transformed into a strongly laterised traditional limonite, however, an extended period of silicification occurred throughout the lower two thirds of the ferruginous saprolite unit cementing Fe-oxides, forming extensive silica vein networks, stockwork alteration and boxwork structures (e.g., Figure 4A,C,D).

This silicification process is suggestive of a wet-to-dry scenario, whereby a change either in weathering conditions from a tropical climate to a seasonal or semi-arid environment promotes silica precipitation [12,14] or a restriction to drainage, promoting extensive silicification at the water table [14]. The extent of the silicification and its impact upon the IOL is evident in Figure 20A, with the seemingly retrograde laterisation step shown by the dashed line, whereby the samples of silicified ferruginous saprolite (SAPSILFE) have an average IOL value of 45.

Considering the importance of MgO and SiO_2 in the weathering of ultramafic bodies and the enrichment of Fe_2O_3 and Al_2O_3 during the laterisation process, [9] developed the quantitative ultramafic index of alteration (UMIA) that has recently been utilised to describe laterite development e.g., [6,9,15,86]. The UMIA is calculated using molar ratios from Equation (2):

$$\text{UMIA} = 100 \times [(\text{Al}_2\text{O}_3 + \text{Fe}_2\text{O}_3)/(\text{SiO}_2 + \text{MgO} + \text{Al}_2\text{O}_3 + \text{Fe}_2\text{O}_3)] \tag{2}$$

The serpentinite at Piauí has an UMIA between 5 and 6 while the saprock evolves from 5 at the lower part of this unit up to 10 close to the Mg-discontinuity at the saprock-SAPSILFE interface. These values are well matched to those for similar rocks at other laterite deposits [9,15,86]. The ferruginous saprolite SAPFE varies from 25 to 55, well below the values calculated for limonite horizons [9,15,86], again reflecting the persistence of phyllosilicates and clays within SAPFE. This developmental pathway can be indicated on the AF-S-M plot in Figure 20B by the dotted line where the loss of MgO and SiO_2, and limited enrichment on Fe_2O_3 is evident. The silicification process is also well highlighted in Figure 20B where the enrichment of the ferruginous saprolite in SiO_2 is indicated by the dashed line and the SAPSILFE samples develop lower UMIA values (13–31) compared to those of the SAPFE samples.

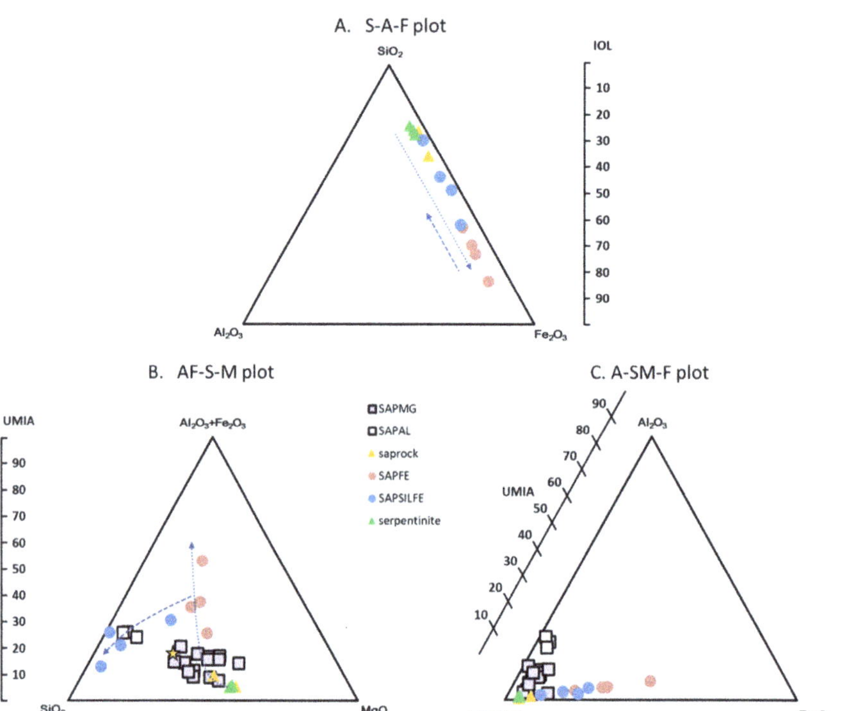

Figure 20. (**A**) Wt% SiO_2-Al_2O_3-Fe_2O_3 (SAF) ternary plot showing the alteration trend during laterisation of the dunite protolith (dotted arrow) and the subsequent impact of late stage silicification (dashed arrow). (**B,C**) Molar ternary plots in UMIA space (SiO_2-MgO-Fe_2O_3-Al_2O_3) showing the weathering trends of the dunite and troctolitic intercalations. (**B**) Highlighting the changed trend due to silicification of the SAPFE unit. Gold star represents average troctolite composition from [27]. (**C**) The bauxitisation trend associated with the formation of SAPMG and SAPAL compared to the laterisation process of the saprock SAPFE and SAPSILFE units.

The bulk chemistry for the ultramafic zone dunite reported by [27] is very similar to that of the serpentinised dunite of the Piauí laterite (Table 4), with the olivine of the ultramafic zone dunite containing 1000–3000 ppm NiO [27], suggesting this is the source of the Ni in the bulk of the laterite profile. The dunite olivine has up to 2800 ppm MnO [27], but the serpentinisation phase does not produce nickeliferous or manganiferous serpentine (Table 5). Instead, the serpentine-I is low in trace elements although Mn-bearing Ni and Co-rich magnetite (Figure 9) formed along with asbolane (Figure 11) containing ~7 wt% of both Co and Ni (Table 5).

The weathering process consumed any remaining olivine, producing Ni-bearing serpentine-II in the saprock. This serpentine-II is richer in Ni and Fe than serpentine-I and thus, is substantially more resilient within the weathering profile than the first formed, trace element-poor serpentine. This process preserved the igneous textures and further increased the porosity with the Si/Mg ratio increasing from 0.75 in the serpentinite to 0.98 in the saprock, just below the Mg-discontinuity e.g., [15]. High Si/Mg ratios during this phase may have induced the formation of the pimelite and nickeliferous talc bearing silica nodules within the ferruginous saprolite (prior to the late-stage conversion of SAPFE to

SAPSILFE) just above the saprock e.g., [15,87]. Continued saprolitisation developed the ferruginous saprolite above the saprock and Mg-discontinuity, within which the less mobile elements Fe, Cr and Al are concentrated (Table 4) relative to the saprock and serpentinite.

In addition to the silicification process, the Piauí laterite profile is further complicated by the presence of the green (SAPMG) and white-green (SAPAL) layers. These have weathered via a different mineralogical pathway to the main parts of the profile based in part on a different starting mineralogy, geochemistry and texture, but perhaps also due to enhanced fluid flow pathways at the interfaces between the ferruginous saprolite and these layers. SAPMG and SAPAL are, in general, substantially richer in Al_2O_3 compared to SAPFE or SAPSILFE, while SAPMG has retained between 13 and 30 wt% MgO, levels approaching those of the saprock and serpentinite (Tables 1, 2 and 4).

Salgado et al. (2016) [27] show that the ultramafic zone dunite has numerous troctolite intercalations or layers, with significant levels of plagioclase associated with the olivine. Moreover, in places, the ultramafic zone dunite possesses layers (from a few cm to several metres thick) characterised by larger amounts of interstitial minerals, particularly plagioclase. Troctolites within the transition and upper zones of the Brejo Seco–Piauí Complex possess 16–21 wt% Al_2O_3, 10–20 wt% MgO and 5–15 wt% Fe_2O_3 [27]. Importantly, however, the analyses of [27] show that these troctolites contain substantially lower Ni (100–750 ppm) than the dunite (1500–3300 ppm). The SAPMG and SAPAL units at Piauí likely derive from similar troctolitic intercalations/layers, and while the geochemistry of SAPMG is very variable (Table 2), average values (Table 4) do relate well to the troctolite chemistry of [27].

Considering the high concentrations of Al_2O_3, the initial serpentinisation of the troctolitic layers may have produced a mixed chlorite and serpentine (serpentine-I) assemblage where both phases had low Ni contents. The onset of saprolitisation provided substantial Ni into these chloritic layers, stabilising the chlorite by Ni for Mg exchange while the more soluble serpentine is lost [14], with the elevated levels of Ni permitting the occasional formation of nimite. During continued saprolitisation of the whole profile fluid flow may have become concentrated around the margins of the intercalated layers, as reported to have occurred at the Siberia Complex, Western Australia [88], further depleting the MgO and SiO_2 in these areas, leaving Al_2O_3-rich layers dominated by kaolinite (with aluminous vermiculite and smectite), associated with the margins of the chloritic SAPMG layers. Moreover, if the Ni content of the chlorite in these margins was less than for the chlorite in the core regions of the layers (as suggested by Table 2) then these would have been preferentially solubilised [14], enhancing the alteration in the margins to produce the SAPAL units (Figure 3D).

Such an alteration pathway can be envisaged in Figure 20B, where an initial troctolite composition evolves into a MgO enriched saprolite (SAPMG), while the highly leached margins of the intercalated troctolite become very depleted in MgO. The alteration pathway of these green and white-green saprolitic layers may be better described as bauxitisation rather than laterisation, as it progresses entirely along a $SiO_2 + MgO - Al_2O_3$ join, as seen in the A-SM-F plot of Figure 20C.

6.2. Element Mobility during the Piauí Laterite Formation

Within the main part of the Piauí profile, the silicified ferruginous saprolites have the highest average ΣREE (36 ppm), while the serpentinite, saprock and SAPFE all have similar abundances with average values of 3.5, 5 and 12, respectively. Although only two saprock samples were analysed, the one closest to the serpentinite has the lowest levels of REE, while the one closest to the Mg-discontinuity has REE values between those of the two ferruginous saprolite units, with a general trend very similar to that displayed by the SAPSILFE samples (Figure 6) and the overlying SAPSILFE. While there is no true limonite unit in the Piauí profile, elevated REE levels in the upper saprock are similar to those reported for the saprolites of Loma de Hierro [15] and Ibra [89], in contrast to the highest levels being reported for the limonites at Punta Gorda, Loma Caribe, Loma Peguera [8] and Wingellina [6]. In general, though not always, the oxide-rich zones have REE trends revealing enrichment in LREE, while the saprolite and protolith/bedrock show flat trends across the REEs. At Piauí, the SAPSILFE and upper part of the saprock show LREE enrichment, while the serpentinite, saprock and SAPFE show no consistent enrichment across the REEs (Figure 6).

The SAPMG and SAPAL units are very much enriched in REE relative to the other parts of the Piauí profile, with maximum ΣREE values of 626 ppm and 1292 ppm, respectively, reflecting their troctolitic source, and all of these samples show a negative slope. There is good correlation between the ΣREE and Mn (r = 0.70) and Co (r = 0.69) for the SAPMG and SAPAL units, suggesting that the REEs may be hosted within Mn oxy-hydroxide phases, but, contrary to previous findings [6,8,15], not across the other lithologies.

Serpentinisation of the dunite mobilised the Co and Ni to magnetite/spinel and into asbolane, with a Co:Ni ratio averaging close to 1 (Table 5; Figure 11). Serpentine-I contains very little Ni and no detectable Co (by EMPA), and the weathering stages of the profile development initially resulted in Ni-bearing serpentine-II and nickeliferous chlorite. We have no evidence for Co hosted within serpentine, contrary to the findings of [53], who used linear combination fitting of XANES data to confirm that serpentine in laterites from the Goro Peninsula, New Caledonia, contain Co. As the saprolitisation process continues, asbolane within the serpentinite is reworked to produce low-Al asbolane-lithiophorite intermediates with an average Ni:Co ratio of about 3.5 (Figure 11), reflecting the increased availability of Ni as the silicates are removed. While Ni-rich goethite is commonplace across the main lateritic profile units, SAPSILFE and SAPFE, we find that Co incorporation into goethite has a strong dependency upon Mn, such that the only Co bearing goethites are those with significant levels of Mn. This may reflect local reworking or dissolution of asbolane within an increasingly Fe-rich environment, or direct formation from Co and Mn-rich magnetite.

The uppermost unit of the profile (SAPFE) has the highest levels of Co, Ni and Mn from the main profile, with these elements hosted primarily within low-Al asbolane-lithiophorite intermediates and goethites, and there is no textural or geochemical evidence to suggest that the asbolane-lithiophorite intermediates within SAPFE are a different generation to those from SAPSILFE. Indeed, there is no strong evidence for multiple generations of Mn oxy-hydroxides within the SAPFE or SAPSILFE profile units. Moreover, it is possible that the remobilisation of Ni and Co downward within the profile, as is common in other deposits, has not occurred at Piauí, as either the lateritisation process has not proceeded far enough, or perhaps due to the change in weathering environment and the extensive late-stage silicification. Limited downward mobilisation and enrichment of Ni was also suggested for the Intex deposit, correlating with the less advanced degree of weathering overall [86].

This observation correlates with the limited evidence found for microbial weathering or a microbial contribution to metal enrichment. The present day Piauí laterite contains a relatively low diversity of prokaryotes and fungi, commonly associated with soil/rhizosphere environments, rather than necessarily reflecting the laterite geochemistry. While it was not possible to identify many of the fungal species, the prokaryotic community is primarily composed of bacteria capable of oxidising simple N and C compounds and Mn(II) and Fe(II) under oxic or suboxic conditions, and a small proportion of sequences are closely related to Fe(III)- and sulphur-reducers, which could suggest there is potential for active microbial metal cycling in the Piauí laterite, and associated redistribution of redox sensitive metals. However, microcosm experiments conducted to stimulate metal-reducing conditions in sediments collected from different horizons of the laterite showed limited evidence for metal mobilisation. Therefore, it is unlikely that active biogeochemical cycling of metals is occurring and contributing to Co enrichment in the present day Piauí laterite. Correlations between Co and Mn are likely to be a relic of parent rock weathering, rather than due to biogeochemical processes; a conclusion that agrees well with the mineralogical associations.

The presence of the SAPMG and SAPAL layers within the Piauí profile have had a profound effect upon the mobility of Co, Ni and Mn. The chlorite dominated SAPMG layers have the highest Ni concentrations, with substantial amounts of Ni hosted within chlorite (Table 7; Figure 14). Additionally, chlorite within SAPMG is also the only silicate phase we identified that hosts levels of Co detectable by EMPA (Table 6). SAPAL has the highest levels of Mn and Co across the entire deposit, more than double the levels of Co and almost double the levels of Mn compared to SAPFE, the next most concentrated unit (Table 4). The elevated levels of Mn in SAPAL are reflected by the high number of Mn oxy-hydroxide grains, and the field observations of regions of concentrated Mn oxy-hydroxide formation as layers or coatings at the SAPAL boundary with the ferruginous saprolite (e.g., Figures 3E and 4E –G). These Mn oxy-hydroxides, along with those in SAPMG, are all rich in Al and are classed as asbolane-lithiophorite intermediates (Figure 11) and have an average Ni:Co ratio of ~4. Within these intermediates, the Ni is incorporated as $Ni(OH)_2$ within the interlayer.

Dublet et al. (2017) [53] proposed that within laterite profiles Co is progressively incorporated in Mn oxy-hydroxides as oxidized Co^{3+} as the serpentinite is weathered to saprolite and further to limonite. Using XANES, EXAFS and STXM we find, however, that in the Al-rich, clay dominated SAPMG and SAPAL, there is substantial Co^{2+} in the asbolane-lithiophorite intermediates hosted within the interlayer, while the Co^{3+} replaces Mn in the MnO_6 layers. There is no evidence for the presence of nanoscale goethite, that can contain Co^{2+} [53], and it is possible that in parts of the asbolane-lithiophorite intermediate grains all the available sites in the MnO_6 layers that are required for the oxidation of Co^{2+} to Co^{3+} are used up. Potentially, the geochemical and Eh/pH properties of the Mn oxy-hydroxide forming fluids within SAPMG and SAPAL are different to those of the ferruginous saprolite at Piauí and in other laterite deposits, promoting the precipitation of asbolane-lithiophorite intermediates as surface coatings and that limit the oxidation of Co^{2+} upon incorporation into the Mn oxy-hydroxide structure e.g., [56,90]. In essence there is the potential for the development of redox fronts at or close to the SAPAL and ferruginous saprolite boundary, similar to the changes in Eh and pH between lateritic clay and limonite zones [88], a mechanism suggested for the similarly silicified, clay rich deposit at Ravensthorpe [91].

Additionally, it may be that the seasonal or reduced rate of the lateritisation process at Piauí related to a change in climatic conditions, silicification and limited drainage promoted more reducing conditions with hydromorphic activity at the SAPAL–ferruginous boundary, e.g., [3], or even precluded the gradual reworking of these Mn oxy-hydroxides and subsequent complete oxidation of Co^{2+} to Co^{3+}. The presence of Al in the asbolane-lithiophorite intermediates may make these Mn oxy-hydroxides more resistant to weathering and subsequent reworking e.g., [92]. Nevertheless, the presence of these SAPMG and SAPAL layers have clearly influenced the local precipitation and nature of the Mn oxy-hydroxides, and moreover, the fluid flow within the adjoining ferruginous saprolite, further supported by the localised precipitation of romanechite phases. The SAPMG and SAPAL layers have effectively created a hydrologic barrier promoting Eh/pH conditions suitable for Mn oxy-hydroxide precipitation further into the ferruginous saprolite and channelling Ni and Co-rich fluids into these regions (e.g., Figure 4E,F), as shown to have occurred in the laterite deposit over the Siberia Complex Laterite, Western Australia [88].

7. Conclusions

The laterite profile at Piauí is essentially an oxide type deposit, though the lack of a true limonite layer and the presence of clay rich units dominated by chlorite, smectite and kaolinite give the profile some clay silicate deposit characteristics. The deposit formed on a serpentinised dunite that contained many troctolitic intercalated layers relatively rich in Al compared to the rest of the ultramafic protolith. The profile was further complicated by a change in climatic and weathering conditions that induced late stage silicification throughout the saprolite units.

Within the ferruginous saprolite layers (SAPSILFE and SAPFE), goethite is volumetrically the most significant Co and Ni bearing mineral, while low-Al asbolane-lithiophorite intermediates possess the highest concentrations of these elements. Within the intercalated green and white-green layers (SAPMG and SAPAL) chlorite is volumetrically the most significant Ni bearing mineral, with concentrations of Ni up to 24 wt% Ni. Al-rich asbolane-lithiophorite intermediates and Ba-rich romanechite with high concentrations of both Co and Ni are associated with these two layers often concentrated at the interface between the SAPAL and ferruginous saprolitic units.

The Co and Ni are structurally bound within the asbolane-lithiophorite intermediates, such that Co^{3+} substitutes for Mn in the MnO_6 layers and Ni is hosted as $Ni(OH)_2$ within the 'brucite-like' layers. Substantial levels of Co^{2+} are identified structurally bound within the asbolane-lithiophorite intermediates, a finding that may be unique to Piauí, associated with its complex petrologic structure and formation history, or may reflect the limited number of equivalent studies on other similar deposits. The formation of the asbolane-lithiophorite intermediates, romanechites and other Mn oxy-hydroxide minerals is associated with redox cycling whereby fluid flow is strongly affected by the presence of the SAPMG and SAPAL layers or where these layers locally impart a defining influence on the redox properties of the metal loaded fluids.

The mineralogy of limonites has a profound impact on their leaching properties [93] and knowledge that Co is predominantly present in Mn oxy-hydroxides and Ni in clays may be used to target mineral processing methods to recover these elements. The mineralogy and geotechnical properties of the Piauí laterite make it very amenable to long term atmospheric heap leaching to extract the Co, Ni and other technologically significant metals. Indeed, the application of tailored short-duration bioreductive technologies, e.g., [48,49], have the potential to specifically target the Mn oxy-hydroxides and the Ni-rich clays at Piauí to recover the Co and Ni hosted within these phases [35,94].

Supplementary Materials: The following supporting information can be downloaded at: https://www.mdpi.com/article/10.3390/min12101298/s1. Figure S1: Schematic of Piauí Test Pit; Figure S2: sampling sites; Figure S3: Sampling sites continued; Figure S4: REE plots; Figure S5: XRD patterns of high Ni samples and Fe oxides; Figure S6: FTIR spectrum of F050; Figure S7: Glucose stimulated microcosm results; Figure S8: Acetate/lactate stimulated microcosm results; Table S1: Additional elements from the bulk chemical analyses; Table S2: Additional elements from the bulk chemical analyses; Table S3: Fitting parameters for the Co K-edge microfocus XAS spectra; Table S4: Fitting parameters for the Ni K-edge microfocus XAS spectra; Table S5: Details of samples analysed for microbial communities; Table S6: Closest phylogenetic relatives of the five most abundant prokaryotic OTUs; Table S7: Closest phylogenetic relatives of the five most abundant fungal OTUs; Methods for microbiological characterisation: Prokaryotes; Methods for microbiological characterisation: Fungi [95–117].

Author Contributions: Conceptualisation: A.D., P.F.S., L.N., R.J.H. and A.O; Methodology: A.D., P.F.S., L.N., R.J.H., J.F.W.M., B.K., M.K., T.A., J.K. and J.R.L.; Analytical Investigation: A.D., P.F.S., L.N., T.J.S., J.F.W.M., B.K., T.A., M.K. and R.L.N.; Field Investigation: P.F.S., L.N., R.J.H., J.K. and A.O.; Resources: P.F.S., L.N., R.J.H., J.F.W.M., B.K. and J.R.L.; Data Curation: A.D., P.F.S. and L.N.; Original Draft Preparation: A.D., P.F.S. and L.N.; Review and Editing: A.D., P.F.S., L.N., J.F.W.M., R.J.H., B.K., T.A., M.K., T.J.S., J.K., A.O., R.L.N. and J.R.L.; Funding Acquisition: R.J.H., P.F.S., J.F.W.M., B.K. and J.R.L. All authors have read and agreed to the published version of the manuscript.

Funding: This work was funded by the Natural Environment Research Council, UK (COG3: The geology, geometallurgy and geomicrobiology of cobalt resources leading to new product streams) Grant references NE/M011488/1, NE/M011518/1 and NE/M011127/1. The work was also supported by Diamond Light Source, proposal references NT14882, SP14908 and SP17882. JK received funding from the People Programme (Marie Curie Actions) of the European Union's Seventh Framework Programme FP7/2007–2013/under REA grant agreement No. [608069].

Data Availability Statement: Data supporting reported results can be found in the Supplementary Materials.

Acknowledgments: This work was supported by the Natural Environment Research Council, UK (COG3: The geology, geometallurgy and geomicrobiology of cobalt resources leading to new product streams NE/M011488/1, NE/M011518/1 and NE/M011127/1). The authors are grateful to Mike Oxley and all the staff of Brazilian Nickel for their support, guidance and hospitality, particularly during the sample collecting at the Piauí deposit. Synchrotron work was carried out with the support of Diamond Light Source, instruments I18 (proposals NT14882 and SP14908) and I08 (proposal SP17882). JK received funding from the People Programme (Marie Curie Actions) of the European Union's Seventh Framework Programme FP7/2007–2013/under REA grant agreement No. [608069]. We are grateful to Rachael James, Jens Najorka, John Spratt, Tobias Salge, Simon Kocher, Ashley King, Christopher Boothman, Paul Lythgoe and Alastair Bewsher for their help during this study.

Conflicts of Interest: The authors declare no conflict of interest.

References

1. Gleeson, S.A.; Herrington, R.J.; Durango, J.; Velásquez, C.A.; Koll, G. The mineralogy and geochemistry of the Cerro Matoso S.A. Ni laterite deposit, Montelíbano, Colombia. *Econ. Geol.* **2004**, *99*, 1197–1213. [CrossRef]
2. Thorne, R.L.; Roberts, S.; Herrington, R. Climate change and the formation of nickel laterite deposits. *Geology* **2012**, *40*, 331–334. [CrossRef]
3. Lambiv Dzemua, G.L.; Gleeson, S.A.; Schofield, P.F. Mineralogical characterization of the Nkamouna Co–Mn laterite ore, southeast Cameroon. *Miner. Depos.* **2013**, *48*, 155–171. [CrossRef]
4. Butt, C.R.M.; Cluzel, D. Nickel Laterite Ore Deposits: Weathered Serpentinites. *Elements* **2013**, *9*, 123–128. [CrossRef]
5. Mudd, G.M.; Weng, Z.; Jowitt, S.M.; Turnbull, I.D.; Graedel, T.E. Quantifying the recoverable resources of by-product metals: The case of cobalt. *Ore Geol. Rev.* **2013**, *55*, 87–98. [CrossRef]
6. Putzolu, F.; Boni, M.; Mondillo, N.; Maczurad, M.; Pirajno, F. Ni-Co enrichment and High-Tech metals geochemistry in the Wingellina Ni-Co oxide-type laterite deposit (Western Australia). *J. Geochem. Explor.* **2019**, *196*, 282–296. [CrossRef]
7. Samouhos, M.; Godelitsas, A.; Nomikou, C.; Taxiarchou, M.; Tsakiridis, P.; Zavasnik, J.; Gamaletsos, P.N.; Apostolikas, A. New insights into nanomineralogy and geochemistry of Ni-laterite ores from central Greece (Larymna and Evia deposits). *Geochemistry* **2018**, *79*, 268–279. [CrossRef]
8. Teitler, Y.; Cathelineau, M.; Ulrich, M.; Ambrosi, J.P.; Munoz, M.; Sevin, B. Petrology and geochemistry of scandium in New Caledonian Ni-Co laterites. *J. Geochem. Explor.* **2019**, *196*, 131–155. [CrossRef]
9. Aiglsperger, T.; Proenza, J.A.; Lewis, J.F.; Labrador, M.; Svojtka, M.; Rojas-Purón, A.; Longo, F.; Durisova, J. Critical metals (REE, Sc, PGE) in Ni-laterites from Cuba and the Dominican Republic. *Ore Geol. Rev.* **2016**, *73*, 127–147. [CrossRef]
10. Brand, N.W.; Butt, C.R.M.; Elias, M. Nickel laterites: Classification and features. *J. Aust. Geol. Geophys.* **1998**, *17*, 81–88.
11. Freyssinet, P.; Butt, C.R.M.; Morris, R.C.; Piantone, P. Ore-forming processes related to lateritic weathering. *Econ. Geol.* **2005**, *100*, 681–722. [CrossRef]
12. Thorne, R.; Herrington, R.; Roberts, S. Composition and origin of the Çaldağ oxide nickel laterite, W. Turkey. *Miner. Depos.* **2009**, *44*, 581–595. [CrossRef]
13. Mudd, G.M. Global trends and environmental issues in nickel mining: Sulfides versus laterites. *Ore Geol. Rev.* **2010**, *38*, 9–26. [CrossRef]
14. Golightly, J.P. Progress in understanding the evolution of nickel laterites. *Soc. Econ. Geol. Spec. Publ.* **2010**, *15*, 451–475.
15. Domènech, C.; Galí, S.; Soler, J.M.; Ancco, M.P.; Meléndez, W.; Rondón, J.; Villanova-de-Benavent, C.; Proenza, J.A. The Loma de Hierro Ni-laterite deposit (Venezuela): Mineralogical and chemical composition. *Boletín Soc. Geológica Mex.* **2020**, *72*, A050620. [CrossRef]
16. Herrington, R.; Mondillo, N.; Boni, M.; Thorne, R.; Tavlan, M. Bauxite and Nickel-Cobalt Lateritic Deposits of the Tethyan Belt. *Soc. Econ. Geol. Spec. Publ.* **2016**, *19*, 349–387.
17. Trescases, J.J.; Melfi, A.J.; de Oliveira, S.M.B. Nickeliferous laterites of Brazil. In *Proceedings of the Laterisation Processes, Proceedings of the International Seminar on Laterisation Processes, Trivandrum, India, 11–14 December 1979*; Balkema: Rotterdam, The Netherlands, 1981; pp. 170–184.
18. de Oliveira, S.M.B.; Trescases, J.J.; Melfi, A.J. Lateritic nickel deposits of Brazil. *Miner. Depos.* **1992**, *27*, 137–146. [CrossRef]
19. Melfi, A.J.; Trescases, J.J.; de Oliveira, S.M.B. Les "laterites" nickeliferes du Brésil. *Cah. ORSTOM Série Géologie* **1979**, *11*, 15–42.
20. Dino, R. Gênese do Minério de Niquel de São João do Piauí Por Alteração Intempérica. Ph.D. Thesis, Instituto do Geosciéncias, Universidade de Sao Paulo, Sao Paulo, Brazil, 1984.
21. Oxley, A.; Smith, M.E.; Caceres, O.Y. Why heap leach nickel laterites? *Miner. Eng.* **2016**, *88*, 53–60. [CrossRef]
22. Oxley, A.; Smith, M.E.; Caceres, O.Y. The Piauí Nickel heap leach project, Brazil. *Proc. Heap Leach Solut.* **2015**, *2015*, 123–133.
23. BRN. The Piauí Nickel Project Fact Sheet. 2021. Available online: https://www.braziliannickel.com/piaui-nickel-project/ (accessed on 30 December 2021).
24. De Almeida, F.F.M.; Hasui, Y.; Brito Neves, B.B.; Fuck, R.A. Brazilian Structural Provinces: An introduction. *Earth Sci. Rev.* **1981**, *17*, 1–29. [CrossRef]

25. Salgado, S.S.; Ferreira Filho, C.F.; Uhlein, A.; de Andrade Caxito, F. Geologia, Estratigrafia e Petrografia do Complexo de Brejo Seco, Faixa Riacho do Pontal, sudeste do Piauí. *Geonomos* **2014**, *22*, 10–21. [CrossRef]
26. Oliveira, R.G. Geophysical Framework, Isostasy and Causes of the Cenozoic Magmatism of the Province of Borborema and Its Continental Margin (Northeastern Brazil). Ph.D. Thesis, Instituto do Geosciéncias, Universidade Federal do Rio Grande do Norte, Natal, Brazil, 2008.
27. Salgado, S.S.; Ferreira Filho, C.F.; de Andrade Caxito, F.; Uhlein, A.; Dantas, E.L.; Stevensen, R. The Ni-Cu-PGE mineralized Brejo Seco mafic-ultramafic layered intrusion, Riacho do Pontal Orogen: Onset of Tonian (ca. 900 Ma) continental rifting in Northeast Brazil. *J. S. Am. Earth Sci.* **2016**, *70*, 324–339. [CrossRef]
28. Devlin, R.; (CSA Global, Perth, Australia). Mineral Resource Estimate, Brazilian Nickel Ltd, Piaui Nickel Project Sao Joao do Piauí, Piauí, Brazil. Unpublished Report No: R259.2013. 2013.
29. Pouchou, J.L.; Pichoir, F. A New Model for Quantitative X-ray Microanalysis, Part I: Application to the Analysis of Homogeneous Samples. *Rech. Aerosp.* **1984**, *3*, 13–38.
30. Pouchou, J.L.; Pichoir, F. A New Model for Quantitative X-ray Microanalysis, Part II: Application to In-depth Analysis of Heterogeneous Samples. *Rech. Aerosp.* **1984**, *5*, 47–65.
31. Mosselmans, J.F.W.; Quinn, P.D.; Roque Rosell, J.; Atkinson, K.D.; Dent, A.J.; Cavill, S.I.; Hodson, M.E.; Kirk, C.A.; Schofield, P.F. The first environmental science experiments on the new microfocus spectroscopy beamline at Diamond. *Mineral. Mag.* **2008**, *72*, 197–200. [CrossRef]
32. Mosselmans, J.F.W.; Quinn, P.D.; Dent, A.J.; Cavill, S.A.; Diaz-Moreno, S.; Peach, A.; Leicester, P.J.; Keylock, S.J.; Gregory, S.; Atkinson, K.D.; et al. I18-the microfocus spectroscopy beamline at the Diamond Light Source. *J. Synchrotron Radiat.* **2009**, *16*, 818–824. [CrossRef]
33. Ravel, B.; Newville, M. ATHENA, ARTEMIS, HEPHAESTUS: Data analysis for X-ray absorption spectroscopy using IFEFFIT. *J. Synchrotron Radiat.* **2005**, *12*, 537–541. [CrossRef]
34. Lerotic, M.; Mak, R.; Wirick, S.; Meirer, F.; Jacobsen, C. MANTiS: A program for the analysis of X-ray spectromicroscopy data. *J. Synchrotron Radiat.* **2014**, *21*, 1206–1212. [CrossRef]
35. Newsome, L.; Solano Arguedas, A.; Coker, V.S.; Boothman, C.; Lloyd, J.R. Manganese and cobalt redox cycling in laterites; Biogeochemical and bioprocessing implications. *Chem. Geol.* **2020**, *531*, 119330. [CrossRef]
36. Lovley, D.R.; Phillips, E.J. Organic matter mineralization with reduction of ferric iron in anaerobic sediments. *Appl. Environ. Microbiol.* **1986**, *51*, 683–689. [CrossRef] [PubMed]
37. Lovley, D.R.; Phillips, E.J.P. Rapid assay for microbially reducible ferric iron in aquatic sediments. *Appl. Environ. Microbiol.* **1987**, *53*, 1536–1540. [CrossRef] [PubMed]
38. Caldeira, J.F.; (Vale, Belo Horizonte, Brazil). Relatorio Preliminar De Pesquisa Integrado, Processos DNPM 803.156/2009 E 803.158/2009, Municipio: Capitao Gervasio Oliveira—PI, Belo Horizonte. VALE Unpublished Internal Report. July 2012.
39. Al-Khirbash, S. Geology, mineralogy, and geochemistry of low grade Ni-lateritic soil (Oman Mountains, Oman). *Chem. Erde* **2016**, *76*, 363–381. [CrossRef]
40. Chassé, M.; Griffin, W.L.; O'Reilly, S.Y.; Calas, G. Australian laterites reveal mechanisms governing scandium dynamics in the critical zone. *Geochim. Cosmochim. Acta* **2019**, *260*, 292–310. [CrossRef]
41. Chassé, M.; Griffin, W.L.; O'Reilly, S.Y.; Calas, G. Scandium speciation in a world-class lateritic deposit. *Geochem. Perspect. Lett.* **2017**, *3*, 105–114. [CrossRef]
42. McDonough, W.F.; Sun, S.-S. The composition of the Earth. *Chem. Geol.* **1995**, *120*, 223–253. [CrossRef]
43. Villanova-de-Benavent, C.; Domènech, C.; Tauler, E.; Galí, S.; Tassara, S.; Proenza, J.A. Fe-Ni-bearing serpentines from the saprolite horizon of Caribbean Ni-laterite deposits: New insights from thermodynamic calculations. *Miner. Depos.* **2017**, *52*, 979–992. [CrossRef]
44. Schwartz, S.; Guillot, S.; Reynard, B.; Lafay, R.; Debret, B.; Nicollet, C.; Lanari, P.; Auzende, A.L. Pressure-temperature estimates of the lizardite/antigorite transition in high pressure serpentinites. *Lithos* **2013**, *178*, 197–210. [CrossRef]
45. Lambiv Dzemua, G.L.; Gleeson, S.A. Petrography, mineralogy, and geochemistry of the Nkamouna serpentinite: Implications for the formation of the cobalt-Manganese laterite deposit, southeast Cameroon. *Econ. Geol.* **2012**, *107*, 25–41. [CrossRef]
46. Putzolu, F.; Balassone, G.; Boni, M.; Maczurad, M.; Mondillo, N.; Najorka, J.; Pirajno, F. Mineralogical association and Ni-Co deportment in the Wingellina oxide-type laterite deposit (Western Australia). *Ore Geol. Rev.* **2018**, *97*, 21–34. [CrossRef]
47. Roqué-Rosell, J.; Mosselmans, J.F.W.; Proenza, J.A.; Labrador, M.; Galí, S.; Atkinson, K.D.; Quinn, P.D. Sorption of Ni by "lithiophorite–asbolane" intermediates in Moa Bay lateritic deposits, eastern Cuba. *Chem. Geol.* **2010**, *275*, 9–18. [CrossRef]
48. Santos, A.L.; Dybowska, A.; Schofield, P.F.; Herrington, R.J.; Johnson, D.B. Chromium (VI) Inhibition of low pH Bioleaching of Limonitic Nickel-Cobalt Ore. *Front. Microbiol.* **2022**, *12*, 802991. [CrossRef]
49. Santos, A.L.; Dybowska, A.; Schofield, P.F.; Herrington, R.J.; Johnson, D.B. Sulfur-enhanced reductive bioprocessing of cobalt-bearing materials for base metals recovery. *Hydrometallurgy* **2020**, *195*, 105396. [CrossRef]
50. Carvalho e Silva, M.L.M.d.; de Oliveira, S.M.B. As fases portadoras de níquel do minério laterítico de níquel do Vermelho, Serra dos Carajás (PA). *Rev. Bras. Geociências* **1995**, *25*, 69–78. [CrossRef]
51. Ratié, G.; Garnier, J.; Calmels, D.; Vantelon, D.; Guimarães, E.; Monvoisin, G.; Nouet, J.; Ponzevera, E.; Quantin, C. Nickel distribution and isotopic fractionation in a Brazilian lateritic regolith: Coupling Ni isotopes and Ni K-edge XANES. *Geochim. Cosmochim. Acta* **2018**, *230*, 137–154. [CrossRef]

52. Andersen, J.C.Ø.; Rollinson, G.K.; Snook, B.; Herrington, R.; Fairhurst, R.J. Use of QEMSCAN for the characterization of Ni-rich and Ni-poor goethite in laterite ores. *Miner. Eng.* **2009**, *22*, 1119–1129. [CrossRef]
53. Dublet, G.; Juillot, F.; Brest, J.; Noël, V.; Fritsch, E.; Proux, O.; Olivi, L.; Ploquin, F.; Morin, G. Vertical changes of the Co and Mn speciation along a lateritic regolith developed on peridotites (New Caledonia). *Geochim. Cosmochim. Acta* **2017**, *217*, 1–15. [CrossRef]
54. Burns, R.G. The uptake of cobalt into ferromanganese nodules, soils and synthetic manganese (IV) oxides. *Geochim. Cosmochim. Acta* **1976**, *40*, 95–102. [CrossRef]
55. Manceau, A.; Llorca, S.; Calas, G. Crystal chemistry of cobalt and nickel in lithiophorite and asbolane from New Caledonia. *Geochim. Cosmochim. Acta* **1987**, *51*, 105–113. [CrossRef]
56. Manceau, A.; Drits, V.A.; Silvester, E.; Bartoli, C.; Lanson, B. Structural mechanism of Co^{2+} oxidation by the phyllomanganate buserite. *Am. Mineral.* **1997**, *82*, 1150–1175. [CrossRef]
57. Manceau, A.; Combes, J.M. Structure of the Mn and Fe oxides and oxyhydroxides: A topological approach by EXAFS. *Phys. Chem. Miner.* **1988**, *15*, 283–295. [CrossRef]
58. Cressey, G.; Henderson, C.M.B.; van der Laan, G. Use of L-edge X-ray absorption spectroscopy to characterize multiple valence states of 3d transition metals; a new probe for mineralogical and geochemical research. *Phys. Chem. Miner.* **1993**, *20*, 111–119. [CrossRef]
59. Schofield, P.F.; Henderson, C.M.B.; Cressey, G.; van der Laan, G. 2p X-ray absorption spectroscopy in the Earth sciences. *J. Synchrotron Radiat.* **1995**, *2*, 93–98. [CrossRef]
60. Smith, A.D.; Schofield, P.F.; Cressey, G.; Cressey, B.A.; Read, P.D. The development of X-ray photo-emission electron microscopy (XPEEM) for valence-state imaging of mineral intergrowths. *Mineral. Mag.* **2004**, *68*, 859–869. [CrossRef]
61. Schofield, P.F.; Smith, A.D.; Scholl, A.; Doran, A.; Covey-Crump, S.J.; Young, A.T.; Ohldag, H. Chemical and oxidation-state imaging of mineralogical intergrowths: The application of X-ray photo-emission electron microscopy (XPEEM). *Coord. Chem. Rev.* **2014**, *277–278*, 31–43. [CrossRef]
62. Mulroy, D.S.J. The Microbiology of Lateritic Co-Ni-Bearing Manganese Oxides. Ph.D. Thesis, University of Manchester, Manchester, UK, 2020.
63. Dublet, G.; Juillot, F.; Morin, G.; Fritsch, E.; Fandeur, D.; Ona-Nguema, G.; Brown, G.E., Jr. Ni speciation in a New Caledonia lateritic regolith: A quantitative X-ray absorption spectroscopy investigation. *Geochim. Cosmochim. Acta* **2012**, *95*, 119–133. [CrossRef]
64. Perez, O.C.; Coto, J.M.; Schippers, A. Quantification of the microbial community in lateritic deposits. *Integr. Sci. Ind. Knowl. Biohydrometall.* **2013**, *825*, 33–36. [CrossRef]
65. Algora, C.; Vasileiadis, S.; Wasmund, K.; Trevisan, M.; Krüger, M.; Puglisi, E.; Adrian, L. Manganese and iron as structuring parameters of microbial communities in Arctic marine sediments from the Baffin Bay. *FEMS Microbiol. Ecol.* **2015**, *91*, fiv056. [CrossRef]
66. Carmichael, S.K.; Bräuer, S.L. Microbial Diversity and Manganese Cycling: A Review of Manganese-oxidizing Microbial Cave Communities. In *Microbial Life of Cave Systems*; Engel, A.S., Ed.; Walter de Gruyter Gmbh: Berlin, Germany; Boston, MA, USA, 2015; Chapter 3; pp. 137–160.
67. Daims, H. The Family Nitrospiraceae. In *The Prokaryotes*; Springer: Berlin/Heidelberg, Germany, 2014; pp. 733–749. [CrossRef]
68. Colmer, A.R. Relation of the iron oxidizer, *Thiobacillus ferrooxidans*, to thiosulfate. *J. Bacteriol.* **1962**, *83*, 761–765. [CrossRef]
69. Hallberg, K.B.; Hedrich, S.; Johnson, D.B. *Acidiferrobacter thiooxydans*, gen. nov. sp. nov.; an acidophilic, thermo-tolerant, facultatively anaerobic iron- and sulfur-oxidizer of the family Ectothiorhodospiraceae. *Extremophiles* **2011**, *15*, 271–279. [CrossRef]
70. Norris, P.R. Acidimicrobiia class. nov. In *Bergey's Manual of Systematics of Archaea and Bacteria*; Trujillo, M.E., Dedysh, S., De Vos, P., Hedlund, B., Kämpfer, P., Rainey, F.A., Whitman, W.B., Eds.; John Wiley & Sons, Ltd.: Chichester, UK, 2015. [CrossRef]
71. Pester, M.; Brambilla, E.; Alazard, D.; Rattei, T.; Weinmaier, T.; Han, J.; Lucas, S.; Lapidus, A.; Cheng, J.-F.; Goodwin, L.; et al. Complete genome sequences of *Desulfosporosinus orientis* DSM765T, *Desulfosporosinus youngiae* DSM17734T, *Desulfosporosinus meridiei* DSM13257T, and *Desulfosporosinus acidiphilus* DSM22704T. *J. Bacteriol.* **2012**, *194*, 6300–6301. [CrossRef] [PubMed]
72. Senko, J.M.; Zhang, G.; McDonough, J.T.; Bruns, M.A.; Burgos, W.D. Metal Reduction at Low pH by a *Desulfosporosinus* species: Implications for the Biological Treatment of Acidic Mine Drainage. *Geomicrobiol. J.* **2009**, *26*, 71–82. [CrossRef]
73. Dimkpa, C.; Svatos, A.; Merten, D.; Büchel, G.; Kothe, E. Hydroxamate siderophores produced by *Streptomyces acidiscabies* E13 bind nickel and promote growth in cowpea (*Vigna unguiculata* L.) under nickel stress. *Can. J. Microbiol.* **2008**, *54*, 163–172. [CrossRef] [PubMed]
74. Dorsey, C.W.; Tomaras, A.P.; Connerly, P.L.; Tolmasky, M.E.; Crosa, J.H.; Actis, L.A. The siderophore-mediated iron acquisition systems of *Acinetobacter baumannii* ATCC 19606 and *Vibrio anguillarum* 775 are structurally and functionally related. *Microbiology* **2004**, *150*, 3657–3667. [CrossRef]
75. Kloepper, J.W.; Leong, J.; Teintze, M.; Schroth, M.N. Enhanced plant growth by siderophores produced by plant growth-promoting rhizobacteria. *Nature* **1980**, *286*, 885–886. [CrossRef]
76. Podschun, R.; Fischer, A.; Ullmann, U. Siderophore production of *Klebsiella* species Isolated from Different Sources. *Zent. Bakteriol.* **1992**, *276*, 481–486. [CrossRef]
77. Chen, W.-M.; Moulin, L.; Bontemps, C.; Vandamme, P.; Béna, G.; Boivin-Masson, C. Legume symbiotic nitrogen fixation by beta-proteobacteria is widespread in nature. *J. Bacteriol.* **2003**, *185*, 7266–7272. [CrossRef]

78. Videira, S.S.; de Araujo, J.L.S.; da Silva Rodrigues, L.; Baldani, V.L.D.; Baldani, J.I. Occurrence and diversity of nitrogen-fixing *Sphingomonas* bacteria associated with rice plants grown in Brazil. *FEMS Microbiol. Lett.* **2009**, *293*, 11–19. [CrossRef]

79. Prosser, J.I.; Head, I.M.; Stein, L.Y. The Family Nitrosomonadaceae. In *The Prokaryotes*; Rosenberg, E., De Long, E.F., Lory, S., Stackebrandt, E., Thompson, F., Eds.; Springer: Berlin/Heidelberg, Germany, 2014; pp. 901–918. [CrossRef]

80. Stieglmeier, M.; Klingl, A.; Alves, R.J.E.; Rittmann, S.K.-M.R.; Melcher, M.; Leisch, N.; Schleper, C. *Nitrososphaera viennensis* gen. nov.; sp. nov.; an aerobic and mesophilic, ammonia-oxidizing archaeon from soil and a member of the archaeal phylum Thaumarchaeota. *Int. J. Syst. Evol. Microbiol.* **2014**, *64*, 2738–2752. [CrossRef]

81. Baldani, J.I.; Rouws, L.; Cruz, L.M.; Olivares, F.L.; Schmid, M.; Hartmann, A. The Family Oxalobacteraceae. In *The Prokaryotes*; Springer: Berlin/Heidelberg, Germany, 2014; pp. 919–974. [CrossRef]

82. Du, Y.; Yu, X.; Wang, G. *Massilia tieshanensis* sp. nov.; isolated from mining soil. *Int. J. Syst. Evol. Microbiol.* **2012**, *62*, 2356–2362. [CrossRef]

83. Alori, E.T.; Glick, B.R.; Babalola, O.O. Microbial phosphorus solubilization and its potential for use in sustainable agriculture. *Front. Microbiol.* **2017**, *8*, 971. [CrossRef]

84. Ahmed, E.; Holmström, S.J.M. Siderophores in environmental research: Roles and applications. *Microb. Biotechnol.* **2014**, *7*, 196–208. [CrossRef]

85. Babechuk, M.G.; Widdowson, M.; Kamber, B.S. Quantifying chemical weathering intensity and trace element release from two contrasting basalt profiles, Deccan Traps, India. *Chem. Geol.* **2014**, *363*, 56–75. [CrossRef]

86. Tupaz, C.A.J.; Watanabe, Y.; Sanematsu, K.; Echigo, T.; Arcilla, C.; Ferrer, C. Ni-Co Mineralization in the Intex Laterite Deposit, Mindoro, Philippines. *Minerals* **2020**, *10*, 579. [CrossRef]

87. Galí, S.; Soler, J.M.; Proenza, J.A.; Lewis, J.F.; Cama, J.; Tauler, E. Ni-enrichment and stability of Al-free garnierite solid-solutions: A thermodynamic approach. *Clays Clay Miner.* **2012**, *60*, 121–135. [CrossRef]

88. Elias, M.; Donaldson, M.J.; Giorgetta, N. Geology, Mineralogy, and Geochemistry of Lateritic Nickel-Cobalt Deposits near Kalgoorlie, Western Australia. *Econ. Geol.* **1981**, *76*, 1775–1783. [CrossRef]

89. Al-Khirbash, S.; Semhi, K.; Richard, L.; Nasir, S.; Al-Harthy, A. Rare earth element mobility during laterization of mafic rocks of the Oman ophiolite. *Arab. J. Geosci.* **2014**, *7*, 5443–5454. [CrossRef]

90. Murray, K.J.; Webb, S.M.; Bargar, J.R.; Tebo, B.M. Indirect oxidation of Co(II) in the presence of the marine Mn (II)-oxidizing bacterium *Bacillus* sp. Strain SG-1. *Appl. Environ. Microbiol.* **2007**, *73*, 6905–6909. [CrossRef]

91. Miller, G.W.; Sampson, D.; Fleay, J.; Conway-Mortimer, J.; Roche, E. Ravensthorpe nickel project beneficiation prediction MLR and interpretation of results. In Proceedings of the International Laterite Nickel Symposium, Charlotte, NC, USA, 14–18 March 2004; pp. 121–136.

92. Dowding, C.E.; Fey, M.V. Morphological, chemical and mineralogical properties of some manganese-rich oxisols derived from dolomite in Mpumalanga province, South Africa. *Geoderma* **2007**, *141*, 23–33. [CrossRef]

93. Stanković, S.; Martin, M.; Goldmann, S.; Gäbler, H.-E.; Ufer, K.; Haubrich, F.; Moutinho, V.F.; Giese, E.C.; Neumann, R.; Stropper, J.L.; et al. Effect of mineralogy on Co and Ni extraction from Brazilian limonitic laterites via bioleaching and chemical leaching. *Miner. Eng.* **2022**, *184*, 107604. [CrossRef]

94. Smith, S.L.; Grail, B.M.; Johnson, D.B. Reductive bioprocessing of cobalt-bearing limonitic laterites. *Miner. Eng.* **2017**, *106*, 86–90. [CrossRef]

95. Lane, D.J. 16S/23S rRNA sequencing. In *Nucleic Acid Techniques in Bacterial Systematics*; Stackebrant, E., Goodfellow, M., Eds.; John Wiley & Sons Ltd.: London, UK, 1991; pp. 115–175.

96. Caporaso, J.G.; Lauber, C.L.; Walters, W.A.; Berg-Lyons, D.; Huntley, J.; Fierer, N.; Owens, S.M.; Betley, J.; Fraser, L.; Bauer, M.; et al. Ultra-high-throughput microbial community analysis on the Illumina HiSeq and MiSeq platforms. *ISME J.* **2012**, *6*, 1621–1624. [CrossRef]

97. Caporaso, J.G.; Lauber, C.L.; Walters, W.A.; Berg-Lyons, D.; Lozupone, C.A.; Turnbaugh, P.J.; Fierer, N.; Knight, R. Global patterns of 16S rRNA diversity at a depth of millions of sequences per sample. *Proc. Natl. Acad. Sci. USA* **2010**, *108*, 4516–4522. [CrossRef]

98. Kozich, J.J.; Westcott, S.L.; Baxter, N.T.; Highlander, S.K.; Schloss, P.D. Development of a dual-index sequencing strategy and curation pipeline for analyzing amplicon sequence data on the MiSeq Illumina sequencing platform. *Appl. Environ. Microbiol.* **2013**, *79*, 5112–5120. [CrossRef]

99. Martin, M. Cutadapt removes adapter sequences from high-throughput sequencing reads. *EMBnet J.* **2011**, *17*, 10. [CrossRef]

100. Joshi, N.A.; Fass, J.N. Sickle: A Sliding-Window, Adaptive, Quality-Based Trimming Tool for FastQ Files. 2011. Available online: http://github.com/najoshi/sickle (accessed on 31 August 2017).

101. Nurk, S.; Bankevich, A.; Antipov, D.; Gurevich, A.A.; Korobeynikov, A.; Lapidus, A.; Prjibelski, A.D.; Pyshkin, A.; Sirotkin, A.; Sirotkin, Y.; et al. Assembling single-cell genomes and mini-metagenomes from chimeric MDA products. *J. Comput. Biol.* **2013**, *20*, 714–737. [CrossRef]

102. Masella, A.P.; Bartram, A.K.; Truszkowski, J.M.; Brown, D.G.; Neufeld, J.D. PANDAseq: Paired-end assembler for Illumina sequences. *BMC Bioinf.* **2012**, *13*, 31. [CrossRef]

103. Haas, B.J.; Gevers, D.; Earl, A.M.; Feldgarden, M.; Ward, D.V.; Giannoukos, G.; Ciulla, D.; Tabbaa, D.; Highlander, S.K.; Sodergren, E.; et al. Chimeric 16S rRNA sequence formation and detection in Sanger and 454-pyrosequenced PCR amplicons. *Genome Res.* **2011**, *21*, 494–504. [CrossRef]

104. Edgar, R.C. UPARSE: Highly accurate OUT sequences from microbial amplicon reads. *Nat. Methods* **2013**, *10*, 996–998. [CrossRef]

105. Edgar, R.C. Search and clustering orders of magnitude faster than BLAST. *Bioinformatics* **2010**, *26*, 2460–2461. [CrossRef]
106. Caporaso, J.G.; Kuczynski, J.; Stombaugh, J.; Bittinger, K.; Bushman, F.D.; Costello, E.K.; Fierer, N.; Peña, A.G.; Goodrich, J.K.; Gordon, J.I.; et al. QIIME allows analysis of high-throughput community sequencing data. *Nat. Methods* **2010**, *7*, 335–336. [CrossRef] [PubMed]
107. Wang, Q.; Garrity, G.M.; Tiedje, J.M.; Cole, J.R. Naive Bayesian classifier for rapid assignment of rRNA sequences into the new bacterial taxonomy. *Appl. Environ. Microbiol.* **2007**, *73*, 5261–5267. [CrossRef] [PubMed]
108. Brown, A.E.; Muthumeenakshi, S.; Sreenivasaprasad, S.; Mills, P.R.; Swinburne, T.R. A PCR primer-specific to Cylindro-carpon heteronema for detection of the pathogen in apple wood. *FEMS Microbiol. Lett.* **1993**, *108*, 117–120. [CrossRef] [PubMed]
109. Gardes, M.; Bruns, T.D. ITS primers with enhanced specificity for basidiomycetes–application to the identification of my-corrhizae and rusts. *Mol. Ecol.* **1993**, *2*, 113–118. [CrossRef]
110. Taylor, D.L.; Walters, W.A.; Lennon, N.J.; Bochicchio, J.; Krohn, A.; Caporaso, J.G.; Pennanen, T. Accurate estimation of fun-gal diversity and abundance through improved lineage-specific primers optimized for Illumina amplicon sequencing. *Appl. Environ. Microbiol.* **2016**, *82*, 7217–7226. [CrossRef]
111. Gweon, H.S.; Oliver, A.; Taylor, J.; Booth, T.; Gibbs, M.; Read, D.S.; Griffiths, R.I.; Schonrogge, K. PIPITS: An automated pipeline for analyses of fungal internal transcribed spacer sequences from the Illumina sequencing platform. *Methods Ecol. Evol.* **2015**, *6*, 973–980. [CrossRef]
112. Edgar, R.C.; Haas, B.J.; Clemente, J.C.; Quince, C.; Knight, R. UCHIME improves sensitivity and speed of chimera detection. *Bioinformatics* **2011**, *27*, 2194–2200. [CrossRef]
113. Borneman, J.; Triplett, E.W. Molecular microbial diversity in soils from eastern Amazonia: Evidence for unusual microorganisms and microbial population shifts associated with deforestation. *Appl. Environ. Microbiol.* **1997**, *63*, 2647–2653. [CrossRef]
114. Wirth, F.; Goldani, L.Z. Epidemiology of Rhodotorula: An emerging pathogen. *Interdiscip. Perspect. Infect. Dis.* **2012**, 465717. [CrossRef]
115. MacGillivray, A.R.; Shiaris, M.P. Biotransformation of polycyclic aromatic hydrocarbons by yeasts isolated from coastal sediments. *Appl. Environ. Microbiol.* **1993**, *59*, 1613–1618. [CrossRef]
116. Jurjevic, Z.; Peterson, S.W.; Horn, B.W. Aspergillus section Versicolores: Nine new species and multilocus DNA sequence based phylogeny. *IMA Fungus* **2012**, *3*, 59–79. [CrossRef]
117. Giraldo, A.; Gené, J.; Sutton, D.A.; Madrid, H.; Cano, J.; Crous, P.W.; Guarro, J. Phylogenetic circumscription of Arthrographis (Eremomycetaceae, Dothideomycetes). *Persoonia* **2014**, *32*, 102–114. [CrossRef]

Review

Origin of Critical Metals in Fe–Ni Laterites from the Balkan Peninsula: Opportunities and Environmental Risk

Maria Economou-Eliopoulos [1],* , Magdalena Laskou [1], Demetrios G. Eliopoulos [2],†, Ifigeneia Megremi [1], Sofia Kalatha [1] and George D. Eliopoulos [3]

[1] Department of Geology and Geoenvironment, National and Kapodistrian University of Athens, 15784 Athens, Greece; laskou@geol.uoa.gr (M.L.); megremi@geol.uoa.gr (I.M.); spkalatha@gmail.com (S.K.)
[2] Institute of Geology and Mineral Exploration (IGME), Olympic Village, 13677 Acharnae, Greece
[3] Department of Chemistry, University of Crete, 70013 Heraklion, Greece; chem1723@edu.chemistry.uoc.gr
* Correspondence: econom@geol.uoa.gr
† Demetrios passed away in April 2019.

Abstract: As the global energy sector is expected to experience a gradual shift towards renewable energy sources, access to special metals in known resources is of growing concern within the EU and at a worldwide scale. This is a review on the Fe–Ni ± Co-laterite deposits in the Balkan Peninsula, which are characterized by multistage weathering/redeposition and intense tectonic activities. The ICP-MS analyses of those laterites indicated that they are major natural sources of Ni and Co, with ore grading from 0.21 to 3.5 wt% Ni and 0.03 to 0.31 wt% Co, as well as a significant Sc content (average 55 mg/kg). The SEM-EDS analyses revealed that fine Fe-, Ni-, Co-, and Mn-(hydr)oxides are dominant host minerals and that the enrichment in these elements is probably controlled by the post-formation evolution of initial ore redeposition. The paucity of rare earth element (REE) within the typical Fe–Ni laterite ore and the preferential occurrence of Co (up to 0.31 wt%), REE content (up to 6000 mg/kg ΣREE), and REE-minerals along with Ni, Co, and Mn (asbolane and silicates) towards the lowermost part of the Lokris (C. Greece) laterite ore suggest that their deposition is controlled by epigenetic processes. The platinum-group element (PGE) content in those Fe–Ni laterites, reaching up to 88 µg/kg Pt and 26 µg/kg Pd (up to 186 µg/kg Pd in one sample), which is higher than those in the majority of chromite deposits associated with ophiolites, may indicate important weathering and PGE supergene accumulation. Therefore, the mineralogical and geochemical features of Fe–Ni laterites from the Balkan Peninsula provide evidence for potential sources of certain critical metals and insights to suitable processing and metallurgical methods. In addition, the contamination of soil by heavy metals and irrigation groundwater by toxic Cr(VI), coupled with relatively high Cr(VI) concentrations in water leachates for laterite samples, altered ultramafic rocks and soils neighboring the mining areas and point to a potential human health risk and call for integrated water–soil–plant investigations in the basins surrounding laterite mines.

Keywords: Fe–Ni–Co laterites; critical metals; REE; PGE; Cr(VI) contamination

Citation: Economou-Eliopoulos, M.; Laskou, M.; Eliopoulos, D.G.; Megremi, I.; Kalatha, S.; Eliopoulos, G.D. Origin of Critical Metals in Fe–Ni Laterites from the Balkan Peninsula: Opportunities and Environmental Risk. *Minerals* **2021**, *11*, 1009. https://doi.org/10.3390/min11091009

Academic Editors: Cristina Domènech and Cristina Villanova-de-Benavent

Received: 4 August 2021
Accepted: 13 September 2021
Published: 16 September 2021

1. Introduction

As gradual shift towards renewable energy sources on a global scale, access to particular raw materials or critical metals (CM), such as platinum-group elements (PGE), rare earth elements (REE), and scandium (Sc), with potential presence in known resources is a growing concern within the European Union (EU) and in a worldwide scale [1,2]. Although the production for these metals is mostly derived from magmatic deposits, volcanogenic massive sulfide (VMS) deposits, the black shale-hosted deposit in Finland and Sweden, and deposits of supergene origin associated with the release of major and trace elements from the alteration of related rocks [1,3–7], deposits of Fe–Ni ± Co laterite may be a potential resource for critical metals as well [8,9]. Ni laterites contain about 60–70% of the world's nickel resources and account for about 40% of the world's nickel production, with the remainder coming from sulfide ores [10]. The Ni–Co laterite deposits provide one of the two major natural sources of nickel and cobalt and became of economic importance with the recent increasing industrialization of developing countries [11]. The type of Ni–Co laterites developed by chemical weathering of ultramafic rocks with potential post redeposition enrichment of weathering products may occur above weathered bedrock. Among others, they have been described in the Philippines (Taganito/Adlay), in Western Australia (The Murrin Murrin deposit), in New Caledonia, Indonesia, and the Dominican Republic [11–13]. Recently, the Fe–Ni ± Co type of laterites is an attractive research topic due to their large tonnage, easy exploitation (open pit mining), a significant development of a Ni and Co demand, and the technological innovations implemented by exploration companies [11,12]. With respect to known classifications, three mineralogical subtypes of Fe–Ni ± Co laterite deposits are recognized: the oxide, hydrous Mg–silicate, and clay types, with median Ni and Co grades, 1.14% and 0.09% of the oxide type, 1.44 and 0.06% for the Mg type, and 1.27 and 0.06% for the clay type [11,13]. The Fe–Ni laterite deposits in the Balkan Peninsula are associated with ophiolites, which represent a remnant of the Tethyan oceanic lithosphere, located in the Mirdita–Sub-Pelagonian and Pelagonian geotectonic zones [14]. They provide some 2–3% of the World's total Ni and include deposits in Serbia (Topola), in N. Macedonia, former F.Y.R.O.M. (Rzanovo), Albania (Bitincka, Gouri-Perjuegjiun, and Katjeli) and Greece (Lokris, C. Evia, Kastoria, Vermion, Paleochori, Edessa, Olympos) (Figure 1) [15–21].

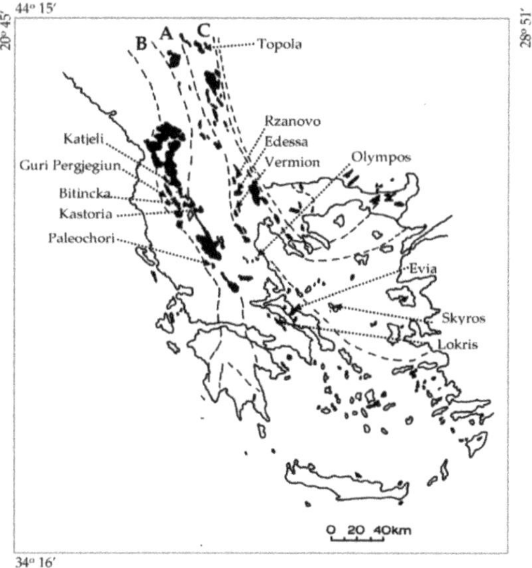

Figure 1. Sketch map showing the Pelagonian (A), Sub-Pelagonian (B), and Axios (C) geotectonic zones in the Balkan Peninsula, and the distribution of ophiolites and the associated Fe–Ni ± Co laterite deposits.

Furthermore, the weathering/alteration of laterite deposits and parent ultramafic rocks, mining, and large volumes of smelting residues (slag) are potential sources of environmental hazards for terrestrial and aquatic ecosystems, food quality, and socioeconomic problems [1–4,10,11]. An evaluation of the environmental risks associated with a possible exploitation of the deposits may contribute to the potential ways forward to protect soil and groundwater for human health and ecosystems. The present study focuses on the combination of Ni–Co–Mn mineral chemistry with geochemical characteristics of laterites from the Balkan Peninsula, which offer a variety of laterite types. The aim is the delineation of geochemical constraints as a contribution from the origin, genesis and exploration of critical metals (REE, PGE, Co, Sc) in laterites to the environmental impact from mining and plant processes of laterite ores.

2. Geological Outline of Fe–Ni ± Co Deposits and Occurrences in the Balkan Peninsula

Based on national mineral resource agencies, approximately 500 Co-bearing deposits and occurrences have been identified in 25 countries in Europe [9]. The measured mineral resources in Greece are 220,000,000 tons [9,11]. The Ržanovo deposit, located close to the contact between the Axios zone and the Pelagonian massif, is one of the largest Fe–Ni ± Co deposits in the Axios zone [20]. In general, large Fe–Ni–Co laterite deposits, which have been exploited for many years, are mainly related to the extensive lateritization of ultramafic ophiolites in the Balkan Peninsula (Shebenik–Pogradec Massif at the central and southern part of the Mirdita ophiolites—Sub-Pelagonian and Pelagonian geotectonic zones) into Greece and by extension the Anatolides zone of western Turkey [20,22]. Major deposits show a complete weathering profile that comprises: (1) a lower zone of saprolitized peridotite with or without partially weathered core stones; (2) a transition zone which may be dominated by quartz or smectite clays; (3) a ferruginous saprolitic limonite zone; (4) a recrystallized and locally transported limonite zone; and (5) a goethite–hematite zone [23]. Although Fe–Ni ± Co laterites in the Balkan Peninsula, related to Upper Jurassic–Lower Cretaceous serpentinized ultramafic ophiolites, may be partially preserved in situ, they are mostly allochthonous, redeposited as marine sediments, and buried by later sedimentary rocks [22]. The laterite ores have commonly been transported and redeposited either onto peridotites, such as at Kastoria, W. Vermion, and Palaiochori (Greece), Bitincka and Guri–Pergjegjum (Albania), or onto limestones, such as at Lokris (Aghios Ioannis), Katjeli (Albania) (Figures 1 and 2A). Those laterite deposits are often overlain by a sequence of mudstone, limestones, and organic carbon-rich mudstones, which have been attributed to a stagnant reducing marine environment during tectonic down warping [15]. In particular, the Katjeli Fe–Ni deposit lies on karstified Jurassic limestone, and it is conformably overlain by a Tertiary series consisting of small beds of lignite and sandstone intercalation and molasses. The exploited ores have been reworked due to repeated marine transgression and regression, deposited in a shallow marine environment, and affected by thrust faulting [15,17,22,24]. They can be divided into those which are unconformably overlain by Early Eocene shallow marine sediments, such as the Kastoria and Katjeli deposits in the Balkan Peninsula, and those overlain by Cretaceous limestones, such as the Lokris [15,17]. Specifically, the Kastoria Fe–Ni deposit is located in northern Greece (Figure 1), NNW of Kastoria town, is developed on Upper Jurassic–Lower Cretaceous serpentinized ultramafic ophiolites of the Mirdita–Sub-Pelagonian zone, and is overlain by Tertiary molasses. A goethite zone, 5-m thick, consisting mainly of goethite and hematite overlies a weathering crust [18,25]. At the W. Vermion (Profitis Elias), ophiolites are mainly comprised of large peridotite masses of harzburgite, dunite, and orthopyroxene–dunite. The weathering crust is overlain by a highly silicified zone, while the ore is pelitic at its lowest part and becomes pisolitic towards the top. The Fe–Ni ore is mainly composed of goethite, hematite, Ni-bearing chlorite, quartz, calcite, and chromite. Talc is common in the lowest part of the ore, while illite is dominant in its upper part. In a distance of less than 1 km north of the Profitis Elias Ni laterite deposit, the Parhari deposit of bauxite laterite is located in an area where mafic rocks (diabases) are the dominant rock types [18]. More attention is paid to the Lokris (Aghios Ioannis) Fe–Ni laterite deposits (Figure 2) of allochthonous origin because of the association of Fe–Ni ore with the bauxite laterite at the Nissi–Patitira location, and the relatively high contents of critical metals [22,24,26].

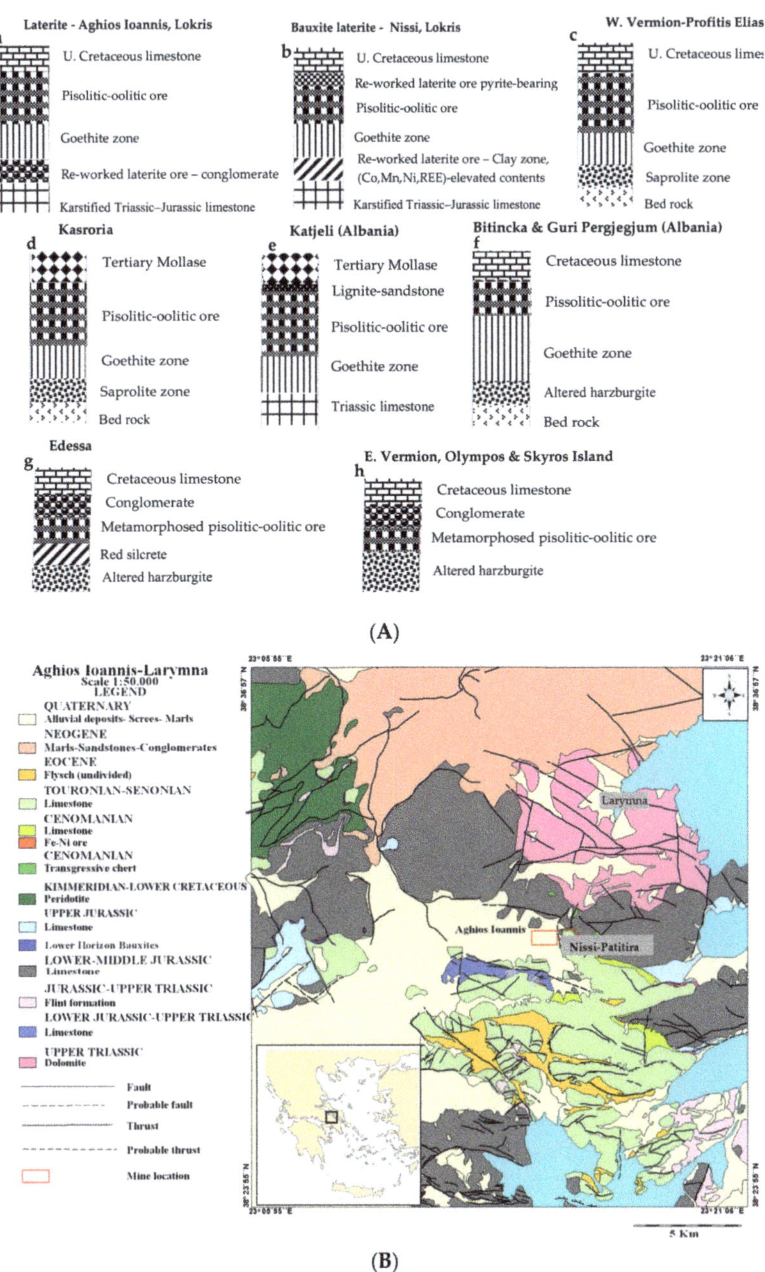

Figure 2. (**A**) Schematic stratigraphic sections of the most representative laterite deposits in the Balkan Peninsula. (**B**) Simplified geological map, after LARCO GMM (General Mining and metallurgical Company), cited in ProMine Project [27] showing the location of the main Fe–Ni laterite deposits at Lokris (Aghios Ioannis).

Small (1 m × 15 m) lens-like Fe–Ni occurrences, metamorphosed to amphibolite facies (probably erosional remnants of redeposited laterites), are found in the E. Vermion, Edessa, Olympos, and Skyros islands of Greece (Figures 1 and 2A) and are overlain by

Cretaceous limestones [19]. The dismembered ophiolite masses of Upper Jurassic–Lower Cretaceous age, which consist of mainly serpentinized harzburgite and, to a lesser extent, crustal magmatic rocks (pyroxenites and gabbros), outcrop along the eastern margin of the Pelagonian massif [28]. Due to intense tectonism, the Fe–Ni laterite occurrences are often entirely enclosed within serpentinized harzburgites [19].

3. Methods of Investigation

Due to the heterogeneous character of most Ni laterite deposits, samples of a minimum 2 kg weight were collected from surface exposures. Although the analytical methods applied for the determination of major and trace elements in rocks, laterite ores, groundwater, and water leachates, including Cr stable isotopes and As speciation, is provided in the relative publications [29–31], a brief outline is given here. Two samples of metallurgical residue (slag samples, Slag96 and Slag14) from the Larymna plant, kindly provided by the mining company GMM LARCO (March 1996 and November 2014, respectively), and representative samples of ultramafic altered rocks were analyzed. Major and trace elements were determined by atomic absorption at the Institute of Geology and Mineral Exploration, Greece, and minor and trace elements were obtained by Inductively Coupled Plasma Mass Spectrometry (ICP–MS) analysis, after multi-acid digestion (HNO_3–$HClO_4$–HF–HCl) at the ACME Laboratories Ltd., Vancouver, BC, Canada. Platinum-group element (PGE) analyses were carried out using the Ni-sulfide fire-assay pre-concentration technique, with the nickel fire-assay technique from large (30 g) samples at Genalysis Laboratory Services, Perth, Australia. This method allows for complete dissolution of samples. The detection limits were 1 ppb for Pd, 10 ppb for Pt, and 5 ppb Au. CDN–PGMS–23 was used as standard.

Representative surface (up to 20 cm) soil samples from the rhizosphere of plants and corresponding plants/crops covering some sites surrounding laterite deposits have been analyzed by inductively coupled plasma mass spectroscopy (ICP/MS) after aqua regia digestion at ACME Laboratories Ltd., Vancouver, BC, Canada.

A series of batch leaching experiments have been carried out, using natural water in order to study the long-term leaching responses of Cr under atmospheric conditions [31]. For these experiments, 20 g of a crushed ore, metallurgical residue (slag), rock, or soil sample and 150 mL of natural water were transferred in a 200 mL Erlenmeyer flask at room temperature. The reaction flask was shaken at approximately 120 rpm by a reciprocal shaker for 5 weeks. After the period of shaking, the slurries were filtered through a 0.45 μm polyamide membrane filter. Furthermore, following the same methodology, except the shaker time, which was 24 h in this case, water leaching experiments were carried out in the present study.

Since oxidative weathering of Cr-bearing ultramafic rocks facilitate the oxidation of Cr(III) into water-soluble Cr(VI) and back again, the chromium isotopes have been applied in the investigation of contaminated groundwater and water leachates. Water and water leachates in an appropriate amount to have approximately 1 μg of Cr_{total} were used for the determination of the Cr isotope composition following the method described by [29]. Both Cr concentrations and isotope ratios were analyzed using an IsotopX/GV IsoProbe T thermal ionization mass spectrometer (TIMS) equipped with eight Faraday cups at the University of Copenhagen, Denmark. Four Cr beams ($^{50}Cr+$, $^{52}Cr+$, $^{53}Cr+$, and $^{54}Cr+$) were analyzed simultaneously with $^{49}Ti+$, $^{51}V+$, and $^{56}Fe+$ beams, which were used to monitor interfering ions. The final isotope composition of a sample was determined as the average of the repeated analyses and reported relative to the certified SRM 979 standard as follows: $\delta^{53}Cr(‰) = [(^{53}Cr/^{52}Cr_{sample}/^{53}Cr/^{52}CrSRM_{979}) - 1] \times 1000$. The raw data were corrected for naturally and instrumentally induced isotope fractionation using the double spike routine. To assess the precision of the analyses, a double spike-treated, certified standard reference material (NIST SRM_{979}) was used.

A representative ore sample from the Nissi (Lokris) bauxite laterite deposit with remarkably high As content (350 mg/kg in bulk) was investigated by Synchrotron Radiation (SR). The SR micro-X-ray Fluorescence (μ-XRF) elemental mapping and micro-X-ray

Absorption Fine Structure (μ–XAFS) spectra were both obtained in the X-ray beamline of the Laboratory for Environmental Studies (SUL-X) of the ANKA Synchrotron Radiation Facility (Karlsruhe Institute of Technology/KIT, Karlsruhe, Germany). For this purpose, solid fragments of the ore were embedded into resin and polished, whereas powders of reference minerals and compounds were pressed with cellulose to pellets. The SR μ-XRF elemental maps and the As K-edge of the μ–XAFS spectra revealed that As is exclusively correlated to Fe, occurring as As^{5+} in the form of arsenate anions (AsO_4^{3-}) [30].

Thin polished sections of Fe–Ni laterites were investigated using a reflected light microscope, a scanning electron microscope (SEM), and energy dispersive spectroscopy (EDS). The SEM–EDS semi-quantitative analyses were carried out at the Department of Geology and Geoenvironmemt, National and Kapodistrian University of Athens (NKUA), using a JEOL JSM 5600 SEM (JEOL, Tokyo, Japan), equipped with the ISIS 300 OXFORD automated energy dispersive X-ray analysis system. Analytical conditions were 20 kV accelerating voltage, 0.5 nA beam current.

4. Mineralogical Characteristic Features

The texture characteristics of Fe–Ni $\pm$ Co ores from the Balkan Peninsula reflect a multistage evolution of the mineralogy and mineral chemistry of the mineralization. Specifically, the Lokris Fe–Ni $\pm$ Co laterite deposits are mostly composed of goethite, limonite, hematite, Ni-bearing chlorite, Fe-chlorite, rutile, quartz, calcite, and chromite, while AlOOH polymorphs boehmite and diaspore are dominant minerals in the bauxite laterite; gibbsite, illite, kaolinite, montmorillonite, smectite, takovite, and Ni–Co lithiophorite (13 wt% Ni and 9.2 wt% Co) replaced commonly by asbolane, manganomelane (6.2 wt% Ni and 4.9 wt% Co), halloysite, and Al–Ni–Co–Mn silicates are more abundant towards fractures and the lowest parts of the deposit [15,17,18,24–26]. A characteristic feature of the transitional zone between the overlying sequence of limestones and the carbonate basement of the laterite ores and organic carbon-rich mudstones is the presence of organic matter (Figure 3c,d) [24]. In addition, the lowermost part of the Lokris bauxite laterite deposit lying on karstified Jurassic limestone is characterized by the presence of abundant rare earth minerals (Figure 3e,f), such as authigenic (hydroxyl)bastnaesite-(Nd) and (La,Nd,Y)-bastnaesite, [26,29]. The Kastoria laterites, overlying a weathering crust, are composed of a pelitic matrix (goethite, hematite), clastic grains of quartz and chromite, as well as silicates (Fe–serpentine, talc), carbonates (calcite, siderite) and Mn-oxides (pyrolusite, lithiophorite), occurring as more than one type: rounded fragments of goethite occur in a subsequently formed matrix of (Fe, Mn, Ni)-hydrous oxides associated with detrital quartz and chromite, while siderite and calcite cross-cut all previous formations (Figure 3g,h). Fine-grained goethite, growing over the vestiges of microorganism cells (Figure 3i) suggests the involvement of microorganisms in the deposition of goethite [24,26].

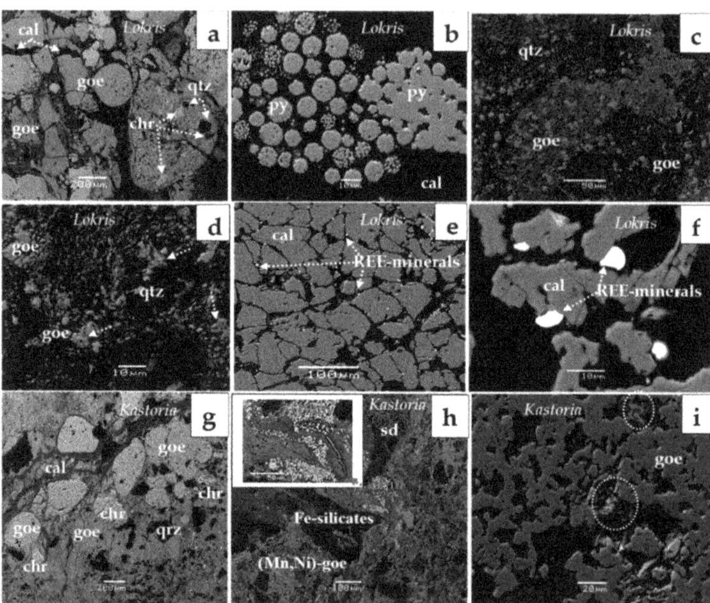

Figure 3. Selected back scattered electron (BSE) images from the Lokris and Kastoria laterites. A general overview from Lokris (**a**); the transitional zone between ore and overlying limestone showing neoformed clusters of framboidal pyrite with fine crystals at the contour of early framboidal pyrite (**b**); isolated and aggregates of bacteriomorphic goethite (goe) associated with quartz (qtz) (**c,d**) and REE-minerals (**e,f**) from the contact of the lowest part of the Nissi–Patitira deposit with the carbonate basement. Similarly, from the Kastoria Fe–Ni laterites showing different morphologies: Rounded fragments of goethite in a matrix of (Mn,Ni)-goethite with goethite fragments of an earlier stage, detrital quartz, and chromite (**g**); close-up view of morphology of fine-grained hematite crosscutting earlier mineral phases (**h**); Mn-bearing siderite cutting all previous phases (**h**) and neo–formed goethite growing over microorganism cells (**i**). Abbreviations: goe = goethite; chr = chromite; py = pyrite; qtz = quartz; cal = calcite; sd = siderite; hem = hematite; REE = Rare Earth Elements.

In addition, a salient feature of the Nissi–Patitira deposit at Lokris is the existence of transitional zones between a typical brown-red laterite ore, gradually grading to grey-white, grey-black, pinkish-white, and pale green zones (Figure 4a), as well as different goethite morphologies, with sub-micron spherulitic aggregates being the dominant micro-textures (Figure 4b–f). The occurrence of As-bearing bacteriomorphic goethite in samples with significant organic matter and the presence of C, N, Fe, Al, K, Si, Ca, Mn, and Ni in goethite, probably due to the presence of microorganisms, has been emphasized in previous studies [24,26].

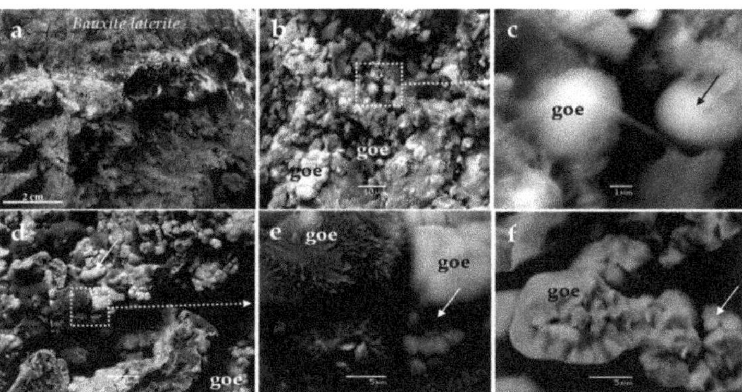

Figure 4. Field photograph (**a**) and BSE images (**b**–**f**) from unpolished parts of the Nissi–Patitira bauxite laterite ore. Highly tectonized reddish to deep red laterite gradually grading to gray, pinkish white, and pale green laterite ores along fractures and towards the carbonate basement limestones (**a**); goethite (goe) of different morphologies, with sub-micron spherulitic aggregates being the dominant microtextures (**b**–**f**); close-up view morphology of goethite (**c**,**e**) of the (**b**,**d**) corresponding, resembling bacterial cells coated by goethite (black and white arrows).

The groundmass of pisolitic and pelitomorphic ore in the Palaiochori and Katjeli deposits is fine grained, and it is mainly composed of iron oxides (goethite, magnetite, and hematite), clastic grains of chromite, quartz, Fe-chlorite (chamosite), calcite, Mn-bearing siderite, and Mn-oxides (pyrolusite) (Figure 4) [19]. Chromite grains are present in a relatively small proportion, whereas the pisolitic ore overlies the pelitomorphic ore [32]. The occurrence of fossilized organic matter and different morphologies of goethite and/or carbonate minerals (siderite) are common in the Lokris, Kastoria, Palaiochori, Katjeli, and other deposits (Figures 3 and 5). The occurrence of chromite grains, inherited from the parent ophiolitic rocks, is a common feature of all Fe–Ni laterite ores and bauxite-laterites. They usually show a cataclastic texture, cemented by magnetite, or occur as chromite cores surrounded by a zone of Fe chromite, showing a gradual decrease of Cr, Al, and Mg as Fe increases outward from the core [33,34]. However, the alteration zone along cracks and peripheral parts of chromite from the Palaiochori and Katjeli deposits is very limited, and it is commonly darker in the BSE images (Figure 5c,e,f). A small amount of apatite is also present in the ores from Edessa (Figure 5g), while the presence of organic matter is common in Palaiochori and Katjeli deposits (Figure 5a,b,d).

The Fe–Ni ore in the W. Vermion is mainly comprised of goethite, hematite, Ni-bearing chlorite, quartz, calcite, and chromite. Talc is common in the lowest part of the deposit, while illite is dominant in its upper parts. Although the majority of the W. Vermion laterites are classified as Fe–Ni–Co laterites, bauxite-laterite has been located lying on serpentinized peridotites, at the Parchari area, in a distance of less than 1 km north of the Profitis Ilias Ni laterite deposit [18]. The bauxite laterite ore is mainly comprised of goethite, hematite, boehmite, Ni–Fe chlorite, illite, quartz, calcite, chromite, rutile, and small amounts of pyrite [18]. Apart from the dominant peridotites, mafic rocks (diabases) are also present in the Parchari area.

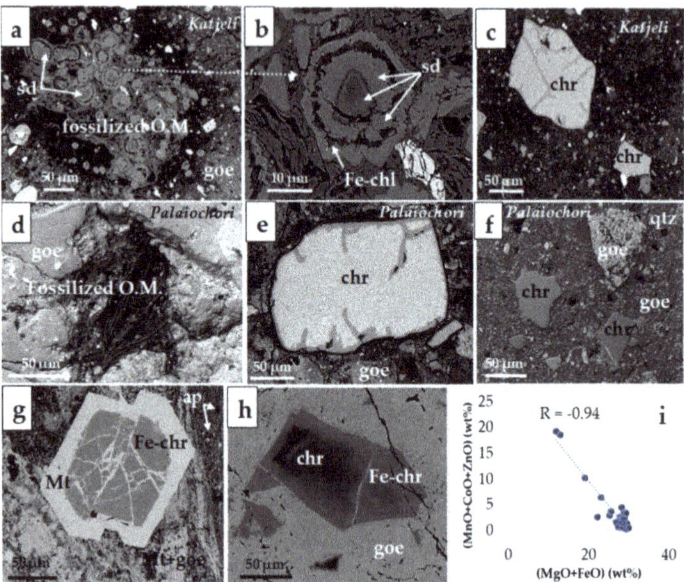

Figure 5. Selected BSE images showing fossilized organic matter and different morphological forms of goethite and/or siderite in Fe–Ni laterite ore from the Palaiochori and Katjeli deposits (**a**,**b**,**d**); very limited alteration zone along cracks and peripheral parts of chromite from the Palaiochori and Katjeli deposits (**c**,**e**,**f**); apatite in the ores from Edessa (**g**); magnetite and Fe-chromite surrounding chromite cores (**g**,**h**); a negative correlation between (MnO + CoO + ZnO) content versus (FeO + MgO) content (**i**) data from [19]. Abbreviations: ap = apatite; Fe-chl = Fe-chlorite; O.M. = organic matter; mt = magnetite; as in Figure 3.

Most Fe–Ni ± Co deposits are dominated by Fe-oxyhydroxides as the main nickel carrier, while the Mn-oxides are commonly enriched in Co and Ni, and zoned chromite grains may contain Mn, Co, and Zn as well [19]. A common feature of the E. Vermion, Olympos, Edessa, and Skyros small laterite occurrences, which are found on serpentinized harzburgites, is the gradual increase of Mn, Co, and Zn outwards of the chromite, attaining the greatest values at the periphery of chromite cores and in the Fe-chromite, reaching values up to 13.0, 4.1 and 2.1 wt%, respectively, and dropping off to negligible values in magnetite, as is exemplified by the plot of (MnO + CoO + ZnO) versus (MgO + FeO) content and the strong negative correlation between them (Figure 5i) [19].

At the area of E. Vermio, there is abundant garnet (grossular) and calcite, while Ni is mainly hosted in chlorite, serpentine, and theophrastite (Figure 6a), the latter containing 80 wt% NiO [35]. A (Co,Mn,Ni)-hydroxide with a wide compositional range occurs in a spatial association with theophrastite and organic matter (Figure 6a,b,c). Hydroxides dominated by Co, occur towards the central parts of the concentring development, while the peripheral parts are dominated by Mn (Figures 6c–f and 7, Table 1). The association of $(Co,Mn,Ni)(OH)_2$ and theophrastite with silicate minerals (mostly Ni-serpentine), garnet, and magnetite, all cross-cutting earlier deformation events (Figure 6), may have a common origin in space. The theophrastite coexists with magnetite and frequently encloses tiny grains of magnetite, both showing an orientation parallel to the general direction of the schistosity (Figure 6a).

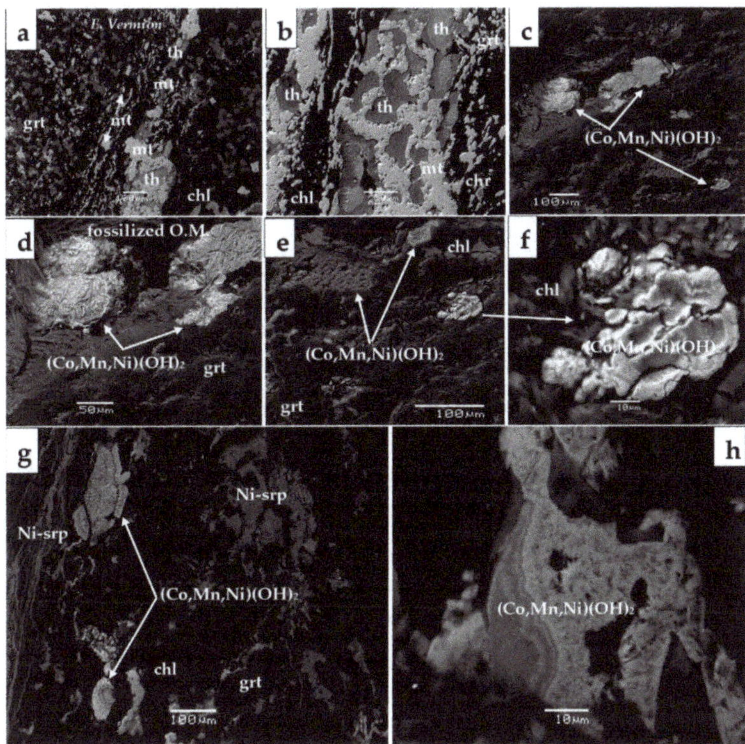

Figure 6. BSE images of metamorphosed Fe–Ni laterites from E. Vermion. Theophrastite associated with magnetite and garnet, in the form of crosscutting veinlets (**a**); the (Co–Mn–Ni)(OH)$_2$ associated with organic matter (O.M.) in a matrix dominated by garnet, well observed in unpolished parts of the studied sections (**b**,**c**); close-up view morphology of (Co–Mn–Ni)(OH)$_2$ (**d**–**h**), showing successive thin layers, composed by fine fibrous crystals (**h**). Abbreviations O.M. = organic matter; mt = magnetite; grt = garnet; chl = chlorite; Ni-srp = Ni-serpentine.

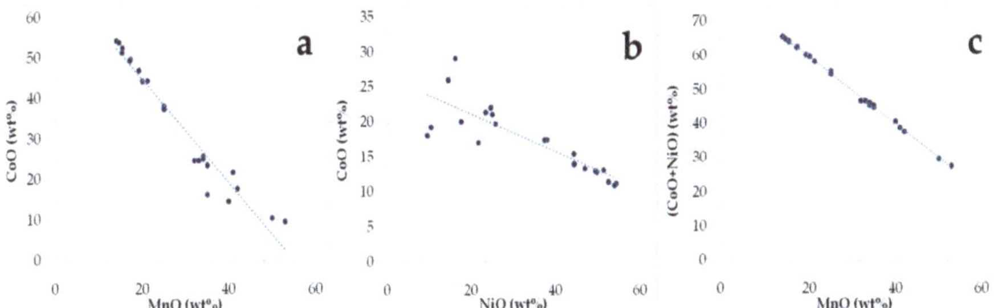

Figure 7. Plots CoO versus MnO, NiO, (**a**,**b**) and (CoO + NiO) versus MnO (**c**). Data from Table 1.

Table 1. Representative SEM–EDS analyses of (Co–Mn–Ni)-hydroxides from metamorphosed Fe–Ni laterites from E. Vermion.

Content in wt%									
No.	MnO	CoO	NiO	Total	No.	MnO	CoO	NiO	Total
V.1	25.0	38.0	17.4	80.4	V.12	34.0	25.7	19.6	79.3
V.2	40.0	14.5	26.0	80.5	V.13	14.6	53.9	10.9	79.4
V.3	14.0	54.3	11.2	79.5	V.14	20.0	44.3	15.4	79.7
V.4	15.4	52.5	11.4	79.3	V.15	35.0	16.2	29.1	80.3
V.5	15.3	51.3	13.1	79.7	V.16	32.0	24.6	22.0	78.6
V.6	17.3	49.7	12.8	79.8	V.17	33.0	24.6	22.0	79.6
V.7	17.1	49.4	12.9	79.4	V.18	34.0	25.0	21.0	80.0
V.8	19.2	46.9	13.3	79.4	V.19	53.0	9.5	18.0	80.5
V.9	21.2	44.4	13.9	79.5	V.20	42.0	17.6	20.0	79.6
V.10	25.0	37.3	17.3	79.6	V.21	41.0	21.7	17.0	79.7
V.11	35.0	23.4	21.3	79.7	V.22	50.0	10.4	19.2	79.6

4.1. Chemical Composition of Fe–Ni ± Co Laterite Ores

Detailed studies on major Fe–Ni laterite deposits of Greece, in the frame of the European GeoNickel project, reported several typical bulk compositions for major and trace elements [36]. Representative analyses of various Fe–Ni laterite deposits and occurrences from Greece, as well as Albania, Serbia, and N. Macedonia (former F.Y.R.O.M.), completed for some critical element contents are shown in the Tables 2–5. Cobalt contents in laterite samples from Fe–Ni laterite deposits of Greece (Lokris, Kastoria, Palaiochori, W, Vermion, Olympos, Edessa, E. Vermion, and Skyros), Albania (Bitincka, Guri Pergjegjum, and Katjeli), Serbia (Topola), and N. Macedonia, former F.Y.R.O.M. (Rzanovo), range from 180 to 1600 mg/kg, reaching values over 3100 mg/kg Co at the lowest contact of the Lokris bauxite laterite deposit with the carbonate basement (Table 2). In addition, it is clear that the REE, U, and Th are mostly associated with bauxite-laterites and to a lesser degree with Fe–Ni laterites of karstic-type, whereas in Fe–Ni laterites with in situ features, such as the Kastoria and Bitincka, the ΣREE content is <10 mg/kg, which appears to be consistent with their mineralogical composition (Figure 3e,f). In particular, the lowest part of the deposit lying on karstified Jurassic limestone is accompanied by an enrichment in ΣREE, reaching values over 6000 mg/kg ΣREE, 66 mg/kg U, 25 mg/kg Th, and 1800 mg/kg As (Table 2). In general, the Co grades in the Balkan Peninsula range between 0.02 and 0.16% wt% Co and show a positive correlation with Ni and Mn (Figure 8). The majority of the Sc contents of lanthanides were 32–90 mg/kg in the Lokris deposits; 41–77 mg/kg Sc in the Kastoria, Palaiochori, and W. Vermion deposits; 30–82 mg/kg Sc in the Olympos, Edessa, E. Vermion, and Skyros Fe–Ni laterite occurrences; and 12–96 mg/kg Sc in the Albania, Serbia, and N. Macedonia, former F.Y.R.O.M., deposits (Tables 2–4). Increasing Fe contents are accompanied by elevated Sc, the best positive correlation between Sc and Fe being evidenced in the Lokris deposit and W. Vermion (Tables 2–5 and Figure 9).

Table 2. Major and trace element contents in representative Fe–Ni ± Co laterite and bauxite laterite samples from the Lokris and Evia deposits.

	LOKRIS								LOKRIS						LOKRIS					EVIA	
	Profile from top to underlying base of carbonates								Bauxitic laterite						Fe–Ni laterite					Fe–Ni laterite	
wt%	N-3H	N-3D	N-3C	N-3Ba	N-3Bb	N-2B	N-1Ba	N-1Bb	LAR4	N-18	N-20	N-22	N-32	N-36	N-11	N-14	N-15	N16	N-17	Ev.1	Ev.2
SiO_2	1.85	24.2	23.1	11.5	19.6	15.9	19.5	20.2	11.6	5.1	7.4	5.8	5.6	20.9	34.0	24.1	23.8	31.2	19.5	26.8	15.9
Al_2O_3	44.1	20.3	18.8	43.8	22.4	16.1	23.6	24,1	36.4	33.6	40.5	29.9	31.2	25.2	8.6	6.7	16.9	7.1	6.25	5.8	7.1
Fe_2O_3	40.6	35.3	34.6	1.6	21.1	39.8	19.1	20.1	30.1	46.1	34.8	50.6	47.1	35.9	42.3	53.9	33.6	42.8	61.8	45.1	57.3
TiO_2	2.7	1.1	1.25	0.03	0.55	0.35	0.4	0.38	1.9	2.25	2.6	1.95	1.95	1.25	0.4	0.3	1.2	0.37	0.3	0.13	0.12
CaO	0.15	0.2	0.41	3.3	1.6	0.95	0.55	0.6	2.8	0.6	0.15	0.1	0.2	0.3	0.35	0.5	3.1	2.2	0.3	1.4	0.5
MgO	0.25	3.6	3.4	0.25	2.14	1.25	1.2	1.6	2.4	0.7	0.85	0.95	1.2	3.7	3.4	2.8	3.8	4.4	2.3	0.03	0.1
MnO	0.17	0.3	0.1	0.8	0.4	0.55	0.75	0.69	0.2	0.17	0.12	0.15	0.25	0.22	0.4	0.2	0.18	0.3	0.32	0.12	0.4
mg/kg																					
Ni	800	18,000	16,000	35,000	28,000	21,000	30,000	35,000	25,000	2700	2700	4600	4300	9700	4700	5500	22,000	5900	4400	8100	12,000
Cr	4900	11,000	13,000	340	5700	3500	3800	3900	9100	7000	8700	8300	5300	4400	19,000	16,000	10,000	23,000	17,000	20,000	22,000
Co	57	2200	430	3100	1400	1400	2200	2200	580	280	230	300	560	1400	510	470	750	520	410	590	830
As	230	140	200	12	790	1800	680	130	140	240	130	250	300	2600	2.0	2.0	6.0	12	5.0	2.0	2.0
Sc	90	58	49	13	34	32	45	44	45	74	69	79	76	68	46	56	47	41	55	60	67
La	56	144	172	2510	1570	2740	2420	2370	73	115	72	79	176	129	15	9.0	39	8.0	19	4.0	5.0
Ce	207	126	84	401	242	292	351	348	107	164	132	128	313	1030	50	25	72	44	67	57	40
Nd	<10	110	160	2040	1460	2740	2250	2690	50	70	60	50	80	140	10	<10	30	<10	10	<10	<10
Sm	3.6	24	34	401	233	404	391	385	9.9	12	11	9.7	11	30	2.7	1.2	5.6	1	1.8	0.9	0.8
Eu	1.2	6.5	9.2	89	54	91	97	95	2.4	3.6	3.9	4.1	2.8	8.7	0.7	0.3	1.2	0.5	0.5	0.4	0.3
Tb	1.0	4.0	4.9	33	30	38	35	32	0.9	1.8	1.8	1.0	2.1	4.5	<0.5	<0.5	<0.5	0.6	<0.5	<0.5	<0.5
Yb	3.7	20.5	14.1	27	24	27	33	33	5.5	8.8	8.7	6.4	9.7	21	1.8	0.9	3.6	0.7	1.3	0.5	0.5
Lu	0.86	3.3	2.2	4.6	4.2	4.9	4.1	4.9	0.9	1.6	1.6	1.1	1.6	2.9	0.3	0.18	0.6	0.11	0.19	0.07	0.09
Th	25	12	<10	<10	<10	<10	<10	<10	9.5	28	31	21	15	10	2.2	1.1	7.0	1.0	2.2	1.7	1.2
U	6.9	13	18	66	41	46	58	58.4	5.6	15	14	17	12	7	<0.5	<0.5	4.2	<0.5	2.3	<0.5	<0.5
ΣREE	273	438	480	5506	3617	6337	5581	5958	265	377	291	279	596	1366	81	37	152	55	100	63	47
µg/kg																					
Pt	<10	22	23	<10	<10	<10	<10	<10	<10	<10	<10	<10	<10	12	17	20	15	86	<10	19	13
Pd	1	19	9	5	5	5	3	3	<1	<1	<1	<1	<1	1	3	4	2	6	1	17	1
Au	18	14	36	<5	<5	<5	<5	<5	<5	<5	<5	<5	25	<5	<5	<5	<5	<5	<5	<5	5

Table 3. Major and trace element contents in representative Fe–Ni $\pm$ Co laterite and bauxite laterite samples from deposits of northern Greece.

| | KASTORIA | | | | | | PALAIOCHORI | | | | W. VERMION | | | | | | | | | | |
| | Goethite zone | | Pisolitic zone | | | | Fe–Ni laterite | | Saprolite | | Fe–Ni laterite | | | | | Saprolite | | Bauxitic laterite | | | |
wt%	Ka-4a	Ka-4b	Ka-5	Ka-6	Ka-7	Ka-8	P.G.2	P.G.3	PR1	PR2	PR5	PR6	WV4	WV5	WV6	U.R.	PV1	PV2	PV3	PW4	PW5
SiO_2	36.6	22.0	3.8	3.8	4.5	4.05	12.1	17.8	45	30	9.0	16.5	38	39	15.2	37	17.5	14.5	13	14	13
Al_2O_3	1.2	2.2	3	5.1	1.8	3.4	4.7	5.4	0.5	0.2	3.34	13	3.6	1.8	12.5	1.0	29	32	31.9	32	37.4
Fe_2O_3	44.1	54.8	77.6	77.6	62.3	78.2	67.5	62.6	18.6	7	65.2	78	42.8	31.6	58.9	11	36	36.8	32.4	39.6	33.5
CaO	5.25	0.56	0.2	0.3	0.6	0.1	3.6	5.4	0.3	18	0.5	0.4	0.3	0.7	0.38	0.5	0.8	0.8	0.6	0.9	0.9
TiO_2	0.01	0.02	0.03	0.03	0.03	0.1	0.18	0.21	0.1	0.1	0.2	0.4	0.1	0.1	0.11	0.1	1.28	1.7	1.65	1.69	1.15
MgO	1.1	3.25	0.6	0.55	0.8	0.6	2.86	5.11	25.9	23.2	1.7	1.9	7.4	20	4.65	37	3.6	2.4	2.1	2.7	2.1
MnO	0.95	0.88	0.25	1.7	0.79	0.3	0.25	0.42	0.24	0.11	0.97	0.61	0.41	0.33	0.62	0.1	0.37	0.41	0.24	0.36	0.26
P_2O_5	—	—	—	—	—	—	—	—	0.01	0.01	0.01	0.1	0.05	0.01	0.01	0.06	0.48	0.31	0.27	0.32	0.2
mg/kg																					
Ni	8500	10,500	8000	9100	6800	4400	9000	17,800	5500	2200	15,000	6900	10,200	9500	7800	3100	5200	5200	6400	6400	4200
Cr	16,000	23,000	20,000	15,000	23,000	17,000	13,000	10,500	5800	2500	22,000	12,900	11,500	11,400	11,000	3800	5500	4400	3800	4100	3600
Co	1100	1200	330	820	420	330	800	790	300	200	1300	700	500	400	82O	340	400	320	400	400	400
Sc	23	41	66	65	55	42	77	75	2.0	5.0	68	60	50	56	58	6	29	64	68	65	70
As	<2	<2	3	15	14	13	4	3	<2	<2	7	6	10	8	6	<2	7	11	12	8	10
La	<1	<1	<1	<1	3.0	<1	19	10	2.0	3.0	18	29	19	15	170	2.0	103	115	114	117	212
Ce	<3	<3	<3	<3	<3	<3	32	3	3	4	31	62	38	20	15	4	140	145	145	150	230
Nd	<10	<10	<10	<10	<10	<10	<10	<10	<10	<10	10	20	10	10	38	<10	33	35	36	33	75
Sm	<0.5	<0.5	<0.5	<0.5	<10	<0.5	1.9	<0.5	<0.5	11	3.8	0.4	2.9	0.9	20	<0.5	12	14	17	18	29
Eu	0.4	0.5	0.3	0.3	0.3	<0.2	0.8	<0.2	<0.2	0.3	0.2	2.1	0.8	1.6	6.5	<0.2	1.8	2.2	2.8	3.4	4.7
Tb	<0.5	<0.5	<0.5	<0.5	<0.5	<0.5	<0.5	<0.5	<0.5	<0.5	<0.5	0	<0.5	<0.5	1.4	0.7	<0.5	<0.5	<0.5	2.9	6.3
Yb	<0.2	<0.2	<0.2	<0.2	<0.5	<0.2	1.5	<0.2	<0.5	0.6	2.5	4.2	3.3	2.4	1.4	<0.5	6	7	8	9	16
Lu	<0.05	<0.05	<0.05	<0.05	<0.05	<0.05	0.15	<0.05	<0.05	0.1	0.45	0.65	0.37	0.45	5.5	<0.05	1.1	1.3	1.5	1.6	2.5
Th	<1	<1	<1	<1	<1	<1	6.3	<1	0.7	0.8	3.1	5.6	3.1	1.9	1.4	<0.5	19	20	17	18	28
U	<1	<1	<1	<1	1.3	<1	<1	<1	0.9	0.9	3.6	4.8	1.6	1.6	3.2	<0.5	4.2	4.0	3.9	4.2	5.0
µg/kg																					
Pt	12	12	15	88	40	10	25	25	11	<10	82	56	34	28	39	<10	25	26	54	48	29
Pd	3	6	<1	<1	<1	<1	<1	6	5	<1	45	38	8	8	186	<1	5	4	10	7	5
Au	<5	<5	<5	<5	9	<5	<5	<5	<5	<5	5	5	4	5	17	<5	5	5	15	16	7

Table 4. Major and trace element contents from metamorphosed laterite occurrences of northern Greece.

	OLYMPOS			E. VERMION			EDESSA			SKYROS	
wt%	OL.1	Ol.3	OL.5x	A.K.10	AL2	ST.B	Ed.P10	Ed.P.11	Ed.P.3	Sk.3	Sk.4
SiO_2	4.2	3.9	7.0	22.3	10.1	15.1	17.5	22.7	14.2	11	11.1
Al_2O_3	3.4	2.85	5.8	5.5	11.1	7.5	8.65	9.5	1.4	7.4	9.4
Fe_2O_3	58.34	59.7	65.64	44.4	67.8	54.8	59.5	52.2	67.65	63.52	59.63
TiO_2	0.07	0.09	0.06	1.28	1.8	0.15	0.27	0.31	0.2	0.21	0.27
CaO	0.07	0.05	0.05	7.0	0.05	1.9	0.46	2.4	0.1	<1	<1
MgO	3.4	2.87	4.35	3.6	1.05	11.5	3.25	3.5	7.05	2.4	2.05
MnO	1.05	0.77	0.27	0.37	0.05	0.49	0.58	0.38	1.1	0.36	0.16
P_2O_5	0.21	0.15	0.17	0.31	0.26	0.28	0.45	0.38	0.41	–	–
mg/kg											
Ni	8600	9300	16,000	5200	1600	8900	9500	6300	11,000	9800	10,000
Cr	91,000	82,000	20,000	10,000	25,000	17,000	9600	7400	39,000	22,000	23,000
Co	300	320	1200	400	180	1100	800	430	1200	360	320
Zn	340	430	850	500	89	500	60	70	290	60	110
Sc	31	30	28	57	67	59	82	80	30	64	61
As	7.0	8.0	19	57	16	22	4.0	2.0	17	2.0	24
La	1.0	1.0	10	11	44	18	75	40	13	6.0	8.0
Ce	3.0	3.0	19	32	95	28	70	76	40	14	23
Nd	<10	<10	<10	10	40	10	49	30	10	<10	<10
Sm	<0.5	<0.5	1.5	2.9	8.2	4.4	7.8	5.5	4	1.1	1.2
Eu	0.6	0.9	0.6	0.5	3.0	0.8	0.3	1.5	0.5	<0.2	0.6
Tb	<0.5	<0.5	<0.5	0.6	1.1	3.5	0.5	0.6	0.6	<0.5	<0.5
Yp	1	0.5	0.9	1.3	3.7	9.0	1.8	1.6	0.6	0.9	1.1
Lu	0.19	0.05	0.05	0.18	0.6	1.22	0.22	0.23	0.05	0.16	0.15
Th	0.6	<0.5	1.0	19	20	<0.5	4.0	3.7	2.6	3.6	2.8
U	0.5	2.3	0.6	4.2	3.9	2.8	2.1	2.6	1.8	1.0	1.0
µg/kg											
Pt	<10	59	38	58	33	10	10	10	48	<10	48
Pd	1.0	2.0	2.0	9.0	1.0	6.0	26	23	22	25	22
Au	2.0	8.0	19	11	13	20	3.0	3.0	15	5.0	15

Table 5. Major and trace element contents in representative Fe–Ni ± Co laterite deposits from Albania, Serbia, and N. Macedonia (former F.Y.R.O.M.).

	ALBANIA									SERBIA				N. MACEDONIA	
	Bitincka					Gouri–Perjuegjiun				Katjeli	Katjeli	Topola	Topola	former F.Y.R.O.M.,	
	Ser. Hartz.	Fe–Ni ore				Ser. Hartz.	Fe–Ni ore			Fe–Ni ore		Fe–Ni ore		Rzanovo	Rzanovo
	Bi.52	Bi.46	Bi.45	Bi.44	Bi.43	G.P.42	G.P.39	G.P.40	G.P.41	K.01	K.02	Yu3	Yu4	RZ1	RZ2
wt%															
SiO_2	37.8	17	5.9	7.9	6.0	37.5	3.5	4.0	4.0	28.44	26.32	19.5	20.8	29.6	12.8
$A_{12}O_3$	0.7	2.25	6.1	6.75	6.5	0.15	2.75	4.0	2.9	7.15	6.05	5.4	5.2	5.4	4.4
CaO	1.65	2.15	0.3	0.35	0.4	0.2	0.55	0.15	0.15	3.57	8.92	0.05	0.1	0.4	0.3
Fe_2O_3	7.8	56.6	71	66.5	69.8	7.6	80	77.3	81.4	44.62	42.66	51.77	61.65	41.3	58.9
TiO_2	0.01	0.03	0.08	0.05	0.05	0.01	0.15	0.1	0.13	0.18	0.21	0.1	0.3	0.4	0.15
MgO	34.1	5.56	1.15	1.45	1.35	37.2	1.0	0.7	0.75	2.08	1.27	5.7	3.8	8.8	10.2
MnO	0.1	0.19	0.17	0.81	0.27	0.1	0.36	0.26	0.29	0.45	0.27	0.05	0.1	0.2	0.45
mg/kg															
Ni	2000	16,000	7200	13,000	9800	2000	13,000	8600	6600	11,000	9100	5500	9200	5100	8200
Cr	3100	26,000	17,000	19,000	21,000	2500	22,000	35,000	26,000	22,500	20,000	21,000	17,000	25,000	15,000
Co	120	960	650	800	410	100	570	550	390	600	600	250	250	260	990
Zn	70	650	310	680	410	100	510	400	180	370	350	330	90	150	430
Sc	12	14	52	70	89	8.0	71	96	68	41	31	60	57	30	56
As	2.0	11	10	5.0	3.0	<2	32	31	25	4.0	5.0	12	2.0	3.0	2.0
La	<1	5.0	3.0	1.0	1.0	<1	5.0	1.0	1.0	9.0	6.0	6.0	4.0	6.0	3.0
Ce	<3	28	27	<3	<3	<3	53	<3	<3	7.0	9.0	9.0	5.0	12	4.0
Nd	<10	<10	<10	<10	<10	<10	<10	<10	<10	<10	<10	<10	<10	<10	<10
Sm	<0.5	<0.5	0.6	<0.5	<0.5	<5	1.2	<0.5	<0.5	1.5	1.1	0.6	0.7	1.1	0.5
Eu	0.3	0.3	0.4	<0.2	<0.2	<2	0.6	0.6	1.0	<0.2	<0.2	<0.2	<0.2	0.3	<0.2
Tb	<0.5	<0.5	<0.5	<0.5	<0.5	<2	0.7	0.3	1.0	0.9	0.6	<0.5	<0.5	<0.5	<0.5
Yb	<0.2	<0.2	<0.2	<0.2	<0.2	<0.2	0.7	0.3	<0.2	0.9	0.6	0.5	0.6	0.3	0.3
Lu	<0.05	<0.05	0.13	<0.05	<0.05	<0.05	0.11	0.09	0.05	0.08	0.12	0.06	0.07	0.1	0.08
Hf	<1	<1	<1	<1	<1	<1	<1	<1	<1	<1	<1	<1	<1	1	0.5
Th	<0.5	<0.5	<0.5	1.9	<0.5	<0.5	1.6	<0.5	<0.5	2.4	2.3	0.7	0.6	0.9	<0.5
U	<0.5	4.1	5.0	<0.5	<0.5	<0.5	1.6	4.7	<0.5	<0.5	<0.5	<0.5	0.8	0.5	<0.5
µg/kg															
Pt	<10	26	31	41	27	<10	28	41	40	16	31	76	81	16	19
Pd	3.0	14	8.0	16	12	<1	<1	<1	<1	9.0	13	25.0	23	12	10
Au	<5	<5	<5	<5	<5	11	<5	<5	<5	7.0	6.0	10	10	<5	5.0

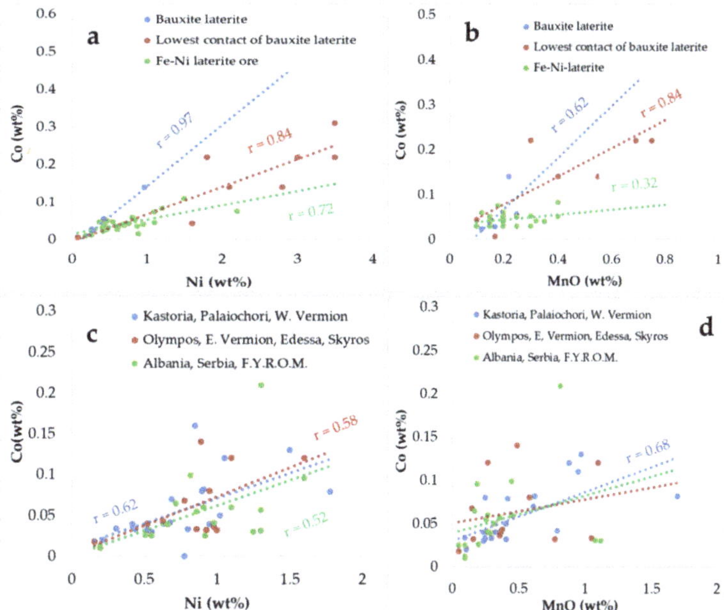

Figure 8. Plots of Co versus Ni and MnO from the Lokris (**a**,**b**), and other Fe–Ni laterite deposits from Greece, Albania, Serbia, and N. Macedonia, former F.Y.R.O.M. (**c**,**d**), revealed the pronounced positive correlation between Co and Ni, Mn at the lowest parts of the Lokris deposit, near the contact with the carbonate basement. Data from Tables 2–5.

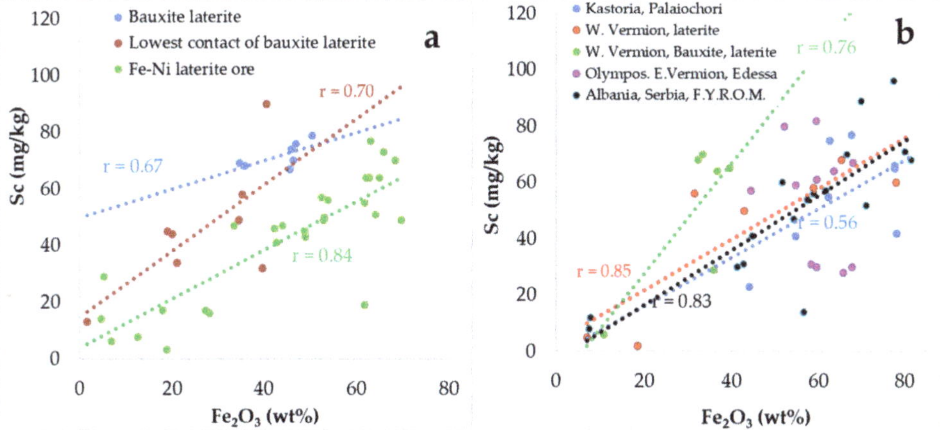

Figure 9. Plots of Sc versus Fe_2O_3 for Fe–Ni laterites and bauxite laterites from the Balkan Peninsula. The best positive correlation between Sc and Fe is observed in the Fe–Ni laterite deposits of Lokris (**a**) and W. Vermion (**b**). Data from Tables 2–5.

Assuming that values >1 and <1 are called positive and negative anomalies, respectively, chondrite normalized REE patterns show in general negative slopes from La–Eu and slight positive slopes from Tb–Lu for the Fe–Ni laterites and bauxite laterite samples from the Lokris and Evia (Figure 10a) and W. Vermion (Figure 10b), except the samples from the lowest part of the Lokris deposit, showing a negative Ce anomaly (Figure 10a). In addition, a positive Nd anomaly in the later samples is observed that seems to be consistent with the occurrence of (hydroxyl)bastnaesite-(Nd) and (La,Nd,Y)-bastnaesite (Figure 3e,f) [26,29,30]. Laterites from W. Vermion are characterized by positive Ce and Eu anomalies. In general, negative slopes are a common feature of the chondrite normalized REE patterns of laterites from northern Greece (Figure 10c), Albania, Serbia, and N. Macedonia, former F.Y.R.O.M. (Figure 10d), with varying positive and negative anomalies.

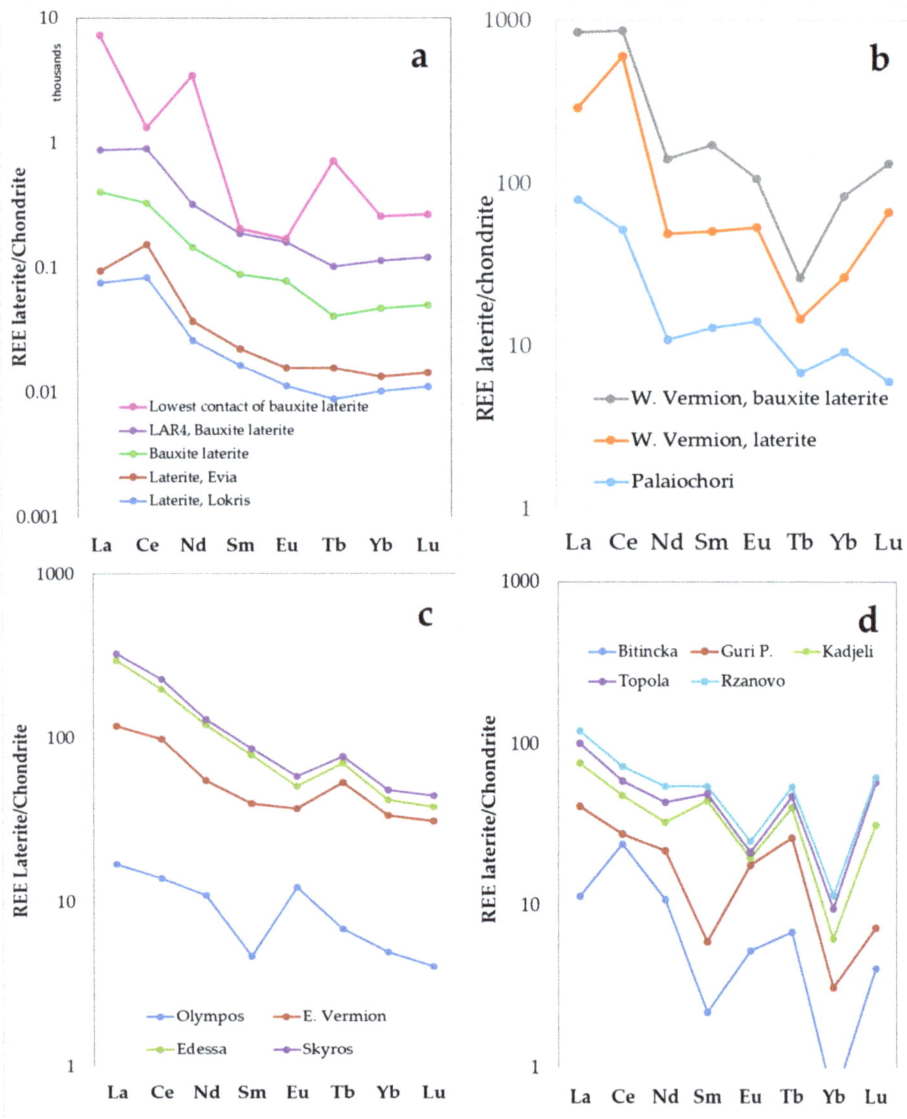

Figure 10. Chondrite normalized REE-patterns for average values of samples from Fe–Ni laterites and bauxite laterites, from the Lokris and Evia (**a**), Palaiochori and W. Vermion (**b**), Olympos, E. Vermion, Edessa, and Skyros (**c**) and Albania, Serbia and N. Macedonia (former F.Y.R.O.M.) (**d**). Data from Tables 2–5.

In general, the PGE content, reaching up to 88 µg/kg Pt and 45 µg/kg Pd (up to 186 µg/kg Pd in one sample), is higher in Fe–Ni ± Co laterites than in bauxite laterites (Tables 2–5 and Figure 11d–f) and bauxites [37]. A common feature of the studied deposits from the Balkan Peninsula, including two Fe–Ni laterite profiles from the Gouri-Perjuegjiun and Bitincka deposits of Albania, the Kastoria, and W. Vermion laterites, is the higher Pt than Pd contents in ores compared to parent peridotites, their concentration mostly occuring in the pelitomorphic and pisolitic–oolitic types (Tables 2–5) [16,18].

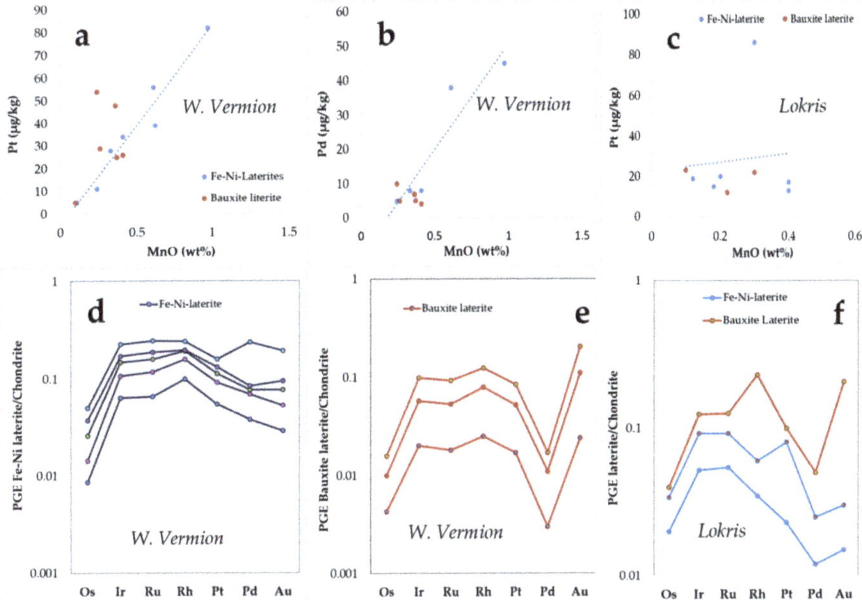

Figure 11. Plots of Pt and Pd versus MnO (**a–c**) and chondrite normalized PGE patterns for Fe–Ni laterites and bauxite from W. Vermion and Lokris (**d–f**). Data from Tables 2–5.

Any relationship between Sc and Pt or Pd is not clear, but there is a positive trend in the laterite samples from W. Vermion (Figure 12a,b), whereas both Pt and Pd exhibit a negative trend with the ΣREE (Figure 12b). In contrast, there is a positive trend between the ΣREE and Th, U, and P_2O_5 contents (Figure 12c–e), as well as between Sc–P_2O_5 contents (Figure 12f).

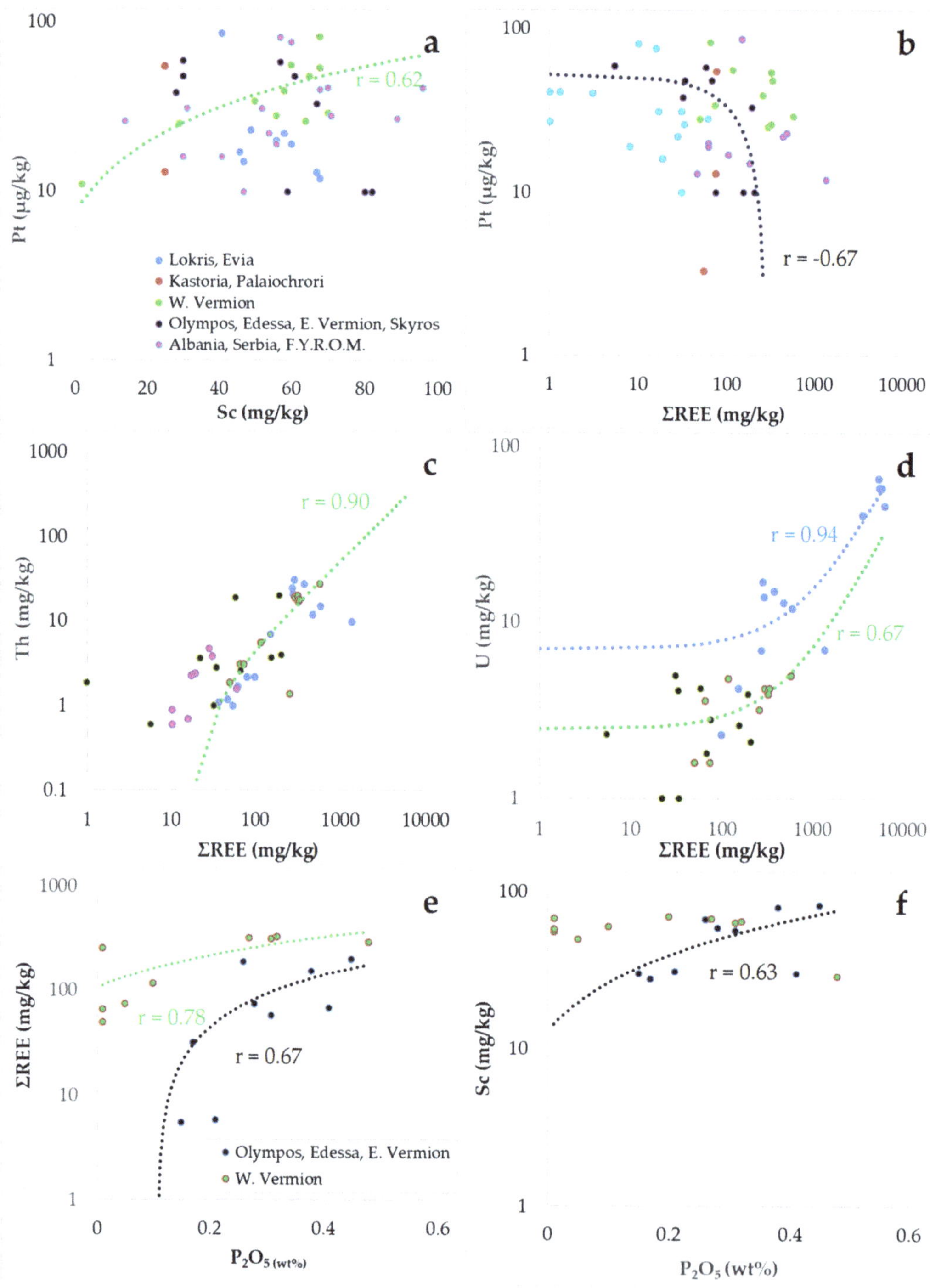

Figure 12. Plots of Pt versus Sc and ΣREE (**a**,**b**); Th and U versus ΣREE (**c**,**d**), and ΣREE, Sc versus P_2O_5 (**e**,**f**). Data from Tables 2–5.

4.2. Environmental Impact from Mining and Smelting of Fe–Ni Laterites

Integrated approaches to the soil–groundwater–plant/crops compositional variations related to the weathering of rocks/ores, mining, and smelting of laterites and the preliminary results of leaching experiments of Fe–Ni laterite have been presented in previous publications and will be discussed. Furthermore, BSE images from the unpolished parts of the Nissi (Lokris) bauxite laterite ore are given in the present study (Figure 4) in order to discuss the role of organic matter into heavy metal biomobilization and biomineralization in soil and the contamination of groundwater affected by rock/ore weathering and mining of laterite deposits.

4.2.1. Metallurgical Residues (Slag)

The result of the reductive smelting of the laterite is the formation of two separate phases, the metallic Fe–Ni and the slag residue. Bulk chemical analyses of representative samples of pyro-metallurgical residue (slag96 and slag14 samples) from the Larymna plant, indicated a significant Cr content (Table 6).

Table 6. Chemical composition of slag (smelting residue) from Fe–Ni–Co laterites, Larymna plant, Lokris.

mg/kg	Slag96	Slag14	Detection limit	mg/kg	Slag96	Slag14	Detection limit
Cu	16	15	0.5	Yb	1.0	1.1	0.5
Pb	5.0	6.2	0.5	Lu	0.5	0.5	0.5
Zn	180	92	5	Nb	4.3	4.1	0.5
Ni	2400	930	0.5	Y	16	15	0.5
Co	210	110	1	Hf	1.3	1.4	0.5
Mn	2700	2820	5	µg/kg			
Cr	19,000	20,600	1	Pt	<10	<10	10
As	12	6	5	Pd	<1	<1	1
U	3.1	2.6	0.5	Au	<5	<5	5
Th	5.8	5.3	0.5	wt%			
V	250	260	10	Si	15.6	14.5	0.01
Zr	39	43	0.5	Fe	29.60	27.88	0.01
Sc	51	50	1	Al	2.81	3.97	0.01
La	20	16	0.5	Ti	0.07	0.17	0.001
Ce	39	27	5	Mg	3.70	5.15	0.01
Nd	10	10	10	Ca	2.58	3.97	0.01
Sm	2.3	3.3	0.5	Na	0.08	0.07	0.01
Eu	1.0	1.3	0.05	K	0.30	0.29	0.01
Tb	<0.5	<0.5	0.5	S	0.16	0.17	0.05

However, preliminary results have shown that the Cr(VI) concentrations in water leachates from the metallurgical residue (slag) of Fe–Ni laterite samples from Lokris were very low (below 1 µg/L) [31].

4.2.2. Concentrations of Cr(VI) in Water Leachates for Fe–Ni Laterites and Bauxite Laterites

Preliminary results of the water leaching experiments for Fe–Ni laterite ores have shown very low Cr(VI) concentrations in the samples from Lokris and up to 1200 µg/L Cr(VI) in samples from the Kastoria deposit [31]. Further leaching experiments carried out in the present study for Fe–Ni laterite deposits from the Balkan Peninsula showed significant Cr(VI) concentrations in certain laterites from W. Vermion, the Bitincka, and Guri– Pergjegjum (Albania) deposits and bauxite laterite samples from the lowest parts of the Nissi deposit at Lokris (Table 7).

Table 7. Concentrations of Cr(VI) in water leachates for Fe–Ni laterites and bauxite laterites from the Balkan Peninsula. Present study and [31]. Symbol * = samples analyzed in a previous study.

Location	Sample I.D.	Cr(VI) (µg/L)	Location	Sample I.D.	Cr(VI) (µg/L)
Kastoria			Guri-Pergjegium		
Pisolitic z.	Ka-7 *	750	Pisolitic z.	G.P.40	40
	Ka-6	460		G.P.41	50
	Ka-5	500			
Goethite z.	Ka-4 *	1200	Goethite z.	G.P.39	60
W. Vermion			Lokris		
Pisolitic z.	W.V5	7	Laterite	LT-18 *	1.8
	W.V.6	26		LN-12 *	1.3
	W.V.P2	12		LN-15 *	0.6
Goethite z.	W.V.4	<4		LN-1 *	0.7
Bitincka			Bauxite laterite	LAR.4	140
Pisolitic z.	Bi43	27		N3A	10
	Bi44	90		N3B	21
	Bi45	17		N-9A	200
				N-9B	60
Goethite z.	Bi47	4		N-P1	24
				N-P2	66

5. Discussion

The development of critical elements' production from local sources is considered to be crucial in view of the future needs of industry in the EU [1]. The investigation of critical metals in the frame of research projects, including the ProMine databases and EURARE provide geological information and mineral resources for mineral deposits, mines, and mining wastes and their assessment in relation to the importance of laterites for the EU economy [1–8,27]. Although the production of Co in Europe is derived from magmatic sulfide deposits, volcanogenic massive sulfides (VMS), hydrothermal deposits, and the black shale-hosted deposit in Finland and Sweden, the Fe–Ni–Co laterite deposits may be a potential resource for Co as well [6,9,11].

5.1. Cobalt Potential in Fe–Ni Laterites and Genesis of Co-Bearing Minerals

In hydrous Mg silicate type of deposits Ni and Co are higher in the lower part of the saprolite, where nickel is hosted by Ni-bearing layer silicates, often referred to as garnierite and Al–Ni–Co–Mn silicates [23,38]. The best-known example of this sub-type is the Goro deposit in New Caledonia, which has a resource of 323 million tons, grading 0.11% Co and 1.48% Ni [11]. In the clay silicate sub-type of laterites, Ni and Co are enriched in clay minerals such as saponite and smectite in the mid to transition zone between the saprolite and the limonite zone. Murrin Murrin in Western Australia is a good example of a large Ni–Co deposit of this type. It has a combined resource of 231 million tons, grading 1.01% Ni and 0.08% Co [39]. However, in the Balkan Peninsula and Turkey, Co grades and tonnages are low compared to other deposits in New Caledonia, Australia, and Cuba [9].

The Fe–Ni laterite deposits in Greece account for about 80% of the nickel production in the the Balkan Peninsula [40], while the Co reserves reported for the Fe–Ni–Co laterite deposits in Greece contain almost 50,000 tons, including the Lokris, Evia, and Kastoria deposits [22,41]. The cobalt contents in Fe–Ni laterite deposits of the Balkan Peninsula rane from 180 to 1600 mg/kg Co and reach values over 3100 mg/kg Co at the lowest contact of the Lokris bauxite laterite deposit with the carbonate basement, which also yield the highest (0.8 wt%) Mn content (Table 2). In the Kastoria laterite deposit and at the lowermost parts of the Lokris Fe–Ni laterite and bauxite laterite deposits, the highest Co, Mn, and Ni contents (Table 2 and Figure 7) are mostly hosted in Ni–Co lithiophorite replaced commonly by asbolane and manganomelane, which are more abundant towards fractures [15,17,18,24,29].

The presence of fragmented (Co, Mn, Ni)-hydroxides cemented by secondary silicates at E. Vermion and the replacement of serpentine by Ni serpentine (Figure 6) suggest that the Ni enrichment event post-dated that of the (Co, Ni, Mn)(OH)$_2$ fragmentation. There is a wide compositional variation in the (Co, Mn, Ni)(OH)$_2$ and an excellent negative correlation between MnO and the sum (CoO + NiO) (Table 1; Figure 7) with the highest Co content (54 wt% CoO) of an unknown Co-hydroxide occurring towards the central parts of the concentric aggregates (Figure 6). Those (Co, Mn, Ni)-hydroxides appear clearly post-dated the transformation (schistosity) of main Fe–Ni laterite ore (Figure 6). The occurrence of the nickel hydroxides [Ni(OH)$_2$] in nature is rare. Theophrastite and Ni(OH)$_2$ were discovered in the metamorphosed laterites of the E. Vermion [35], and subsequently, it was described in chromitites from the Hagdale Quarry, Unst, Shetland Islands [42]; in nickel ores of the Lord Brassey mine, Tasmania, Australia [43]; and elsewhere [44]. Although a significant mutual substitution of Mn^{2+}, Ni^{2+}, and Co^{2+} is facilitated in those hydroxides, due to the similarity in their ionic radius, Co^{2+} (0.72 Å), Ni^{2+} (0.69 Å), and Mn^{2+} (0.80 Å), they are very rare in nature [45]. The association of (Co, Mn, Ni)(OH)$_2$ and Ni(OH)$_2$ in the studied deposits may reflect a common origin, subsequently of a strong late tectonic evolution, which obscures earlier deformation events. The formation processes of the Ni-hydroxide have been attributed to alkaline (in the pH region of 7 to 8) conditions due to the neighboring limestone and to the low temperature (<200 °C) [46]. However, the extremely tiny crystal size of these hydroxides and the unnamed Co-dominant hydroxide require further research and interpretation.

5.2. The Co, Mn–Zn-Bearing Fe Chromite

A (Mn, Zn, Co) enrichment in the Fe chromite of zoned chromite grains (Figure 5g,h), due to the substitution of Mg^{2+} and Fe^{2+} by Mn, Zn, and Co in the chromite lattice (Figure 4i), has been recorded only in certain Ni laterite deposits and has been attributed to the availability of Mn^{2+}, Zn^{2+}, and Co^{2+} in solution and to the Eh and pH conditions, probably during diagenetic and meta-diagenetic stages of the laterite ore [19]. The lack of any significant Mn, Zn, or Co enrichment and Fe chromite along cracks or peripheral parts (Figure 5c,e) of chromite hosted in laterite ores from the Palaochori and Katjeli deposits, hosting abundant fossilized organic matter (Figure 5a,b,d) may reflect a significant role of the organic matter in the reduction of metals (Figure 5i). Nevertheless, the small portion of chromite grains or fragments in Fe–Ni laterites and the occurrence of elevated Co in small Fe-chromite zones of certain laterite occurrences only (E. Vermion, Olympos, Edessa, Skyros Island) suggest that they must be considered a negligible source for Co production.

5.3. REE and Sc Potential and Their Genetic Significance

The highest productive potential of REE in the EU is recognized for alkaline igneous rock deposits (Sweden, Greenland/Denmark), while a smaller potential is known for skarn hydrothermal deposits and iron oxide-apatite deposits (in Sweden). Small carbonatite deposits are known in Finland and Germany, while small placer deposits are located in Spain [1,8]. In general, karst-type Fe–Ni laterites display REE contents ranging from a few tens to thousands of mg/kg, depending on the site of the Ni laterite samples along vertical profiles and on the post-deposition processes that affect their composition. Specifically, the Fe–Ni laterites lying on peridotites such as the Kastoria, Bitincka, and Guri–Pergjegjum (Albania) deposits show a low REE content, in contrast to the lowest part of the Lokris deposit lying on karstified Jurassic limestone, exhibiting an enrichment in ΣREE, reaching values over 6000 mg/kg ΣREE, 66 mg/kg U, 25 mg/kg Th, and 1800 mg/kg As (Table 2). Such elevated REE content, along with Co, Mn, and Ni, may indicate the mobilization of these metals under reducing and acidic conditions, and subsequently, redeposition under alkaline conditions at the lowest parts at the contacts between laterites and carbonate rocks, which is supported by the presence of the authigenic REE minerals (hydroxyl)bastnaesite-(Nd) and (La, Nd, Y) bastnaesite at the Lokris deposit (Figure 3e,f); [26,30].

The difference between the REE patterns of the laterite samples from the lowest part with a negative Ce anomaly and the upper parts of the Lokris deposit with positive Ce anomalies (Figure 10a) may reflect the oxidation of Ce^{3+} to Ce^{4+} and its incorporation into other Ce-rich minerals [26]. In addition, a good positive correlation between the ΣREE and P_2O_5 content (Figure 12e) may suggest the existence of secondary phosphate minerals in the Vermion laterites, which have a significant P_2O_5 content (Tables 3 and 4). Thus, Fe–Ni ± Co laterites in the Balkan Peninsula may contain a significant REE potential, but it is probably controlled by the redox and pH conditions during the mobilization and reprecipitation processes that caused metal leaching and residual enrichment throughout deposit profiles and REE concentration at the lowest parts of the deposits [26,30,47].

Scandium, another critical element, is classified as a rare earth element by the International Union of Pure and Applied Chemistry, but although all the other REEs are present in certain deposits, the Sc content may be insignificant [48]. Due to the similarity of the Sc ionic radius with that of Mg^{2+} and Fe^{2+}, which is smaller than that of any other REE, it is mainly hosted in clinopyroxenes, in contrast to the REE-minerals [48]. After pyroxene breakdown, Sc is released into percolating water where it is transported and exhibits a positive correlation with Fe_2O_3 (Figure 9), suggesting its association with the Fe-oxides in laterites. It is low (0.6 to 12 mg/kg) in weathered peridotites and saprolite zones, but in laterite ores, it varies from 41 to 79 mg/kg (average 60 mg/kg) (Tables 2–5) [16,18,49]. Higher Sc contents, about 120 mg/kg (average approximately 70–80 mg/kg Sc) have been recorded in the Fe–Ni laterites from Cuba and the Dominican Republic [12]. Thus, assuming that Sc can be produced as a by-product of metallurgical processes or from tailings and residues [49,50], resources with more than 20 mg/kg, such as Fe–Ni laterites from the Balkan Peninsula can be considered as a Sc source.

5.4. Genetic and Economic Significance of PGE

The total PGE contents in Fe–Ni laterites around the world are in the range of less than 100 μg/kg to up to a few hundred μg/kg [51,52]. However, lateritic crust with 2 mg/kg PGE over the Ora Banda Sill in Western Australia has been reported [53], as well as more than 4 mg/kg PGE from Burundi [54]. In general, the platinum-group element (PGE) content, reaching up to 88 μg/kg Pt and 186 μg/kg Pd is higher in Fe–Ni ± Co laterites than in bauxite laterites (Tables 2–5 and Figure 11d–f), bauxites [37], and chromite concentrates from ultramafic rocks [55]. A common feature of the studied deposits from the Balkan Peninsula, including two Fe–Ni laterite profiles from the Gouri-Perjuegjiun and Bitincka deposits of Albania, the Kastoria, and W. Vermion, is the higher Pt than Pd contents, higher (Pt + Pd) contents in ores compared to parent peridotites, and their concentration mostly in the pelitomorphic and pisolitic–oolitic types (Tables 2–5) [16,18]. Although PGE-bearing or PGE minerals were not identified yet in laterites of the Balkan Peninsula, the elevated PGE content in those laterites, and the presence of PGE-bearing mineral compounds at the border of secondary Fe oxide(s) grains elsewhere [12] support the PGEs' mobility during weathering [56–60]. The higher solubility and mobility of Pd compared to that of Pt [61] may be reflected in the observed decoupling between Pt and Pd in the allochthonous laterites, as is exemplified by the higher value of the Pt/Pd ratio in the laterite deposits than in the lowermost part of the Lokris deposit (Table 2). In addition, it has been suggested that high Eh (0.3–1.11 V) and low pH (3–5) conditions caused the mobility of PGEs with subsequently incorporation into Fe oxides in laterites of southeast Cameroon [52]. The occurrence of secondary PGM in laterite profiles support the mobilization of PGEs during weathering, lateritization, and post-depositional processes [12,18]. Thus, the existence of high Pt and Pd (up to 88 μg/kg Pt and 186 μg/kg Pd) in laterite samples from the Balkan Peninsula, which are much higher compared to those in typical chromitites [62], may be related to the mobilization of PGEs during weathering, lateritization, and post-depositional processes. In addition, the chromium isotope data for laterite are limited, laterite samples from the Lokris and Kastoria deposits have shown highly variable $\delta^{53}Cr$ values ranging from -0.2 in the former to 1.08 ‰ in the latter, while the average (n = 3) of the $\delta^{53}Cr$ values in water leachates was 0.79 ‰ for weathered ultramafic rocks [31], showing a negative correlation with the Pt content (Figure 13).

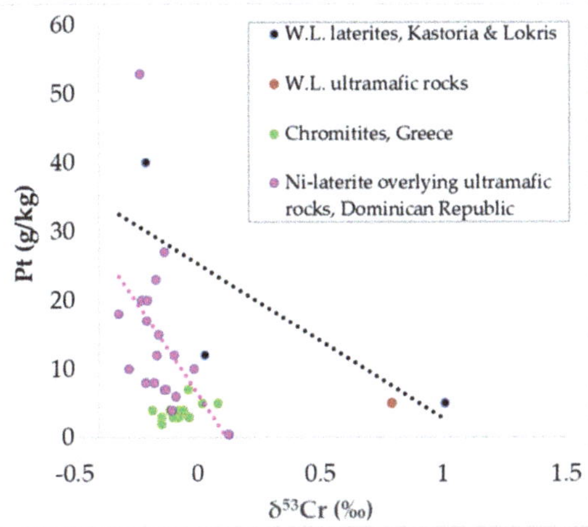

Figure 13. Plot of Pt versus δ^{53}Cr values for Fe–Ni ± Co laterites, chromitites, and weathered ultramafic rocks from Greece and the Dominican Republic. Data from [12,31,63].

Therefore, the redox processes have a significant effect on the Cr isotope fractionation and the Pt distribution in laterites. The low Pt contents and the low δ^{53}Cr values in chromitites may reflect mostly magmatic conditions (less redistribution), while the low δ^{53}Cr values and the relatively high Pt content in the Kastoria laterites suggest oxidation of Cr to Cr(VI) during the redeposition of the laterite ore, the adsorption on goethite, and the limited back-reduction to Cr(III). Assuming that during the reduction of the Cr(VI) to Cr(III) the lighter chromium isotopes (^{52}Cr) are preferentially reduced, an enrichment of ^{53}Cr relative to the values of ^{52}Cr is recorded [29]. Thus, the lower Pt content and higher δ^{53}Cr values in the Lokris deposits may indicate that Cr released from parent ultramafic rocks and oxidized to Cr(VI) due to the presence of Mn-oxides [64] was subsequently back-reduced to Cr(III) under more reductive conditions [29,65].

6. Environmental Risk from Laterite Ores

The environmental impact from metals/metalloids may be related to the mineralogical and geochemical characteristics of the ore, the effect during mining, pre-concentration, and smelting processes of Fe–Ni laterite ores. High Co concentrations have been reported in mine waters and groundwater related to the weathering and dissolution of Co-bearing sulfide ores and secondary minerals in regions of acidic pH, while in water flowing from the leach plant at pH 2.9 may occur up to 34,400 mg/L Cr(VI) [66]. With exception of the Bou Azzer deposit, where Co is the chief commodity [66], Co is almost always mined as a by-product, including the Fe–Ni laterite deposits, the main product being Ni and Cu, depending on the deposit type [66,67]. Cobalt's aqueous geochemistry is dominated by +2 and +3 oxidation states, with Co^{3+} being thermodynamically unstable and changing under Eh–pH conditions [67]. Mining activity and the associated Co-bearing wastes are potential sources of contamination by Co. However, knowledge gaps include the mechanism and kinetics of secondary Co-bearing mineral transformation, the extent at which such environmental cycling is facilitated by microbial activity, the nature of Co speciation across different Eh–pH conditions and Co toxicity for human/ecosystems [66–68].

6.1. Environmental Impact from Mining/Smelting Processes

The soil contamination by heavy metals (Cr, Fe, Mn, and Co) due to the weathering/alteration of laterite deposits and parent ultramafic rocks, mining, and large volumes of smelting residues (slag) and their transfer into groundwater and plants/crops is common in many countries of the world [31,51,63–65,69].

Available data on the bioavailability and bioaccumulation of metal in C. Evia and the Assopos–Thiva Basins contaminated by weathered ophiolites/laterite mining, agricultural, and industrial activities have shown that the Cr content in plants/crops (dry weight) ranges from <1 to tens of mg/kg, and it is higher in the roots compared to being poorly transported to the shoots [69]. Another threat from the exploitation of Fe–Ni laterites may be fine particles with a diameter of less than 10 microns (PM10) and of 2.5 microns or less (2.5 PM), which are considered to be among the greatest health risk since they are deposited and remain within the respiratory system [70]. The measuring and monitoring of PM10 and PM2.5 particles with distance from the dust source is required for the establishment of background particulate levels to be used in the air quality modelling and health protection. The processing for the production of Fe–Ni alloys at the Larymna plant is accompanied by the production of millions of tons of slag and the emission of tons of fly ash and fine dust collected in the anti-pollution systems [40]. The detailed investigation of slag and fly ash from laterite smelting (Niquelandia, Brazil) has shown that leaching was highly pH-dependent for both slag and fly ash, the highest releases of harmful elements occurred between pH 3 and 7, and the fly ash was significantly more reactive, suggesting that the disposal sites for the fly ash can represent a significant source of local pollution [70,71].

6.2. The Cr(VI) Concentrations in Water Leachates from Fe–Ni ± Co Laterites

Although Cr(III) is a required nutrient, the form of oxidized Cr(VI) is highly toxic, very soluble, and causes serious health problems [72]. Thus, further Cr(VI) concentrations in water leachates for the majority of Fe–Ni laterites were determined. The groundwater from Neogene shallow aquifers (10–100 m) throughout C. Evia (Messapia), characterized by the extensive presence of ultramafic rocks and F–Ni laterite deposits, contains over the maximum acceptable level for Cr_{total} in drinking water (50 µg/L), reaching up to 360 µg/L Cr(VI) in irrigation water [63,69]. In addition, relatively high Cr(VI) concentrations are known in the leachates of Fe–Ni laterites from the Kastoria deposit [31]. The present study revealed significant Cr(VI) concentrations in leachates from the Gouri-Perjuegjiun and Bitincka deposits of Albania and the W. Vermion laterites, all lying on weathered peridotites, as well as in bauxite laterites from the Nissi–Patitira deposit (Table 7). The latter bauxite laterite samples come from highly tectonized zones, towards the contact of the deposit with carbonate basement, commonly grading from red to a gray, pinkish-white color (Figure 4a). They are characterized by hundreds of mg/kg of As, elevated contents of REE and Co (Table 2), while in laterites from Kasroria, W. Vermion, Gouri-Perjuegjiun, and Bitincka deposits, the REE and As contents are relatively low (Tables 3 and 5). Since As in soil, plant/crops, and drinking water is a crucial contaminant for human health and ecosystems [72–74], the As oxidation state was studied in a bauxite laterite sample, by Synchrotron Radiation (SR) spectroscopic techniques, and it was found to be adsorbed exclusively onto goethite-type phases, occurring in the form of arsenate anions (AsO_4^{3-}) [30]. Although there is still debate on whether the mineral is of direct organic or inorganic origin [75,76], recent studies of the internal structure of framboids within a microbial biofilm suggest that the organic matrix may play a central role in the nucleation of pyrite microcrystals that form framboids [77,78].

6.3. The Role of Organic Matter into Biomobilization and Biomineralization

In general, the presence of organic matter and microorganism traces in highly tectonized zones of the metamorphosed Fe–Ni laterites (Figures 3–6) may suggest the direct involvement of microorganisms and appropriate conditions for metal bioleaching and biomineralization [24,79–87]. It is well known that microorganisms of varying morphological forms produce enzymes which are considered to be a powerful factor to catalyze redox reactions and act as nucleation sites for the precipitation of secondary minerals [83–87]. The negative $\delta^{34}S$ values ($-25.0‰$ to $-26.2‰$) [24] for sulfide-bearing samples and the presence of bacterio-morphic goethite in the Lokris deposit (Figure 3) have been suggested to reflect the existence of the appropriate conditions for metal bioleaching and biomineralization [24]. This review highlights that the occurrence of organic matter is a common feature in many laterite deposits of the Balkan Peninsula, such as a sequence of carbon-rich mudstones in Lokris and elsewhere [15], the Kastoria deposit (Figure 3), the Paleochori and Katjeli deposits (Figure 5), and the E. Vermion metamorphosed laterite occurrence (Figure 6). In addition, the Katjeli Fe–Ni deposit is conformably overlain by a Tertiary series consisting of small beds of lignite and sandstone intercalations (Figure 2A). In the reworking of laterites, due to the repeated marine transgression/regressions and its deposition in a shallow marine environment [15,17,18,24], microorganisms which reduce Mn(IV) and Fe(III) and use the gained energy for their growth [88,89] may facilitate the dissolution of Mn(IV) and Fe(III) minerals under reducing and slightly acidic conditions, forming aqueous Mn(II) and Fe(II) [90].

The negatively charged surfaces of bacteria cells offer extensive surfaces for biosorption of metals where elements with a higher positive charge are preferentially adsorbed [80]. Both the metabolic activity and the biomineral precipitation by organisms may play an important role in trace element co-precipitation [77,81,82]. Thus, the described bacterio-morphic goethite at the Nissi–Patitira deposits, coupled with the negative $\delta^{34}S$ values for sulfide-bearing samples towards the top of the deposit [24] may suggest the existence of the appropriate conditions for iron bioleaching, the top-down remobilization of the laterite deposit, and the biomineralization at the contact with the carbonate basement [24]. The existence of significant As only in bacteriomophic goethite type in the Patitira–Nissi deposits [26] and the presence of Cr(VI) in water leachates from such ore samples may reflect that both bacteria and chemical processes may be invoked in the As and Cr redox reactions, and the need for monitoring of the Cr(VI) content in groundwater aquifers affected by the mining of Fe–Ni laterites. In addition, a quantitative relationship between the Cr(VI) production rates and Cr(III) solubility of $CrxFe_{1-x}(OH)_3$ has been established in an attempt to predict Cr(VI) production rates at different conditions [81]. These authors showed that soluble Cr(III) released from $CrxFe_{1-x}(OH)_3$ solids to aqueous solution can migrate to MnO_2 surfaces and provide a major geochemical pathway for Cr(VI) occurrence from Cr(III) in groundwater, soils, or sub-seafloor environments.

Therefore, the association on (Mn, Ni, Co)-hydroxides with fossilized organic material (Figures 3–6) may indicate that microbial metal cycling play an important role in controlling the mobility of Co, Ni, REE, Sc, and PGE downward and their redeposition under alkaline conditions, probably during post-depositional and diagenesis stages. Based on such evidence on the biogeochemical cycling of Co, Mn, and Ni in several laterite deposits, new bioprocessing techniques have been proposed to recover these metals [68,70,71]. In addition, the investigation of the ability of organic matter in reducing Cr uptake by crops irrigated with Cr(VI) water provides a way for the restriction of Cr transfer to plants/crops from contaminated soils and irrigation water. Specifically, using organic matter in the form of leonardite (an oxidized form of lignite), due to its high content of humic acid, is considered to be a useful organic fertilizer that provides possibilities for combining food production with soil protection [91].

7. Conclusions

The presented data combined with the review of the available geological, mineralogical, and geochemical features of Fe–Ni $\pm$ Co laterites from the Balkan Peninsula led us to the following conclusions:

- The bulk ore analyses indicated that the Fe–Ni $\pm$ Co laterites from the Balkan Peninsula are a potential source of nickel ranging from 0.21 to 3.5 wt% Ni, and of cobalt ranging from 0.03 to 0.31 wt% Co. In addition, an average content of 55 mg/kg Sc,

- which is more than 20 mg/kg, and is considered as a Sc source suggests that laterites are a potential source of Sc as well.
- In general, the REE potential in the Fe–Ni ± Co laterites of the Balkan Peninsula is relatively low, but metal-leaching and reprecipitation of REE at the lowest parts of the karst-type deposits, controlled by the redox and pH conditions, may result in a significant REE content.
- The existence of up to 88 μg/kg Pt and 186 μg/kg Pd in laterite samples from the Balkan peninsula, which are much higher compared to those in typical chromitites (the main collector of PGE in ophiolites) and the positive Pt–Fe_2O_3 correlation in the goethite and pisolitic–oolitic zones of laterites, support the PGE mobilization during weathering, lateritization, and post-depositional processes.
- Present textural, mineralogical, and geochemical data on the Ni–Co laterite deposit from the Balkan Peninsula with elevated Co, Mn, and Ni have evolved through the redeposition and post-deposition processes rather than during lateritization.
- The association on (Mn, Ni, Co)-hydroxides with fossilized organic material and the interelement relationships may indicate that microbial metal cycling play an important role in controlling their downward mobility and their redeposition under alkaline conditions, probably during post-depositional and diagenesis stages.
- The relatively high Pt content versus low δ^{53}Cr values and high Cr(VI) concentrations in leachates from the Kastoria laterite samples suggest the oxidation of Cr to Cr(VI) during the redeposition of the laterite ore, the adsorption onto goethite, and the limited back-reduction to Cr(III). In contrast, the Lokris laterite deposits with lower Pt content exhibit higher δ^{53}Cr values, suggesting that the oxidation of the released Cr from ultramafic rocks into Cr(VI) is followed by a back-reduction to Cr(III) under more reductive conditions.
- The bioaccumulation of Cr in contaminated soil and irrigation groundwater coupled with the elevated Cr(VI) concentrations in natural water leachates for Fe–Ni laterite samples from the Kastoria, Bitincka, Guri–Pergjegjum, and Nissi–Patitira (Lokris) deposits point to the potential human health risk and environmental significance of an integrated water–soil–plant investigation of the Cr contamination in basins surrounding laterite mines.

Author Contributions: Conceptualization and methodology: M.E.-E., M.L. and D.G.E.; software and validation of data, writing—original draft preparation: M.E.-E., M.L., I.M., S.K. and G.D.E. All authors have read and agreed to the published version of the manuscript.

Funding: The financial support in the frame of the "Integrated Technologies for Minerals Exploration: Pilot Project for Nickel Ore Deposits" (Part of an EU BRITE EURAM III Project—Contract No. BE 1117 CT95-0052) is greatly acknowledged once again.

Acknowledgments: Many thanks are expressed to E. Michaelidis, University of Athens, for his assistance with the electron probe analyses. The criticism and constructive suggestions by two anonymous reviewers and the Editors Cristina Domènech and Cristina Villanova-de-Benavent are much appreciated.

Conflicts of Interest: The authors declare no conflict of interest.

References

1. European Commission. *A New Industrial Strategy for Europe. Communication from the Commission to the European Parliament, the Council, the European Economic and Social Committee and the Committee of the Regions, COM(2020) 102 Final;* European Commission: Brussels, Belgium, 2020.
2. Lewicka, E.; Guzik, K.; Galos, K. On the Possibilities of Critical Raw materials Production from the EU's Primary Sources. *Resources* **2021**, *10*, 50. [CrossRef]
3. Cassard, D.; Bertrand, G.; Billa, M.; Serrano, J.J.; Tourlière, B.; Angel, J.M.; Gaál, G. ProMine Mineral Databases: New Tools to Assess Primary and Secondary Mineral Resources in Europe. In *3D, 4D and Predictive Modelling of Major Mineral Belts in Europe;* Weihed, P., Ed.; Springer: Cham, Switzerland, 2015.

4. Cassard, D.; Tertre, F. Improvement of KDPs' Applications and Interaction with the RMIS and the GeoERA Information Platform. 2021. Available online: https://geoera.eu/wp-content/uploads/2021/03/D5.5-KDP%E2%80%99s-applications-delivery-to-RMIS.pdf (accessed on 30 May 2021).

5. Lauri, L.S.; Eilu, P.; Brown, T.; Gunn, G.; Kalvig, P.; Sievers, H. Identification and Quantification of Primary CRM Resources in Europe. 2018. Available online: http://scrreen.eu/wp-content/uploads/2018/03/SCRREEN-D3.1-Identification-and-quantification-of-primary-CRM-resources-in-Europe.pdf (accessed on 22 March 2021).

6. Sadeghi, M.; Arvanitidis, N.; Ladenberger, A. Geochemistry of Rare Earth Elements in Bedrock and Till, Applied in the Context of Mineral Potential in Sweden. *Minerals* **2020**, *10*, 365. [CrossRef]

7. Minerals4EU. Available online: http://www.minerals4eu.eu/ (accessed on 17 March 2021).

8. Goodenough, K.M.; Schilling, J.; Jonsson, E.; Kalvig, P.; Charles, N.; Tuduri, J.; Deady, E.A.; Sadeghi, M.; Schiellerup, H.; Mueller, A.; et al. Europe's rare earth element resource potential: An overview of REE metallogenic provinces and their geodynamic setting. *Ore Geol. Rev.* **2016**, *72*, 838–856. [CrossRef]

9. Horn, S.; Gunn, A.G.; Petavratzi, E.; Shaw, R.A.; Eilu, P.; Törmänen, T.; Bjerkgård, T.; Sandstad, J.S.; Jonsson, E.; Kountourelis, S.; et al. Cobalt resources in Europe and the potential for new discoveries. *Ore Geol. Rev.* **2021**, *130*, 103915. [CrossRef]

10. Mudd, G.M. Global trends and environmental issues in nickel mining: Sulfides versus laterites. *Ore Geol. Rev.* **2010**, *38*, 9–26. [CrossRef]

11. Berger, V.I.; Singer, D.A.; Bliss, J.D.; Moring, B.C. *Ni-Co Laterite Deposits of the World—Database and Grade and Tonnage Models*; USGS Open-File Report: Menlo Park, CA, USA, 2011; Volume 1058, p. 26.

12. Aiglsperger, T.; Proenza, J.A.; Zaccarini, F.; Lewis, J.F.; Garuti, G.; Labrador, M.; Longo, F. Platinum-group minerals (PGM) in the Falcondo Ni-laterite deposit, Loma Caribe peridotite (Dominican Republic). *Mineral. Depos.* **2015**, *50*, 105–123. [CrossRef]

13. Freyssinet, P.; Butt, C.R.M.; Morris, R.C.; Piantone, P. Ore-forming processes related to lateritic weathering. *Econ. Geol. 100th Anniv. Vol.* **2005**, *2005*, 681–722.

14. Milushi, I. An Overview of the Albanian Ophiolite and Related Ore Minerals. *Acta Geol. Sin.* **2015**, *89*, 61–64. [CrossRef]

15. Valeton, I.; Biermann, M.; Reche, R.; Rosenberg, F. Genesis of Nickel laterites and bauxites in Greece during the Jurassic and Cretaceous, and their relation to ultrabasic parent rocks. *Ore Geol. Rev.* **1987**, *2*, 359–404. [CrossRef]

16. Tashko, A.; Laskou, M.; Eliopoulos, D.; Economou-Eliopoulos, M. The behavior of Pt, Pd, and Au during lateritization process of the ultramafic rocks of the Shebenic-Pogradec massif, Albania: Plate tectonic aspects of the Alpine metallogeny in the Carpatho-Balkan region. In Proceedings of the Annual Meeting of IGCP Project 356, Sofia, Bulgaria, December 1996; pp. 121–131.

17. Alevizos, G. Mineralogy, Geochemistry and Origin of the Sedimentary Fe–Ni Ores of Lokris. Ph.D. Thesis, Technical University of Crete, Chania, Greece, 1997; p. 245.

18. Eliopoulos, D.; Economou-Eliopoulos, M. Geochemical and mineralogical characteristics of Fe–Ni and bauxitic–laterite deposits of Greece. *Ore Geol. Rev.* **2000**, *16*, 41–58. [CrossRef]

19. Economou-Eliopoulos, M. Apatite and Mn, Zn, Co-enriched chromite in Ni-laterites of northern Greece and their genetic significance. *J. Geochem. Explor.* **2003**, *80*, 41–54. [CrossRef]

20. Boev, B.; Jovanovski, G.; Makreski, P. Minerals from Macedonia. XX. Geological setting, lithologies, and identification of the minerals from Rzanovo Fe-Ni deposit. *Turk. J. Earth Sci.* **2009**, *18*, 631–652. [CrossRef]

21. Thorne, R.; Herrington, R.J.; Roberts, S. Composition and origin of the Çaldağ oxide nickel laterite, W. *Turkey. Miner. Depos.* **2009**, *44*, 581–595. [CrossRef]

22. Eliopoulos, D.G.; Economou-Eliopoulos, M.; Apostolikas, A.; Golightly, J.P. Geochemical features of nickel-laterite deposits from the Balkan Peninsula and Gordes, Turkey: The genetic and environmental significance of arsenic. *Ore Geol. Rev.* **2012**, *48*, 413–427. [CrossRef]

23. Golightly, J.P. Progress in understanding the evolution of nickel laterites. In *The Challenge of Finding New Mineral Resources—Global Metallogeny, Innovative Exploration, and New Discoveries*; Goldfarb, R.J., Marsh, E.E., Monecke, T., Eds.; Economic Geology. Special Publication 15; Society of Economic Geologists, Inc.: Littleton, CO, USA, 2010; Volume 15, pp. 451–486.

24. Kalatha, S.; Economou-Eliopoulos, M. Framboidal pyrite and bacteriomorphic goethite at transitional zones between Fe-Ni-laterites and limestones: Evidence from Lokris, Greece. *Ore Geol. Rev.* **2015**, *65*, 413–425. [CrossRef]

25. Skarpelis, N.; Laskou, M.; Alevizos, G. Mineralogy and geochemistry of the nickeliferous lateritic iron-ores of Kastoria, N.W. Greece. *Chem. Erde* **1993**, *53*, 331–339.

26. Kalatha, S.; Perraki, M.; Economou-Eliopoulos, M.; Mitsis, I. On the origin of bastnaesite-(La,Nd,Y) in the Nissi (Patitira) bauxite laterite deposit, Lokris, Greece. *Minerals* **2017**, *7*, 45. [CrossRef]

27. Baker, J.; Cassard, D.; Grundfelt, B.; Ehinger, S.; D'Hugues, P.; Gaál, G.; Kaija, J.; Kurylak, W.; Mizera, W.; Sarlin, J.; et al. A New Approach to Mineral Resources in Europe—ProMine. In Proceedings of the Sustainable Production and Consumption of Mineral Resources—Integrating the EU's Social Agenda and Resource Efficiency–FP7 Cooperation Work Programme, 2013, Theme 4, NMP-2008-4.0, Wroclav, Poland, 20–22 October 2011.

28. Saccani, E.; Photiades, A.; Santato, A.; Zeda, O. New evidence for supra-subduction zone ophiolites in the Vardar Zone from the Vermion Massif (northern Greece): Implication for the tectono-magmatic evolution of the Vardar oceanic basin. *Ofioliti* **2008**, *33*, 17–37.

29. Ellis, A.S.; Johnson, T.M.; Bullen, T.D. Chromium isotopes and the fate of hexavalent chromium in the environment. *Science* **2002**, *295*, 2060–2062. [CrossRef]

30. Gamaletsos, P.N.; Kalatha, S.; Godelitsas, A.; Economou-Eliopoulos, M.; Göttlicher, J.; Steininger, R. Arsenic distribution and speciation in the bauxitic Fe-Ni-laterite ore deposit of the Patitira mine, Lokris area (Greece). *J. Geochem. Explor.* **2018**, *194*, 189–197. [CrossRef]
31. Economou-Eliopoulos, M.; Frei, R.; Megremi, I. Potential leaching of Cr(VI) from laterite mines and residues of metallurgical products (red mud and slag): An integrated approach. *J. Geochem. Explor.* **2016**, *162*, 40–49. [CrossRef]
32. Orphanoudakis, A.; Mposkos, E.; Kastritsis, I. A study of the mineralogical and geochemical composition of the Fe-Ni-laterite from the area of Paleochori (Grevena). *Geol. Soc. Greece* **1997**, *31*, 7–22.
33. Paraskevopoulos, G.; Economou, M. Genesis of magnetite ore occurrences by metasomatism of chromite ores in Greece. *Neues. Jahrb. Mineral. Abh.* **1980**, *140*, 29–53.
34. Michailidis, K.M. Zoned chromites with high Mn-contents in the Fe–Ni–Cr-laterite ore deposits from the Edessa area in Northern Greece. *Miner. Depos.* **1990**, *25*, 190–197. [CrossRef]
35. Marcopoulos, T.; Economou, M. Theophrastite, Ni(OH)$_2$, a new mineral from northern Greece. *Am. Mineral.* **1981**, *66*, 1020–1021.
36. Arnisalo, J.; Makela, K.; Bourgeois, B.; Spyropoulos, C.; Varoufakis, S.; Morten, E.; Maglaras, C.; Soininen, H.; Angelopoulos, A.; Eliopoulos, D.; et al. Integrated technologies for minerals exploration; pilot project for nickel ore deposits. *Trans. Instn. Min. Metall. Sect. B Appl. Earth Sci.* **1999**, *108*, B151–B163.
37. Laskou, M.; Economou, M. Palladium, Pt, Rh and Au Contents in Some Bauxite Occurrences of Greece. In Proceedings of the Balkan-Carpathian Congress, Sofia, Bulgaria, 11–13 December 1989; pp. 1367–1371.
38. Brindley, G.W.; Maksimović, Z. The nature and nomenclature of hydrous nickel-containing silicates. *Clay Miner.* **1974**, *10*, 271–277. [CrossRef]
39. Glencore. Annual Report of Glencore Company Plc; Baar, Switzerland, 2019. Available online: https://www.glencore.com/dam/jcr:ae4466b4-7ef4-4407-ae00-6ca55b694028/GLEN_2018_Resources_Reserves_Report-.pdf (accessed on 22 July 2020).
40. INSG. *Nickel in the Balkan Region. Joint Study of the Copper, Lead, Zinc and Nickel Industries in the Balkan Region*; International Nickel Studies Group; INSG: San Diego, CA, USA, 2015; p. 85.
41. Petavratzi, E.; Gunn, A.G.; Kresse, C.C. BGS, London, Commodity Review, 2019; 72 p. Available online: https://www.bgs.ac.uk/mineralsuk/search/home.html (accessed on 13 September 2021).
42. Livingstone, A.; Bish, D.L. On the new mineral thoephrastite, a nickel hydroxide, from Unst, Shetland, Scotland. *Mineral. Mag.* **1982**, *46*, 1–5. [CrossRef]
43. Henry, D.A.; Birch, W.D. Otwayite and theophrastite from the Lord Brassey Mine, Tasmania. *Mineral. Mag.* **1992**, *56*, 252–255. [CrossRef]
44. Arai, S.; Ishimaru, S.; Miura, M.; Akizawa, N.; Mizukami, T. Post-Serpentinization Formation of Theophrastite-Zaratite by Heazlewoodite Desulfurization: An Implication for Shallow Behavior of Sulfur in a Subduction Complex. *Minerals* **2020**, *10*, 806. [CrossRef]
45. Economou-Eliopoulos, M. A new solid solution [(Co, Mn, Ni)(OH)$_2$], in the Vermion Mt (Greece) and its genetic dignificance for the mineral group of hydroxides. In *Chemical Mineralogy, Smelting and Metallization*; McLaughlin, E.D., Braux, L.A., Eds.; Nova Science Publishers: New York, NY, USA, 2009.
46. Economou, M.; Marcopoulos, T. Genesis of the new mineral theophrastite, Ni(OH)2. *Chem. Erde* **1983**, *42*, 53–56.
47. Maksimovic, Z. Genesis of Mediterranean karstic bauxites and karstic nickel deposits. *Bull. Acad. Serbe Sci. Arts* **2010**, *46*, 1–27.
48. Williams-Jones, A.E.; Vasyukova, O.V. The Economic Geology of Scandium, the Runt of the Rare Earth Element Litter. *Econ. Geol.* **2018**, *113*, 973–988. [CrossRef]
49. Jaireth, S.; Hoatson, D.M.; Miezitis, Y. Geological setting and resources of the major rare-earth-element deposits in Australia. *Ore Geol. Rev.* **2014**, *62*, 72–128. [CrossRef]
50. Kim, J.; Azimi, G. An innovative process for extracting scandium from nickeliferous laterite ore: Carbothermic reduction followed by NaOH cracking. *Hydrometallurgy* **2020**, *191*, 105194. [CrossRef]
51. Augé, T.; Legendre, O. Pt-Fe nuggets from alluvial deposits in Eastern Madagascar. *Can. Mineral.* **1992**, *30*, 983–1004.
52. Ndjigui, P.D.; Bilong, P. Platinum-group elements in the serpentinite lateritic mantles of the Kongo–Nkamouna ultramafic massif (Lomié region, South-East Cameroon). *J. Geochem. Explor.* **2010**, *107*, 63–76. [CrossRef]
53. Gray, R.; Owen, D.; Adams, C. *Accounting and Accountability: Changes and Challenges in Corporate Social and Environmental Reporting*; Prentice-Hall: London, UK, 1996.
54. Maier, W.D.; Barnes, S.-J.; Bandyayera, D.; Livesey, T.; Li, C.; Ripley, E. Early Kibaran rift-related mafic–ultramafic magmatism in western Tanzania and Burundi: Petrogenesis and ore potential of the Kapalagulu and Musongati layered intrusions. *Lithos* **2008**, *101*, 24–53. [CrossRef]
55. Konstantopoulou, G.; Economou-Eliopoulos, M. Distribution of Platinun-Group Elements and Gold within the Vourinos chromitite ores, Greece. *Econ. Geol.* **1991**, *86*, 1672–1682. [CrossRef]
56. Aiglsperger, T.; Proenza, J.A.; Font-Bardia, M.; Baurier-Amat, S.; Gali, S.; Lewis, J.F.; Longo, F. Supergene neoformation of Pt-Ir-Fe-Ni alloys: Multistage grains explain nugget formation in Ni-laterites. *Mineral. Depos.* **2017**, *52*, 1069–1083. [CrossRef]
57. Aiglsperger, T.; Proenza, J.A.; Lewis, J.F.; Labrador, M.; Svojtka, M.; Rojas-Purón, A.; Longo, F.; Ďurišová, J. Critical metals (REE, Sc, PGE) in Ni laterites from Cuba and the Dominican Republic. *Ore Geol. Rev.* **2016**, *73*, 127–147. [CrossRef]
58. Bowles, J.F.W. The development of platinum-group minerals in laterites. *Econ. Geol.* **1986**, *81*, 1278–1285. [CrossRef]

59. Bowles, J.F.W.; Gize, A.P.; Vaughan, D.J.; Norris, S.J. Organic controls on platinum group PGE solubility in soils: Initial data. *Chron. Rech. Min.* **1995**, *520*, 65–73.
60. Bowles, J.F.W.; Suárez, S.; Prichard, H.M.; Fisher, P.C. Weathering of PGE-sulfides and Pt-Fe alloys, in the Freetown Layered Complex, Sierra Leone. *Miner. Depos.* **2017**, *52*, 1127–1144. [CrossRef]
61. Grey, J.D.; Schorin, K.H.; Butt, C.R.M. Mineral associations of platinum and palladium in lateritic regolith, Ora Banda Sill, Western Australia. *J. Geochem. Explor.* **1996**, *57*, 245–255. [CrossRef]
62. Economou-Eliopoulos, M. Platinum-group element distribution in chromite ores from ophiolite complexes: Implications for their exploration. *Ore Geol. Rev.* **1996**, *11*, 363–381. [CrossRef]
63. Economou-Eliopoulos, M.; Frei, R.; Atsarou, C. Application of chromium stable isotopes to the evaluation of Cr (VI) contamination in groundwater and rock leachates from central Euboea and the Assopos basin (Greece). *Catena* **2014**, *122*, 216–228. [CrossRef]
64. Oze, C.; Bird, K.D.; Fendorf, S. Genesis of hexavalent chromium from natural sources in soil and groundwater. *Proc. Natl. Acad. Sci. USA* **2007**, *104*, 6544–6549. [CrossRef]
65. Economou-Eliopoulos, M.; Antivachi, D.; Vasilatos, C.; Megremi, I. Evaluation of the Cr(VI) and other toxic element contamination and their potential sources: The case of the Thiva basin (Greece). *Geosci. Front.* **2012**, *3*, 523–539. [CrossRef]
66. Ziwa, G.; Crane, R.; Hudson-Edwards, K.A. Geochemistry, Mineralogy and Microbiology of Cobalt in Mining-Affected Environments. *Minerals* **2021**, *11*, 22. [CrossRef]
67. Roberts, S.; Gunn, G. Cobalt. In *Critical Metals Handbook*; John Wiley & Sons, Inc.: New York, NY, USA, 2014; pp. 122–149.
68. Newsome, L.; Arguedas, A.S.; Coker, V.S.; Boothman, C.; Lloyd, J.R. Manganese and cobalt redox cycling in laterites; Biogeochemical and bioprocessing implications. *Chem. Geol.* **2020**, *531*, 119330. [CrossRef]
69. Megremi, I. Controlling Factors of the Mobility and Bioavailability of Cr and Other Metals at the Environment of Ni-Laterites. Ph.D. Thesis, University of Athens, Athens, Greece, 2010; p. 316.
70. Ettler, V. Soil contamination near non-ferrous metal smelters: A review. *Appl. Geochem.* **2016**, *64*, 56–74. [CrossRef]
71. Arruti, A.; Fernandez-Olmo, I.; Irabien, A. Evaluation of the contribution of local sources to trace metals levels in urban PM2.5 and PM10 in the Cantabria region (Northern Spain). *J. Environ. Monit.* **2010**, *12*, 1451–1458. [CrossRef]
72. World Health Organization (WHO). Guidelines for drinking water quality. In *World Health Organization Geneva*, 4th ed.; World Health Organization (WHO): Geneva, Switzerland, 2017.
73. USEPA. *Literature Review of Contaminants in Livestock and Poultry Manure and Implications for Water Quality*; USEPA: Washington, DC, USA, 2013.
74. Gamaletsos, P.N.; Godelitsas, A.; Filippidis, A.; Pontikes, Y. The Rare Earth Elements Potential of Greek Bauxite Active Mines in the Light of a Sustainable REE Demand. *J. Sustain. Metall.* **2019**, *5*, 20–47. [CrossRef]
75. Maksimovic, Z.J.; Pantó, G. Authigenic rare earth minerals in karst-bauxites and karstic nickel deposits. In *Rare Earth Minerals, Chemistry, Origin and Ore Deposits*; Jones, A.P., Wall, F., Williams, C.T., Eds.; The Mineralogical Society Series 7: Washington, DC, USA, 1996; Chapter 10; pp. 257–279.
76. Ohfuji, H.; Rickard, D. Experimental syntheses of framboids—a review. *Earth Sci. Rev.* **2005**, *71*, 147–170. [CrossRef]
77. Heim, C. An Integrated Approach to the Study of Biosignatures in Mineralizing Biofilms and Microbial Mats. Ph.D. Thesis, University of Göttingen, Göttingen, Germany, 2010; p. 175.
78. MacLean, L.C.W.; Pray, T.J.; Onstott, T.C.; Brodie, E.L.; Hazen, T.C.; Southam, G. Mineralogical, chemical and biological characterization of an anaerobic biofilm collected from a borehole in a deep gold mine in South Africa. *J. Geomicrobiol.* **2007**, *24*, 491–504. [CrossRef]
79. Hudson-Edwards, K.A.; Santini, J.M. Arsenic-Microbe-Mineral Interactions in Mining-Affected Environments. *Minerals* **2013**, *3*, 337–351. [CrossRef]
80. Haferburg, G.; Kothe, E. Microbes and metals: Interactions in the environment. *J. Basic Microbiol.* **2007**, *47*, 453–467. [CrossRef]
81. Pan, C.; Liu, H.; Catalano, J.G.; Qian, A.; Wang, Z.; Giammar, D.E. Rates of Cr(VI) generation from CrxFe1-x(OH)3 solids upon reaction with manganese oxide. *Environ. Sci. Technol.* **2007**, *51*, 12416–12423. [CrossRef]
82. Vithanage, M.; Kumarathilaka, P.; Oze, C.; Karunatilake, S.; Seneviratne, M.; Hseu, Z.Y.; Gunarathne, V.; Dassanayake, M.; Ok, Y.S.; Rinklebe, J. Occurrence and cycling of trace elements in ultramafic soils and their impacts on human health: A critical review. *Environ. Int.* **2019**, *131*, 104974. [CrossRef]
83. Laskou, M.; Economou-Eliopoulos, M. The role of micro-organisms on the mineralogical and geochemical characteristics of the Parnassos-Ghiona bauxite deposits, Greece. *J. Geochem. Explor.* **2007**, *93*, 67–77. [CrossRef]
84. Laskou, M.; Economou-Eliopoulos, M. Bio-mineralization and potential biogeochemical processes in bauxite deposits: Genetic and ore quality significance. *Mineral. Petrol.* **2013**, *107*, 471–486. [CrossRef]
85. Baskar, S.; Baskar, R.; Kaushik, A. Role of microorganisms in weathering of the Konkan-Goa laterite formation. *Curr. Sci.* **2003**, *85*, 1129–1134.
86. Russell, M.J.; Hall, A.J.; Boyce, A.J.; Fallick, A.E. On hydrothermal convection systems and the emergence of life. *Econ. Geol. 100th Anniv. Spec.* **2005**, *100*, 419–438.
87. Southam, G.; Saunders, J.A. The geomicrobiology of ore deposits. *Econ. Geol.* **2005**, *100*, 1067–1084. [CrossRef]
88. Lovley, D.R.; Phillips, E.J.P. Availability of ferric iron for microbial reduction in bottom sediments of the freshwater tidal Potomac River. *Appl. Environ. Microbiol.* **1986**, *52*, 751–757. [CrossRef] [PubMed]

89. Myers, C.R.; Nealson, K.H. Bacterial manganese reduction and growth with manganese oxide as the sole electron acceptor. *Science* **1988**, *240*, 1319–1321. [CrossRef] [PubMed]
90. Perez, J.; Jeffries, T.W. Roles of manganese and organic acid chelators in regulating lignin degradation and biosynthesis of peroxidases by Phanerochaete chrysosporium. *Appl. Environ. Microbiol.* **1992**, *58*, 2402–2409. [CrossRef]
91. Raptis, S.; Gasparatos, D.; Economou-Eliopoulos, M.; Petridis, A. Chromium uptake by lettuce as affected by the application of organic matter and Cr(VI)-irrigation water: Implications to the land use and water management. *Chemosphere* **2018**, *210*, 597–606. [CrossRef]

Article

Metal Mobility in Embryonic-to-Proto-Ni-Laterite Profiles from Non-Tropical Climates

José María González-Jiménez [1,*], Cristina Villanova-de-Benavent [2], Lola Yesares [3], Claudio Marchesi [1,4], David Cartwright [1,4], Joaquín A. Proenza [2], Luis Monasterio-Guillot [4] and Fernando Gervilla [1,4]

[1] Instituto Andaluz de Ciencias de la Tierra (CSIC-UGR), Avda. Palmeras 4, 18100 Armilla, Spain; claudio@ugr.es (C.M.); davidcb94@gmail.com (D.C.); gervilla@ugr.es (F.G.)

[2] Departament de Mineralogia, Petrologia i Geologia Aplicada, Facultat de Ciències de la Terra, Universitat de Barcelona (UB), Martí i Franquès 1, 08028 Barcelona, Spain; cvillanovadb@ub.edu (C.V.-d.-B.); japroenza@ub.edu (J.A.P.)

[3] Departamento de Mineralogía y Petrología, Facultad de Ciencias Geológicas, Universidad Complutense de Madrid, C/José Antonio Nováis 2, 28040 Madrid, Spain; myesares@ucm.es

[4] Departamento de Mineralogía y Petrología, Facultad de Ciencias, Universidad de Granada, Avda. Fuentenueva s/n, 18071 Granada, Spain; luismonasterio@ugr.es

* Correspondence: jm.gonzalez.j@csic.es

Abstract: We evaluated the mobility of a wide suite of economic metals (Ni, Co, REE, Sc, PGE) in Ni-laterites with different maturities, developed in the unconventional humid/hyper-humid Mediterranean climate. An embryonic Ni-laterite was identified at Los Reales in southern Spain, where a saprolite profile of ~1.5 m thick was formed at the expense of peridotites of the subcontinental lithospheric mantle. In contrast, a more mature laterite was reported from Camán in south-central Chile, where the thicker (~7 m) weathering profile contains well-developed lower and upper oxide horizons. This comparative study reveals that both embryonic and mature laterites can form outside the typical (sub)-tropical climate conditions expected for lateritic soils, while demonstrating a similar chemical evolution in terms of major (MgO, Fe_2O_3, and Al_2O_3), minor (Ni, Mn, Co, Ti, Cr), and trace (REE, Y, Sc, PGE, Au) element concentrations. We show that, even in the earliest stages of laterization, the metal remobilization from primary minerals can already result in uneconomic concentration values.

Keywords: PGE; REE; Sc; Co; Ni-rich serpentinite; Mn-oxyhydroxides; laterite

Citation: González-Jiménez, J.M.; Villanova-de-Benavent, C.; Yesares, L.; Marchesi, C.; Cartwright, D.; Proenza, J.A.; Monasterio-Guillot, L.; Gervilla, F. Metal Mobility in Embryonic-to-Proto-Ni-Laterite Profiles from Non-Tropical Climates. *Minerals* **2023**, *13*, 844. https://doi.org/10.3390/min13070844

Academic Editor: Maria Boni

Received: 22 May 2023
Revised: 20 June 2023
Accepted: 21 June 2023
Published: 22 June 2023

1. Introduction

Ni-Co laterite deposits account for about 60% of world's land-based Ni resources [1] and mineable resources of Co and Sc. Furthermore, laterites are steadily gaining ground as potential sources of other critical metals like rare earth elements (REE) and platinum-group elements (PGE) [2,3]. Ni-Co laterites are regoliths formed through the weathering of mafic-to-ultramafic bedrocks under tropical and subtropical (i.e., hot and humid) climates, where high rates of annual average rainfall (>1000 mm/yr) and high temperatures (15–31 °C) are common [4]. In fact, most modern laterites are located at tropical latitudes (±23° Lat.), where these specific conditions are met [5] (Figure 1). Nevertheless, the observation of laterites out of the tropic suggests the existence of paleo-laterites displaced from their original location by tectonic drift, as well as suggesting that other interacting factors, besides temperature and rainfall, also play a role in their genesis, i.e., the lithology of the parental rock, drainage, topography, and the abundance of organic matter [6,7].

Despite the existence of good knowledge about laterite formation, there is still uncertainty about the mechanisms operating when the ultramafic rock is first transformed into an embryonic laterite. In particular, how metals of economic interest behave at these initial stages in the formation of the weathering profile is mostly unknown. This is mainly because most laterite profiles in tropical systems are very evolved, and the initial stages of alteration are usually obliterated.

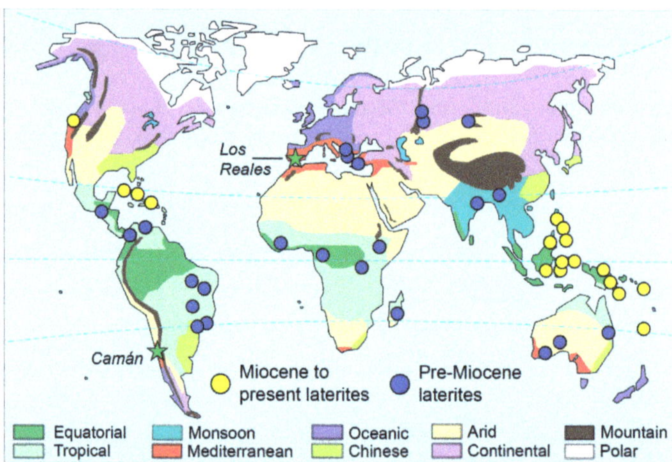

Figure 1. Locations of Los Reales and Camán weathering profiles (green stars) and major Ni-Co laterite districts [5] in the context of the global present-day distribution of climate zones.

This paper aims to fill this gap in knowledge by providing insights on the behavior of economic metals (i.e., Ni, Co, REE, Sc, PGE) in weathering profiles that display contrasting geological settings and alteration histories, while being exposed to an unusual climate regime for laterite formation, i.e., sub-tropical, semi-oceanic, humid/hyper-humid Mediterranean climate (Figure 1). Other researchers [6] have shown that this climate regime is the most optimal for evaluating the initial-to-intermediate stages of the formation of supergene profiles on ultramafic rocks leading to the concentration of economic metals. For this study, we selected: (1) a laterite profile developed from the alteration of ultramafic rocks of the subcontinental lithospheric mantle (SCLM) units exposed in Los Reales, Serranía de Ronda, in the Málaga province of Spain [8]; and (2) a laterite profile formed from the weathering of oceanic lithospheric mantle (OLM) peridotites exposed in Camán, Coastal Cordillera, in the Valdivia province of Chile [9–11]. The mantle peridotites of Ronda are products of partial melting and kilometric scale refertilization by asthenospheric melts in the SCLM [12] and were hydrated at low P–T conditions [13], whereas those from the Coastal Cordillera of Chile originated in an oceanic setting and were later hydrated at higher P–T conditions during subduction [14]. Our study revealed that, outside the typical climate conditions expected for laterization, embryonic-to-proto laterites might also form, and economic metals may be already mobilized to certain degree of intensity.

2. Geology, Climate Setting, and Field Characteristics of the Weathering Profiles

We selected two weathering profiles developed on ultramafic rocks of contrasting origins but exposed to similar humid/hyper humid Mediterranean climate regimes. Their locations as they pertain to climate and regional geology are shown in Figures 1–3.

2.1. The Los Reales Range, Serranía de Ronda, in Southwestern Spain

The first weathering profile is located in the Los Reales range at the highest altitude of Sierra Bermeja, in the westernmost part of the Málaga province in SW Spain (Figure 2). The geology of this area is dominated by the Ronda ultramafic massif, which is the largest outcrop of SCLM peridotites exposed on Earth (~300 km^2). Dated as Proterozoic, this SCLM was emplaced into the continental crust in the early Miocene, likely during the development of a back-arc basin in a suprasubduction setting behind the Betic–Rif orogenic wedge [15].

The Ronda massif mainly consists of lherzolites and harzburgites, with minor dunites and pyroxenite layers (usually <10%). These rocks are arranged in a kilometric-scale petrological and structural zoning, consisting, from the top to the bottom of the mantle section, of the following domains (Figure 2): (1) spinel ($\pm$garnet) tectonite consisting of foliated spinel tectonites, bounded by garnet-spinel mylonites in direct contact with the crustal units and corresponding to the exhumed SCLM roots (~2 GPa-900 °C; [16]); (2) granular peridotite formed through the heating and melting of pre-existing spinel ($\pm$garnet) tectonite due to the upwelling of the asthenosphere during unroofing (~1.5 GPa-1250 °C; [16]); and (3) plagioclase tectonite corresponding to shear zones, originated shortly before or jointly to the crustal emplacement (<1 GPa-1000 °C; [16]). A narrow (ca. 200–400 m wide) and continuous (ca. 20 km long in the Ronda massif) transitional zone, referred to as the recrystallization front, separates the spinel tectonites from the granular peridotite domain. The circulation of fluids through the fracture network produced intensive hydrothermal alteration and the formation of serpentines and deposits of talc, asbestos, and vermiculite [17].

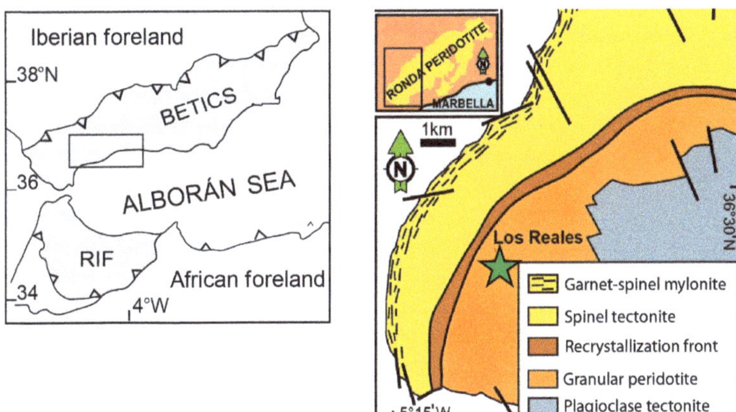

Figure 2. Main geological domains of the SW portion of the Ronda peridotite massif (after [12]) in the western portion of the Betic-Rifean orogenic belt, showing the location of the Los Reales profile.

The Los Reales weathering profile developed under a unique climatic area of the Gibraltar strait. Here, a rainfall regime exceeding 2000 mm per year is controlled by the reliefs of the strait, where flows of humidity from the Atlantic Ocean penetrate through the Gulf of Cadiz in topographic depressions. The Los Reales range faces this humidity to the west and experiences average annual temperatures close to 10 °C at altitudes >1000 m asl [18]. These general climatic features, along with its geological and edaphic singularities, make Sierra Bermeja a paleoecological refuge for an endemic floristic element, i.e., the Spanish fir (Abies pinsapo Boiss) [19].

2.2. The Camán Ultramafic Body, Chilean Coastal Cordillera

This weathering profile developed on the Camán ultramafic body, which is located 27 km southeast of the city of Valdivia in Chile (Figure 3). This is one of the largest known occurrences of ultramafic rocks of oceanic origin preserved in the Late Paleozoic accretionary prism of the Coastal Cordillera of south-central Chile. The latter is a paired metamorphic belt comprising two parallel N–S-trending metamorphic units known as the Western and the Eastern Series, which are characterized by different metamorphic grades and rock assemblages.

The Eastern Series consists of slightly deformed meta-sedimentary rocks (metagreywackes and metapelites), which experienced high-T and low-tointermediate-P metamorphism by the intrusion of the Coastal Batholith ca. 320–300 Ma ago [20,21]. In contrast, the Western Series consists of metapelites and metapsammites of continental origin that locally host metamorphic rocks of oceanic affinity (greenschists, blueschists, amphibolites, metasediments, metavolcanics, metagabbros, and metaperidotites). All of these rocks were affected by high-pressure and low-temperature (HP–LT) metamorphism under greenschist and blueschist facies with maximums of 2.0–2.5 GPa and <600 °C recorded in serpentinitic rocks [22]. One study [14] investigated the geochemistry and mineralogy of the Camán ultramafic body and showed that it is a portion of lherzolitic upper oceanic mantle affected by antigoritization at medium P–T conditions in a subduction channel.

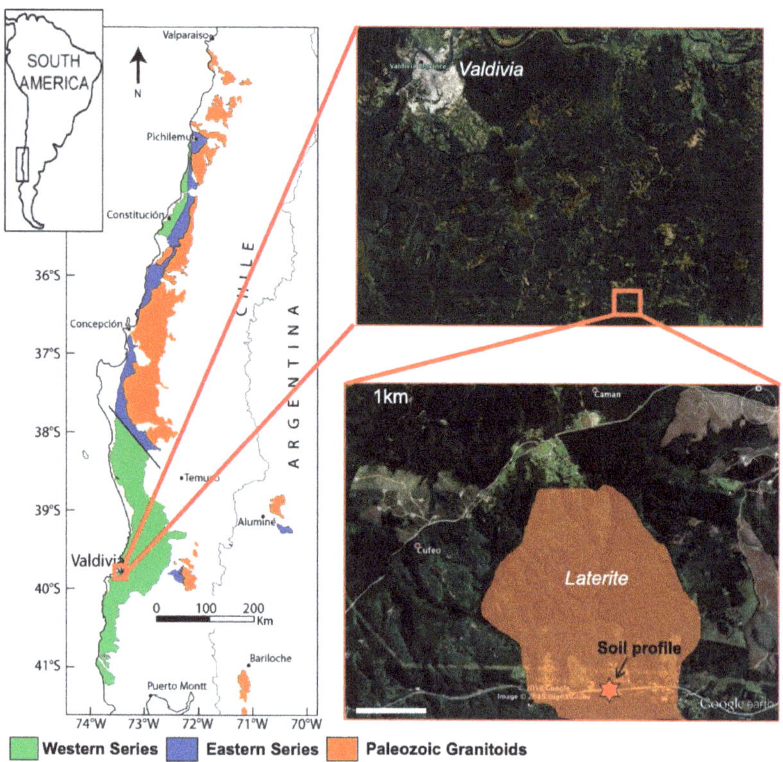

Figure 3. Main geological domains of the Coastal Cordillera (after [22]) of south-central Chile, showing the location of the Camán laterite.

The climate of the Coastal Cordillera at 40° S is characterized by a mean annual temperature of 11 °C and an average annual precipitation above 2000 mm per year in coastal areas and lowlands [23]. Precipitation is more intense during the fall and winter seasons (March–August). This cold and humid climate supports the growth of abundant vegetation in native forests (i.e., the Valdivian rainforest) and patches of exotic anthropic plantation of Pinnus radiata and Eucalyptus. In this region, relatively thick weathering profiles have been developed on the metamorphic rocks, mainly on ultramafic peridotites.

2.3. Local Geology of the Studied Weathering Profiles

The weathering profile at Los Reales covers an area of a few tens of m^2 and is best exposed in the road trench along a segment of ~10 m. Considering the road as a baseline, the soil profile reaches here up to 2 m in thickness and consists, from bottom to top, of floating blocks of the parent bedrock (i.e., partially serpentinized peridotite) of up to 0.5 m high and saprolite of up to ~1.5 m thick (Figure 4).

Figure 4. Field relationships of the studied weathering profiles. (**a**) Saprolite profile derived from the SCLM peridotites exposed at Los Reales in Spain (2 m length of measuring tape as scale). (**b**) Laterite profile of Camán from the weathering of oceanic peridotites cropping out in the Coastal Cordillera of Chile.

The weathering profile at Camán in south-central Chile covers an area of approximately 5 km^2 and is well recognized along a segment of ~60 m in a road trench of Route 206 from Valdivia to Paillaco (Figure 4). Considering the road as a baseline, the soil profile here reaches up to 7 m in thickness and consists, from bottom to top, of isolated blocks of the parent rock (i.e., serpentinite) of up to 2 m high; a lower oxide zone (up to ~5 m high), containing floating blocks of serpentinite; and an upper oxide zone (~1–5 m thick) (Figure 4). These are distinguished in the field by their greenish, pale-yellow, and reddish-yellow colors, respectively.

3. Materials and Methods

3.1. Sampling

A total of five samples, including one from bedrock (Sample 0) and four of saprolite (from Samples 1 to 4 upward), were collected from the weathering profile of Los Reales in southwestern Spain, while two from the bedrock (Samples 0 and 1) and five from oxide zone (from Samples 2 to 6 upward) were collected from the weathering profile of Camán in south-central Chile.

Polypropylene cilindric tubes were hand-drilled through each horizon of the selected weathering profiles. Each one of the samples weighted 1.5–3 kg and was the composite of two centimeter-sized borehole samples extracted within a single horizon. The extraction sites were selected on the basis of changes in the macroscopic appearance of the soil material, taking into account color, texture, and the degree of aggregation. The samples were dried at 60 °C over 48 h, then crushed and pulverized for geochemical and mineralogical characterization. Aliquots were also employed to obtain several thin sections, polished thin sections, and cylindrical polished Epoxy® mounts.

3.2. Geochemistry of Major, Minor, and Trace Elements

3.2.1. Major, Minor, and Trace Elements

Bulk major, minor, and trace elements were analyzed using ICP-OES or ICP-MS at Actlabs (Ancaster, ON, Canada). Sample dissolution was performed using a lithium metaborate/tetraborate fusion, and molten beads were digested in a 5% nitric acid solution. For Cu and Zn, digestion was performed using multiacid solutions (HF-HNO$_3$-HClO$_4$). Element concentrations were determined through external calibration with certified international standards. The compositions of reference materials DNC-1, SY-4, BIR-1a, BCR-2, W-2, OREAS-101b, USZ-25, and ZW-C were analyzed as unknowns during the analytical runs and showed good agreement with the working values of these international standards. Bulk major, minor, and trace element data are reported in the Supplementary Material (Table S1).

3.2.2. Platinum-Group Elements and Gold

Concentrations of platinum-group elements (Ir, Ru, Rh, Pt, and Pd) and gold in bulk rocks and soils were also determined at Actlabs. Noble metals were measured in 30 g of sample by NiS fire assay (NiS-FA) and ICP-MS. A nickel sulfide bead collected noble metals, which were then dissolved in HCl and re-dissolved in aqua regia prior to ICP-MS analysis. Detection limits in this study were 1 ppb for each noble metal. The analyses of the OREAS 13b-certified reference material compared well with the published values of these standards. Table S1 in the Supplementary Material shows the concentrations of PGE and Au in the analyzed weathering profiles. More details on the analytical methods for PGE and Au at Actlabs have been reported by [24].

3.3. Mineralogy
3.3.1. X-ray Powder Diffraction (XRPD)

Powdered samples from Los Reales were analyzed in a Bruker D8 Advance diffractometer with Cu-Kα radiation at 40 kV and 30 mA at the Universidad Complutense de Madrid (Spain). Randomly oriented powders were scanned from 5 to 55° (2θ) at a scanning rate of 2.0°/min. To confirm the presence of expandable clay silicates in the samples, the <2 μm fraction was separated from the coarse sample by ultrasonic dispersion and centrifugation. Oriented aggregate samples were prepared by mounting the suspended solution of the <2 μm fraction on glass slides and drying at room temperature. These oriented samples were solvated with ethylene glycol vapor in a desiccator for 24 h and heated in an oven for one hour at a temperature of 500 °C. The patterns of samples on oriented mounts were obtained by XRPD (model Bruker D8 Advanced) analysis with Cu-Kα radiation at 30 kV and 20 mA. The samples were scanned from 2 to 30° (2θ) at a scanning rate of 0.02°/s.

The powdered samples from Camán were analyzed in a PANalytical X'Pert PRO MPD Alpha1 diffractometer in a Bragg-Brentano h/2h geometry with a 240 mm radius and nickel filtered Cu Kα radiation (k = 1.5418 Å), at 45 kV and 40 mA, at the Scientific and Technological Centers of the Universitat de Barcelona (CCiT-UB, Barcelona, Spain). During analysis, the randomly oriented powder sample was spun at 2 revolutions per second, a variable divergence slit kept an area illuminated constant (10 mm), and a mask was used to limit the length of the beam (12 mm). Axial divergence Soller slits of 0.04 radians were used. Powdered samples were scanned from 4 to 80° 2θ with a step size of 0.017° and a measuring time of 50 s per step, using a X'Celerator detector (active length = 2.122°). The mineral identification of all samples was obtained using the XPert HSP software. Phase identification was performed using the PDF-2 database.

3.3.2. Scanning Electron Microscope and Electron Micro Probe Analyzer

Polished Epoxy® mounts of laterites and the thin sections of protolith were examined by means of transmitted and reflected light petrographic microscopy at the Departament de Mineralogia, Petrologia i Geologia Aplicada de la Facultat de Ciències de la Terra (Universitat de Barcelona, Spain). They were later carbon coated and studied using two field emission scanning electron microscopes (FESEM), the Quanta 200 FEI and the XTE 325/D8395, along with an INCA energy dispersive spectrometer (EDS) microanalysis system and a Jeol JSM-7100 equipped with an INCA Energy 250 EDS, at 20 kV and 5 nA, at the CCiT-UB. Chemical analyses were obtained in an electron probe microanalyzer

(EPMA) at the CCiT-UB using a JEOL JXA-8230 electron microprobe, equipped with five wavelength-dispersive spectrometers and an energy-dispersive spectrometer. The operating conditions were 20 kV accelerating voltage, 15 nA beam current, 2 μm beam diameter, and a counting time of 20 s per element. The calibration standards used were wollastonite (Si, Ca), corundum (Al), orthoclase (K), hematite (Fe), periclase (Mg), rhodonite (Mn), NiO (Ni), metallic Co (Co), Cr_2O_3 (Cr), and barite (Ba).

4. Results

4.1. Los Reales

4.1.1. Geochemistry

The SiO_2 and MgO contents decreased from bedrock (from ~42 to ~35 wt.% and from 37.41 wt.% to 17 wt.%, respectively) to the saprolite (Figure 5). This trend anticorrelated with the parallel increases in Al_2O_3 (~2 to 8.5 wt.%), Fe_2O_3 (from 8.5 to 26.7 wt.%), Cr_2O_3, TiO_2, and MnO (<1.5 wt.%). Cobalt (from 105 ppm to 301 ppm), Cu (from ~16 to ~47 ppm), Zn (from ~38 to ~96 ppm), and V (from 63 to 166 ppm) also normally increased upwards in the profile, while Ni increased from the bedrock to Sample 2 from the saprolite unit (2680 ppm to 5370 ppm) and then decreased closer to the top (Figure 5; Table S1).

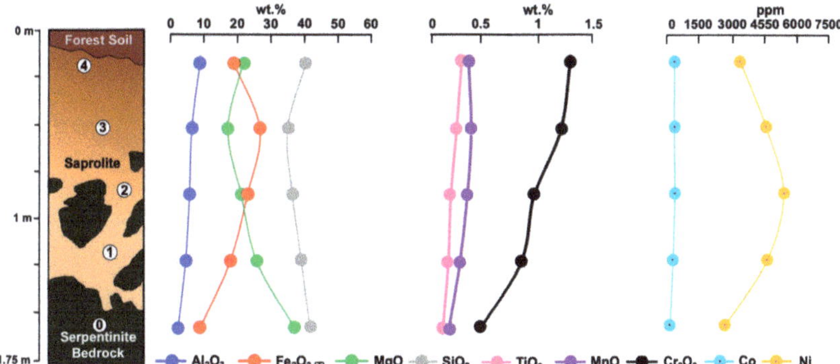

Figure 5. Schematic weathering profile at Los Reales, showing variations of selected major and minor elements. Numbers 0-4 indicate location of the studied samples.

The contents of REE were <40 ppm (Table S1), with light rare earth elements (LREE: La + Ce + Pr + Nd + Sm) more abundant than heavy rare earth elements (HREE: Ho + Er + Tm + Yb + Lu) (Figure 6). Both LREE and HREE experienced a steady enrichment from bottom to top in the profile (Figure 6). Yttrium and Sc also increased from the parent bedrock to the top of the profile (from 1 to 8 ppm and from 13 to 32 ppm, respectively; Figure 6). The concentrations of noble metals were very low, with total PGE (Ir + Ru + Rh + Pt + Pd) ranging from 30 to 55 ppb and Au ≤ 3 ppb (Figure 6). The partially serpentinized bedrock had the lowest PGE contents, while the highest concentrations were detected in the upper middle portion of the profile (Sample 3; Figure 6), which correlated with the Pt/Pd ratios decreasing from 0.70 to 0.23. The Au content remained negligible and almost constant across the weathering profile.

4.1.2. Mineralogy

The parent bedrock at Los Reales was a partially serpentinized lherzolite with olivine, orthopyroxene, and clinopyroxene. Olivine was altered to lizardite with a mesh texture and to magnetite; orthopyroxene was replaced by "bastite" lizardite pseudomorphs. All of them were crosscut by post-dating veinlets of lizardite and/or chrysotile (Figure 7a–d). Chromium spinel, often with alteration rims enriched in Fe and/or magnetite, was the most common accessory mineral.

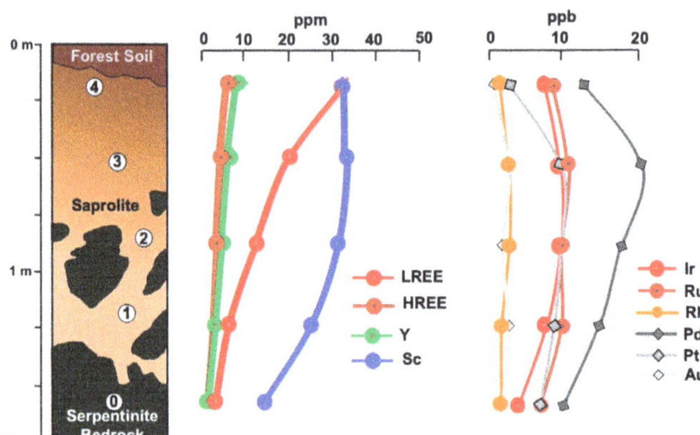

Figure 6. Schematic weathering profile at Los Reales, showing the location of samples and variation in LREE, HREE, Y, Sc, PGE, and Au contents. Numbers 0-4 indicate location of the studied samples.

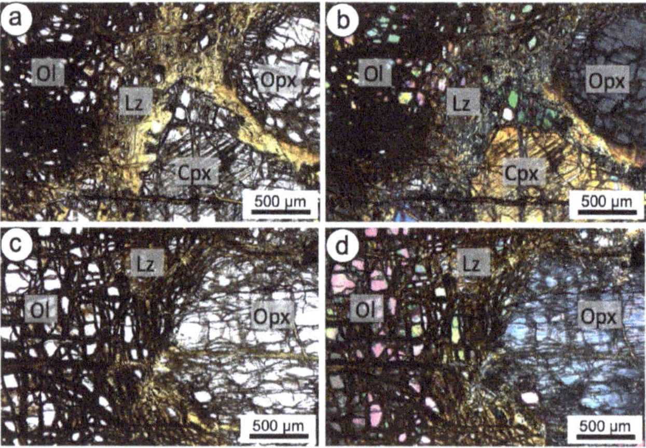

Figure 7. Optical photomicrographs of the typical mineralogy and fabric of the lherzolite parent bedrock of the Los Reales weathering profile. This rock, shown in images (**a**–**d**), consists of olivine (Ol), partially replaced by mesh-textured serpentine (lizardite; Lz) coexisting with porphyroblastic orthopyroxene (Opx) and clinopyroxene (Cpx). (**a,c**) are plane-polarized light images, whereas (**b,d**) are the corresponding cross-polarized micrographs.

The X-ray diffraction patterns of the four samples from the Los Reales profile confirmed that the main relict magmatic minerals were forsterite, enstatite, chromium spinel, and diopside, whereas actinolite, goethite, chlorite, and quartz were the main phases of a secondary (i.e., hydrothermal to supergene) origin (Figure 8). The presence of chlorite as a major clay component was conformed through XRD on clay aggregates (Figure S1).

A combination of optical microscopy, FESEM, and EPMA revealed a fine-grained groundmass of goethite enveloping the relict grains of partially altered olivine, pyroxenes, amphibole, and chromium spinel (Figure 9). Relict olivine was replaced by serpentine (Figure 9a,e,i,m), whereas orthopyroxene and clinopyroxene grains were partially altered along their rims into complex, fine-grained intergrowths of Mg-phyllosilicates as smectite and/or chlorite (Figure 9b,f,j,n).

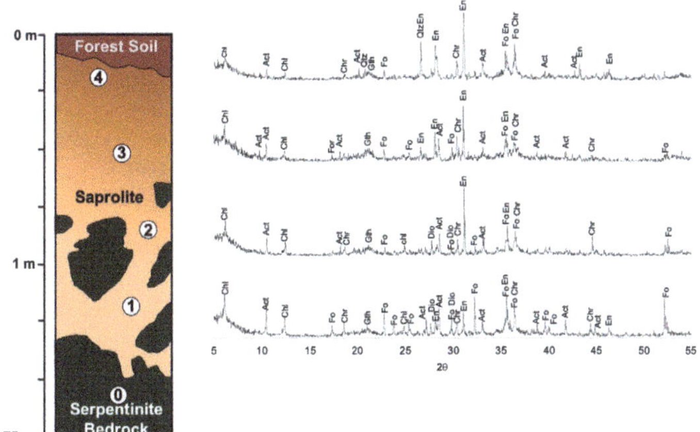

Figure 8. X-ray diffraction patterns of the four analyzed samples collected from the Los Reales weathering profile. Key: Fo: forsterite; En: enstatite; Chr: Chromium spinel; Dio: diopside; Act: actinolite; Gth: goethite; Chl: chlorite; Qtz: quartz. Numbers 0-4 indicate location of the studied samples.

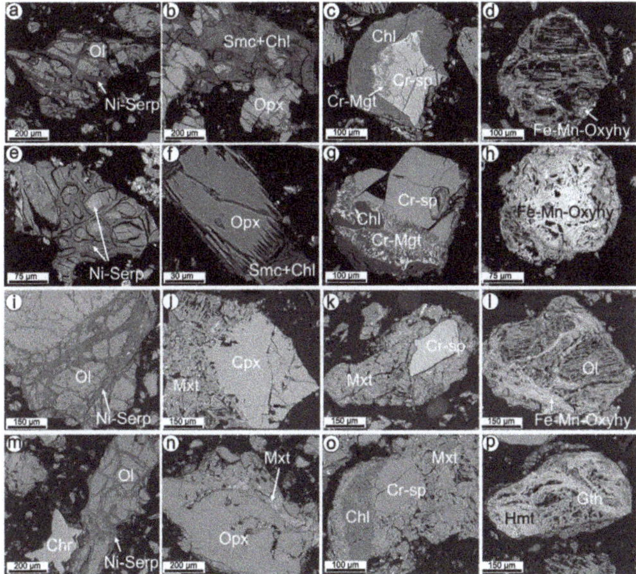

Figure 9. Backscattered scanning electron images (SEM-BSE) showing the textures of minerals of the Los Reales weathering profile. (**a–d**) Sample 1, (**e–h**) Sample 2, (**i–l**) Sample 3, and (**m–p**) Sample 4. Key: Ol: olivine; Opx (orthopyroxene); Cpx (clinopyroxene); Chl: chlorite; Serp: serpentine; Cr-sp: chromium spinel; Cr-Mgt: Cr-magnetite; Fe-Mn-Oxy-hy: Fe-Mn-oxy-hydroxides; Gth: goethite; Hmt: hematite; Mxt: mixture of silicate and oxides.

Chromium spinel grains with Cr-magnetite plus chlorite alteration rims were common at the bottom of the profile close to the parent bedrock, whereas in the uppermost section of the laterite, relict chromium spinel was more rounded and exhibited coatings of chlorite and/or fine-grained mixtures of chlorite with Fe-oxy-hydroxides and Fe-oxides (goethite and hematite; (Figure 9c,g,k,o). Fe- and Mn-oxy-hydroxides were very common across the whole profile, and they were found as pseudomorphs after olivine or irregular concretional aggregates (Figure 9d,h,j,p).

Serpentine had variable NiO contents (0.47–1 wt.%) that were higher than the parental olivine (<0.45 wt.%) and pyroxene (<0.12 wt.%) (Figure 10; Table S2). The average structural formula of serpentine is $Mg_{2.49}Fe^{3+}_{0.30}Ni_{0.03}(Si_{1.99}Al_{0.02}O_5)(OH)_4$. Olivine is typically magnesian (Fo_{88-91}), and pyroxenes consist of enstatite (average structural formula: $Mg_{1.72}Fe_{0.18}Al_{0.09}Cr_{0.01}(Si_{1.91}Al_{0.05})O_6$; En_{90-91}) and diopside (average structural formula: $Ca_{0.87}Mg_{0.91}Fe_{0.08}Al_{0.08}Cr_{0.03}(Si_{1.92}Al_{009})O_6$; Di_{91-93}).

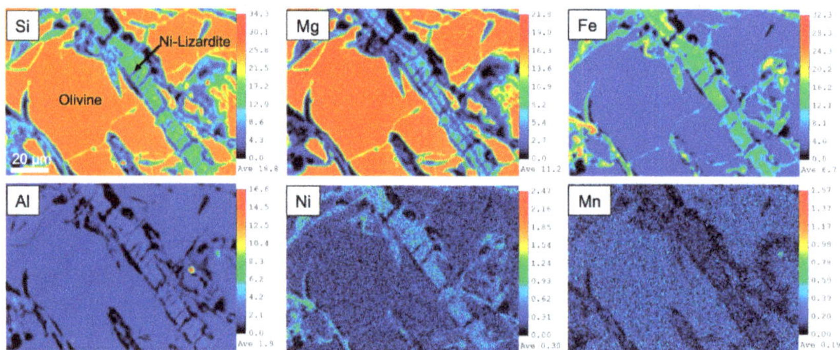

Figure 10. Quantitative (wt.%) wavelength-dispersive spectrometry (WDS) X-ray elemental maps of the mesh-serpentine replacing olivine in the Los Reales weathering profile.

Chromium spinel bore trace amounts of MnO (up to 0.11 wt.%), NiO (up to 0.36 wt.%), and CoO (up to 0.1 wt.%), and larger amounts of Cr_2O_3 (up to 23.59 wt.%) than the aforementioned silicates. Its structural formula is $(Mg^{2+}_{0.72}Fe^{2+}_{0.24})(Al^{3+}_{1.53}Cr^{3+}_{0.37})O_4$, with Cr# (Cr/Cr + Al) ranging from 1.38 to 1.67 and Mg# (Mg/(Mg + Fe^{2+})) from 0.69 to 0.76.

Goethite showed significant contents of MnO (up to 7.4 wt.%), NiO (up to 2.04 wt.%), Cr_2O_3 (up to 0.44 wt.%), and CoO (up to 0.40 wt.%), as well as conspicuous amounts of SiO_2 (up to 9.43 wt.%) and Al_2O_3 (up to 6.4 wt.%); the latter were very likely related with secondary silica intergrowths (Table S2).

The Fe- and Mn-oxy-hydroxides are fine-grained and poorly crystalline phases, difficult to identify by EMPA and/or X-ray powder diffraction. The volume analyzed by electron microprobe represented a mixture (at least two phases) that contained high Ni (2.97–10.4 wt.%) and Co (1.49–3.18 wt.%) contents (Table S2).

4.2. Camán

4.2.1. Geochemistry

SiO_2 and MgO in Camán strongly decreased upwards from the serpentinite bedrock to the lower oxide zone (from ~40 wt.% to ~15 wt.% and from ~38 wt.% to ~1 wt.%, respectively). A further loss in silica occurred at the transition between the upper and lower oxide zones (~5 wt.%), whereas Al_2O_3 and especially Fe_2O_3 increased from the parent bedrock to the lower and upper oxide zones (Al_2O_3 from ~1 to ~14 wt.% and $Fe_2O_3(t)$ from ~8 to ~50 wt.%) (Figure 11).

Cr_2O_3 decreased while MnO and TiO_2 increased from the parent bedrock upwards, with overall contents of < 3 wt.% (Figure 11). Nickel mirrored the distribution of MnO and was more enriched in the upper oxide zone (up to 6860 ppm) relative to the lower oxide zone (~3500 ppm). A similar increase from the bedrock to upper oxide zone was also observed for Co (from 113 to 890 ppm), Cu (from ~9 to 180 ppm), and Zn (from ~40 to 400 ppm) (Table S1).

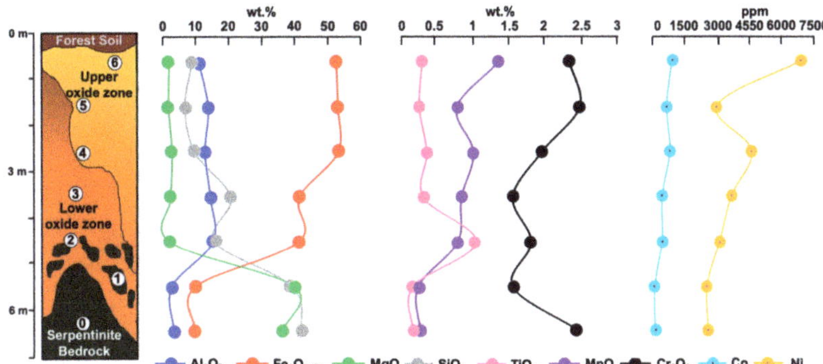

Figure 11. Schematic weathering profile in Camán, showing the location of samples and variations in selected major and minor elements. Numbers 0-6 indicate location of the studied samples.

The REE had total contents of < 240 ppm and showed an overall enrichment in LREE (up to 162 ppm) compared to HREE (up to 31 ppm). Abundances of LREE, HREE, and Y displayed identical distributions within the profile and increased from the bottom to the top, singularly close to the transition between the lower and upper oxide zones (Sample 3) (Figure 12). It is noteworthy that Sc mirrored the distribution pattern of Fe_2O_3, with a sharp increase from the bedrock (~10 ppm) to lower and upper oxide zones (~70 ppm) (Figures 11 and 12).

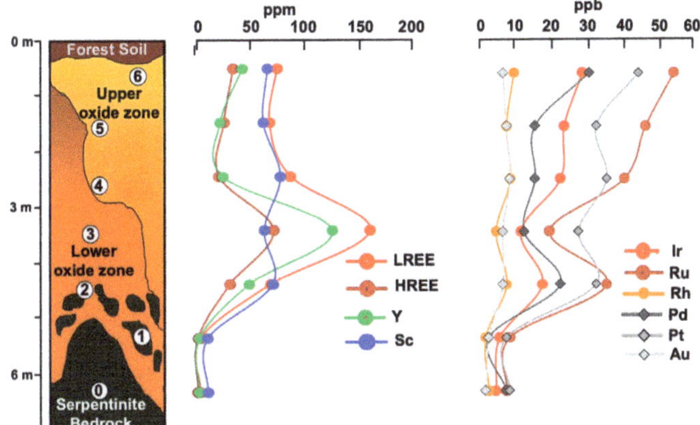

Figure 12. Schematic weathering profile in Camán, showing the locations of samples and variations of LREE, HREE, Y, Sc, PGE and Au contents. Numbers 0-6 indicate location of the studied samples.

The concentrations of noble metals were strongly variable throughout the soil profile, with total PGE = 23–165 ppb and Au = 1–8 ppb, which were higher than the respective figures from the Los Reales profiles (Table S1). The lowest PGE contents were in the serpentinite bedrock (~25 ppb; Pt/Pd = 1.1–3.5), and there was a general trend of enrichment upwards through the lower oxide zone (73–113 ppb; Pt/Pd = 1.5–2.3) and upper oxide zone (120–165 ppb; Pt/Pd = 2.3–1.5) (Figure 12). Gold also increased in content from the serpentinite bedrock (~1 ppb) upwards to the oxide zones (up to 8 ppb) (Figure 12).

4.2.2. Mineralogy

The parent bedrock of the Camán profile was a serpentinite, consisting of randomly oriented and interlocking short, blade-shaped crystals of antigorite (Figure 13a,b), often crosscut by chrysotile veinlets. This rock preserved ghost clinopyroxene morphologies

(Figure 13c,d), accessory chromium spinel partly altered to magnetite, and minor amounts of rutile and ilmenite.

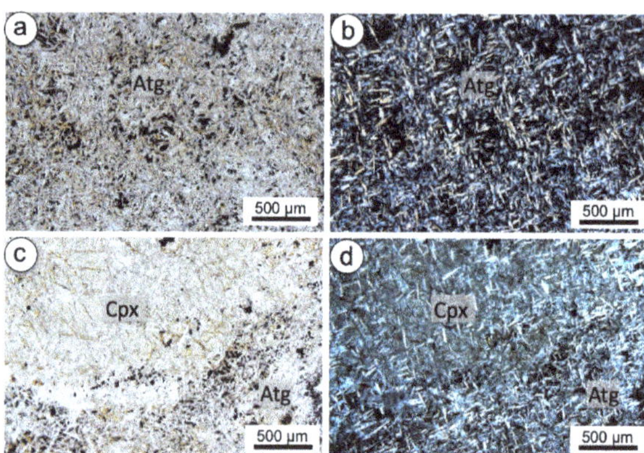

Figure 13. Optical micrographs of the typical mineralogy and fabric of the serpentinite parent rock in the Camán weathering profile. This rock, shown in images (**a–d**), consists of antigorite blade-shaped crystals and clinopyroxene phantoms. (**a,c**) are plane-polarized light images whereas (**b,d**) are the corresponding cross-polarized micrographs. Key: Atg: antigorite; Cpx: clinopyroxene phantom.

The XRPD analyses of the two representative bedrock samples (0 and 1) confirmed the presence of antigorite and, to a lesser extent, chrysotile (Figure S2). Conversely, the lower and upper oxide zones almost completely consisted of goethite, with trace amounts of serpentine at the bottom of the profile (Figure 14).

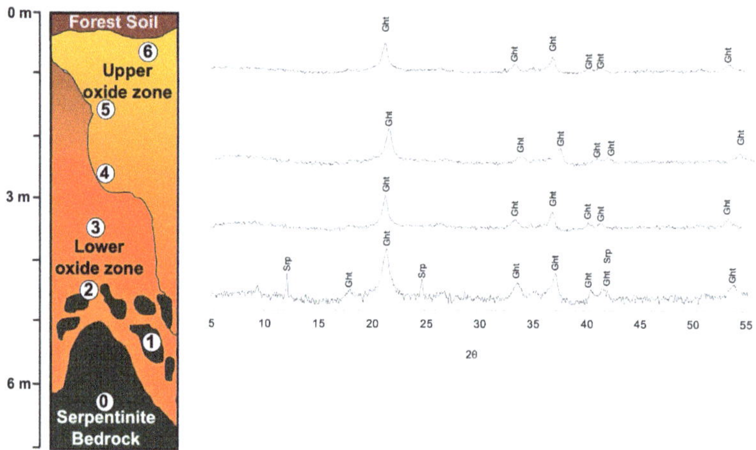

Figure 14. X-ray diffraction patterns of the four analyzed samples collected from the Camán weathering profile. Key: Atg: antigorite; Ctl: chrysotile; Gth: Goethite; Srp: serpentine. Numbers 0-6 indicate location of the studied samples.

Goethite in the upper and lower oxide zones was found as variably crystalline pisolitic and spheroidal aggregates (Figure 15). Other mineral species included chromium spinel, Cr-rich magnetite, zircon, and ilmenite. They had rounded morphologies with partially corroded outlines, locally coated by goethite (e.g., Figure 15c). These relict minerals were progressively less frequent higher in the upper oxide horizon, being replaced by the more abundant Mn-rich oxy-hydroxides of the asbolane–lithiophorite intermediates and minor amounts of Mg-phyllosilicates (chlorite >> smectite+talc). Chlorite and smectite formed fine-grained coatings on Cr- or Ti-rich oxide grains (Cr-rich magnetite and ilmenite; e.g., Figure 15g) and/or Fe- and Mn-rich oxy-hydroxides and oxides (goethite, hematite, asbolane–lithiophorite; e.g., Figure 15o). The Mn-rich oxy-hydroxides also appeared in flaky aggregates of a few micrometers, coexisting with Fe-rich oxy-hydroxides and talc; they may have also formed colloform to botryoidal aggregates of parallel and/or radial fibrous crystals that were tens to a hundreds of micrometers in diameter (Figure 15p).

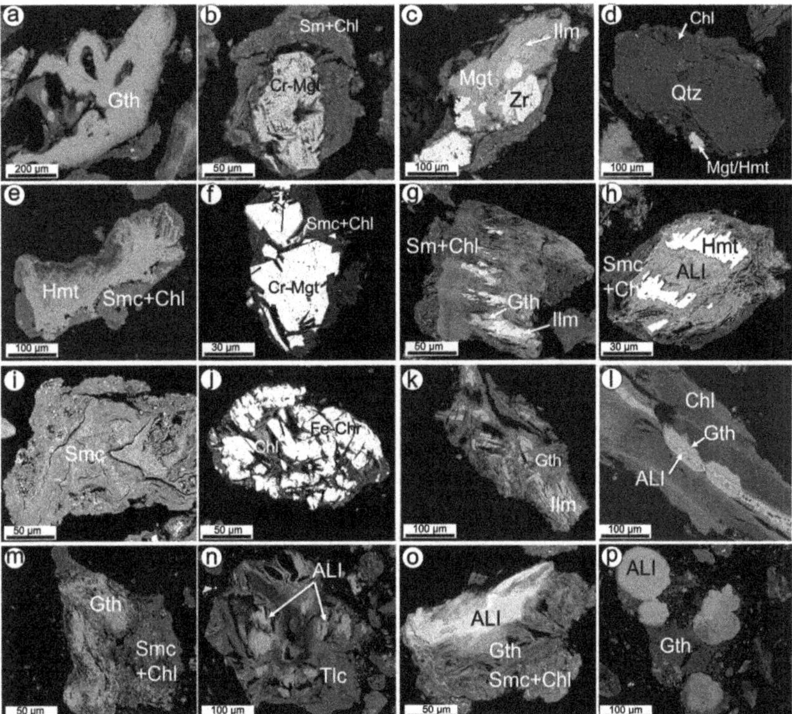

Figure 15. Backscattered electron images (SEM-BSE) showing the textures of minerals in the Camán weathering profile. Samples 2 and 3 from the lower oxide zone are shown in (**a–d**), whereas Samples 4, 5, and 6 from the upper oxide zone are shown in (**e–h**), (**i–l**), and (**m–p**), respectively. Key: Chl: chlorite; Serp: serpentine; Smc: smectite; Tlc: talc; Cr-sp: Chromium spinel; Cr-Mgt: Cr-magnetite; Mgt: magnetite; Ilm: ilmenite; ALI: asbolane–lithiophorite intermediates; Gth: goethite; Hmt: hematite; Zr: zircon.

Goethite had detectable contents of MnO (up to 2.1 wt.%), NiO (1.9 wt.%), Cr_2O_3 (up to 0.73 wt.%), and CoO (up to 0.44 wt.%), as well as high amounts of SiO_2 (up to 9.78 wt.%) and Al_2O_3 (up to 8.73 wt.%), indicative of secondary silica intergrowths (Table S2).

The asbolane–lithiophorite intermediates had the highest contents of NiO (0.6–14.14 wt.%) and CoO (0.56–9.13 wt.%) and bore variable amounts of MnO (23.64 to 43.5 wt.%), Al_2O_3 (4.08 to 15.5 wt.%), and FeO (1.4 to 23.48 wt.%, with most analyses being above 10 wt.%). They also had subordinate SiO_2 (3.3 wt.% on average, with a maximum of 6.53 wt.%) and MgO contents (below 2.73 wt.%). Manganese, Fe, Al, Ni, and Co in the asbolane–lithiophorite from Camán displayed the same correlations as the Fe-Mn-oxy-hydroxides from Los Reales but contained remarkably more Mn, Ni, Co, and Al, and less Fe (Figure 16).

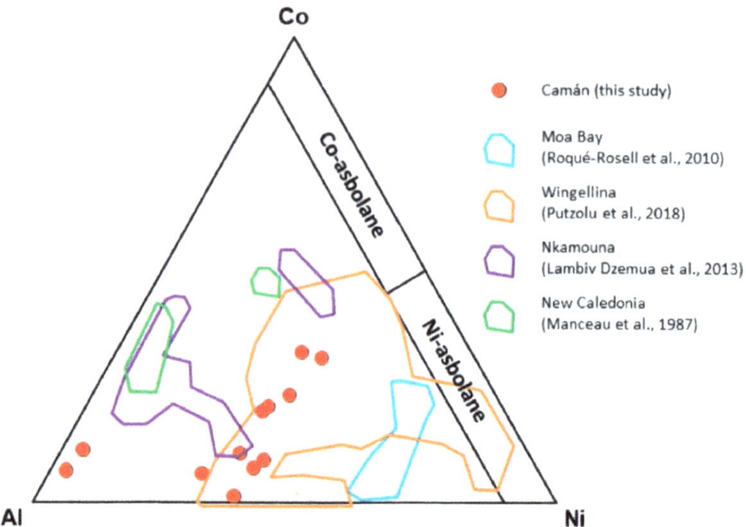

Figure 16. Co–Ni–Al ternary plot (cationic composition wt.%) displaying the mineral chemistry of the lithiophorite–asbolane intermediates from the Camán weathering profile. Compositional fields of other localities worldwide are shown for comparison: Moa Bay, Cuba [25]; Nkamouna, SE Cameroon [26]; New Caledonia [27] and Wingellina, W Australia [28].

5. Discussion

5.1. Maturity of the Ni-Laterite Profiles

In the two studied weathering profiles, Si and Mg decreased whereas Fe, Al, Ni, and Cr increased as the samples approached the top (Figures 5 and 11). However, the intensity of these weathering-related trends differed between the studied profiles. Specifically, these compositional changes can be quantified by using the ultramafic index of alteration (UMIA) proposed by [2]:

$$UMIA = 100 \times [(Al_2O_3 + Fe_2O_{3(T)})/(SiO_2 + MgO + Al_2O_3 + Fe_2O_{3(T)})] \text{ (in molar ratios)}$$

This geochemical index considers MgO and SiO_2 to be the dominant components in the parental rock, while CaO, Na_2O, and K_2O are present in negligible amounts, which is consistent with the compositions of the bedrocks analyzed here (Table S1). According to the UMIA index, a weak weathering rate was envisaged at Los Reales, while a deeper weathering took place at Camán, particularly towards the uppermost part of the profile (Figure 17a).

The parent rocks from the two studied profiles both had an UMIA index of 4. This index reached values between 11 and 18 in Los Reales, whilst it was much higher in the lower (53–60) and upper (73–83) oxide zones of Camán, respectively (Figure 17b; Table S1). The low UMIA values in the Los Reales profile are characteristic of saprolite originating during early laterization. In contrast, the strong Fe enrichment and higher UMIA values in Camán were akin to a more developed laterite produced by a more intense weathering of ultramafic rocks (Figure 17b,c). In the latter profile, the identification of chemically zoned horizons (i.e., lower and upper oxide zones) highlights a differential degree of maturity within the profile itself, like those in oxide-type Ni-laterite deposits dominated by Fe-oxy-hydroxides [29].

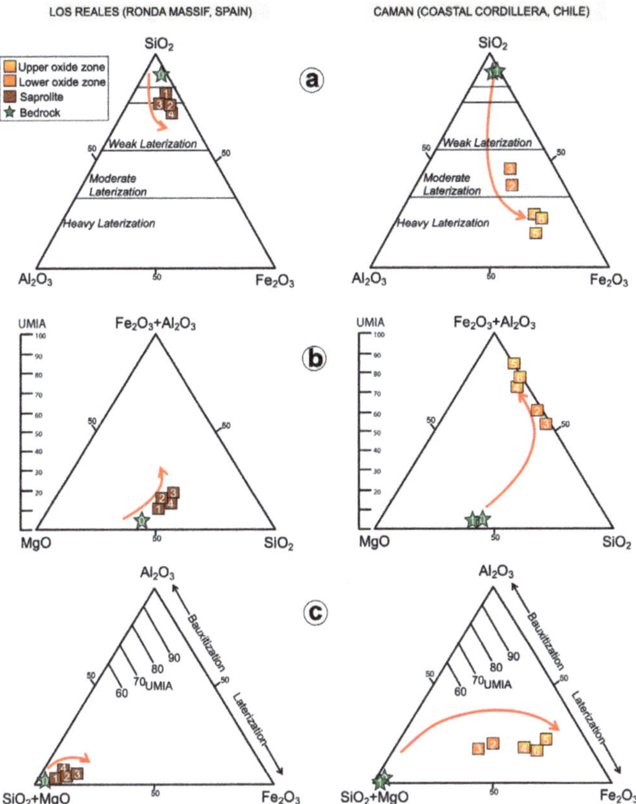

Figure 17. SiO$_2$–Fe$_2$O$_3$–Al$_2$O$_3$–MgO (bulk-rock geochemistry in wt.%) ternary plots showing the weathering trends of Los Reales (left) and Camán (right). (**a**) Ternary SiO$_2$–Fe$_2$O$_3$–Al$_2$O$_3$ diagram illustrating the degree of weathering in relation with decreasing SiO$_2$. (**b**) Ternary Fe$_2$O$_3$+Al$_2$O$_3$–SiO$_2$–MgO diagram displaying the increase in the UMIA [2] with increasing Fe$_2$O$_3$ and Al$_2$O$_3$. (**c**) Ternary Al$_2$O$_3$–Fe$_2$O$_3$–SiO$_2$ + MgO diagram showing the increase in the UMIA with increasing Fe$_2$O$_3$ (laterization) or Al$_2$O$_3$ (bauxitization).

The low UMIA of the Los Reales saprolite (Figure 17b,c; Table S1) supports the presence of abundant mineral relicts (altered olivine, pyroxenes, and chromium spinel) from the partially serpentinized peridotite bedrock. The angular outlines of the broken fragments of orthopyroxene (e.g., Figure 9f) and chromium spinel (e.g., Figure 9c,g,k,o) evidence the residual origin of these grains, resulting from the weathering of the parent rock during incipient laterization. The high resistance of spinel to alteration was very likely due to its relatively slow dissolution kinetics in comparison to other primary phases such as olivine or pyroxene [2,6,30].

Moreover, in the lower portion of the Los Reales profile, Ni-rich serpentine with a mesh texture and "bastite" pseudomorphs, as well as the hydrothermal replacement of chromium spinel by chlorite and/or Cr-magnetite, were preserved. This is also consistent with an incomplete degradation during the early stages of weathering and, to a lesser extent, with mass loss and regolith collapse. In contrast, the observation of fine-grained mixtures of phyllosilicates and oxides coating grains higher up the profile (e.g., Figure 9j–k from Sample 3 and n-o from Sample 4) may have been caused by a more advanced stage of laterization. This transition marks the shift from the pure physical processes of the disaggregation of the parent rock towards more in situ weathering and mass changes involving differential elemental mobility. In this frame, pseudomorphs of secondary Fe- and Mn-oxy-hydroxides and smectite after olivine ($\pm$serpentine) (Figure 9d,j,p) likely formed through acid hydrolysis that partially to completely dissolved the parental rock, according to the equation below [31]:

$2(Mg,Fe)_3Si_2O_5(OH)_4 + 3H_2O =$
Serpentine

$$Mg_3Si_4O_{10}(OH)_2 \bullet 4H_2O + 2Mg^{2+} + FeO(OH) + 3OH^- \tag{1}$$

$$\text{smectite} \qquad\qquad \text{goethite}$$

The serpentine-to-smectite transformation highlighted by Reaction (1) implies a substantial volumetric increase which, in turn, increases porosity. Following this process, a portion of Fe^{2+} structurally founded in olivine and/or serpentinite can be released and easily transported by oxidizing seeping pore waters, being eventually fixed as Fe^{3+} in Fe-oxy-hydroxide-forming, open-space-filling textures (e.g., Figure 9p). Through this reaction, Mg^{2+} was likely leached out of the profile, whereas the Ni released from olivine and/or serpentine could be enriched in the profile through fixation by goethite, thus explaining the slight enrichment in this metal higher up the profile (Figure 5; e.g., [32,33]).

The identification of boxwork Fe-Mn-oxy-hydroxides after olivine grains (Figure 9d,l) indicated the neoformation of these minerals during a very early stage of laterization by the direct replacement of olivine, or as newly formed phases in open fractures.

On the other hand, in the Camán profile, relict minerals (primary silicates and Cr-Ti-Fe oxides) were mainly restricted to the bottom of the lower oxide horizon, whereas higher up, the more abundant minerals were the secondary Fe- and Mn-oxy-hydroxides (goethite >> hematite $\pm$ asbolane–lithiophorite), Mg-phyllosilicates (chlorite >> smectite + talc), and quartz (Figure 15a–p). The colloform and/or botryoidal textures of some Fe- and Mn-oxy-hydroxides (e.g., Figure 15p) suggest that neoformation is the main minerogenetic process, thus explaining the pronounced Fe and Mn gains at the transition between the bedrock and the lower oxide horizon, as well as from the upper portion of the lower oxide zone upwards. The progressive enrichment in Fe and Mn is coupled with a loss in Si and Mg (i.e., the "Mg-discontinuity"; [29]), which could reflect the transition from more acidic conditions at the top towards the bottom of the weathering profiles.

5.2. Mineralogical Siting of Economic Metals: Clues on Enrichment Mechanisms
5.2.1. Nickel and Cobalt

Overall, both studied profiles experienced a Ni enrichment from the bedrock upwards. However, several differences regarding the distribution and enrichment mechanism of economic metals can be outlined. In the Los Reales profile, Ni increased from the bedrock to the overlying saprolite (Samples 1 and 2) and decreased again at the top of the profile (Figure 5), whereas in the Camán profile, Ni was steadily enriched from the bedrock to the overlying lower and upper oxide horizons (Figure 11).

In the Los Reales saprolite, serpentine was the main Ni repository, with concentrations of up to 1 wt.% NiO. This is typical of saprolites derived from ultramafic rocks worldwide [30,34,35]. However, it must be noted that goethite and Mn-oxy-hydroxides showed much higher Ni grades, with concentrations of up to 2.04 wt.% NiO and 10.4 wt.% NiO, respectively. This is coherent with the Ni incorporation in goethite being faster or starting at earlier stages than the Ni enrichment in serpentine. Goethite may capture remarkable amounts of Ni^{2+}, either by adsorption onto their surfaces of by substitution for Fe^{3+} [36–38]. In Mn-oxy-hydroxides, Ni enrichment may be due to the replacement of Al in the gibbsite-like octahedral sites of lithiophorite-like structures, which consequently triggers the development of an asbolane-type layer when Al is entirely replaced [39]. In another study [40], it was suggested that in Al-rich systems with pH > 6, Ni is rapidly incorporated into lithiophorite due to the development of Al–Ni precipitates in the form of gibbsite-like layers. In contrast, if the system is Al-free with a neutral pH, Ni should enter the structure of a Mn-oxide type layer, occupying vacancies [25,41,42]. The chemistry of the asbolane–lithiophorite intermediates in Camán overlapped that of the asbolane–lithophorite intermediates that originated in Al-rich systems with pH > 6 (e.g., type IV intermediates reported by [28] in Wingellina, Australia; Figure 16). It is worth noting that the asbolane–lithiophorite intermediates from Camán have the highest Ni grades in the profile (up to 14.14 wt.% NiO), higher than the more abundant goethite (1.9 wt.% NiO).

The maximum bulk-rock MnO and Co contents of the parent rocks in the two studied profiles were essentially identical: 0.12 wt.% and 105 ppm in Los Reales saprolite, and 0.14 wt.% and 116 ppm in Camán. However, there was a great difference in the enrichment observed in these two elements between the two weathering profiles, with lower average values in Los Reales (MnO 0.30 wt.% and Co up to 269 ppm) than in Camán (MnO up to 0.90 wt.% and Co up to 649 ppm). This suggests that, as observed for Ni, the primary Co and Mn enrichment of the protoliths has a lesser influence than weathering rates.

In the Los Reales saprolite, Mn and Co exhibited identical bulk-rock distribution patterns, characterized by weak enrichment upwards (Figure 5). Their distribution was also similar to those displayed by Al_2O_3 and Cr_2O_3, which were mostly hosted by spinel occurring in the weathering profile. Considering that this oxide contains hundreds to thousands of ppm of both Mn and Co (Table S1), the Mn-Co geochemical signature of the profile is somehow controlled by chromium spinel. Much like the latter, olivine and/or orthopyroxene (or serpentines derived from them) from the peridotite parent rock contained hundreds to thousands of ppm of both Mn and Co (Table S2). During laterization, the olivine dissolution through interaction with acid to neutral fluids is the main process accounting for the Co^{2+} and Mn^{2+} remobilization and uptaking in Fe-Mn-oxy-hydroxides forming pseudomorphs after olivine (e.g., Figure 9d) and/or botryoidal aggregates (Figure 9h). In these neoformed oxides, Co^{2+} would be adsorbed on the surface of the low-crystalline, Mn-rich oxy-hydroxides via its oxidation to Co^{3+} by Mn^{4+} vacancies and the subsequent replacement of Mn^{3+} in the crystal lattice by Co^{3+}, owing to the acidic nature of the solutions [27,43].

In the Camán laterite, the neoformed Fe- and Mn-oxy-hydroxides were the main carriers of Mn at the top of the profile, where chromium spinel was practically absent. This observation confirms previous interpretations that asbolane–lithiophorites formed during laterization are the main carriers of Mn and Co [27,28].

5.2.2. Behavior of REE, Sc, and PGE

The parent rocks of the two studied profiles yielded almost identical concentrations of ΣREE (~3.5–4 ppm). However, the Los Reales saprolite had much lower ΣREE contents (<50 ppm) than the Camán laterite (up to 230 ppm), which again indicates a strong influence of the metal enrichment by the degree of maturity of the profile instead of the composition of the starting bedrock.

The two weathering profiles analyzed here showed a common decoupling between LREE and HREE, characterized by a noticeable enrichment of the former at the tops of the profiles. This trend of enrichment was positively correlated with Al_2O_3, and especially MnO. This is consistent with previous reports that REE enrichments are related to Al_2O_3-rich zones in the oxide orebodies of the Moa Bay in Cuba [2], and in the Wingellina laterite deposit in Australia [28]. Such REE–Al_2O_3 correlation has been ascribed to REE adsorption onto clays, following an ion-adsorption-type concentration process. In contrast, positive correlations between REE and MnO suggest an additional remobilization and accumulation of the lanthanides during chemical weathering into Mn-oxy-hydroxides [2,28,44,45]. This clearly evidences a dual model of the fixation of remobilized REE during weathering.

Another interesting feature of the REE geochemistry of the Camán laterite was the significant REE concentration (highest REE of 230 ppm) observed in Sample 3, marking the transition between the lower and the upper oxide horizons (Figure 12). This abrupt increase in REE coincided with a positive SiO_2 spike and the highest contents of Zr (143 ppm) and Hf

(7 ppm) (Figure 13; Table S1). These observations, along with the identification of detrital zircon in this sample (Figure 16), suggest the concentration of primary REE minerals. It is worth noting that Y displayed a similar distribution to total REE in both profiles, suggesting that its speciation was controlled by the same processes as that of the REE.

On the other hand, the distribution of Sc was identical to that of Fe_2O_3 (Figures 11 and 12). This element tends to accumulate in the upper portion of the profiles where Mn-oxy-hydroxides prevail, reaching maximum concentrations of up to 23 ppm in Los Reales and up to 78 ppm in Camán. Interestingly, other studies [2,46] reported a similar tendency of Sc to concentrate in the uppermost portion of oxide horizons in well-developed laterites from New Caledonia and Cuba.

The maximum total PGE + Au contents of the parent rock of the Los Reales (32 ppb) were similar to those of Caman (25–29 ppb) (Table S1). However, these noble metals were distinctively more enriched in the profile of Camán (79–171 ppm) than in that of Los Reales (36–58 ppb). The positive correlation between the PGE+Au and the UMIA in both profiles ($r^2 = 0.91$; Table S1) indicates again the impact of the degree of laterization in the supergene enrichment of critical metals.

From Figures 6 and 12, it is clear that Rh and Au were the least mobile elements in both profiles, and there were remarkable variations in the enrichment of the other PGE. Thus, in Los Reales, Ir, Ru, Pd, and Pt steadily increased across the whole profile, displaying differential mobility upwards, and all slightly decreased in the uppermost portion. In contrast, in Camán, changes in these PGEs were more irregular, and they were more enriched at the top. Altogether, these observations highlight that these PGE were effectively mobilized in different proportions through the weathering profile, such as reported in other laterites worldwide [2,6,47,48].

In Los Reales, the PGE+Au distribution resembled those of Cr_2O_3, Al_2O_3, MnO, and TiO_2, suggesting that detrital chromium spinel also had a significant role in controlling the bulk-rock content of noble metals. This could be related to the presence of nanometer-to-micrometer-sized platinum-group minerals (PGM) and gold that were initially hosted in chromium spinel and retained in this oxide or eventually released during laterization. Although we have not yet identified PGM in our profiles, previous studies have shown that chromium spinel from the Ronda peridotites contain abundant both PGM and gold particles [49].

The coinciding positive anomalies of noble metals and Cr at the bottom of the Camán profile, where chromium spinel was relatively abundant (i.e., Sample 2; Figures 11 and 12), confirm that this mineral controls the PGE signature of the Camán laterite. Relatively high Ru and Ir contents in the nearby La Cabaña saprolite have been attributed to laurite (RuS_2) and irarsite (IrAsS) hosted in chromium spinel [6], and to laurite and/or Ru-Fe alloys (e.g., hexaferrum) derived from its alteration in laterites from the Dominican Republic by [2,50]. Moreover, the residual accumulation of sperrylite ($PtAs_2$) originally hosted in chromium spinel may explain Ir and Pt enrichments in these aforementioned weathering profiles. Nonetheless, the progressively increasing concentrations of Pd, Ir, Pt, and overall Ru upwards in the Camán profile cannot be exclusively attributed to PGM originally hosted in the residual chromite but can also be explained by the liberation of PGE-rich Ni–Fe–Cu sulfides from serpentinite [14] and/or the supergene neoformation of PGM from the migrating fluids [50].

Studies dealing with elemental geochemistry and Cr ($^{53}Cr/^{52}Cr$) and Re-Os ($^{187}Re/^{188}Os$) isotopes show that PGE are relatively mobile in fluids acting in surficial environments, thus leading to the enrichment of these metals in laterites [6]. Previous studies have shown that acidic conditions (pH < 6 and Eh > 0.4) could promote the dissolution of most PGE species, especially of Pt and Pd. In these conditions, Pt may form aqueous hydroxide species but is less mobile than Pd in surface environments due to the higher oxidation potential to enter into solution [51–53]. The relatively soluble behavior of Pd (and to a lesser extent Pt) in near-surface conditions suggests that Pd may be liberated during the initial stages of weathering from host minerals (PGM, Fe-Ni sulfides, and native Au) to be incorporated into porewater as, e.g., $Pd^{2+}(OH)_{2(aq)}$ and then partially mobilized throughout the profile [54–58]. This may result in Pd (and Pt) being incorporated or trapped onto clay and oxy-hydroxide-rich horizons or being reduced by Mn or organic matter compounds, due to their high surface area and highly negatively charged interfaces [2,59–63]. For example, Pt–Ir–Fe–Ni alloys have been documented as inclusions and packed nanoparticles within pore spaces of Fe- and Mn-oxy-hydroxides in the nearby La Cabaña saprolite [6] and in many other laterites worldwide, such as the Loma Peguera Ni-laterite deposit in the Dominican Republic [50,62–64] and the Planeta Rica Ni-laterite profiles of Northern Colombia [47].

6. Conclusions

For most elements analyzed, protolith rock has a lesser influence on supergene enrichment of these metals than weathering intensity. Thus, despite the similar climate, higher enrichments of metals are related to higher UMIA rather than higher initial concentrations in bedrock. The whole-rock geochemical data show that at the earliest stages of laterization, metals are already mobilized, but in concentrations below cut-off levels. Therefore, a relatively high degree of maturity in the profile is necessary for a laterite to achieve economic interest.

Supplementary Materials: The following supporting information can be downloaded at: https://www.mdpi.com/article/10.3390/min13070844/s1. Table S1: Whole-rock major, minor, and trace element data from parent rocks and soil samples from the Los Reales and Caman laterites; Table S2: Selected electron microprobe analyses of representative minerals from soil samples from the Los Reales and Caman laterites. Figure S1: XRDP oriented_Los Reales; Figure S2: XRDP parent rock Caman.

Author Contributions: Conceptualization, J.M.G.-J. and C.M.; fieldwork, J.M.G.-J., C.M., F.G. and L.Y.; methodology, C.M., L.Y. and C.V.-d.-B.; formal analysis, C.M., L.Y., C.V.-d.-B., D.C. and L.M.-G.; resources J.M.G.-J. and C.M.; data curation C.M., L.Y., C.V.-d.-B. and J.A.P.; writing—original draft preparation J.M.G.-J., C.V.-d.-B. and J.A.P.; writing—review and editing, all the authors; visualization, all the authors; supervision, F.G. and J.A.P.; project administration, C.M.; funding acquisition, J.M.G.-J. and C.M. All authors have read and agreed to the published version of the manuscript.

Funding: This research was fully funded by the MECRAS Project A-RNM-356-UGR20 "Proyectos de I+D+i en el marco del Programa Operativo FEDER Andalucía 2014-2020" of the Consejería de Economía, Conocimiento, Empresas y Universidad de la Junta de Andalucía (Spain).

Data Availability Statement: Not applicable.

Acknowledgments: Authors would like to acknowledge Xavier Arroyo from CAI of Complutense University of Madrid for sample preparation and Xavier Llovet from UB for assisting in electron microprobe analyses. LMG thank the Spanish University Ministry and the European Union (Next Generation EU).

Conflicts of Interest: The authors declare no conflict of interest.

References

1. Available online: https://pubs.usgs.gov/periodicals/mcs2023/mcs2023-nickel.pdf (accessed on 1 January 2020).
2. Aiglsperger, T.; Proenza, J.A.; Lewis, J.F.; Labrador, M.; Svotjka, M.; Rojas-Purón, A.; Longo, F.; Durišova, J. Critical metals (REE, Sc, PGE) in Ni lateritas from Cuba and the Dominican Republic. *Ore Geol. Rev.* **2016**, *73*, 127–147. [CrossRef]
3. Cocker, M. 48th Forum on the Geology of Industrial Minerals, 9th ed. Arizona Geological Survey Special Paper; Conway, F.M., Ed.; Arizona Geological Survey: Tucson, AZ, USA, 2014; Chapter: 4; pp. 1–8.
4. Thorne, R.L.; Herrington, R.; Roberts, S. Climate change and the formation of nickel laterite deposits. *Geology* **2012**, *40*, 331–334. [CrossRef]
5. Berger, V.I.; Singer, D.A.; Bliss, J.D.; Moring, B.C. Ni-Co laterite deposits of the world—Database and grade and tonnage models: U.S. *Geol. Surv. Open File Rep.* **2011**, *2011–1058*, 26p. Available online: http://pubs.usgs.gov/of/2011/1058/ (accessed on 1 January 2020).
6. Rivera, J.; Reich, M.; Scoeneberg, R.; González-Jiménez, J.M.; Barra, F.; Aiglsperger, T.; Proenza, J.A.; Carretier, S. Platinum-group element and gold enrichment in soils monitorted by chromium stable isotopes during weathering of ultramafic rocks. *Chem. Geol* **2018**, *499*, 84–99. [CrossRef]

7. Maurizot, P.; Sevin, B.; Iseppi, M.; Giband, T. Nickel-Bearing Laterite Deposits in Accretionary Context and the Case of New Caledonia: From the Large-Scale Structure of Earth to Our Everyday Appliances. *GSA Today* **2019**, *29*, 4–10. [CrossRef]

8. Yusta, A.; Berahona, E.; Huertas, F.; Reyes, E.; Yáñez, J.; Linares, J. Geochemistry of soils from peridotite in Los Reales, Málaga. *Acta Miner. Petrogr.* **1985**, *29*, 439–498.

9. Zamarsky, V.V.; Salmon, H.O.; Tabok, M. Estudio geoquímico de los productos de intemperismo de las rocas ultrabásicas (serpentinitas) en la provincia de Valdivia, Chile. *Rev. Geológica De Chile: Int. J. Andean Geol.* **1974**, *1*, 81–102.

10. Salinas, S. Formación de la Laterita de Camán XIV Región Chile. Master's Thesis, Univeridad de Chile, Santiago, Chile, 2016.

11. Navarro-Hasse, E. Análisis Mineralógico y Concentración de Cromo, Níquel y Cobalto en el Suelo Serpentinitico del Sector de Camán, Región de los Ríos, Chile. Master's Thesis, Universidad Austral de Chile, Valdivia, Chile, 2020.

12. Lenoir, X.; Garrido, C.J.; Bodinier, J.-L.; Dautria, J.-M.; Gervilla, F. The recrystallization front of the Ronda peridotite: Evidence for melting and thermal erosion of subcontinental lithospheric mantle beneath the Alborán basin. *J. Petrol.* **2001**, *42*, 141–158. [CrossRef]

13. Bucher, K.; Stober, I.; Muller-Sigmund, H. Weathering crusts on peridotite. Contrib. *Miner. Petrol.* **2015**, *169*, 52. [CrossRef]

14. Faundez, R. Distribución de Metales Nobles en Rocas Ultramáficas Serpentinizadas del Centro-Sur de Chile. Master's Thesis, Univeridad de Chile, Santiago, Chile, 2016.

15. Hidas, K.; Booth-Rea, G.; Garrido, C.J.; Martínez-Martínez, J.M.; Padrón-Navarta, J.A.; Konc, Z.; Giaconia, F.; Frets, E.; Marchesi, C. Back-arc basin inversion and subcontinental mantle emplacement in the crust: Kilometre-scale folding and shearing at the base of the proto-Alborán lithospheric mantle (Betic Cordillera, southern Spain). *J. Geol. Soc.* **2013**, *170*, 47–55. [CrossRef]

16. Garrido, C.J.; Gueydan, F.; Booth-Rea, G.; Precigout, J.; Hidas, K.; Padron-Navarta, J.A.; Marchesi, C. Garnet lherzolite and garnet-spinel mylonite in the Ronda peridotite: Vestiges of Oligocene backarc mantle lithospheric extension in the western Mediterranean. *Geology* **2011**, *39*, 927–930. [CrossRef]

17. Luque del Villar, F.J.; Rodas-González, M.; Doval-Montoya, D. Mineralogía y génesis de los yacimientos de vermiculita del macizo de Ojén (Serranía de Ronda, Málaga). *Boletín De La Soc. Española De Mineral.* **1985**, *8*, 229–238.

18. Gómez-Zotano, J.A.; Cunill-Artigas, R.; Martínez-Ibarra, E. Descubrimiento y caracterización geográfica de una depresión ultramáfica en Sierra Bermeja: Nuevos datos geomorfoedáficos, fitogeográficos y paleoecológicos. *Pirineos* **2017**, *172*, 26. [CrossRef]

19. Gutiérrez-Hernández, O.; Cámara Artigas, R.; García, L.V. Regeneración de los pinsapares béticos. Análisis de tendencia interanual y estacional del NDVI. *Pirineos* **2018**, *173*, e035. [CrossRef]

20. Willner, A.P.; Thomson, S.N.; Kröner, A.; Wartho, J.A.; Wijbrans, J.R.; Hervé, F. Time markers for the evolution and exhumation history of a Late Palaeozoic paired metamorphic belt in North–Central Chile (34–35 30' S). *J. Petrol.* **2005**, *46*, 1835–1858. [CrossRef]

21. Deckart, K.; Hervé, F.; Fanning, M.; Ramírez, V.; Calderón, M.; Godoy, E. U-Pb geochronology and Hf-O isotopes of zircons from the Pennsylvanian coastal batholith, south central Chile. *Andean Geol.* **2014**, *41*, 49–82. [CrossRef]

22. González-Jimenez, J.M.; Plissart, G.; Garrido, L.; Padrón-Navarta, J.; Aiglsperger, T.; Romero, R.; Marchesi, C.; Moreno-Abril, A.; Reich, M.; Barra, F.; et al. Ti-clinohumite and Ti-chondrodite in antigorite serpentinites from Central Chile: Evidence for deep and cold subduction. *Eur. J. Miner.* **2017**, *29*, 959–970. [CrossRef]

23. Available online: http://explorador.cr2.cl (accessed on 1 January 2020).

24. Hoffman, E.L.; Dunn, B. Sample preparation and bulk analytical methods for PGE. In *The Geology, Geochemistry and Mineral Beneficiation of Platinum Group Elements*; Cabri, L.J., Ed.; Canadian Institute of Mining, Metallurgyand Petroleum: Montreal, QC, Canada, 2002; Volume 54, pp. 1–11.

25. Roqué-Rosell, J.; Mosselmans, J.F.W.; Proenza, J.A.; Labrador, M.; Galí, S.; Atkinson, K.D.; Quinn, P.D. Sorption of Ni by "lithiophorite–asbolane" intermediates in Moa Bay lateritic deposits, eastern Cuba. *Chem. Geol.* **2010**, *275*, 9–18. [CrossRef]

26. Lambiv Dzemua, G.; Gleeson, S.A.; Schofield, P.F. Mineralogical characterization of the Nkamouna Co-Mn laterite ore, southeast Cameroon. *Miner. Depos.* **2013**, *48*, 155–171. [CrossRef]

27. Manceau, A.; Llorca, S.; Calas, G. Crystal chemistry of cobalt and nickel in lithiophorite and asbolane from New Caledonia. *Geochim. Cosmochim. Acta* **1987**, *51*, 105–113. [CrossRef]

28. Putzolu, F.; Boni, M.; Mondillo, N.; Maczurad, M.; Pirajno, F. Ni-Co enrichment and High-Tech metals geochemistry in the Wingellina Ni-Co oxide-type laterite deposit (Western Australia). *J. Geochem. Explor.* **2019**, *196*, 282–296. [CrossRef]

29. Brand, N.W.; Butt, C.R.M.; Elias, M. Nickel laterites: Classification and features. *AGSO J. Aust. Geol. Geoph.* **1998**, *17*, 81–88.

30. Ulrich, M.; Cathelineau, M.; Muñoz, M.; Boiron, M.C.; Teitler, Y.; Karpoff, A.M. The relative distribution of critical (Sc, REE) and transition metals (Ni, Co, Cr, Mn, V) in some Ni-laterite deposits of New Caledonia. *J. Geochem. Explor.* **2019**, *197*, 93–113. [CrossRef]

31. Brand, N.W.; Butt, C.R.M. Weathering, element distribution and geochemical dispersion at Mt Keith, Western Australia: Implication for nickel sulphide exploration. *Geochem. Explor. Environ. Anal.* **2001**, *1*, 391–407. [CrossRef]

32. Freyssinet, P.; Butt, C.R.M.; Morris, R.C. Ore-forming processes related to lateritic weathering. *Econ. Geol. One Hundredth Anniv. Volume* **2005**, 681–722. [CrossRef]

33. Golightly, J.P. Progress in understanding the evolution of nickel laterites. *Econ. Geol. Spec. Pub.* **2010**, *15*, 451–485.

34. Golightly, J.P.; Arancibia, O.N. The chemical composition and infrared spectrum of nickel- and iron-substituted serpentine form a nickeliferous laterite profile, Soroako, Indonesia. *Can. Miner.* **1979**, *17*, 719–728.

35. Villanova-de-Benavent, C.; Domènech, C.; Tauler, E.; Galí, S.; Tassara, S.; Proenza, J.A. Fe-Ni-bearing serpenties from the saprolite horizon of Caribbean Ni-laterite deposits: New insights from thermodynamic calculations. *Miner. Depos.* **2017**, *52*, 979–992. [CrossRef]
36. Beukes, J.P.; Giesekke, E.W.; Elliot, W. Nickel retention by goethite and haematite. *Miner. Eng.* **2000**, *13*, 1573–1579. [CrossRef]
37. Singh, B.; Sherman, D.M.; Gilkes, R.J.; Wells, M.A.; Mosselmans, J.E.W. Incorporation of Cr, Mn and Ni into goethite (aFeOOH): Mechanism from extended X-ray absorption fine structure spectroscopy. *Clay Min.* **2002**, *37*, 639–649. [CrossRef]
38. Trivedi, P.; Axe, L. Ni and Zn sorption to amorphous versus crystalline iron oxides: Macroscopic studies. *J. Coll. Int. Sci.* **2001**, *244*, 221–229. [CrossRef]
39. Ostwald, J. Two varieties of lithiophorite in some Australian deposits. *Miner. Mag.* **1984**, *48*, 383–388. [CrossRef]
40. Roberts, D.R.; Scheidegger, A.M.; Sparks, D.L. Kinetics of mixed Ni-Al precipitate formation on a soil clay fraction. *Environ. Sci. Technol.* **1999**, *33*, 3749–3754. [CrossRef]
41. Peacock, C.L. Physiochemical controls on the crystal-chemistry of Ni in birnessite: Genetic implications for ferro-manganese precipitates. *Geochim. Cosmochim. Acta* **2009**, *73*, 3568–3578. [CrossRef]
42. Cui, H.; You, L.; Feng, X.; Tan, W.; Qiu, G.; Liu, F. Factors governing the formation of lithiophorite at atmospheric pressure. *Clays Clay Miner.* **2009**, *57*, 353–360. [CrossRef]
43. McKenzie, R.M. Manganese oxides and hydroxides. In *Minerals in Soil Environments*, 2nd ed.; Dixon, J.B., Weed, S.B., Eds.; SSSA: Madison, WI, USA, 1989; pp. 439–465.
44. Laskou, M.; Economou-Eliopoulos, M. The role of microorganisms on the mineralogical and geochemical characteristics of the Parnassos-Ghiona bauxite deposits, Greece. *J. Geochem. Explor.* **2007**, *93*, 67–77. [CrossRef]
45. Vodyanitskii, Y.N. Geochemical fractionation of lanthanides in soils and rocks: A review of publications. *Eurasian Soil Sci.* **2012**, *45*, 56–67. [CrossRef]
46. Audet, M.A. Le Massif du Koniambo-Nouvelle Calédonie: Formation et Obduction d'un Complexe Ophiolitique de Type SSZ. Enrichissement en Nickel, Cobalt et Scandium Dans les Profils Résiduels. Ph.D. Thesis, Université de Nouvelle Calédonie, Nouméa, France, 2008; p. 327. (In French).
47. Tobón, M.; Weber, M.; Proenza, J.A.; Aiglsperger, T.; Betancur, S.; Farré-de-Pablo, J.; Ramírez, C.; Pujol-Solà, N. Geochemistry of Platinum-Group Elements (PGE) in Cerro Matoso and Planeta Rica Ni-Laterite deposits, Northern Colombia. *Boletín De La Soc. Geológica Mex.* **2020**, *72*, A201219. [CrossRef]
48. Al-Khirbash, S.A.; Ahmed, H. Distribution and mobility of platinum-group elements in the Late Cretaceous Ni-laterite iin the Northern Oman Mountains. *JGR Solid Earth* **2021**, *126*, e2021JB022363. [CrossRef]
49. Gervilla, F.; González-Jiménez, J.M.; Hidas, K.; Marchesi, C.; Piña, R. Geology and Metallogeny of the Upper Mantle Rocks from the 939 Serranía de Ronda. *Mineral. Span. Soc. Ronda* **2019**, *122*. Available online: https://digital.csic.es/handle/10261/206926 (accessed on 1 January 2020).
50. Aiglsperger, T.; González-Jiménez, J.M.; Proenza, J.A.; Galí, S.; Longo, F.; Griffin, W.L.; O'Reilly, S.Y. Open System Re-OsIsotope Behavior in Platinum-Group Minerals during Laterization? *Minerals* **2021**, *11*, 1083. [CrossRef]
51. Sassani, D.C.; Shock, E.L. Solubility and transport of platinum-group elements in supercritical fluids: Summary and estimates of thermodynamic properties for ruthenium, rhodium, palladium, and platinum solids, aqueous ions, and complexes to 1000 °C and 5 kbar. Geochim. *Cosmochim. Acta* **1998**, *62*, 2643–2671. [CrossRef]
52. Colombo, C.; Oates, C.J.; Monhemius, A.J.; Plant, J.A. Complexation of platinum, palladium and rhodium with inorganic ligands in the environment. *Geochem. Explor. Environ. Anal.* **2008**, *8*, 91–101. [CrossRef]
53. Reith, F.; Campbell, S.G.; Ball, A.S.; Pring, A.; Southam, G. Platinum in Earth surface environments. *Earth-Sci. Rev.* **2014**, *131*, 1–21. [CrossRef]
54. Fuchs, W.A.; Rose, A.W. The geochemical behavior of platinum and palladium in the weathering cycle in the Stillwater Complex, Montana. *Econ. Geol.* **1974**, *69*, 332–346. [CrossRef]
55. Locmelis, M.; Melcher, F.; Oberthür, T. Platinum-group element distribution in the oxidized main sulfide zone, Great Dyke, Zimbabwe. *Miner. Depos.* **2010**, *45*, 93–109. [CrossRef]
56. Oberthür, T. The fate of platinum-group minerals in the exogenic environment—From sulfide ores via oxidized ores into placers: Case studies bushveld complex, South Africa, and Great Dyke, Zimbabwe. *Minerals* **2018**, *8*, 581. [CrossRef]
57. Junge, M.; Oberthür, T.; Kraemer, D.; Melcher, F.; Piña, R.; Derrey, I.T.; Manyeruke, T.; Strauss, H. Distribution of platinum-group elements in pristine and near-surface oxidized Platreef ore and the variation along strike, northern Bushveld Complex, South Africa. *Miner. Depos.* **2019**, *54*, 885–912. [CrossRef]
58. Korges, M.; Junge, M.; Borg, G.; Oberthür, T. Supergene mobilization and redistribution of platinum-group elements in the Merensky Reef, eastern Bushveld Complex, South Africa. *Can. Miner.* **2021**, *59*, 1381–1396. [CrossRef]
59. Bowles, J.F.W. The development of platinum-group minerals in laterites. *Econ. Geol.* **1986**, *81*, 1278–1285. [CrossRef]
60. Cabral, A.R.; Radtke, M.; Munnik, F.; Lehmann, B.; Reinholz, U.; Riesemeier, H.; Tupinambá, M.; Kwitko-Ribeiro, R. Iodine in alluvial platinum-palladium nuggets: Evidence for biogenic precious-metal fixation. *Chem. Geol.* **2011**, *281*, 125–132. [CrossRef]
61. Kubrakova, I.V.; Fortygin, A.V.; Lobov, S.G.; Koscheeva, I.Y.; Tyutyunnik, O.A.; Mironenko, M.V. Migration of platinum, palladium, and gold in the water systems of platinum deposits. *Geochem. Int.* **2011**, *49*, 1072–1084. [CrossRef]
62. Aiglsperger, T.; Proenza, J.A.; Zaccarini, F.; Lewis, J.F.; Garuti, G.; Labrador, M.; Longo, F. Platinum group minerals (PGM) in the Falcondo Ni-laterite deposit, Loma Caribe peridotite (Dominican Republic). *Miner. Depos.* **2015**, *50*, 105–123. [CrossRef]

63. Aiglsperger, T.; Proenza, J.A.; Galí, S.; Longo, F.; Roqué-Rosell, J. Geochemical and mineralogical survey of critical elements (PGE, REE, Sc and Co) in Ni laterites from the Caribbean. *EGU Gen. Assem.* **2019**, *21*, 1.
64. Aiglsperger, T.; Proenza, J.A.; Font-Bardia, M.; Baurier-Amat, S.; Galí, S.; Lewis, J.F.; Longo, F. Supergene neoformation of Pt-Ir-Fe-Ni alloys: Multistage grains explain nugget formation in Ni-laterites. *Miner. Depos.* **2017**, *52*, 1069–1083. [CrossRef]

Article

Evaluation of Sc Concentrations in Ni-Co Laterites Using Al as a Geochemical Proxy

Yoram Teitler [1,*], Sylvain Favier [1], Jean-Paul Ambrosi [2], Brice Sevin [3], Fabrice Golfier [1] and Michel Cathelineau [1]

1   Université de Lorraine, CNRS, GeoRessources, 54506 Vandoeuvre-lès-Nancy, France; sylvain.favier@univ-lorraine.fr (S.F.); fabrice.golfier@univ-lorraine.fr (F.G.); michel.cathelineau@univ-lorraine.fr (M.C.)
2   Aix-Marseille Université, CNRS, IRD, INRAE, CEREGE, CEDEX 4, 13545 Aix-en-Provence, France; ambrosi@cerege.fr
3   Service de la Géologie de Nouvelle-Calédonie, Direction de l'Industrie, des Mines et de l'Énergie, 1ter Rue Unger, BP 465, CEDEX, 98845 Nouméa, New Caledonia; brice.sevin@gmail.com
*   Correspondence: yoram.teitler@univ-lorraine.fr

Citation: Teitler, Y.; Favier, S.; Ambrosi, J.-P.; Sevin, B.; Golfier, F.; Cathelineau, M. Evaluation of Sc Concentrations in Ni-Co Laterites Using Al as a Geochemical Proxy. *Minerals* **2022**, *12*, 615. https://doi.org/10.3390/min12050615

Academic Editors: Cristina Domènech and Cristina Villanova-de-Benavent

Received: 23 March 2022
Accepted: 10 May 2022
Published: 12 May 2022

Publisher's Note: MDPI stays neutral with regard to jurisdictional claims in published maps and institutional affiliations.

Abstract: Scandium (Sc) is used in several modern industrial applications. Recently, significant Sc concentrations (~100 ppm) were reported in some nickel-cobalt lateritic ores, where Sc may be valuably co-produced. However, Sc is typically not included in routine analyses of Ni-Co ores. This contribution examines the relevance of using routinely analysed elements as geochemical proxies for estimating Sc concentration and distribution. Three Ni-Co lateritic deposits from New Caledonia were investigated. In each deposit, Sc is well correlated with Al_2O_3. The slopes of deposit-scale Sc-Al_2O_3 regression lines are remarkably controlled by the composition of enstatite from the parent peridotite. In all deposits, maximum Sc enrichment occurs in the yellow limonite, above the highest Ni and Co enrichment zones. Sc- and Al-bearing crystalline goethite is predominant in the oxide-rich zones, though Sc shows a higher affinity for amorphous iron oxides than Al. We propose that, in already assayed Ni-Co lateritic ores, the concentration and distribution of Sc can be estimated from that of Al. Deposit-scale Sc-Al_2O_3 correlations may be determined after analysing a limited number of spatially and chemically representative samples. Therefore, mining operators may get a first-order evaluation of the Sc potential resource in Ni-Co lateritic deposits at low additional costs.

Keywords: scandium; laterite; Ni-laterite; sequential extraction; New Caledonia

1. Introduction

Scandium (Sc) is mainly used as a hardening additive to aluminium to form Al-Sc alloys for aerospace industries and high-quality sports equipment [1,2]. In addition, Sc is notably used in high-temperature lights, lasers, and ceramics manufacturing and finds promising applications in the development of Solid Oxide Fuel Cells (SOFCs). The global Sc supply and consumption remain marginal (~15 to 25 t/yr Sc_2O_3, [3]). Sc is solely recovered as a by-product through titanium, zirconium, uranium, cobalt, and nickel process streams. Nevertheless, the development of energy-saving technologies and Sc extraction techniques have raised interest in this metal. In the last decade, Sc-rich occurrences with economically attractive grades and tonnages have been identified in some oxide-rich laterites developed after mafic and ultramafic rocks [4–14]. There, Sc enrichment is largely residual and results from the intense leaching of mobile cations during the lateritisation of the parent rock. Scandium is, thus, trapped and concentrated in neo-formed goethite, and moderate remobilisation can occur during repeated stages of goethite dissolution-recrystallisation and transformation into hematite. The chemical speciation of Sc in laterites from Australia and the Philippines has been investigated using a combination of X-ray absorption near-edge structure (XANES) spectroscopy and sequential extractions [7,8,11]. These studies further support the preferential affinity of Sc for goethite compared to hematite or smectite. However, the relative importance of adsorption and incorporation processes and the impact of goethite crystallinity in the formation of Sc-rich goethite zones remain debated. They may vary from one deposit to another. In New Caledonia, significant Sc concentrations were reported in several Ni-Co lateritic oxide ores [13,14]. There, maximum Sc grades reach ~100 ppm in the yellow limonite horizon. Although such concentrations are too low to be economically attractive as primary resources, Sc could be a valuable by-product of Ni and Co processing, provided that Sc-rich zones sufficiently overlap Ni- and Co-rich zones [13].

Scandium is, however, not routinely analysed during exploration and resource estimation of Ni-Co laterites. Consequently, mining operators do not evaluate the potential Sc resources hosted in Ni-Co lateritic ores. In contrast, Fe, Al, and Cr are routinely analysed. As suggested by previous investigations, they could serve as geochemical proxies for inferring the distribution of Sc concentrations in lateritic Ni-Co ores [13]. Although Fe is the predominant constituent of lateritic oxide ores with up to 80 wt% Fe_2O_3, Teitler et al. [13] suggested that Fe may be correlated to Sc only in the lower to intermediate sections of the lateritic profiles and that the Sc-Fe_2O_3 correlation is usually not valid in the Sc-rich, yellow limonite horizon. In contrast, Sc is better correlated with Al in all the facies of the lateritic profiles, except for the uppermost ferruginous duricrust. The similar distribution patterns of Sc and Al in nickel laterites may result from a similar behaviour during weathering. Indeed, as for Sc, Al is mostly immobile during lateritic weathering though it can be, to some extent, expelled from iron oxides-oxyhydroxides during repeated stages of dissolution-recrystallisation [6,13].

Nevertheless, the Sc-Al_2O_3 correlation proposed by Teitler et al. [13], based on the combined analysis of individual vertical profiles from several deposits, shows significant dispersion and does not assess how reliable the Sc-Al_2O_3 proxy is at the deposit scale. In this contribution, we discuss the relevance of the Sc-Al_2O_3, Sc-Fe_2O_3 and Sc-Cr_2O_3 correlations at the deposit scale in three Ni-Co deposits from the Koniambo, Cap Bocage, and Tiébaghi massifs, New Caledonia, based on the geochemical analysis of spatially and lithologically representative samples from each deposit. Using laser ablation inductively coupled mass spectroscopy (LA-ICP-MS), we then examine the dependency of bedrock mineral compositions on the Sc-Al_2O_3 regression coefficients and the variability of Al and Sc contents in secondary-formed minerals. Third, we conduct selective chemical leaching on some Sc-bearing samples to provide further insights into the speciation of Sc in the investigated deposits. These results are used to discuss the potential of the Al proxy for the evaluation of Sc concentrations and distribution in peridotite-hosted laterites in the perspective of Ni-Co-Sc co-valorisation.

2. Regional Geology

The "Grande Terre" island of New Caledonia (Figure 1a) consists of a 300 km long allochthonous peridotite ophiolite referred to as the "Peridotite Nappe". The Peridotite Nappe, formed at ca. 35 Ma [15–17], is primarily composed of harzburgite locally interlayered with dunite, except in the northernmost klippes where lherzolite dominates [18]. The extensive lateritisation of the peridotite ophiolite led to the formation of world-class Ni-Co(-Sc) lateritic resources representing about 10% of the world's nickel reserves [19]. The development of Ni-Co(-Sc) laterites in New Caledonia resulted in the formation of two main ore types, (i) the saprolite ore dominated by Ni-bearing phyllosilicates and (ii) the oxide ore dominated by Ni-bearing ochreous goethite [20–27]. The Ni-rich (>2.0 wt% Ni) saprolite ore reserves are rapidly being depleted as they have been actively mined since the late 19th century. Lower-grade (1.0–2.0% Ni) oxide ore ("limonite") reserves will therefore represent the bulk of Ni reserves of New Caledonia in the future. Moreover, Ni oxide ores often yield elevated Co (>2000 ppm) and Sc (60–100 ppm) concentrations adding significant value to the ore (Figure 1b, [6,13]). Maximum Co concentrations (>2000 ppm) typically occur at the interface between saprolite and limonite (referred to as transition zone), where Co is mainly associated with Mn-oxides. Upwards in the limonitic horizons,

Co concentrations decrease, partially scavenged and trapped in ochreous goethite [28]. Maximum Sc grades are reached in the yellow limonite horizon (Figure 1b). Therefore, Ni-, Co-, and Sc-rich zones in limonite may partly overlap, and the degree of such overlap is critical for co-valorising Ni, Co, and Sc from lateritic Ni ores [6,13].

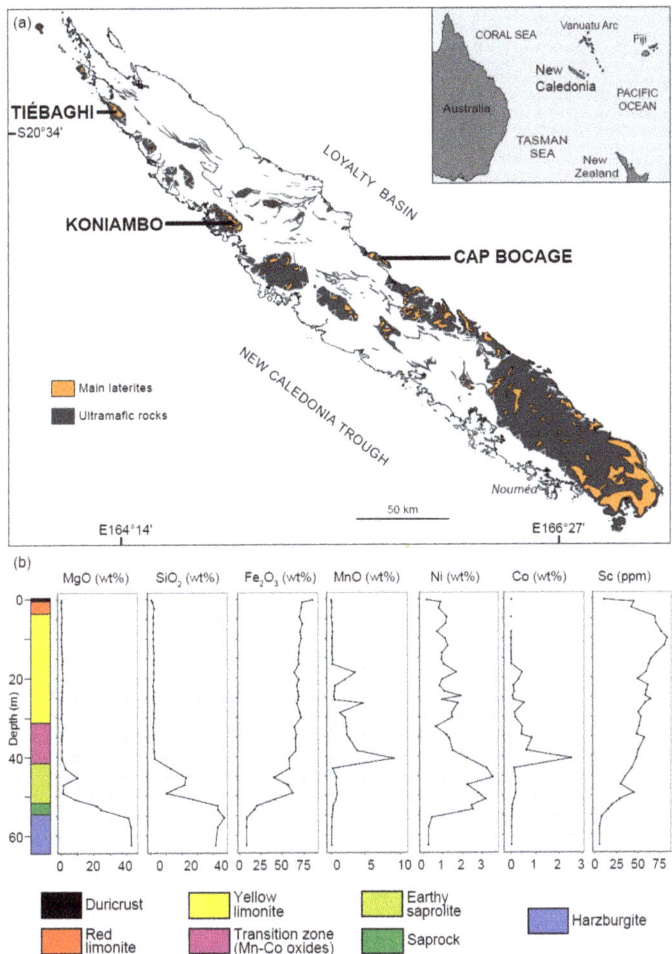

Figure 1. (**a**) Geological map of New Caledonia and location of the investigated mining massifs. Modified from Maurizot and Vendé-Leclerc [29]. (**b**) Geochemical evolution along a typical Ni-lateritic profile in New Caledonia (modified from Bailly et al. [6]).

3. Methodology

3.1. Sampling Strategy

Sampling was conducted in multiple locations within the investigated deposits, along several drillcores and pit walls, encompassing the diversity of representative lithofacies. The objective was to check whether Sc-Al$_2$O$_3$ correlations may be generalised at the deposit scale. At the Ma-Oui deposit (Koniambo massif), 39 samples were collected along six drillcores and one pit wall profile (Figure 2a). At the Coquette Red tenement (Cap Bocage massif), 27 samples were collected along two drillcores and one outcrop profile (Figure 2b). At the East Alpha deposit (Tiébaghi massif), 48 samples were collected along nine pit wall profiles, including three samples of saprolitised gabbro (Figure 2c).

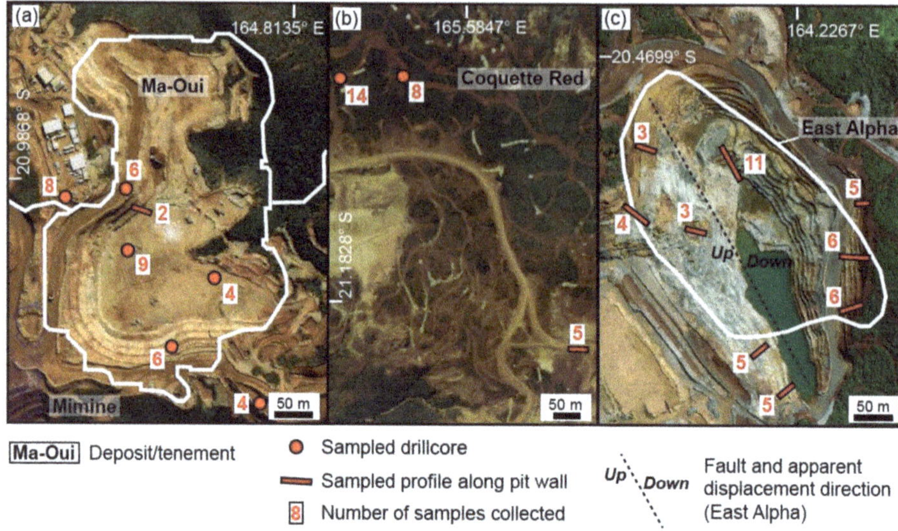

Figure 2. Outline of the investigated deposits and location of collected samples. (**a**) Ma-Oui deposit (Koniambo massif). (**b**) Coquette Red tenement (Cap Bocage massif). (**c**) East Alpha deposit (Tiébaghi massif).

3.2. Analytical Strategy

Among the 114 samples collected for whole-rock geochemical analysis, 49 samples (batch 1) were analysed at the SARM analytical service of the CRPG (France), and 65 samples (batch 2) were analysed at the NILAB laboratory (New Caledonia). Batch 1 sample pulps were analysed for whole-rock major- and trace-element geochemistry using the procedure of Carignan et al. [30]. Major element oxides and Sc were analysed using an iCap6500 ICP-OES with Li borate fusion. Trace elements were analysed using an iCapQ ICP-MS with Li borate fusion followed by nitric acid digestion. Analytical accuracy (2 s standard deviation) lies within the typical uncertainty of the analytical data for both major element oxides and trace elements, which is less than 1% for major oxides and less than 5% for most trace elements. Batch 2 samples were analysed for major elements, Ni, Co, and Sc, using an Axios FAST Wavelength Dispersive XRF spectrometer with LiF200 crystal. The detection limit for Sc is 12 ppm, and analytical accuracy is less than 7% in the 50–120 ppm concentration range. The relative analytical precision between the two datasets, tested by duplicating the analysis of three samples in both laboratories, is better than 2% for Sc.

Principal Component Analysis (PCA) was conducted on whole-rock geochemical data using the atomic weight percentages of Mg, Al, Si, Sc, Cr, Mn, Fe, Co, Ni, and LOI following centred log-ratio (CLR) transformation to avoid spurious proportionality and boundary effects [31,32]. PCA consists in transforming a multivariate dataset—here, each elemental concentration represents a variable—into a lower-dimensional dataset while preserving as much of the data's variation as possible. The dataset is projected into a new coordinate system based on the principal components, which are orthogonal linear combinations of original variables corresponding to the eigenvectors of the covariance matrix. The advantage of such analysis is that mineral stoichiometry likely controls principal components, providing a more realistic representation of geological variability. Centred log-ratio transformation of Mg, Al, Si, Sc, Cr, Mn, Fe, Co, and Ni concentrations and LOI performed before PCA has the benefit of removing the proportionality effect (i.e., the predominance of major elements on the geochemical variability) and the boundary effects (negative bias and spurious correlation effect [33]). Saprolitised gabbro samples from East Alpha were excluded from PCA calculations to better explore the impact of weathering on peridotite bedrocks alone. In a covariance biplot of log-ratio transformed data, a short distance between the ray tips of two variables (i.e., coincident rays) indicates that these variables are highly proportional [32].

Sixteen polished, thin sections were prepared and examined using reflected/transmitted light and scanning electron microscopy with a JEOL JSM7600F at Georessources Laboratory and SCMEM. In situ mineral chemistry analysis for major/minor elements (Mg, Al, Si, Ca, Cr, Mn, Fe, Co, and Ni) was conducted using a CAMECA SX100 electron probe micro analyser (EPMA) at the SCMEM with typical beam conditions of 15 kV and 10 nA. In situ analysis for Sc and minor/trace elements was conducted using LA-ICP-MS (193 nm MicroLas ArF Excimer coupled with Agilent 7500c quadrupole ICP-MS). LA-ICP-MS analysis of Sc and Al in silicates and phyllosilicates was performed with the NIST610 reference standard using Si concentration obtained from EPMA as the internal standard. For oxides, analysis was performed with the StdGoe 1.1 pelletised goethite standard developed and validated by Ulrich et al. [14] and using Fe concentration as an internal standard. Reference standards were analysed before and after every 15 ablations conducted on samples. Before each analysis, the background signal, or gas blank, was measured for 25 s. Ablation time, spot sizes, and laser pulse frequency were 45 s, 60 μm, and 5 Hz, respectively. Data processing was conducted using the SILLS program [34]. The detection limit was below 1 ppm for most trace elements.

Sequential extractions were performed at the CEREGE laboratory on seven Sc-bearing samples from Coquette Red and East Alpha, encompassing earthy saprolite, smectitic saprolite, and yellow and red limonite facies. Sequential extractions were not conducted on Ma-Oui samples as their mineralogy and chemistry are similar to that of Coquette Red samples. Four reagents were successively used to extract major and trace elements selectively: (i) ultrapure water for easily soluble elements; (ii) 0.1 mol·L^{-1} hydroxylamine hydrochloride NH_2OH-HCl at pH = 3.5, which is particularly efficient and selective for manganiferous phases; (iii) 0.2 mol·L^{-1} ammonium oxalate $(NH_4)_2C_2O_4$ at pH = 3 for amorphous and poorly crystallised iron oxides; and (iv) citrate-bicarbonate-dithionite $Na_2S_2O_4$ (22% Na-citrate and 1 g Na-dithionite) for well-crystallised iron oxides [6]. For each extraction, approximately 1 g of sample powder (<50 μm) was weighed into 50 mL of reagent solution and agitated for 76 h. Each tube was then centrifuged at 10,000 rpm, and the supernatant liquid was filtered at 0.2 μm before ICP-AES analysis. The residue was then rinsed with Milli-Q$^®$ water and used for subsequent extraction after re-suspension.

4. Lithofacies and Mineral Assemblages

The investigated Ni-Co deposits exhibit a continuum of alteration facies with specific mineral assemblages and textures. At Ma-Oui, moderately serpentinised harzburgite is the predominant parent rock of Ni-Co laterites (Figures 3a and 4). At Coquette Red, the laterite is developed both on harzburgite and dunite. At East Alpha, lherzolite dominates over harzburgite, although the intensive serpentinisation of the peridotite complicates its univocal recognition in the field. In the saprock, weathering initiates through the progression of the mantle silicate alteration front, marked by the onset of forsterite and enstatite hydrolysis and the development of Ni-rich (up to 15–20 wt% Ni), greenish talc-like (kerolite) veins that root into the unweathered peridotite (Figure 4). Serpentine (mostly lizardite) is preserved from dissolution in the saprock. The evolution from the saprock to the saprolite, wherein weathering is more pronounced but remains isovolumetric, is characterised by the complete disappearance of mantle silicate and the partial replacement of serpentines by goethite. In contrast, at East Alpha, smectite extensively develops from primary silicates and serpentine (Figures 3b and 4). It remains preserved throughout most of the saprolite until it is eventually replaced by ochreous goethite. The interface between the saprolite and the overlying limonite, referred to as the transition zone, is marked by the progressive accumulation of Mn-Co-Ni oxides (lithiophorite, asbolane). Upwards, the transition zone evolves into the yellow limonite, wherein ochreous goethite predominates. The yellow limonite then grades into the red limonite, wherein hematite forms at the expense of ochreous goethite resulting in a mineral assemblage dominated by goethite but containing significant amounts (>5 vol%) of hematite (Figure 4). Lateritic profiles are capped by a ferruginous duricrust that is either directly developed after the underlying limonite or derived from the ferrugination of transported material (Figures 3b–d and 4). Following Anand et al. [35], the term duricrust describes regolith materials cemented by Fe, irrespective of the substrate origin. When it is residual, the iron-cemented material is called a lateritic residuum, and when it is formed and indurated in a transported cover, ferricrete. Lateritic residuum exhibits, at the microscopic scale, some locally preserved textural relics (Figure 3e). There, both goethite and hematite are largely recrystallised into coarser crystallites. In contrast, ferricrete shows angular goethite and hematite nodules with goethite pisolitic envelopes. They are cemented by vitreous goethite (Figure 3f). Subvertical dykes of amphibolitic gabbros locally exposed at East Alpha alter to form a mineral assemblage composed of kaolinite-gibbsite-hematite (Figure 3g,h and Figure 4).

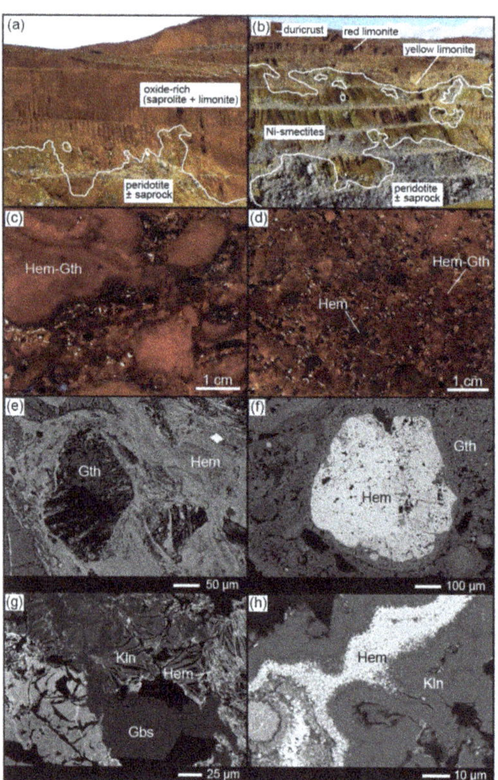

Figure 3. (**a**) Typical sharp transition from silicate-dominated bedrock and saprock to oxide-dominated saprolite and limonite (Ma-Oui). (**b**) Well-developed Ni smectite-rich zone (up to >15 m in vertical thickness) above lherzolitic serpentinite at East Alpha. (**c**) Vesicular hematite-goethite-bearing lateritic residuum (East Alpha). (**d**) Nodular/pisolitic ferricrete (East Alpha). (**e**) Recrystallisation of goethite and hematite with local preservation of inherited textures at the microscopic scale in the lateritic residuum, East Alpha (BSE imaging). (**f**) Goethitic pisolitic cortexes overgrowing onto nodular hematite in a goethite-bearing cement, ferricrete horizon, East Alpha (BSE imaging). (**g**,**h**) Kaolinite-gibbsite-hematite mineral assemblage in saprolitised gabbro at East Alpha (BSE imaging). Mineral abbreviations: Gth = goethite, Hem = hematite, Kln = kaolinite, Gbs = gibbsite. Onset (**b**) is modified after [13].

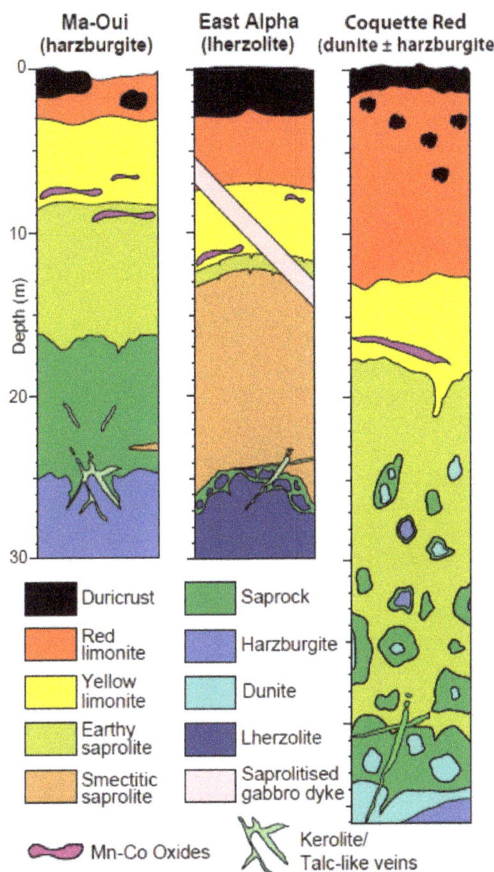

Figure 4. Stratigraphic logs of representative Ni-Co laterites from the Ma-Oui deposit, Coquette Red tenement, and East Alpha deposit. Modified after [13].

5. Whole-Rock Geochemical Correlations at the Deposit Scale

Whole-rock geochemical analysis of the investigated lateritic profiles shows that the lateritisation of unweathered peridotites is associated with the progressive enrichment of poorly mobile elements. In particular, Fe, Al, Cr, and Sc concentrations co-increase during weathering, exhibiting positive correlation trends (Figure 5, Table S1). In the Ma-Oui deposit, Al_2O_3, Fe_2O_3, Cr_2O_3, and Sc concentrations progressively increase from the bedrock to the oxide-rich horizons up to about 6 wt%, 80 wt%, 4.5 wt%, and 80 ppm, respectively. The Sc-Al_2O_3, Sc-Fe_2O_3, and Sc-Cr_2O_3 linear regression models encompassing variable lithofacies (unweathered harzburgite, saprock, saprolite, Mn-Co-rich transition zone, and yellow and red limonite) provide a particularly relevant fit of the data (Figure 5a). Scandium concentrations may be approximatively estimated from the Al_2O_3, Fe_2O_3, and Cr_2O_3 concentrations from the bedrock to the red limonite as follows:

$$Sc\ (ppm) = 12.03 \times Al_2O_3\ (wt\%) + 2.42;\ R^2 = 0.93 \tag{1}$$

$$Sc\ (ppm) = 0.92 \times Fe_2O_3\ (wt\%) + 0.07;\ R^2 = 0.86 \tag{2}$$

$$Sc\ (ppm) = 16.27 \times Cr_2O_3\ (wt\%) + 0.31;\ R^2 = 0.91 \tag{3}$$

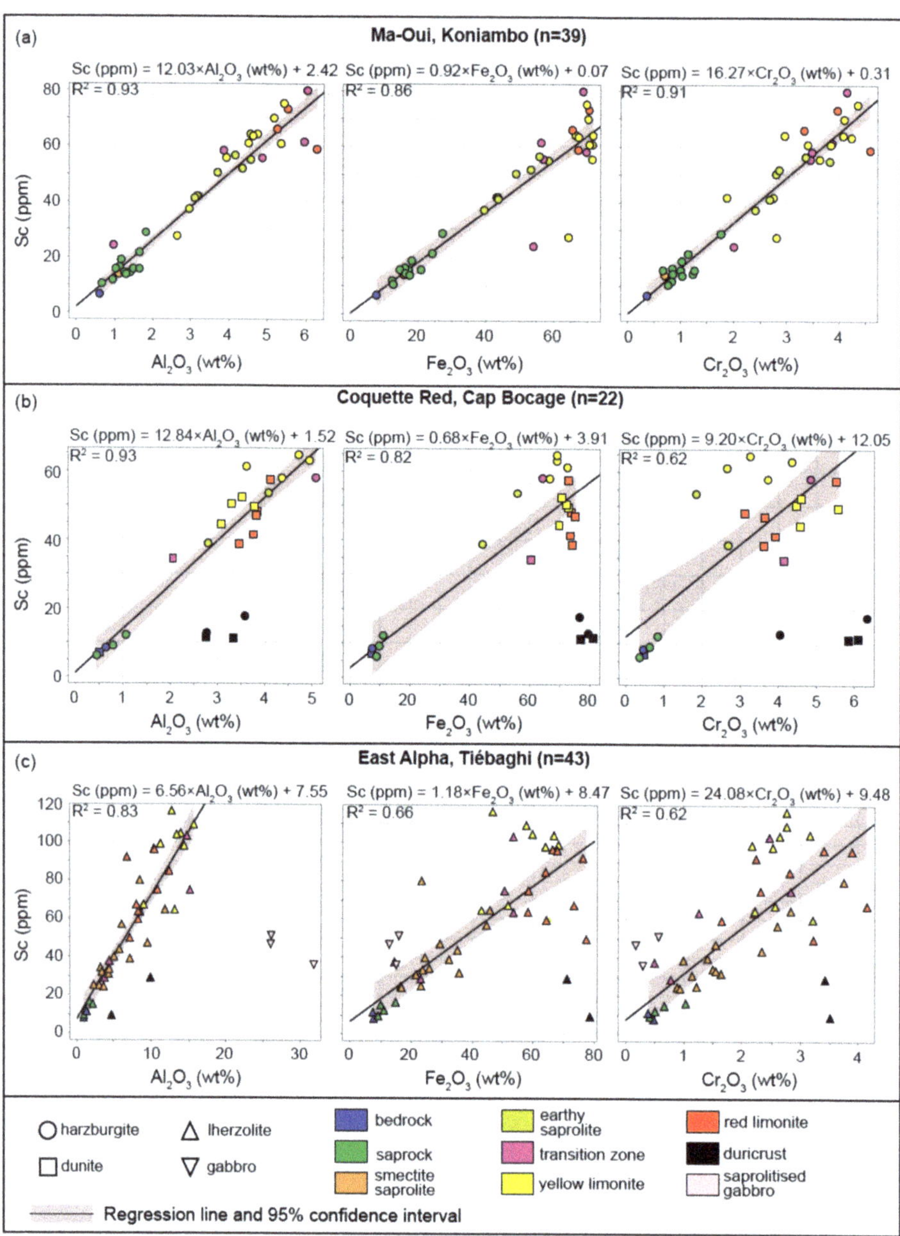

Figure 5. Sc-Al$_2$O$_3$, Sc-Fe$_2$O$_3$, and Sc-Cr$_2$O$_3$ scatter plots and regression lines obtained for samples collected at (**a**) the Ma-Oui deposit (Koniambo massif), (**b**) the Coquette Red tenement (Cap Bocage massif), and (**c**) the East Alpha deposit (Tiébaghi massif). Peridotite-derived duricrust and gabbro-derived saprolite samples lie out of the correlation trends and are therefore excluded from the regression line modelling.

The Sc-Fe$_2$O$_3$ correlation trend shows low dispersion in the low-Sc concentration range (0–50 ppm) but significant dispersion in the higher-Sc concentration range (50–80 ppm). In addition, two samples collected in the transition and yellow limonite zone, respectively, fall well out of the Sc-Fe$_2$O$_3$ regression line. The Sc-Cr$_2$O$_3$ correlation trend is better defined, yet it presents moderate data dispersion. Comparatively, the best fit is obtained for the Sc-Al$_2$O$_3$ regression line, which exhibits the lowest dispersion of the data. Notably, the decrease in Al and Sc concentrations from the yellow to red limonite, as commonly identified by Teitler et al. (2019) in New Caledonian lateritic profiles, is not observed in the Ma-Oui geochemical dataset. In the Coquette Red tenement (Cap Bocage massif), Al$_2$O$_3$, Fe$_2$O$_3$, Cr$_2$O$_3$, and Sc concentrations increase to about 5 wt%, 80 wt%, 5.5 wt%, and 65 ppm in the limonite, respectively. Like the Ma-Oui deposit, the Sc-Al$_2$O$_3$ regression line efficiently fits the data from the bedrock to the red limonite, although the laterite is developed on dunite and harzburgite (Figure 5b). Analysed duricrust samples exhibit a large offset from the trend, with a substantial depletion in Sc relative to Al$_2$O$_3$. It is worth noting that the Sc-Al$_2$O$_3$ regression lines obtained at Ma-Oui and Coquette Red have similar regression coefficients (respectively 12.03 and 12.84; Equations (1) and (4)).

In contrast, Fe$_2$O$_3$, Cr$_2$O$_3$ and Sc concentrations do not correlate when considering the limonitic horizons. Therefore, Al$_2$O$_3$ here constitutes the only relevant explanatory variable for Sc concentrations among major elements. More specifically, Sc and Al$_2$O$_3$ are significantly enriched compared to Fe$_2$O$_3$ in the yellow limonite and, on the opposite, depleted in the red limonite. Regression lines obtained from Coquette Red samples, accounting for all lithologies apart from the duricrust, may be expressed as follows:

$$\text{Sc (ppm)} = 12.84 \times \text{Al}_2\text{O}_3 \text{ (wt\%)} + 1.52; R^2 = 0.93 \qquad (4)$$

$$\text{Sc (ppm)} = 0.68 \times \text{Fe}_2\text{O}_3 \text{ (wt\%)} + 3.91; R^2 = 0.82 \qquad (5)$$

$$\text{Sc (ppm)} = 9.20 \times \text{Cr}_2\text{O}_3 \text{ (wt\%)} + 12.05; R^2 = 0.62 \qquad (6)$$

The East Alpha deposit (Figure 5c), developed after strongly serpentinised lherzolite, exhibits Sc concentrations up to 115 ppm in the limonite, i.e., significantly higher than those observed at Ma-Oui and Coquette Red. Moreover, Al$_2$O$_3$ concentrations reach about 15 wt% in the lherzolite-derived limonite, i.e., about three times more elevated than in the harzburgite $\pm$ dunite-derived limonite at Ma-Oui and Coquette Red. In comparison, gabbro-derived saprolite yields Al$_2$O$_3$ concentrations up to about 30 wt%. Maximum Fe$_2$O$_3$ and Cr$_2$O$_3$ concentrations at East Alpha are similar to those observed in the other investigated sites, at 80 and 4 wt%, respectively. The Sc-Al$_2$O$_3$ regression line modelled for East Alpha samples fits the data relatively well, except for duricrust and gabbro samples that fall well out of the correlation trend. The regression coefficient of the Sc-Al$_2$O$_3$ line is about two times lower at East Alpha than at Ma-Oui and Coquette Red (Equation (7)). It is worth noting that smectitic saprolite samples from East Alpha exhibit Al, Sc, Fe, and Cr co-variations and a quite extensive concentration range (Figure 5c).

Nevertheless, significant dispersion from the regression line is also observed in the yellow and red limonite horizons compared to the Ma-Oui deposit and the Coquette Red tenement (Figure 5c). Similar to Coquette Red, Sc and Al$_2$O$_3$ are significantly enriched in the yellow limonite and depleted in the red limonite compared to Fe$_2$O$_3$. The Sc-Fe$_2$O$_3$ and Sc-Cr$_2$O$_3$ correlations in the East Alpha deposit are poorly defined, especially in the yellow and red limonite, wherein Fe$_2$O$_3$, Cr$_2$O$_3$, and Sc concentrations do not correlate. Regression lines obtained for East Alpha, modelled after exclusion of duricrust and gabbro-derived samples, may be expressed as follows:

$$\text{Sc (ppm)} = 6.56 \times \text{Al}_2\text{O}_3 \text{ (wt\%)} + 7.55; R^2 = 0.83 \qquad (7)$$

$$\text{Sc (ppm)} = 1.18 \times \text{Fe}_2\text{O}_3 \text{ (wt\%)} + 8.47; R^2 = 0.66 \qquad (8)$$

$$\text{Sc (ppm)} = 24.08 \times \text{Cr}_2\text{O}_3 \text{ (wt\%)} + 9.48; R^2 = 0.62 \qquad (9)$$

Principal Component Analysis (PCA) allows us to explore further the overall compositional variability of the samples and the strength of the colinearity between Sc and poorly mobile elements. The PCA, performed after centred log-ratio transformation of the data, shows that the cumulated explained variance in the PC1-PC2 plane reaches 78.4% of the overall variability of the initial dataset (respectively, 53.0 and 25.4% for PC1 and PC2, Figure 6). All of the elements constitutive of the samples, except Mn and Co, predominantly contribute to PC1, with Mg, Si, and Ni having a positive contribution to PC1. At the same time, Fe, Al, Cr, and Sc negatively contribute to PC1 (Figure 6b). PC1 underlines a contrast between Mg-Si-Ni and Al-Sc-Fe-Cr, and PC2 describes the Mn and Co positive correlation. LOI exhibits a strong negative contribution to PC2. It is worth noting that the angles between Mg, Si, and Ni rays are low, yet these rays are not coincident. Similarly, Sc and Al, on the one hand, and Fe and Cr, on the other hand, form two pairs of strongly coincident rays. These two pairs of coincident rays are at a relatively low angle, yet they are not colinear with one another. Sc is, therefore, better correlated with Al than with Fe or Cr. Regarding sample scores, bedrock samples fall close to the Mg ray. Saprock samples, in contrast, appear globally shifted towards the Si and Ni rays. Most smectitic saprolite samples still display a positive PC1 score, but lower than that of bedrock and saprock samples, with a low PC2 score. Earthy saprolite samples exhibit small positive and negative PC1 scores and a significant shift towards positive PC2 scores. Earthy saprolite samples represent, therefore, intermediate compositions between the Mg-Si-Ni and the Al-Sc-Fe-Cr groups along PC1, together with the onset of an Mn-Co compositional signature along PC2. Comparatively, transition zone samples are slightly shifted towards negative PC1 scores, indicative of an increasing contribution of the Al-Sc-Fe-Cr group. Moreover, transition zone samples exhibit the highest PC2 scores, consistent with their enrichment in Mn and Co. Yellow limonite samples are further shifted towards negative PC1 scores while still displaying positive PC2 scores indicating the subsistence of an Mn-Co signature. Red limonite samples also show negative PC1 scores but appear to be forming two subgroups, one with positive PC2 scores similar to the yellow limonite and one with negative PC2 scores, identical to duricrust samples.

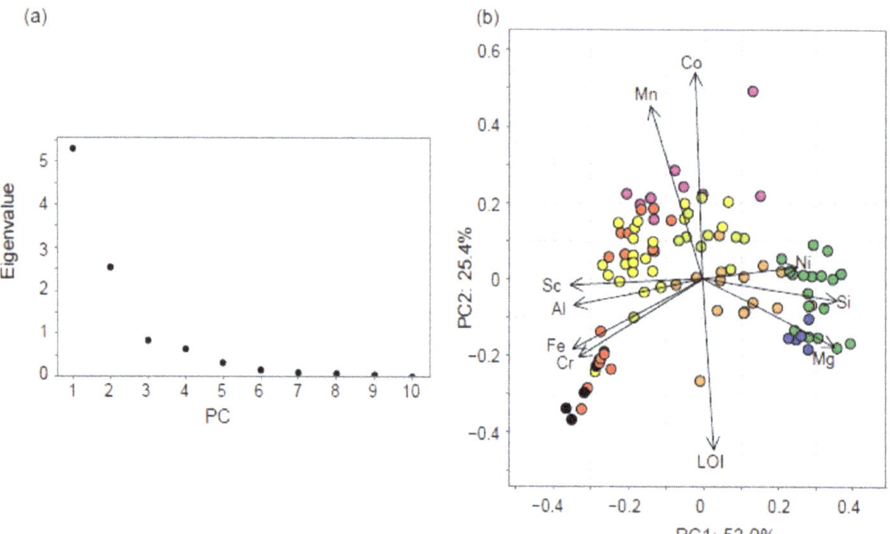

Figure 6. Results of PCA performed after centred log-ratio transformation of Mg, Al, Si, Sc, Cr, Mn, Fe, Co, and Ni concentrations and LOI from the global whole-rock geochemical dataset of peridotite-derived laterites (Ma-Oui, Coquette Red and East Alpha). Saprolitised gabbro samples from East Alpha were excluded from PCA calculations. (**a**) Scree plot. (**b**) PC1-PC2 biplot showing variable loadings (arrows) and sample scores (circles). See Figure 5 for lithology colour coding.

6. Mineral Chemistry

The mineral chemistry data obtained from LA-ICP-MS analyses (Table S2) allow us to assess the contribution of primary silicates and serpentine to the Al and Sc budget in the unweathered peridotites. Forsterite has a low Sc content in all investigated sites, about 3–6 ppm at Ma-Oui and Coquette Red, about 7 ppm at East Alpha (Figure 7), and Al below detection limits. Enstatite yields higher Sc and Al concentrations, which vary from one site to another. Sc concentrations in enstatite from Ma-Oui and Coquette Red are similar (24 and 22.5 ppm, respectively). Al_2O_3 concentrations in enstatite are slightly higher at Ma-Oui (about 1.8 wt%) than at Coquette Red (about 1.4 wt%). In contrast, enstatite from East Alpha is enriched in Sc (about 32 ppm) and Al_2O_3 (about 3.4 to 3.8 wt%). Diopside yields an Al_2O_3 content similar to that of enstatite, with a higher Sc content (about 50 ppm in Ma-Oui and 65 ppm in Coquette Red). The mineral composition of lizardite developed after mantle silicates (forsterite, enstatite, and rarer diopside) commonly reflects the composition of their precursor mineral. Indeed, lizardite developed after forsterite yields low Sc and very low Al_2O_3 contents, whereas lizardite developed after enstatite and diopside yields higher Sc and Al_2O_3 contents (Figure 7). Lizardite forming the mesh network typically has Sc and Al_2O_3 contents intermediate between forsterite and enstatite. Chromiferous spinel yields elevated Al_2O_3 contents (>12 wt%) together with very low Sc contents (<3 ppm, Table S2). Importantly, Sc-Al_2O_3 regression lines obtained from major bedrock minerals (forsterite, enstatite, lizardite after enstatite and forsterite, and lizardite mesh) in each investigated site (Figure 7) closely match those obtained from whole-rock geochemical analysis throughout the weathering sequences (Figure 5). More specifically, regression coefficients estimated from bedrock mineral chemistry vs. global whole-rock geochemistry, respectively, are (i) 10.95 vs. 11.02 at Ma-Oui, (ii) 13.53 vs. 12.84 at Coquette Red, and (iii) 6.60 vs. 6.56 at East Alpha.

Further support for the intimate relationship between Sc-Al_2O_3 contents of peridotite mantle silicates and Sc-Al_2O_3 contents of peridotite-derived laterite samples is provided by the mineral chemistry data of secondary mineral phases along laterite profiles as analysed at East Alpha (Figure 8, Table S2). Together with forsterite and enstatite, Sc and Al_2O_3 concentrations in smectite, ochreous goethite from the limonite zone, and goethite-hematite from the lateritic residuum appear to be strongly proportional, with a regression line closely similar to that obtained from whole-rock geochemistry. Smectite in the smectitic saprolite and ochreous goethite in the limonite exhibit variable yet proportional Sc and Al_2O_3 concentrations. Indeed, Sc and Al_2O_3 concentrations in smectite vary from 20 to 80 ppm and from 2.5 to 10 wt%, respectively, while Sc and Al_2O_3 concentrations in limonitic, ochreous goethite vary from 55 to 122 ppm and from 7 to 16 wt%, respectively. Goethite-hematite from the lateritic residuum is depleted both in Sc and Al_2O_3. In contrast, goethite and hematite nodules and pisolitic cortexes from the ferricrete are depleted mostly in Sc and therefore fall below the regression line.

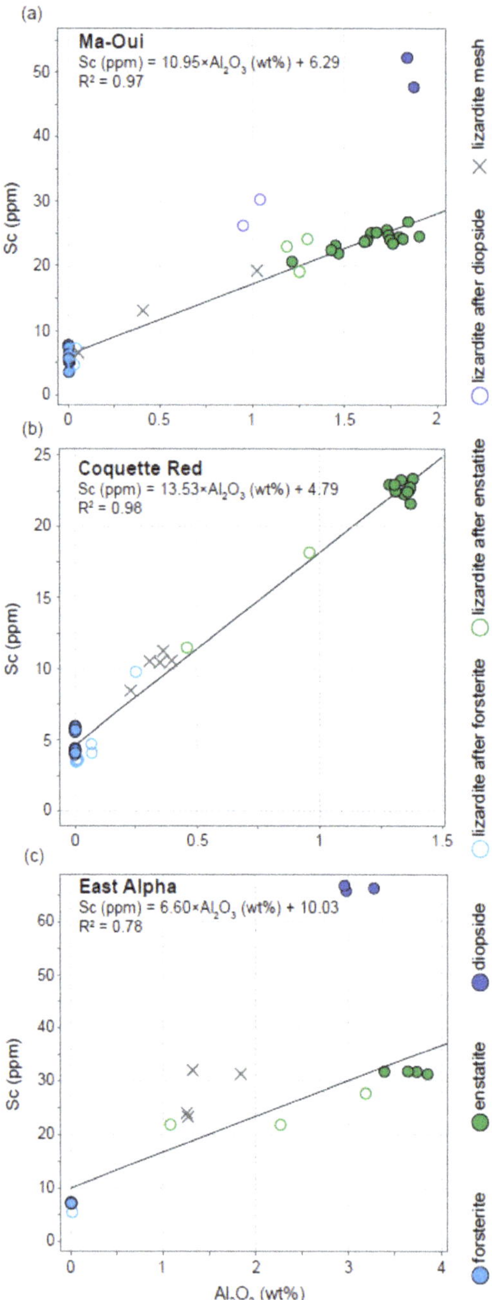

Figure 7. Sc-Al$_2$O$_3$ scatterplots and regression lines obtained for mantle silicates (forsterite, enstatite, diopside) and lizardite (developed after forsterite, enstatite, and diopside, or as mesh) in peridotite bedrocks from (**a**) Ma-Oui, (**b**) Coquette Red, and (**c**) East Alpha from LA-ICP-MS analyses. The regression line modelling excluded accessory phases (diopside, lizardite developed after diopside, and chromiferous spinel).

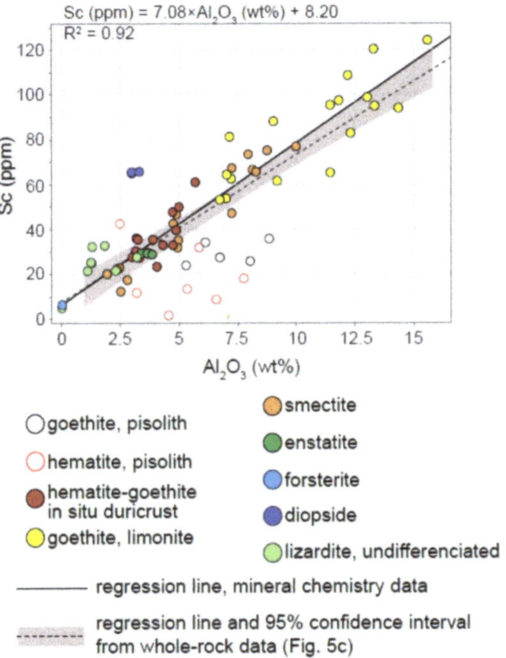

Figure 8. Sc-Al$_2$O$_3$ scatterplot and regression line obtained for silicates, smectite, and iron oxides in the East Alpha deposit (Tiébaghi massif) from LA-ICP-MS analyses. The regression line calculation excluded spot analyses of diopside from the bedrock and of goethite and hematite from the nodular/pisolitic duricrust.

7. Sequential Extractions

The sequential extractions for Fe, Al, Ni, and Sc are given in Figure 9. During the first extraction step, the ultrapure water did not permit solubilising these elements in any investigated samples. With regards to Fe, the hydroxylamine hydrochloride extraction also proved unsuccessful. In iron-rich samples (earthy saprolite, and yellow and red limonite), about 10–15% of the total Fe mass was extracted during the subsequent extraction step involving ammonium oxalate, while about 50–60% was removed during the last extraction step involving citrate-bicarbonate-dithionite (CBD). In iron-rich lithologies, the overall procedure permits the extraction of 60 to 80% of the total Fe content, and CBD appears to be the most effective reagent to solubilise Fe. In contrast, the general approach only resulted in the extraction of about 25% of the total Fe in the smectitic saprolite sample (TIEA-08) from East Alpha, mainly through the action of CBD and, to a lesser extent, ammonium oxalate. With regards to Al, the overall procedure succeeded in solubilising 40–50% and 10% of the total Al (Figure 9) in iron-rich and smectite-rich samples, respectively. The relative amount of extracted Al is higher in iron-rich samples from Coquette Red (PZ1B-06, PZ1B-09, PZ1B-13 and PZ1B-17) than in their equivalents from East Alpha (TIEA-07 and TIEA-12). Among this fraction, a small portion (less than 5% of the total Al extracted) was solubilised using hydroxylamine hydrochloride. The relative amount of Al removed by ammonium oxalate is similar to that of Fe, about 10% or less. CBD remains the most efficient reagent for Al extraction but seems less effective for solubilising Al than Fe. In addition, Al appears more efficiently extracted by ammonium oxalate in iron-rich samples from Coquette Red (about 10% of the total Al or 15–30% of the whole extracted Al) than in their equivalent from East Alpha (about 5% of the total Al or 10% of the extracted Al).

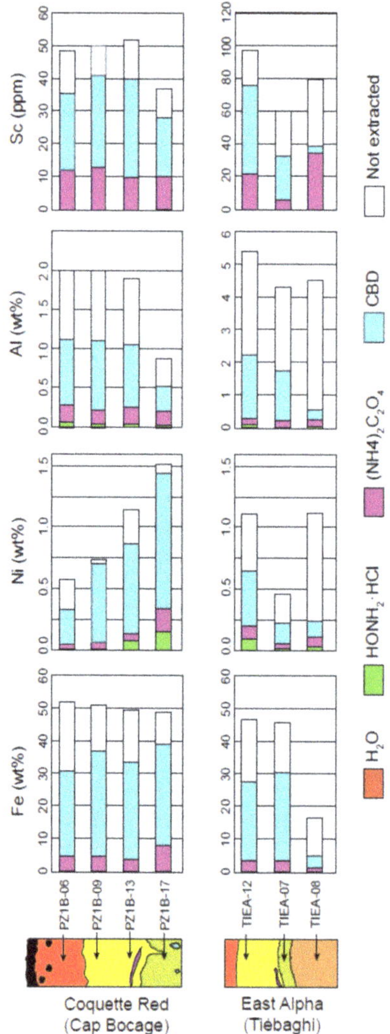

Figure 9. Results of sequential extractions. Cumulative histograms showing the total quantities of Fe (wt%), Ni (wt%), Al (wt%), and Sc (ppm) extracted with the different reagents, as well as the quantities not extracted. H_2O = Ultrapure water (step 1), $HONH_2$-HCl = hydroxylamine hydrochloride 0.1 mol·L^{-1} at pH 3.5 (step 2), $(NH_4)_2C_2O_4$ = ammonium oxalate 0.2 mol·L^{-1} at pH 3 (step 3), CBD = citrate-bicarbonate-dithionite $Na_2S_2O_4$ (22% Na-citrate and 1 g Na-dithionite) (step 4). See Figure 5 for lithology colour coding.

Bulk Ni extraction is similar to or higher than Fe extraction. It is, in particular, higher in iron-rich samples from Coquette Red than in their equivalents from East Alpha. More specifically, two samples from Coquette Red (PZ1B-09 and PZ1B-17) show almost complete Ni extraction, while Ni extraction in East Alpha does not exceed 60%. In contrast, Ni is weakly extracted from the smectitic saprolite sample from East Alpha. The relative proportion of Ni removed by hydroxylamine hydrochloride remains low (5 to 15% of the total Ni extracted), yet higher than that observed for Al. The efficiency of Ni extraction by ammonium oxalate appears similar to that of Fe (about 5 to 15% of the total Ni extracted) and slightly lower than Al. As for Fe and Al, Ni is best extracted through CBD's last extraction step. Regarding Sc, the overall procedure proves efficient in oxide-rich samples, resulting in the extraction of about 75% of the total Sc amount, except in one sample (60% extraction efficiency in TIEA-07). Like Fe, neither the ultrapure water nor the hydroxylamine hydrochloride successfully extracted Sc from any samples. Sc extraction using ammonium oxalate, which accounts for about 20 to 30% of the total Sc contained, appears significantly more efficient than Fe, Al, and Ni in iron-rich samples.

Nevertheless, in these samples, CBD remains the most efficient reagent for Sc extraction. In contrast, in the smectitic saprolite sample from East Alpha, Sc is almost exclusively extracted using ammonium oxalate. However, the total amount of extracted Sc from this sample remains moderate (about 50% of the bulk Sc content). To summarise, it should be noted that, in oxide-rich facies, (i) the extraction procedure is globally efficient for solubilising Fe, Al, Ni, and Sc, though significant differences in global extraction efficiencies are observed depending on the considered element and the sample provenance; (ii) only Al and Ni can be slightly extracted using hydroxylamine hydrochloride; (iii) Fe, Al, Ni, and Sc are best solubilised using CBD and, to a lesser extent, ammonium oxalate; and (iv) the efficiency of Sc extraction using ammonium oxalate is significantly greater than that of Fe, Al, and Ni. In contrast, the overall procedure is less well suited for the sequential extraction in smectitic saprolite, so the greater extraction efficiency of ammonium oxalate relative to other reagents must be considered cautiously.

8. Discussion and Implications for the Evaluation of Sc in Ni-Co Laterites

This study aims to assess in which conditions Al may be reliably used as a geochemical proxy for a first-order evaluation of Sc grades and distribution in Ni-Co lateritic ores. In the following, we discuss the causes of the Al-Sc co-variation regarding the mineralogical evolution in lateritic profiles and the relevance of using deposit-scale Sc-Al_2O_3 correlations to assess the distribution of Sc in Ni-Co laterites.

8.1. Co-Evolution of Al and Sc Concentration and Speciation through Weathering

Analytical data documented in the present contribution reveal significant correlations, at the deposit scale, between Sc and Al_2O_3 concentrations in some lateritic Ni deposits of New Caledonia. The high proportionality between Sc and Al_2O_3, observed in the investigated lateritic deposits from the bedrock to the red limonite, is identified from whole-rock geochemistry (scatterplots of untransformed data and PCA of centred log-ratio-transformed data) and mineral chemistry data (Figures 5, 6 and 8). This correlation has already been documented [13], but not at the deposit scale. Furthermore, the study conducted by Santoro et al. [36] on trace element concentrations in goethite from various Ni-laterite deposits showed a strong association between the Sc and Al contents of goethite from the Wingellina deposit (Australia). The authors interpreted such association as being primarily controlled by the composition of the parent rock, highlighting the influence of Al- and Sc-bearing pyroxenite lenses on the composition of their goethite-bearing, weathered derivatives. The Sc-Al_2O_3 proportionality, documented in the present contribution throughout the weathering sequences of various Ni-laterites from New Caledonia (i.e., from the bedrock to the red limonite), argues for a similar behaviour of Al and Sc during weathering.

In the unweathered mantle silicates, Ni is exclusively bound to forsterite. In contrast, virtually all of the Al is hosted in pyroxenes (enstatite and diopside), and, to a lesser extent, in chromiferous spinel, as it is highly incompatible with forsterite. Sc is also preferentially hosted in pyroxenes, as forsterite only incorporates Sc up to a few ppm. Diopside contains Sc concentrations up to about 65 ppm (Table S2, [13,14]) but is not abundant (Table S1) and has a minor influence on the bulk Sc content of unweathered lherzolite. It has been proposed that serpentinisation of the peridotite most likely has a marginal effect on Sc concentrations in the bedrock [13]. The mineral chemistry data documented on lizardite in the present contribution supports such an assumption, as lizardite composition primarily reflects the composition of its mineral precursor. Although Cr-spinel yields elevated Al_2O_3 contents, it is an accessory phase in the bedrock. Considering (i) the mineral composition of

Cr-spinel, (ii) the bulk Cr content of bedrock samples, and (iii) conservatively assuming that all Cr is hosted in Cr-spinel in the bedrock, the contribution of Cr-spinel on the bulk Al_2O_3 content does not exceed 10% in the bedrock. Therefore, the global Sc and Al_2O_3 contents in unweathered peridotite largely depend on the relative proportion of enstatite and its chemical composition. Nevertheless, forsterite being the predominant mantle silicate in peridotite, it also significantly contributes to the global Sc budget of the bedrock.

In the lower part of the weathering sequence (saprock and smectitic saprolite), Sc and Al concentrations co-increase with Fe and Cr concentrations, suggesting that Sc and Al enrichments are residual. Co-variations, along an extensive concentration range, of Al, Sc, Fe, and Cr in smectitic saprolite samples from East Alpha (Figure 5), result from inherent co-variation of smectite composition, as evidenced by mineral chemistry data (Figure 8). The sequential extraction procedure proved relatively ineffective in solubilising Fe, Ni, and Al and from the smectitic saprolite. These elements are typically present in octahedral and tetrahedral positions in nickeliferous smectite and are not leached easily from smectite [37,38]. However, the significant amounts of Sc extracted from the smectite-rich saprolite using ammonium oxalate raise questions regarding the speciation of Sc in this horizon. Together with the elevated Sc concentrations measured in smectite from LA-ICP-MS analysis, the lack of Fe extraction by ammonium oxalate rules out any significant contribution of amorphous iron oxides/oxyhydroxides to the Sc budget in the smectite-rich zone. Based on the scandium K-edge XANES analysis, Chassé et al. [8] proposed that Sc in the plasmic horizon of the Syerston-Flemington Sc laterite (Australia) is efficiently trapped in smectite through incorporation into octahedral sites. Such interpretation seems unable to account for the high quantities of Sc extracted using ammonium oxalate. Further investigations remain to evaluate the Sc speciation in the smectitic saprolite from the East Alpha deposit.

Contrasting with the Fe-Al-Sc-Cr collinearity identified in the lower portion of the weathering sequences, Sc and Al exhibit a relative enrichment compared to Fe in the earthy saprolite/yellow limonite and a relative depletion in the red limonite. This distribution pattern, observed at Coquette Red and East Alpha (Figure 5b,c), as well as in several New Caledonian Ni-Co laterites [13], but not at Ma-Oui (Figure 5a), supports specific mobility of Sc and Al. It is suggested that these elements are, to some extent, remobilised from the red limonite and accumulated downwards in the yellow limonite. The remobilisation of Sc and Al possibly results from the release of Sc and Al after the dissolution/recrystallisation of increasingly crystallised goethite during the maturation of the lateritic profile [13,39]. Such a model is commonly accepted for explaining the distribution pattern of Ni concentrations, which generally decrease gradually from the earthy saprolite upwards in conjunction with an increase of the mean coherent domain (MCD) size of the goethite crystallites [39,40]. This typical Ni distribution pattern is, for instance, well-evidenced at Coquette Red, where a gradual decrease of Ni concentrations occurs from the earthy saprolite (1.5 wt% Ni) to the red limonite (0.5 wt% Ni). There, Sc and Al distribute differently from Ni, as Sc and Al concentrations are higher in the yellow limonite than in the earthy saprolite. Although the release of Sc during dissolution/recrystallisation of goethite has been proposed as a relevant model for explaining some elevated Sc concentrations in the yellow limonite [8,13], the mobility of Sc released during this process must therefore be lower than that of Ni. In addition, the formation of hematite at the expense of goethite in the red limonite likely results in the downward redistribution of Sc and Al, as both Sc and Al substitute more easily for Fe in goethite than in hematite [41–43], thus contributing to Sc enrichment in the yellow limonite.

Sequential extraction provides further insights on the speciation of Sc and its association with other elements in the earthy saprolite, yellow and red limonite. It is worth noting that the extraction procedure here applied differs from the method previously used on ultramafic-derived laterites from the Syerston-Flemington (Australia) and the Berong (Philippines) deposits [8,11]. The extraction procedure is adapted from Hall et al. [44] and Sanematsu et al. [45] in these two studies. It includes sodium acetate as the first reagent to assess the adsorbed and exchangeable species. In contrast, we used ultrapure water, which only allows us to extract the easily exchangeable species and, therefore, does not estimate the adsorbed species.

Nevertheless, the extracted Sc amounts obtained from sodium acetate treatment are low (10–15%) in the Syerston–Flemington laterites to very low (<5%) in the Borong laterites [8,11]. These two studies combine sequential extraction with XANES analysis. Chassé et al. [8] observed that the low proportion of exchangeable goethite-hosted Sc obtained from sequential extraction (10–15%) is at odds with the results of the XANES analysis. The latter reflects an elevated contribution of Sc adsorbed on goethite (up to 80%) in the global Sc budget, with only a small proportion of Sc being substituted in iron oxides and oxyhydroxides. The authors favour better reliability of the XANES spectra interpretation than the sequential extraction results and propose reconciling these data by arguing for the high stability of the Sc adsorption complex on goethite, thus preventing the extraction of adsorbed Sc through sodium acetate treatment. Similarly, in the Gorong deposit [11], the amounts of Sc adsorbed on goethite appear more significant when estimated from XANES spectra (24–49%) than from sequential extraction (below 5%).

Both of these studies then use hydroxylamine hydrochloride to extract amorphous iron oxides. In contrast, we used hydroxylamine hydrochloride at lower concentrations to extract manganiferous species. No Fe or Sc, and only a very slight fraction of Ni and Al, are extracted using hydroxylamine hydrochloride, highlighting the marginal presence of Mn oxides in the investigated samples and the absence of detectable Sc in Mn oxides. Following hydroxylamine hydrochloride treatment, we used ammonium oxalate to extract amorphous iron oxides. Ammonium oxalate has been reported as an efficient and selective dissolving agent for amorphous and poorly crystalline ferric oxides/oxyhydroxides without significant crystalline goethite/hematite or silicate dissolution [46,47]. Chassé et al. [8] and Qin et al. [11] did not pursue sequential extraction further than the amorphous iron oxides/oxyhydroxides extraction step, hypothesizing that the residue is mostly representative of crystalline iron oxides. In the present study, the last extraction step with CBD unambiguously demonstrates that crystalline iron oxides/oxyhydroxides predominates over amorphous iron oxides/oxyhydroxides in oxide-rich horizons. Along with Fe, the elevated Ni, Al, and Sc fractions extracted using CBD confirm that these elements are mainly bound to crystalline goethite. At Coquette Red, a slight decrease in the total extracted Fe is observed from the earthy saprolite to the red limonite, from about 80 to 60% (Figure 9). Such a decrease, resulting from a slight lowering of both CBD and ammonium oxalate extraction efficiencies (from about 60 to 50% and 15 to 8%, respectively), may indicate an increase in goethite crystallinity. Compared to Fe, Al extraction appears globally less effective, especially at East Alpha, wherein the Al contents of oxide-rich facies are higher than at Coquette Red. Such a discrepancy between the extraction efficiencies of Al and Fe possibly results from the presence of extraction-resistant, Al-bearing minerals in the oxide-rich facies. Chromiferous spinel, an accessory phase in the bedrock, is residually enriched in the oxide-rich facies and contains about 15 wt% Al [13]. Therefore, even in minor amounts, the presence of extraction-resistant Cr-spinel implies that a non-negligible portion of the whole-rock Al content cannot be extracted through the used procedure. However, such quantities of unextractable Al are likely insufficient to account for the apparent lower extractability of Al, or for the differences in Al extraction rates between Coquette Red and East Alpha, as both sites have similar Cr contents. Although not observed in any of the peridotite-derived limonite samples, trace amounts of kaolinite in the lherzolite-derived limonite cannot be ruled out at East Alpha. Similar to Cr-spinel, the used extraction procedure is ineffective in solubilising kaolinite, whose potential presence in the East Alpha limonite may explain the lower extraction rate for Al. Regarding Sc, total extraction rates in the Fe-rich horizons are similar to that of Fe (and therefore higher than that of Al), supporting the strong association of Sc with iron oxides/oxyhydroxides and, in particular, with crystalline goethite. Nevertheless, ammonium oxalate appears relatively more efficient in extracting Sc (about 20 to 35% of total Sc) than Fe (about 10% of the total Fe). The proportion of Sc extracted from amorphous iron oxides in the present study is, therefore, slightly higher than that obtained by Chassé et al. [8], that is about 15–25%, and significantly higher than that obtained by Qin et al. [11], that is below 3%. This discrepancy possibly results from a better efficiency of ammonium oxalate to extract Sc from amorphous iron oxides/oxyhydroxides than hydroxylamine hydrochloride. Our results suggest that Sc may have a higher affinity for amorphous iron oxides/oxyhydroxides than crystalline goethite. Thus significant amounts of Sc can be concentrated in amorphous iron oxides/oxyhydroxides despite the predominance of crystalline goethite throughout the lateritic sequence. The speciation of Sc in oxide-rich facies differs probably in part from that of Al. The latter is mainly incorporated in the lattice of crystalline goethite and preserved in weathering-resistant Cr-spinel. However, despite second-order differences between the speciation of Sc and Al_2O_3, these elements show strong proportionality from the bedrock to the red limonite. As these elements (i) are both mainly immobile during the weathering of peridotite, (ii) have a preferential affinity for goethite, and (iii) are only moderately remobilised following recrystallisation and goethite replacement by hematite, the composition of the parent rock remains the first-order control on their concentrations throughout the weathering sequences. Thus, the Sc-Al_2O_3 regression lines obtained from forsterite-enstatite-lizardite mineral compositions are close to those obtained from weathering-related mineral compositions and whole-rock geochemistry along weathering profiles. The slightly lower regression coefficients obtained from whole-rock geochemistry than those obtained from bedrock mineral composition may result from the second-order contribution of Cr-spinel to the global Al budget in whole-rock geochemical compositions. These results indicate that (i) the Sc-Al_2O_3 content of enstatite in a given peridotite bedrock drives the slope of the Sc-Al_2O_3 regression line within its weathered derivatives, and (ii) the relative proportion of enstatite together with its Sc content largely control the maximum Sc concentrations reached in the yellow limonite. It is worth noting that the positive intercepts observed on whole-rock Sc-Al_2O_3 regression lines likely reflect the contribution of Al-free forsterite.

Contrasting with the proportionality observed between Sc and Al_2O_3 from the bedrock up to the red limonite, samples from the duricrust are out of the Sc-Al_2O_3 whole-rock correlation trends. Such offset, previously documented by Teitler et al. [13], possibly results from the contribution of allochthonous material to the duricrust. Indeed, in situ Sc analysis of goethite-hematite in lateritic residuum matches the Sc-Al_2O_3 regression line obtained on saprolitic and limonitic minerals. On the opposite, nodular and pisolitic goethite and hematite from the ferricrete exhibit a relative depletion in Sc compared to Al, similar to the offsets identified in the whole-rock geochemical dataset. In the East Alpha deposit, the offset of gabbro-derived saprolite from the Sc-Al_2O_3 correlation trend also results from the allochthonous nature of the gabbro compared to the peridotite-derived Ni-Co laterite, together with the specific mineral assemblage of the gabbro-derived saprolite. The formation of kaolinite during the weathering of gabbro is related to its Al content [13,48,49]. As Sc is typically poorly concentrated into kaolinite [7,13,14], gabbro weathering may lead to Sc remobilisation and trapping into nearby yellow limonite. At the same time, Al remains concentrated in saprolitised gabbro as kaolinite. Consequently, Sc and Al_2O_3 may be positively correlated in lateritic deposits if kaolinite, or other Al-bearing phases such as gibbsite, are mostly absent from the lateritic profiles. The formation of kaolinite (or gibbsite) during the lateritisation process requires that the parent rock contains significant amounts of Al, so that Sc-Al_2O_3 correlation trends may only be observed in ultramafic-derived laterites, wherein the Al content is low.

8.2. Implications for the Assessment of Sc in Ni-Laterites

Using Al as a geochemical proxy to conduct a first-order estimation of Sc concentration and distribution in a given Ni-Co deposit depends on the specific reliability of the Sc-Al_2O_3 correlation at the deposit scale and on the method used to characterise Al concentration and distribution. The reliability of the Sc-Al_2O_3 correlation may be influenced by (i) potential heterogeneities in the parent rock lithology, (ii) the occurrence of alteration facies containing Al-rich phases such as kaolinite or gibbsite, and (iii) the inherent data dispersion in Sc-Al_2O_3 scatterplots. Interestingly, the alternate presence of harzburgite and dunite in the bedrock does not seem to significantly affect the Sc-Al_2O_3 regression coefficient nor the dispersion of the data, providing that the composition of enstatite remains similar in both facies. Nevertheless, lithological heterogeneities involving a change in the composition of enstatite, elevated amounts of diopside and plagioclase (e.g., in some lherzolite facies), or the occurrence of mafic intrusive dykes, may cause significant variations of the Sc/Al_2O_3 concentration ratio both in the parent rocks and in their weathered derivatives. These potential variations of the parent rock lithology at the deposit scale may be tested by geological characterisation and geochemical assay data analysis. In addition, alteration facies containing significant amounts of Al-rich phases (e.g., kaolinite, gibbsite) are commonly characterised by a substantial deviation from the Sc-Al_2O_3 correlation trend established from smectite- and Fe-oxide-dominated lithologies. The possible occurrence of Al-rich phases, typically poor in Sc, must be examined to prevent over-estimation of the Sc content. Al-rich phases are commonly associated with weathered intrusive rocks such as gabbros. Still, they may also form during the weathering of some peridotites that yield significant Al concentrations. In particular, some plagioclase-bearing lherzolites (not investigated in the present study) can exhibit Al_2O_3 contents up to ~4 wt% [18,50] and may consequently alter to kaolinite or gibbsite.

Therefore, geochemical homogeneity and low Al-content of the parent rock are necessary conditions for a reliable Sc-Al_2O_3 correlation trend at the deposit scale. Determination of a reliable, deposit-scale Sc-Al_2O_3 correlation requires assessing the range of Al_2O_3 concentrations for which Sc is well correlated with Al_2O_3 through an adequate sampling strategy that encompasses the whole range of Al_2O_3 concentrations and the mineralogical diversity throughout the deposit. Finally, the inherent data dispersion in Sc-Al_2O_3 biplots may be variable depending on the investigated deposit. Therefore, the number of samples used for establishing the Sc-Al_2O_3 correlation must be adapted to the inherent data dispersion to develop a reliable correlation with a good correlation coefficient. Once verified, deposit-scale Sc-Al_2O_3 correlations may first prove useful to estimate Sc concentrations on surface outcrops or pit walls using portable devices such as pXRF. Indeed, the relatively low Sc concentrations observed in Ni-laterites (<100 ppm) limit the use of pXRF to directly assess Sc [51], whereas Al concentrations, typically about a few wt%, can be confidently estimated using such a device. The slightly lower accuracy obtained on Al analysis from pXRF than from a classical assay analysis would only marginally affect the estimation of Sc. More importantly, applying this approach to deposits where Al is routinely assayed could provide a first-order Sc concentration and distribution estimate. The reliability of such estimation would depend on the Sc-Al_2O_3 correlation and the reliability of the Al distribution model. Indeed, block models are developed firstly to estimate the resource and distribution of Ni both in saprolitic and limonitic facies. They rely on statistical variograms set explicitly for several metals (e.g., Si, Mg, Fe, and Ni) but not necessarily Al. In such a case, a specific evaluation of the variability of Al in the limonitic facies may be beneficial.

9. Conclusions

This contribution examines the relevance of using Al as a geochemical proxy for first-order Sc distribution and concentration estimates in some peridotite-hosted, Ni-Co laterites from New Caledonia. Apparent correlations are identified at the deposit scale between Sc and Al_2O_3 concentrations from the bedrock to the red limonite. These correlations put forward the similar behaviour of Al and Sc in weathered peridotite as their concentrations are primarily issued from residual enrichment. Local remobilisation from the uppermost horizons is shown. Al and Sc are predominantly hosted in crystalline goethite, but Sc has a relatively higher affinity for amorphous iron oxides than Al. In all investigated deposits, the Sc-Al_2O_3 regression coefficient remarkably depends on the Sc content in enstatite. Since the parent lithology is homogeneous and relatively depleted in Al, reliable Sc-Al_2O_3 correlations may thus be determined at the deposit scale after analysing a limited number of spatially and chemically representative samples. An adequate sampling strategy is required to cover the range of Sc and Al_2O_3 concentrations throughout the deposit. In addition, it is necessary to consider potential occurrences of specific alteration facies that may affect the relevance of the deposit-scale correlation (ferricrete, weathered intrusive rocks, and kaolinite- or gibbsite-bearing alterite). Although the zones of maximum Sc enrichment are situated above the top Ni and Co enrichment zones, Sc-rich limonite may overlap the Co-rich transition zone. The base of the limonite zone may therefore contain exploitable Co and Sc together with sub-economic Ni grades.

Supplementary Materials: The following supporting information can be downloaded at: https://www.mdpi.com/article/10.3390/min12050615/s1, Table S1: Whole-rock geochemical data; Table S2: Mineral chemistry (LA-ICP-MS) data.

Author Contributions: Conceptualization, Y.T., J.-P.A., B.S. and M.C.; methodology, Y.T., S.F., J.-P.A. and M.C.; software, Y.T. and S.F.; validation, Y.T., J.-P.A., B.S. and M.C.; formal analysis, Y.T. and J.-P.A.; investigation, Y.T. and J.-P.A.; writing—original draft preparation, Y.T., S.F., F.G. and M.C.; supervision, B.S., F.G. and M.C.; project administration, F.G. and M.C.; funding acquisition, Y.T., F.G. and M.C. All authors have read and agreed to the published version of the manuscript.

Funding: This work has been funded and logistically supported by the French National Research Agency through the national program "Investissements d'avenir" of the Labex Ressources 21 with the reference ANR-10-LABX-21-RESSOURCES21 and by the National Centre for Technological Research CNRT "Nickel et son environnement" based in Nouméa, New Caledonia (Project grants: 8PS2013-CNRT.CNRS/SCANDIUM and 9PS2017-CNRT.GEORESSOURCE/TRANSNUM).

Data Availability Statement: Data supporting reported results can be found at Supplementary Materials.

Acknowledgments: The authors would like to thank Stephane Lesimple, Bernard Robineau, and Olivier Monge from the Geological Survey of New Caledonia, together with France Bailly, Laurence Barriller, and Fabien Trotet, from the National Centre for Technological Research, for their technical and logistical support, as well as for sharing their extensive knowledge of the geology of New Caledonia. We are grateful to Société Le Nickel (SLN) and Koniambo Nickel SAS (KNS), and, in particular, to Clément Marcaillou, Matthieu Monnerais, Hubert Dumon, and Ludovic Levy, for support and access to mines and drill cores. We also thank Christophe Cloquet (SARM analytical centre, CRPG, Vandœuvre-lès-Nancy, France), Andreï Leconte (GeoRessources, Vandœuvre-lès-Nancy, France), and Claude Douchement (NiLab laboratory, Pouembout, New Caledonia) for technical support in providing analytical data.

Conflicts of Interest: The authors declare no conflict of interest.

References

1. Royset, J.; Ryum, N. Scandium in aluminium alloys. *Int. Mater. Rev.* **2005**, *50*, 19–44. [CrossRef]
2. Toropova, L.S.; Eskin, D.G.; Kharakterova, M.L.; Dobatkina, T.V. Advanced Aluminium Alloys Containing Scandium. In *Structure and Properties*; Gordon and Breach Science Publishers: Amsterdam, The Netherlands, 1998; p. 175.
3. US Geological Survey. Scandium. In *Mineral Commodity Summaries*; U.S. Geological Survey: Reston, VA, USA, 2022; pp. 146–147.
4. Aiglsperger, T.; Proenza, J.A.; Lewis, J.F.; Labrador Svojtka, M.; Rojas-Puron, A.; Longo, F.; Durisova, J. Critical metals (REE, Sc, PGE) in Ni laterites from Cuba and the Dominican Republic. *Ore Geol. Rev.* **2016**, *73*, 127–147. [CrossRef]
5. Audet, M. Le massif du Koniambo, Nouvelle-Calédonie. Formation et Obduction d'un Complexe Ophiolitique du Type SSZ. Enrichissement en Nickel, Cobalt et Scandium Dans Les Profils Résiduels. Ph.D. Thesis, Université de la Nouvelle Calédonie, Noumea, France, Université du Quebec à Montréal, Montréal, QC, Canada, 2008; 326p.
6. Bailly, L.; Ambrosi, J.P.; Barbarand, J.; Beauvais, A.; Cluzel, D.; Lerouge, C.; Prognon, C.; Quesnel, F.; Ramanaïdou, E.; Ricordel-Prognon, C.; et al. Nickal—Typologie des Latérites de Nouvelle-Calédonie. Gisements de Nickel Latéritique, Volume II. CNRT « Nickel & Son Environnement». 2014. Available online: https://horizon.documentation.ird.fr/exl-doc/pleins_textes/divers19-02/010075036.pdf (accessed on 20 March 2022).
7. Chassé, M.; Griffin, W.L.; O'Reilly, S.Y.; Calas, G. Scandium speciation in a world-class lateritic deposit. *Geochem. Perspect. Lett.* **2017**, *3*, 105–114. [CrossRef]
8. Chassé, M.; Griffin, W.L.; O'Reilly, S.Y.; Calas, G. Australian laterites reveal mechanisms governing scandium dynamics in the critical zone. *Geochim. Et Cosmochim. Acta* **2019**, *260*, 292–310. [CrossRef]
9. Hoatson, D.M.; Jaireth, S.; Miezitis, Y. *The Major Rare-Earth-Element Deposits of Australia: Geological Setting, Exploration, and Resources*; Geoscience Australia: Camberra, Australia, 2011; 204p.

10. Maulana, A.; Sanematmatmatsu, K.; Sakakibara, M. An Overview on the Possibility of Scandium and REE Occurrence in Sulawesi, Indonesia. *Indones. J. Geosci.* **2016**, *3*, 139–147. [CrossRef]
11. Qin, H.-B.; Yang, S.; Tanaka, M.; Sanematsi, K.; Arcilla, C.; Takahashi, Y. Chemical speciation of scandium and yttrium in laterites: New insights into the control of their partitioning behaviors. *Chem. Geol.* **2020**, *552*, 119771. [CrossRef]
12. Sun, J.; Liu, Y.; Liu, X. Iron isotope constraints on the mineralization process of Shazi Sc-rich laterite deposit in Qinglong county, China. *Minerals* **2021**, *11*, 737. [CrossRef]
13. Teitler, Y.; Cathelineau, M.; Ulrich, M.; Ambrosi, J.P.; Muñoz, M.; Sevin, B. Petrology and geochemistry of scandium in New Caledonian Ni-Co laterites. *J. Geochem. Explor.* **2019**, *196*, 131–155. [CrossRef]
14. Ulrich, M.; Cathelineau, M.; Muñoz, M.; Boiron, M.C.; Teitler, Y.; Karpoff, A.M. The relative distribution of critical (Sc, REE) and transition metals (Ni, Co, Cr, V) in some Ni-laterite deposits of New Caledonia. *J. Geochem. Explor.* **2019**, *197*, 93–113. [CrossRef]
15. Cluzel, D.; Aitchison, J.C.; Picard, C. Tectonic accretion and underplating of mafic terranes in the Late Eocene intraoceanic fore-arc of New Caledonia (Southwest Pacific): Geodynamic implications. *Tectonophysics* **2001**, *340*, 23–59. [CrossRef]
16. Cluzel, D.; Maurizot, P.; Collot, J.; Sevin, B. An outline of the Geology of New Caledonia; from Permian–Mesozoic Southeast Gondwanaland active margin to Cenozoic obduction and supergene evolution. *Episodes* **2012**, *35*, 72–86. [CrossRef] [PubMed]
17. Paquette, J.L.; Cluzel, D. U–Pb zircon dating of post-obduction volcanic-arc granitoids and a granulite-facies xenolith from New Caledonia. Inference on Southwest Pacific geodynamic models. *Int. J. Earth Sci.* **2007**, *96*, 613–622. [CrossRef]
18. Ulrich, M.; Picard, C.; Guillot, S.; Chauvel, C.; Cluzel, D.; Meffre, S. Multiple melting stages and refertilization as indicators for ridge to subduction formation: The New Caledonia ophiolite. *Lithos* **2010**, *115*, 223–236. [CrossRef]
19. Maurizot, P.; Sevin, B.; Lesimple, S.; Bailly, L.; Iseppi, M.; Robineau, B. Mineral resources, prospectivity of the ultramafic rocks of New Caledonia. *Geol. Soc. Lond. Mem.* **2020**, *51*, 247–277. [CrossRef]
20. Butt, C.R.M.; Cluzel, D. Nickel laterite ore deposits: Weathered serpentinites. *Elements* **2013**, *9*, 123–128. [CrossRef]
21. Cathelineau, M.; Quesnel, B.; Gautier, P.; Boulvais, P.; Couteau, C.; Drouillet, M. Nickel dispersion and enrichment at the bottom of the regolith: Formation of pimelite target-like ores in rock block joints (Koniambo Ni deposit, New Caledonia). *Miner. Depos.* **2016**, *51*, 271–282. [CrossRef]
22. Cathelineau, M.; Myagkiy, A.; Quesnel, B.; Boiron, M.-C.; Gautier, P.; Boulvais, P.; Ulrich, M.; Truche, L.; Golfier, F.; Drouillet, M. Multistage crack seal vein and hydrothermal Ni enrichment in serpentinized ultramafic rocks (Koniambo massif, New Caledonia). *Miner. Depos.* **2017**, *52*, 945–960. [CrossRef]
23. Freyssinet, P.H.; Butt, C.R.M.; Morris, R.C.; Piantone, P. Ore-Forming Processes Related to Lateritic Weathering. In *Economic Geology 100th Anniversary Volume*; Hedenquist, J.W., Thomson, J.F.H., Goldfarb, R.J., Richards, J.P., Eds.; Economic Geology Publishing Company: New Haven, CT, USA, 2005; pp. 681–722.
24. Golightly, J.P. Progress in understanding the evolution of nickel laterites. In *The Challenge of Finding New Mineral Resources: Global Metallogeny, Innovative Exploration, and New Discoveries, Volume II*; Goldfarb, R.J., Marsh, E.E., Monecke, T., Eds.; Society of Economic Geologists Special Publication: Littleton, CO, USA, 2010; Volume 15, pp. 451–485. [CrossRef]
25. Manceau, A.; Schlegel, M.L.; Musso, M.; Sole, V.A.; Gauthier, C.; Petit, P.E.; Trolard, F. Crystal chemistry of trace elements in natural and synthetic goethite. *Geochim. Et Cosmochim. Acta* **2000**, *64*, 3643–3661. [CrossRef]
26. Trescases, J.-J. L'evolution Géochimique Supergène des Roches Ultrabasiques en Zone Tropicale: Formation des Gisements Nickélifères de Nouvelle-Calédonie. Edited, O.R.S.T.O.M.: Paris, France. 1975. Available online: https://horizon.documentation.ird.fr/exl-doc/pleins_textes/pleins_textes_6/Mem_cm/07526.pdf (accessed on 20 March 2022).
27. Wells, M.A.; Ramanaidou, E.R.; Verrall, M.; Tessarolo, C. Mineralogy and crystal chemistry of "garnierites" in the Goro lateritic nickel deposit, New Caledonia. *Eur. J. Mineral.* **2009**, *21*, 467–483. [CrossRef]
28. Dublet, G.; Julliot, F.; Brest, J.; Noël, V.; Fritsch, E.; Proux, O.; Olivi, L.; Ploquin, F.; Morin, G. Vertical changes of the Co and Mn speciation along a lateritic regolith developed on peridotites (New Caledonia). *Geochim. Et Cosmochim. Acta* **2017**, *217*, 1–15. [CrossRef]
29. Maurizot, P.; Vendé-Leclerc, M. New Caledonia Geological Map, Scale 1/500 000; Direction de l'Industrie, des Mines et de l'Énergie–Service de la Géologie de Nouvelle-Calédonie, Bureau de Recherches Géologiques et Minières. 2009.
30. Carignan, J.; Hild, P.; Mevelle, G.; Morel, J.; Yeghicheyan, D. Routine Analyses of Trace Elements in Geological Samples using Flow Injection and Low Pressure On-Line Liquid Chromatography Coupled to ICP-MS: A Study of Geochemical Reference Materials BR, DR-N, UB-N, AN-G and GH. *Geostand. Geoanalytical Res.* **2001**, *25*, 187–198. [CrossRef]
31. Aitchison, J. The Statistical Analysis of Compositional Data. In *Monographs on Statistics and Applied Probability*; Chapman & Hall Ltd.: London, UK, 1986.
32. Tolosana-Delgado, R.; McKinley, J. Exploring the joint compositional variability of major components and trace elements in the Tellus soil geochemistry survey (Northern Ireland). *Appl. Geochem.* **2016**, *75*, 263–276. [CrossRef]
33. Chayes, F. On correlation between variables of constant sum. *J. Geophys. Res.* **1960**, *65*, 4185–4193. [CrossRef]
34. Guillong, M.M.; Maier, D.L.; Allan, M.M.; Heinrich, C.A. SILLS: A MATLAB based program for the reduction of laser ablation ICPMS data of homogeneous materials and inclusions. In *Laser Ablation ICPMS in the Earth Sciences. Current Practices and Outstanding Issues*; Sylvester, P., Ed.; Mineralogical Association of Canada: Vancouver, BC, Canada, 2008; pp. 328–333.

35. Anand, R.R.; Smith, R.E.; Paine, M.D. *Genesis, Classification and Atlas of Ferruginous Materials, Yilgarn Craton*; Australian Mineral Industries Research Association: Parkville, VIC, Australia; CRC LEME: Millaa Millaa, QLD, Australia; CSIRO: Canberra, ACT, Australia, 2002. Available online: http://crcleme.org.au/Pubs/OPEN%20FILE%20REPORTS/OFR%20073/OFR%20073%20 mid%20res.pdf (accessed on 20 March 2022).
36. Santoro, L.; Putzolu, F.; Mondillo, N.; Boni, M.; Herrington, R. Trace element geochemistry of iron-(oxy)-hydroxides in Ni(Co)-laterites: Review, new data and implications for ore forming processes. *Ore Geol. Rev.* **2022**, *140*, 104501. [CrossRef]
37. Mano, E.S.; Caner, L.; Petit, S.; Chaves, A.P.; Mexias, A.S. Mineralogical characterization of Ni-bearing smectites from Niquelandia, Brazil. *Clays Clay Miner.* **2014**, *62*, 324–335. [CrossRef]
38. Ratié, G.; Garnier, J.; Calmels, D.; Vantelon, D.; Giumaraes, E.; Monvoisin, G.; Nouet, J.; Ponzevera, E.; Quantin, C. Nickel distribution and isotopic fractionation in a Brazilian lateritic regolith: Coupling Ni isotopes and Ni K-edge XANES. *Geochim. Et Cosmochim. Acta* **2018**, *230*, 137–154. [CrossRef]
39. Dublet, G.; Julliot, F.; Morin, G.; Fritsch, E.; Fandeur, D.; Brown, G.E., Jr. Goethite aging explains Ni depletion in upper units of ultramafic lateritic ores from New Caledonia. *Geochim. Et Cosmochim. Acta* **2015**, *160*, 1–15. [CrossRef]
40. Dublet, G.; Juillot, F.; Morin, G.; Fritsch, E.M.; Fandeur, D.; Ona-Nguema, G.; Brown, G.E., Jr. Ni speciation in a New Caledonian lateritic regolith: A quantitative X-ray absorption spectroscopy investigation. *Geochim. Et Cosmochim. Acta* **2012**, *95*, 119–133. [CrossRef]
41. Levard, C.; Borschneck, D.; Grauby, O.; Rose, J.; Ambrosi, J.P. Goethite, a tailor-made host for the critical metal scandium: The $Fe_xSc_{(1-x)}OOH$ solid solution. *Geochem. Perspect. Lett.* **2018**, *9*, 16–20. [CrossRef]
42. Schwertmann, U.; Latham, M. Properties of iron oxides in some New Caledonian oxisols. *Geoderma* **1986**, *39*, 105–123. [CrossRef]
43. Trolard, F.; Bourrie, G.; Jeanroy, E.; Herbillon, A.J.; Martin, H. Trace metals in natural iron oxides from laterites: A study using selective kinetic extraction. *Geochim. Et Cosmochim. Acta* **1995**, *59*, 1285–1297. [CrossRef]
44. Hall, G.E.M.; Vaive, J.E.; Beer, R.; Hoashi, M. Selective leaches revisited, with emphasis on the amorphous Fe oxyhydroxide phase extraction. *J. Geochem. Explor.* **1996**, *56*, 59–78. [CrossRef]
45. Sanematsu, K.; Moriyama, T.; Sotouky, L.; Watanabe, Y. Mobility of Rare Earth Elements in Basalt-Derived Laterite at the Bolaven Plateau, Southern Laos. *Resour. Geol.* **2011**, *61*, 140–158. [CrossRef]
46. Leermakers, M.; Mbachou, B.E.; Husson, A.; Lagneau, V.; Descostes, M. An alternative sequential extraction scheme for the determination of trace elements in ferrihydrite rich sediments. *Talanta* **2019**, *199*, 80–88. [CrossRef]
47. Poulton, S.W.; Canfield, D.E. Development of a sequential extraction procedure for iron: Implications for iron partitioning in continentally derived particulates. *Chem. Geol.* **2005**, *214*, 209–221. [CrossRef]
48. Schwertmann, U.; Friedl, J.; Stanjek, H.; Schulze, D.G. The effect of Al on Fe oxides. XIX. Formation of Al-substituted hematite from ferrihydrite at 25 °C and pH 4 to 7. *Clays Clay Miner.* **2000**, *48*, 159–172. [CrossRef]
49. Trolard, F.; Tardy, Y. A model of Fe^{3+}-kaolinite-Al^{3+}-goethite-Al^{3+}-hematite equilibria in laterites. *Clay Miner.* **1989**, *24*, 1–21. [CrossRef]
50. Marchesi, C.; Garrido, C.J.; Godard, M.; Belleyd, F.; Ferré, E. Migration and accumulation of ultra-depleted subduction-related melts in the Massif du Sud ophiolite (New Caledonia). *Chem. Geol.* **2009**, *266*, 171–186. [CrossRef]
51. Lacroix, E.; Cauzid, J.; Teitler, Y.; Cathelineau, M. Near real-time management of spectral interferences with portable X-ray fluorescence spectrometers: Application to Sc quantification in nickeliferous laterite ores. *Geochem. Explor. Environ. Anal.* **2021**, *21*, 1–13. [CrossRef]

On the Influence and Correction of Water Content on pXRF Analysis of Lateritic Nickel Ore Deposits in the Context of Open Pit Mines of New-Caledonia

Valérie Laperche [1],*, Cyrille Metayer [2], Julien Gaschaud [3], Philippe Wavrer [4] and Thomas Quiniou [2]

1   Bureau de Recherches Géologiques et Minières (BRGM), 45060 Orleans, France
2   Institut des Sciences Exactes et Appliquées-EA 7484, Université de la Nouvelle-Calédonie, CEDEX, 98851 Noumea, France; cyrille.metayer@unc.nc (C.M.); thomas.quiniou@unc.nc (T.Q.)
3   Antea Group, 2–6, Place du Générale De Gaulle, 92160 Antony, France; julien.gaschaud@anteagroup.fr
4   CASPEO, 45009 Orleans, France; p.wavrer@caspeo.onmicrosoft.com
*   Correspondence: v.laperche@brgm.fr; Tel.: +33-2-38-64-36-34

Abstract: In a number of applications, the use of portable X-ray fluorescence (pXRF) instruments offers a time and cost-saving alternative to standard laboratory instruments. This is particularly true in a mining context where decisions must be taken quickly in the field. However, pXRF is a technique known to be efficient, provided that samples are well prepared, i.e., dried and finely ground. On the mine face, little-to-no sample preparation is conceivable as mining vehicles must be able to operate continuously. Therefore, solutions have to be found even for raw materials and one of the most critical problems is the sample water content, in particular in the context of open pit mines in a tropical area. A large number of analysis shows that knowledge of humidity enables the measured concentration to be effectively corrected for the three instruments used (Niton, X-met, Titan). It is possible to overcome the difficulty of measuring water content in the field by fixing it to its maximum value (saturation). The results show that the saturation method is reliable, or at least, promising.

Keywords: pXRF; mine; nickel; iron; water content; dilution law; Beer-Lambert law; saturation

Citation: Laperche, V.; Metayer, C.; Gaschaud, J.; Wavrer, P.; Quiniou, T. On the Influence and Correction of Water Content on pXRF Analysis of Lateritic Nickel Ore Deposits in the Context of Open Pit Mines of New-Caledonia. *Minerals* **2022**, *12*, 415. https://doi.org/10.3390/min12040415

Academic Editors: Cristina Domènech, Cristina Villanova-de-Benavent and Jordi Ibanez-Insa

Received: 23 December 2021
Accepted: 25 March 2022
Published: 29 March 2022

Publisher's Note: MDPI stays neutral with regard to jurisdictional claims in published maps and institutional affiliations.

1. Introduction

The advantage of pXRF, compared to other chemical analysis methods, is that it allows the analysis of elements in concentrations ranging from a few tens of mg kg^{-1} to several percent, on raw samples, whether in the field or in the laboratory, in a very short time, thus allowing a rapid diagnosis, without delay. Thus, this method makes it possible to identify and quantify chemical elements at an acceptable cost and within acceptable deadlines and to delimit areas more or less concentrated in these elements. Because of fluctuations in measurements due to the matrix effects, particle size, heterogeneity of materials and moisture content, it is not intended to replace the usual techniques of chemical analysis with their precision and standards. On the other hand, knowledge of the different parameters that can influence the measurement and their consideration allows an improvement in the quantification of certain chemical elements.

The use of portable X-ray fluorescence (pXRF) devices in mining contexts, whether during the exploration, the borehole mining or the mining phases [1–7], has shown its value even if the on-site results are out of step with the results of chemical analyses carried out in the laboratory and the dispersions can sometimes be significant.

In New Caledonia, mining companies have been using this type of equipment for a few years but are struggling to obtain reliable results. The calibrations are based on empirical laws rather than physical laws. The main advance is to make corrections a posteriori according to the water content of the material. Currently, these devices are mainly used on finely ground materials, saprolitic materials have too much variability. This equipment is used profitably at the level of prospecting which is satisfied with relative indications that are not very precise and where they make it possible to guide the taking of a representative sample. On cored or destructive soundings and on the operating front, difficulties arise in terms of the accuracy of the measurements and their reproducibility.

Finally, on ore stock where certified analyses are needed, these devices cannot replace conventional analysis methods. It is therefore in the field up to the sounding workshop, then logging, that progress can be envisaged. The need to improve the analytical protocol based on real scientific bases was expressed by the college of industrialists of the CNRT (Centre National de Recherche Technologique). However, this research must not lead to solutions that are too cumbersome to implement in the field, making the technology lose all logistical advantage.

Obtaining an average value in accordance with the reference value requires a correction of the local measurement (water content) as well as an adequate sampling strategy. There arc several quantitative XRF analysis methods, being either compensation methods (dilution, internal standard, standard addition or Compton scatter) or correction of the matrix effects methods (fundamental parameters, empirical influence coefficient or theoretical influence coefficient) [3,7–15]. The implementation of compensation methods on mine site is not possible since a complex preparation before is necessary (drying and grinding). The methods used in the mining context are matrix correction methods.

It is therefore necessary to work on improving the accuracy of the measurement and the sampling strategy by focusing on three points: water content, particle size and sampling. The control of these different parameters should eventually make it possible to obtain chemical analyses "close" to the concentrations determined by conventional laboratory methods. Although three points are important, this study focuses more on the water content in the quantification of chemical elements.

Soils can naturally contain high water content, especially in humid tropical environments. Within the sample, water replaces ambient air that fills porosities or fractures [16]. On the surface of the sample, pressure due to contact of pXRF device can induce release of water from macro pores and then can form a thin layer of water [17]. The protective film conventionally used during pXRF measurements can lead to formation of a layer of water on the surface of the sample [18]. Whatever its origin, water influences the intensity of X-rays in two ways. First, water absorbs X-rays more than air, so the absorption of the sample increases with the water content. Secondly, the water scatters the primary X-ray from the source and thus increases the intensity of the background. Both effects will decrease the net area of fluorescence peaks. While the presence of water will greatly affect results for light elements, concentrations measured for heavy elements ($Z > 40$ or $Z > 26$) [17] remain almost constant.

Whatever the element considered, it is generally accepted that moisture contents up to 20% do not significantly influence XRF intensity and the quantification by this method [19–21]. For higher water contents, several correction laws have been proposed to obtain concentration in the dry sample from wet sample measurement. Ge and coauthors [16] assume that reduction of XRF intensity is proportional to the increment of water content and propose a relationship derived from Beer Lambert's law. Ge and coauthors also show that the sum of the intensities resulting from coherent and incoherent scattering of primary X-ray from the source is a linear function of water content. By using this second relationship, measurement of the water content is no longer necessary and only measurement of the intensities is required. Bastos and coauthors [22] retained Ge and coauthors hypothesis but uses the background intensity at low energy instead of scattered intensity [22]. This correction law can give satisfactory results for water contents up to 136.8% [23]. Phedorin and Goldberg [24] and Kido and coauthors [18] have also proposed correction laws based on Beer Lambert's law. The correction proposed by [24] requires an iterative algorithm and to know all macroscopic cross sections. The law proposed by [18] requires knowledge of the mass absorption coefficient of the dry sample. These constraints make their use more complex and these laws are therefore rarely used. More recently, Ribeiro and coauthors [25] applied several correction laws (linear, second-degree polynomial and power) to measurements made on Brazilian soils. Finally, they proposed to use a power law to fit measured data.

In this paper, two water content compensation laws were applied to Caledonian ore samples. The dilution law which compensates the variation of the mass concentration induced by the addition of water and the classical law introduced in Ge and co-authors [16]. On the other hand, this study shows a strong correlation between the concentrations measured for the dry sample and for the sample whose water content is close to the concentration of the saturated ore. By saturating the sample with water, after a calibration phase as it is classically required for dry samples, it is possible to obtain directly the Fe and Ni contents of the dry sample.

2. Materials and Methods

2.1. Portable XRF (pXRF)

The characteristics of the devices used during this project are very similar (Table 1). They are all equipped with a tube capable of operating at high voltages of 45 to 50 kV, allowing measurement of the heaviest elements. Their multi-element detection range is also close, at about 30–35 elements. On the other hand, the technologies of detection systems, hardware and software ergonomics, as well as the implementation of quantification algorithms (FP, coefficients of influence, etc.) may vary from one manufacturer to another. The standardization of the XL3t mining mode is based on fundamental parameters but also uses the Compton scatter (inelastic collisions) to standardize to 100%. The information on how the normalize of the intensities was done, was not available for the two other instruments. Rousseau [26] showed that the normalization conditions may introduce a bias in the results.

Table 1. Specifications of the instruments used in this work.

Manufacturer	Thermo Fisher® Boston, MA, USA	Oxford Instrument® Abingdon, Oxfordshire, UK	Bruker® Billera, MA, USA
Model	Niton GOLDD + 900	XMET 7500	S1 Titan 800
Anode	Ag	Rh	Rh
Tube voltage (kV)	50	45	50
Tube current (μA)	200	50	200
Spot size (mm)	7	9	5
Resolution (eV)	<185	<150	<145
Detector	SDD GOLDD	SDD	Fast SDD
Element range	Mg to U	Mg to U	Mg to U
Application mode	Mining Cu/Zn	mining_fp	Ni-Co Ore Rock (NiOreRock method)

In the laboratory, the devices were operated in a benchtop stand using an AC adapter to create the ideal measurement conditions for the sample cups. The devices were allowed to warm up for a minimum period of 30–45 min before measurements.

In all situations, in the laboratory or in the field, the measurement time was set to 10 s. This measurement time is deliberately short and it is a compromise, obtaining an acceptable accuracy and minimizing the muscular tension of the operator. In situ, the operator will have to use the pXRF instrument several times a day and sometimes in uncomfortable positions, so it is essential to reduce the measurement time as much as possible.

As in situ conditions can be very hard for a pXRF in an open mine, X-ray tubes and detectors can easily be damaged. In order to protect nozzle instruments, all mining operators cover their instruments with a protective tape, e.g., Kapton® for Niton and 3M scotch® (ref. E5016C) for Xmet (Figure 1). In the case of Titan pXRF, as it has a built-in protective shield, no other protection was added. Adding a film will attenuate and scatter the radiation. It will strongly attenuate the low energy radiation typical of the fluorescence of light elements while it will have a negligible influence on the fluorescence of the heavier elements [27].

Figure 1. Protective tape, Kapton on the **left**, 3M scotch on the **right**.

2.2. Samples Description

2.2.1. Geological Setting

The obduction of the peridotite layer on sedimentary formations and the basaltic unit of Poya is attributed to the Late Eocene [28–32]. Originally covering a large part of Grande-Terre in New-Caledonia, the formation was gradually stripped by a succession of episodes of chemical and mechanical alteration. Currently, the remains of this formation cover about a third of Grande-Terre and are represented by:

- the ultrabasic massif to the south and its extension on the east coast of New Caledonia (mines of Goro, Tontouta, Camp des Sapins, Thio, Nakéty, Boa Kaine, Kouaoua, Poro, Monéo);
- a sequence of klippes on the edge of the west coast (Kopéto-Boulinda, Koniambo, Ouazangou-Taom, Tiébaghi, Poum, Bélep island, etc.).

Peridotites, rocks of the Earth's mantle that rarely appear on the surface of the earth, consist mainly of silicates rich in metallic elements including iron, magnesium, manganese, chromium, nickel and cobalt. The alteration of peridotites in hot and humid tropical environments is a supergene process that causes the hydrolysis of the components of the rock and its dissolution. Some elements are leached (Si, Mg), while others remain in place (Fe, Cr, Ni, Co, Mn). Under the action of a humid tropical climate, the peridotites were gradually altered by hydrolysis of ferromagnesian silicates leading to a typical weathering profile [33] composed, from the bottom to the top: fractured peridotic fresh rock, saprolites, yellow limonite, red limonite, nodular layer and ferricrete.

This basic succession has many variations and gaps. While in source rock, Ni and Co are in low concentrations (0.3% and 0.01% respectively), these same elements are concentrated in saprolite and limonite horizons with contents of the order of one percent.

In saprolites, the distribution of Ni is very variable, as is the physical heterogeneity of the material. Economically sized ore bodies are currently mined at an average grade greater than 2% Ni + Co. Ni is hosted by Ni-bearing serpentines that make up the majority of the saprolite horizon, and by less abundant, but high grade, garnierites (green, fine grained mixtures of serpentine, talc, chlorite, sepiolite and smectite).

In the limonite horizon, the content of metals is less fluctuating but overall lower, less than 2% Ni + Co, because these levels almost entirely devoid of magnesia and silica, mainly consisting of iron oxy-hydroxides partially crystallized in fine grained goethite, have a lower Ni retention capacity. Cobalt is often concentrated at the base of yellow limonite in the form of concretions of asbolane, a complex manganese oxide.

A very complete article on the mineral resources and prospectivity of ultramafic rocks in New Caledonia has recently been published and covers all the knowledge acquired on the subject [32].

The Ni ores currently mined in New Caledonia are of two types:

- Saprolitic or garnieritic silicate ores of high contents (>2%);
- Limonitic ores of lower grades (<2%).

Although the two types can coexist in the same mining area, the mining techniques and industrial processes to which they are subject are different. The Société le Nickel (SLN) operates mainly saprolitic ores in five centres. Koniambo Nickel SAS (KNS) will mine saprolitic and limonitic ore by pyrometallurgical treatment. The Valé Company, in the South, mainly processes limonitic ore by a hydrometallurgical process. The other "small miners" are divided between conventional exploitation and the export of saprolites and limonites.

2.2.2. Reference Samples

During this two years project, we have used a database made of 27 reference samples (14 limonites and 13 saprolites) that have passed a round Robin test. This base is shared by all the major mining companies and the Geological Survey (SGNC), a department of New Caledonia's Direction for Industry, Mining and Energy (DIMENC). Minimum and maximum elemental concentrations are given in Table 2.

Table 2. Minimum (min) and maximum (max) values of reference samples concentrations.

	Fe (%)	Ni (%)	Mg (%)	Si (%)	Cr (%)	Al (%)	Mn (%)	Ca (%)
min value	4.96	0.09	0.13	0.52	0.17	0.10	0.04	0.01
max value	53.96	3.29	22.26	31.55	3.54	12.75	4.10	0.43

Reference concentrations of these samples were established by XRF in a New-Caledonian accredited laboratory (Ni, Lab).

All samples were prepared by the loose powder technique in plastic cups covered with a protective 6.0 μm Mylar® polyester film (FluXana, Bedburg-Hau, Germany). Cups were then filled with powders of at least 1 cm thickness, eventually covered with cotton fibers and finally closed in order to hold the setup firmly (Figure 2).

Figure 2. On the **left**: an empty cup and on the **right**: a filled cup with a powder.

Each sample was analyzed with the three pXRF, three times or more during 10 s, and results were then averaged.

2.2.3. Field Sampling

Two sampling campaigns have been conducted in different mines across the territory during this project. The first one took place at the beginning in order to collect samples for the laboratory analysis, the second at the end to test the developed method. All samples, about 5 kg each, were initially sent to a NATA/ISO/IEC17025 accredited laboratory to be weighed, dried at 105 °C until no mass change, weighed again, crushed at 3 mm and split with a rifle sampler. An aliquot of 1 kg was then crushed at 75 μm, split again in a rifle and one part was finally analyzed by XRF on fused disks. This first step allowed us to have access to the concentration of eleven elements/compounds (Ni, Co, Fe_2O_3, MgO, SiO_2, Cr_2O_3, Al_2O_3, MnO, CaO, CuO and ZnO) and to the water content of samples in the field. During the first campaign, 30 samples (11 limonites and 19 saprolites) were collected. Concentrations range from 0.27 wt% to 3.36 wt% and from 4.7 wt% to 52.9 wt% for Ni and Fe, respectively. Water content ranges from 0.75 wt% to 48.03 wt%. Figure 3 shows two panels, a limonitic one on the left and a saprolitic one on the right; their width is about 3.5 m and their height about 2 m. Circles represent the sampling points, and each one was first measured with the three pXRF before being collected in a bag and sent to the accredited laboratory. pXRF measurements are not presented in this paper, as samples were primarily collected for water analysis in the lab.

(a)   (b)

Figure 3. Examples of Ni ore panels sampled during the project: (**a**) Limonite; (**b**) Saprolite.

During the second campaign, 18 samples (12 limonites and 6 saprolites) were collected. Concentrations range from 1.31 wt% to 4.47 wt% for Ni and from 8.8 wt% to 52.4 wt% for Fe.

2.3. Methods

2.3.1. Laboratory Analysis

In order to evaluate how water content affects pXRF analyses on crushed samples, a series of measurements has been made at different water contents with the 30 samples collected during the first field campaign (Figure 4). The process is as follows:

- Weighing of the empty cup (only one side is covered with a thin Mylar® film, the other is left open to facilitate water evaporation during drying);
- Sample saturation with water;
- Filling the cup with the wetted powder;
- Weighing and analysis with the 3 pXRF (one measurement of 10 s);
- Drying in a ventilated oven at 70 °C during one hour;
- Weighing and analysis with the 3 pXRF.

The two last steps were repeated seven or eight times in a day. After the last measurement, each sample was gently dried at 30 °C during 16 h and 2 more hours at 105 °C, then finally weighed and analyzed. The last step gives us element concentrations and masses of the dry samples.

The saturation is achieved by using a vacuum filtration system. The ore is placed on a filter and then water is added in excess to the Büchner funnel. The mass of added water is approximately 8 g and the mass of the dry sample is approximately 6 g. This initial volume of water corresponds to approximately 3 times the volume of the solid phase and 2 times the mass of water contained in the wettest sample. A vacuum is then created in the Büchner flask using a water aspirator vacuum pump to remove excess water. After approximately one minute, the water reaches the upper surface of the sample. The pumping system is then stopped and the sample is immediately placed in the cup. This method does not guarantee that we really reach saturation, as samples could be over-saturated, but it has the advantage of being reproducible. We are probably closed to the saturation, however, and we will not be able, in the field, to precisely saturate the sample.

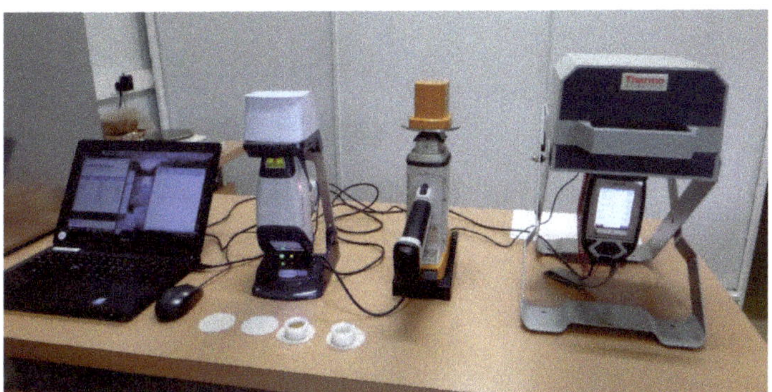

Figure 4. Presentation of the three pXRF devices and their stands during the measurement phase of wetted samples in the lab.

Furthermore, as the measured concentrations may vary depending on the position of the measuring head on the cup, a machined mechanical support with the footprint of the cup was designed and used in order to ensure correct positioning.

2.3.2. Field Measurements (Saturation)

Water content of Ni-ore in humid climate as in New Caledonia can vary from a few percent to fifty percent and even more in the lateritic profiles. Iron-rich limonitic layers are made of soft porous and permeable material that can easily retain water. Saprolitic layers are less weathered than lateritic ones, they can be very rocky or a mix of rock fragments of different size and soft material.

During the second field campaign, and in order to test our saturation method in both limonitic and saprolitic layers, we have used two sampling processes:

- In the presence of a wet, cohesive and homogeneous material, e.g., in a limonitic horizon, we have drawn a 5 × 5 grid directly on the panel as shown in the Figure 5a;
- In other situations, sampling points were randomly distributed over the area of interest as shown in Figure 5b.

Each measurement point (cell of the grid or random point) was saturated with water with a hand spray as shown in Figure 6, measured with the three pXRF and a sample was taken over the grid or around each random point, collected in a bag and sent to the lab. At the end, for a given sample, all local pXRF data were averaged in order to be compared to the reference value issued from the bag. A total of 13 samples measured and collected in this way have been used to calibrate the instruments on saturated samples.

This process has the advantage, from the miner's point of view, of being quick and simple, but sample saturation cannot be strictly ensured. In practice, we added water until the wetted area stopped absorbing but this process is, for sure, operator dependent. However, the calibration phase, which consists in confronting pXRF measurements with reference values obtained on XRF fused disks, will allow us to correct for systematic errors if there are some, and above all, to determine measurement errors, including those induced by the proposed saturation process.

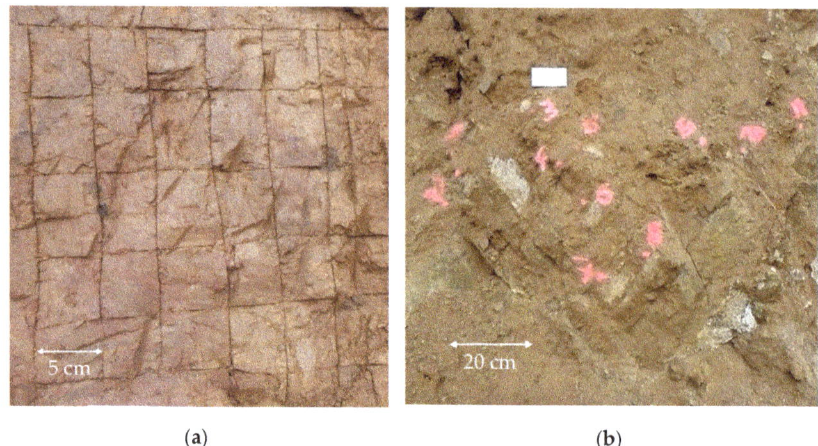

(a)   (b)

Figure 5. Two sampling processes depending on panel humidity and cohesion: (**a**) a 5 × 5 grid in a limonitic panel; (**b**) random sampling in a saprolitic panel.

Figure 6. Water saturation in the field.

2.4. Statistical Analyses

To validate the correction step between measured concentrations and reference concentrations or to compare the correction models according to the water content, the mean error (ME), the root-mean-square error (RMSE) and the coefficient of determination (R^2) have been calculated.

$$\text{ME} = \frac{1}{n} \sum_{i=1}^{n} (e_i - r_i) \tag{1}$$

$$\text{RMSE} = \sqrt{\frac{1}{n} \sum_{i=1}^{n} (e_i - r_i)^2} \tag{2}$$

where n is the number of observations, e_i is the corrected pXRF value and r_i is the reference value for the calibration step or the value measured at zero water content during the water correction step. A value of ME close to zero indicates that corrected values are centered around the reference values and that there is no systematic error. The RMSE quantifies the accuracy of the correction. If the correction is perfect, RMSE is equal to 0. The coefficient of determination, also called r-squared (R^2), is the square of correlation coefficient R. The coefficient of determination is a measure of the scatter about the regression line and is a measure of the strength of the linear association. It ranges in value from 0.0 (no linear association) to 1.0 (perfect linear association).

3. Results

3.1. Calibration with Reference Samples

3.1.1. Single Linear Regression

Even on dry and finely ground samples, a fine-tuning of concentrations is always needed to compensate for the systematic biases induced by cups tape and nozzle protective tape, i.e., Kapton® or Scotch®. Systematic biases can be overcome through a calibration process that consists of finding, for a given element, the relationship between pXRF data (raw data C_r) and XRF, ICP-AES or AAS laboratory data (reference data C). A simple linear regression analysis can be performed for this purpose. Linear regression produces the slope, a, and the y-intercept, b, of the regression line. A slope of 1.0 and a y-intercept of 0.0 indicate that pXRF is accurate. If not, the values of a and b can be entered directly into the analyser. The simple regression is defined by:

$$[C] = a \times [C_r] + b \tag{3}$$

In addition to calibration factors, linear regression analysis produces statistical parameters such as the coefficient of determination, which can be used to evaluate the goodness of the fit. Table 3 shows the parameters of the simple regression for the two elements and for the three devices.

Table 3. Parameters of the simple regression for the two elements and for the three devices (a: Slope, b: y-intercept, R^2: Coefficient of determination, ME: Mean error, RMSE: Root-mean-square error).

	Element	Fe	Ni
Niton	a	1.235	1.282
	b	−1.020	0.235
	R^2	0.9992	0.9563
	ME	6.3×10^{-15}	2.6×10^{-16}
	RMSE	0.46	0.159
Xmet	a	1.039	0.842
	b	−1.825	0.247
	R^2	0.9989	0.9652
	ME	4.8×10^{-15}	5.0×10^{-16}
	RMSE	0.56	0.142
Titan	a	1.079	1.134
	b	0.246	0.009
	R^2	0.9987	0.9950
	ME	3.6×10^{-15}	-5.8×10^{-17}
	RMSE	0.61	0.054

The Figure 7 shows the results obtained with the three pXRF devices, Fe on the left and Ni on the right for the 27 reference samples. The coefficient of determination R^2 is close to one for Fe regardless of instrument model. Titan pXRF is also very good for Ni, the Niton and Xmet show more dispersion.

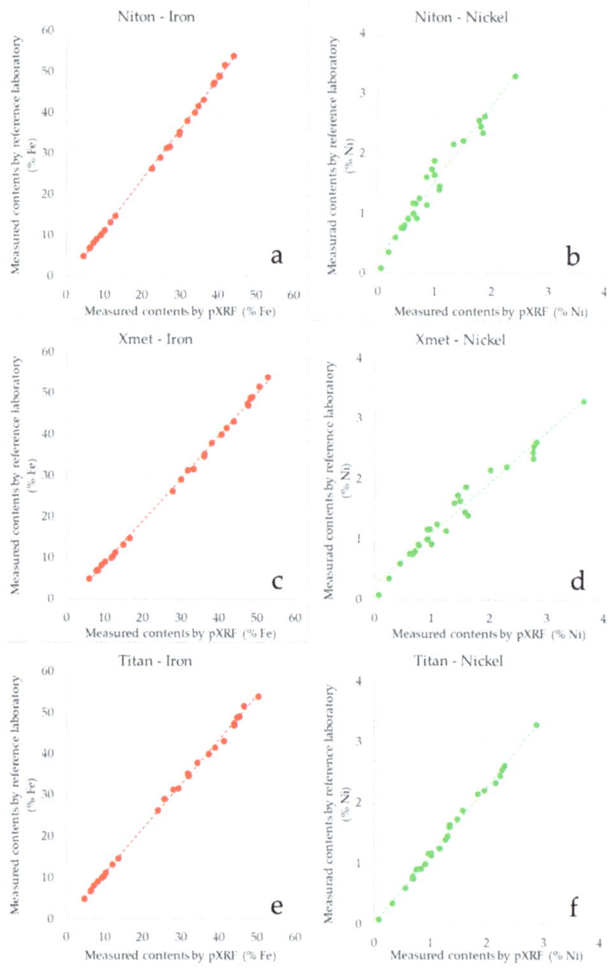

Figure 7. Linear regression for Fe and Ni in the three devices used for measuring: (**a**,**b**) Niton, (**c**,**d**) Xmet and (**e**,**f**) Titan.

3.1.2. Multiple Linear Regression

In a sample containing both Ni and Fe, it is well known that Ni may be underestimated. Ni, after being excited by the primary X-ray source, emits a Kα radiation that may be absorbed by Fe. This two steps phenomenon is called secondary fluorescence Secondary fluorescence effects are considered in the Fundamental Parameter (FP) calculations but our samples contain a very high concentration of Fe compared to Ni, especially in the limonite layer. Although we do not have precise information on how FP algorithms are implemented by each manufacturer, we can here suspect that for the Niton and the Xmet devices, they cannot deal with such high concentrations. However, we can perform a posteriori calibration of Ni concentration with a multiple linear regression that includes both measured Ni and Fe.

In order to improve results for the determination of Ni concentration in the presence of Fe, we propose to apply the following multiple linear regression:

$$[Ni] = a \times [Ni_r] + b \times [Ni_r] \times [Fe_r] + c \tag{4}$$

where Ni is the corrected Ni concentration and Ni_r and Fe_r are the raw concentration of Ni and Fe, respectively.

When we separate our initial set in two subsets, one for the limonites and the other for the saprolites, results on the Ni are greatly improved for both instruments. However, this would imply management of two calibration sets. With a multilinear regression, no distinction has to be made, as shown in Figure 8.

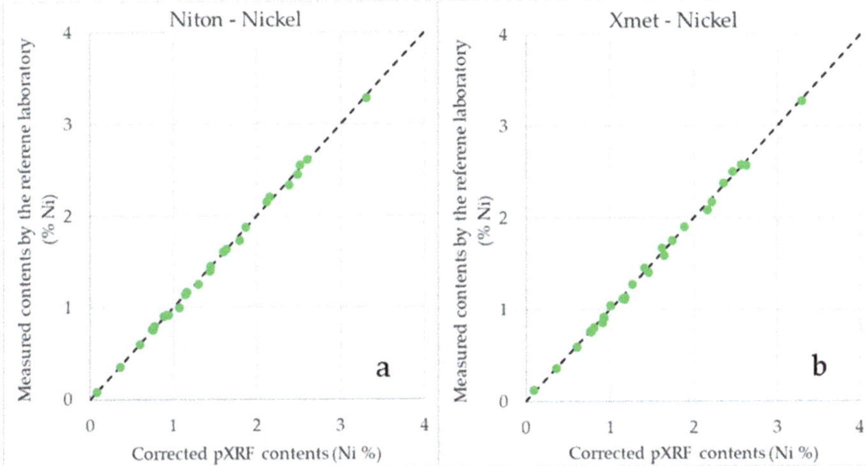

Figure 8. Results from multiple linear regression corrections for Ni, (**a**) for the Niton device and (**b**) for the Xmet device.

Table 4 gives the regression parameters for the Niton and Xmet devices.

Table 4. Parameters of the multiple linear regression for the Ni and for Niton and Xmet (a: Regression coefficient for $[Ni_r]$, b: Regression coefficient for $[Ni_r] \times [Fe_r]$, c: y-intercept, R^2: Coefficient of determination, ME: Mean error, RMSE: Root-mean-square error).

	Element	Ni
	a	1.2166
	b	0.0164
	c	0.0068
Niton	R^2	0.9985
	ME	3.8×10^{-16}
	RMSE	0.029
	a	0.7913
	b	0.0077
	c	0.0645
Xmet	R^2	0.9977
	ME	1.2×10^{-16}
	RMSE	0.037

Using multiple linear regression instead of simple linear regression greatly improves the goodness of fit. The coefficient of determination is closer to 1 and the root-mean-square error is divided by 4. When concentration of an element is highly variable, it may be necessary to correct matrix effects related to this element to improve the quantification.

3.2. Laboratory Study of Water Content Influence

The data were obtained from samples artificially moistened in the laboratory. Figure 9 shows how water content influences Ni and Fe concentration for the three instruments.

Ni dry is the raw concentration of Ni measured on the dry sample; no other correction has been applied. One sample is represented on a vertical line. The most wetted sample (–saturation) is the lower point and the dry one is located on the bisector, represented by a dashed line on the following figures.

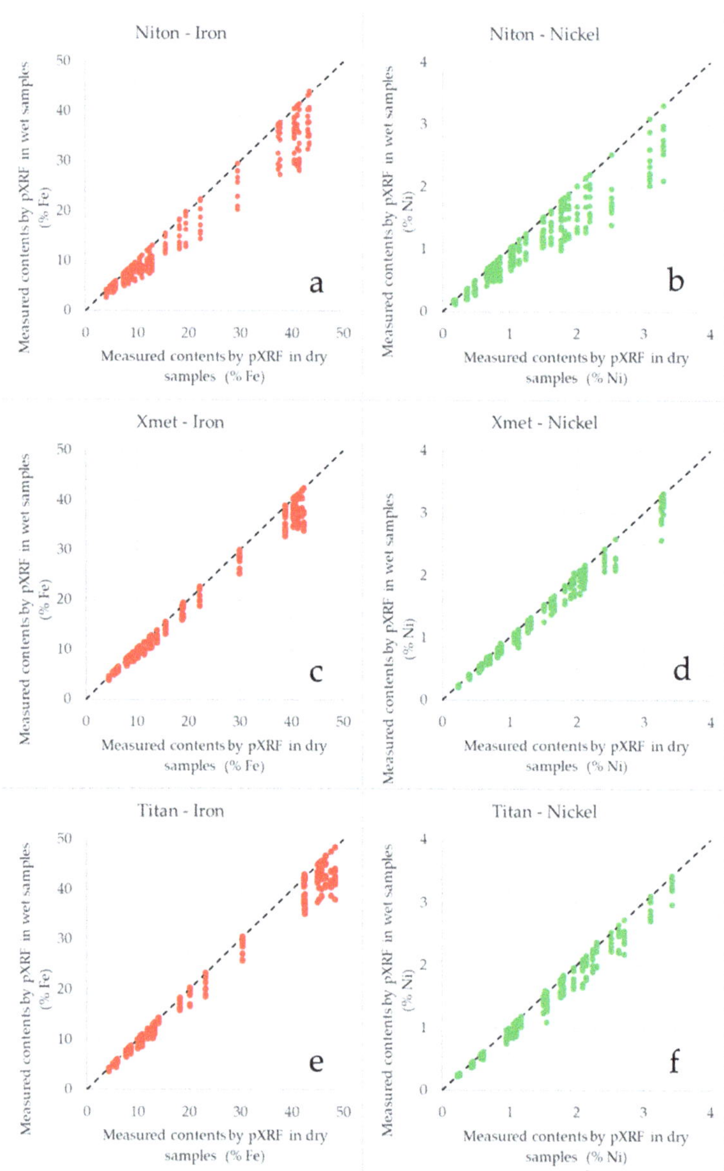

Figure 9. Influence of water content on Fe and Ni concentrations: (**a,b**) Niton, (**c,d**) Xmet and (**e,f**) Titan. The bisector is represented by the dashed line on the figures.

Figure 9 shows two different behaviors between the Niton on one side and both Xmet and Titan on the other side. Water has more influence in the Niton device, the spread is more pronounced compared to the other two. This does not mean one instrument is better than another, only that they do not have the same response when samples are wet. At this step, we do not have enough information on how internal algorithms work but Figure 9 clearly shows that different approaches are used.

3.2.1. Correction Using the Dilution Law

The dilution law can be express as follow [34]:

$$C_c = C_w \frac{m_w}{m_d} \tag{5}$$

where m_w and m_d are wet and dry masses of the sample, respectively, C_c is the corrected element concentration and C_w is the measured element concentration under wet condition. This law is a general one, meaning that it can be applied regardless of the element under consideration (Ni, Fe, Co, etc.).

The sample set from the first field campaign is made of 19 saprolites and 11 limonites. Figures 10 and 11 show the concentrations ratio versus the masses ratio for Ni (up) and Fe (down) in both profiles (left and right) for the Niton (Figure 10) and Xmet pXRF (Figure 11).

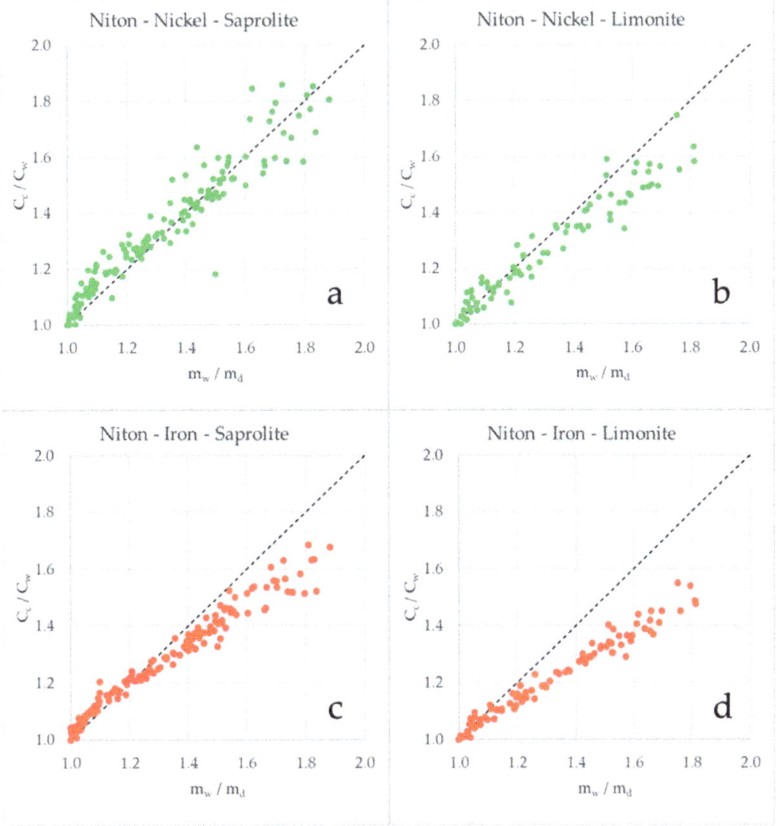

Figure 10. Data from the Niton after correction with the dilution law: (**a**) Ni in the saprolite layer; (**b**) Ni in the limonite layer; (**c**) Fe in the saprolite layer; (**d**) Fe in the limonite layer.

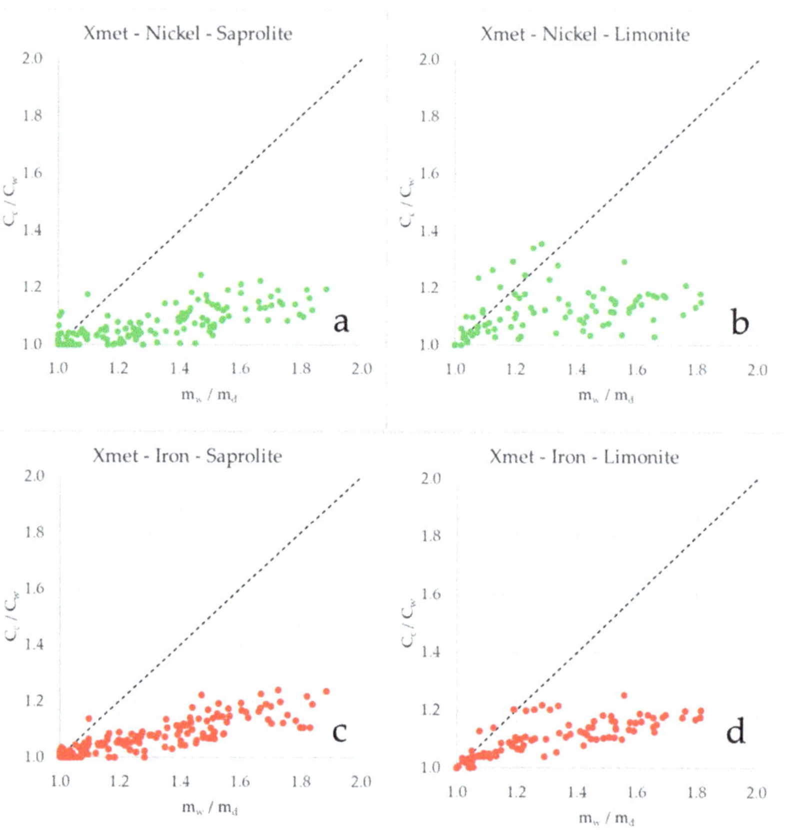

Figure 11. Data from the Xmet after correction with the dilution law: (**a**) Ni in the saprolite layer; (**b**) Ni in the limonite layer; (**c**) Fe in the saprolite layer; (**d**) Fe in the limonite layer.

If the situation was totally ideal (dilution law fully adapted and no sample preparation and measurement errors), all the points should be located on the dashed line (bisector). Figure 10 shows that Fe is farther from the bisector than Ni. The dilutions law could be a candidate only for the Ni in the saprolite layer because in this situation, data are centered on the bisector. In the limonite layer, data for the Ni are slightly under this line. For Fe, data are also more distant to the bisector in the limonite layer than in the saprolite one. This difference of behavior is probably due to physical characteristics (porosity, density and particle size) of each layer. Because element concentrations are derived from an algorithm that run in the Niton, we believe this algorithm is somehow influenced by one or more physical parameters of the sample.

Figure 11 shows that data are far away from the bisector, the dilution law is not a good candidate for the Xmet. The same conclusion applied to for the Titan. However, some practical considerations can be derived from Figure 11. We have seen that the Niton device clearly behaves differently between the saprolite and the limonite layers. The Xmet behaves the same way in both layers, so we can conclude their quantification algorithms are implemented in a different way.

3.2.2. Correction Using a Method Derived from Beer-Lambert Law

In presence of pore water in the sample and if the composition of the matrix remains unchanged, the reduction in X-rays intensity associated with the analyte (dI_x) is directly proportional to the increment of the water content in the sample (dw) [16,22]:

$$dI_x = -\mu_m I_x dw \tag{6}$$

where the correlation coefficient μ_m is constant and I_x is the X-ray intensity typical of the analyte at the water content w. If I_x is equal to I_0 when $w = 0$, the integration of the previous equation gives:

$$I_x = I_0 \cdot e^{-\mu_m w} \tag{7}$$

Instruments do not give the intensities, so mass concentrations are used in the correction law as in a few studies [27,35,36].

$$\frac{C_0}{C_w} = e^{\sigma w} \tag{8}$$

where C_w and C_0 are the elemental concentrations in wet (water content w express in %) and dry conditions, respectively, and σ is an attenuation coefficient (due to soil content, [23]).

Table 5 gives, for the three devices, the attenuation coefficients obtained after fitting data with the Beer-Lambert law (Equation (8)) and parameters characterizing the quality of this correction.

Table 5. Parameters of correction by Beer-Lambert law proposed by Ge and coauthors [16] for the two elements and for the three devices (σ: Attenuation coefficient, ME: Mean error, RMSE: Root-mean-square error).

	Element	Fe	Ni
Niton	σ	0.0063	0.0075
	ME	0.056	−0.023
	RMSE	1.05	0.083
Xmet	σ	0.0027	0.0024
	ME	−0.22	−0.015
	RMSE	1.07	0.079
Titan	σ	0.0027	0.0026
	ME	−0.29	−0.004
	RMSE	1.19	0.067

Figure 12 shows corrected concentrations by the Beer-Lambert law as function of dry sample concentrations. Comparison with Figure 9 shows that concentrations after correction are closer to dry sample concentrations. However, this correction is not perfect and a significant difference may still exist between the corrected concentration and the reference concentration.

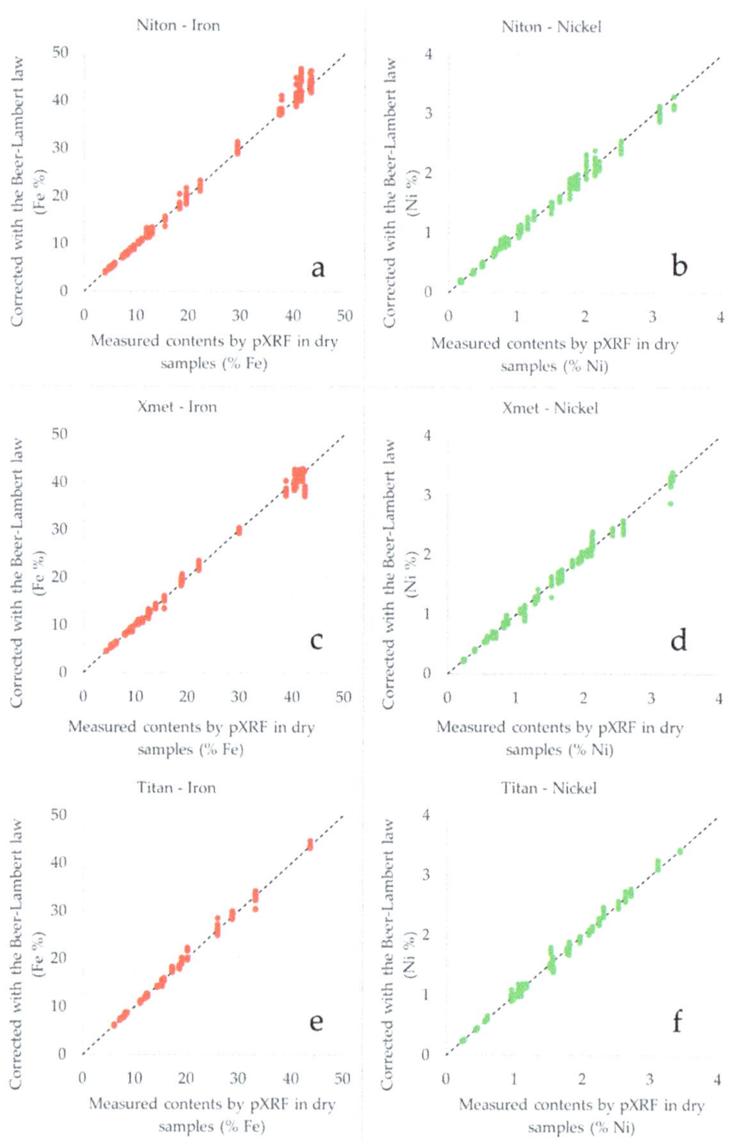

Figure 12. Ni and Fe concentrations after correction with the Beer-Lambert law: (**a**,**b**) Niton, (**c**,**d**) Xmet and (**e**,**f**) Titan.

3.2.3. pXRF Measurements at Water Saturation in the Laboratory

In Figure 9, it can be seen that measurements made at highest values of water content (lower point) vary almost linearly with concentrations in dry samples. Table 6 shows parameters of the linear fit. Values of the coefficients of determination confirm the veracity of this observation ($R^2 > 0.98$).

Using a linear function is therefore possible to estimate concentrations in the dry sample from measurements made in water-saturated samples as shown in Figure 13.

Root-mean-square error values remain relatively large but mean errors are close to 0, so quantification can certainly be improved by increasing the number of measurements.

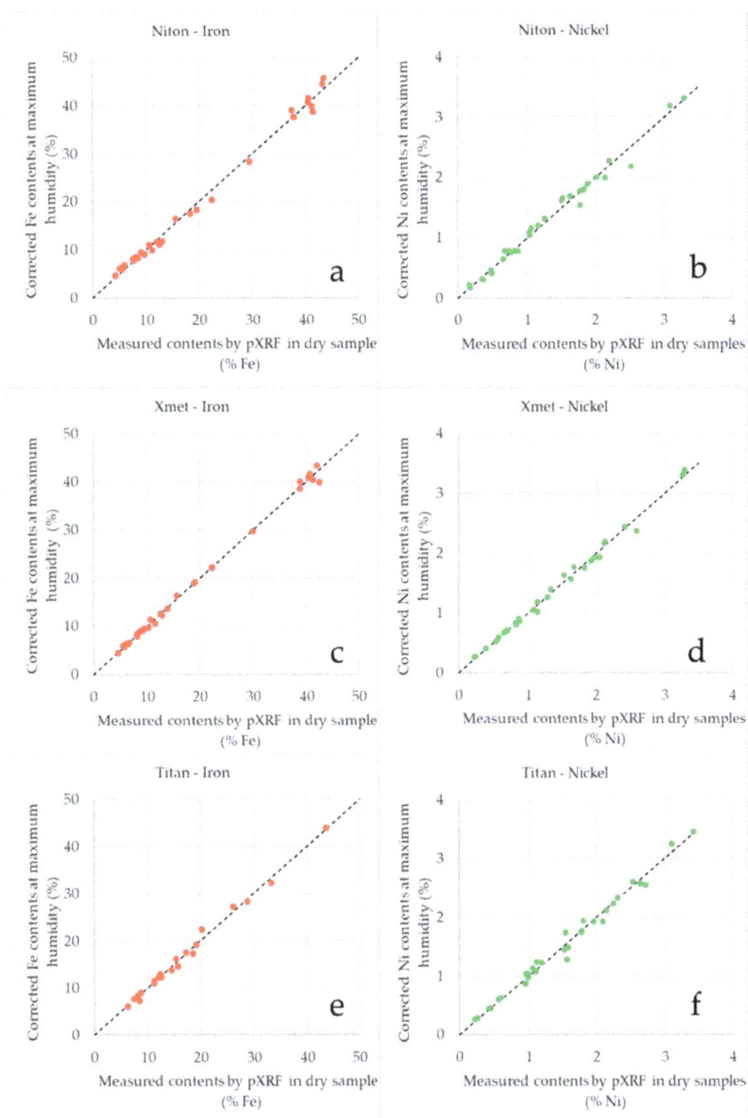

Figure 13. The relationship between corrected Ni and Fe concentrations measured at maximum humidity and concentrations on dry samples: (**a**,**b**) Niton, (**c**,**d**) Xmet and (**e**,**f**) Titan.

For laboratory samples, the particle size is less than 75 μm and sample wetting is well controlled. It is questionable whether this linear relationship is still valid for a direct measurement on the mining front. This method will be usable anyway if the sample is finely ground.

Table 6. Parameters of the linear regression between concentrations at higher water content and concentrations in dry sample for the two elements and for the three devices (a: Slope, b: y-intercept, R^2: Coefficient of determination, ME: Mean error, RMSE: Root-mean-square error).

	Element	Fe	Ni
Niton	a	1.335	1.576
	b	1.142	−0.001
	R^2	0.9936	0.9845
	ME	1.9×10^{-15}	-1.3×10^{-16}
	RMSE	1.13	0.101
Xmet	a	1.177	1.128
	b	−0.072	0.022
	R^2	0.9977	0.9935
	ME	-6.2×10^{-15}	9.0×10^{-17}
	RMSE	0.69	0.067
Titan	a	1.190	1.160
	b	−0.098	0.019
	R^2	0.9966	0.9869
	ME	-5.3×10^{-15}	-1.9×10^{-16}
	RMSE	1.36	0.095

3.3. pXRF Measurements at Water Saturation in the Field

In Section 3.1.1, we have shown that concentrations measured on dry samples are a linear function of the reference concentrations. On the other hand, concentrations measured for saturated samples are a linear function of the concentrations measured for the dry samples (Section 3.2.3). The overall correction law between measurements on saturated samples and reference ones is therefore also linear. As such, we can expect that concentrations measured by pXRF on water-saturated soil are related to reference concentrations by a linear relationship.

Rigorously, the calibration of Ni concentration depends on the Fe concentration for Xmet and Niton (Section 3.1.2). The correction should therefore include terms depending on Fe concentration. The dilution induced by water saturation decreases Fe concentration and its variability, and limits the importance of these terms. These terms do not significantly modify the quality of the correction and therefore we keep a linear correction for its simplicity. The linear regression and results of the correction shown in Figure 14 were established on the same set of samples which contains 13 samples (7 limonites and 6 saprolites).

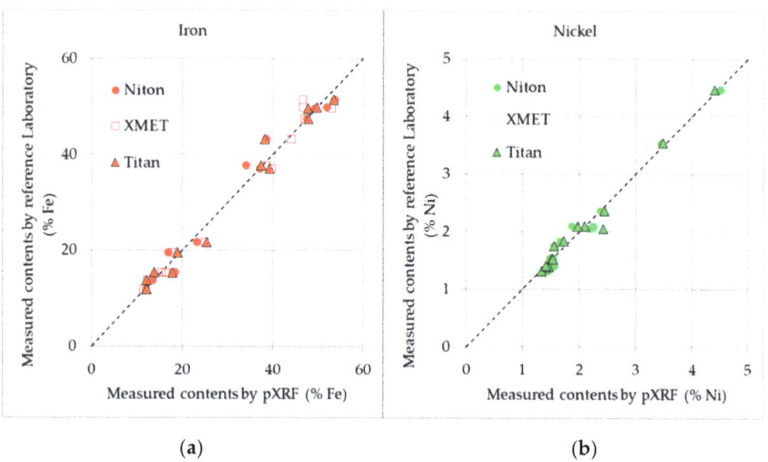

Figure 14. Results of the saturation method after calibration (linear regression): (**a**) Fe oxide; (**b**) Ni.

The Table 7 quantifies the mean error and the root-mean-square error obtained by a linear correction law.

Table 7. Parameters of the linear regression between in situ measurements of water-saturated samples and contents measured by reference laboratory for the two elements and for the three devices (a: Slope, b: y-intercept, R^2: Coefficient of determination, ME: Mean error, RMSE: Root-mean-square error).

	Element	**Fe**	**Ni**
	a	1.978	1.283
	b	−4.739	0.880
Niton	R^2	0.9767	0.9781
	ME	-2.7×10^{-15}	4.1×10^{-16}
	RMSE	2.31	0.131
	a	1.527	0.776
	b	−7.742	0.753
Xmet	R^2	0.9762	0.9711
	ME	1.6×10^{-15}	1.9×10^{-16}
	RMSE	2.33	0.151
	a	1.617	1.215
	b	−2.851	0.434
Titan	R^2	0.9787	0.9789
	ME	1.5×10^{-15}	1.5×10^{-16}
	RMSE	2.21	0.129

The Relative Standard Deviation has also been calculated. It is less than 8% for the Ni and less than 10% for the Fe whatever the pXRF device.

4. Discussion

Concentrations measured by pXRF depend on surface heterogeneities, on the size of particles and thickness of the sample. The values given by the different devices cannot therefore be directly compared with the reference values and a correction step is necessary.

A classic linear regression is used for this step. After correction, results are satisfactory for Fe. RMSE is approximately 0.5% when the range of Fe concentrations varies from 5% to 55%. Corrected measurements are also centered around reference values (ME around 5.10^{-15}). Regarding Ni, for Titan, the precision remains satisfactory. RMSE is around 0.05% when Ni concentrations are included between 1 wt% and 3.3 wt%. Values remain centered around references values (ME of the order of -6.10^{-17}). For Niton and Xmet, RMSE is three times higher than for Titan so corrected measurements are much more dispersed around references values for Niton and Xmet.

It is possible to minimize dispersion of corrected values if two linear regressions are used, one for the limonites and another one for the saprolites. The major drawback of this method is the necessity to determine the nature of the ore to be analyzed. This determination can sometimes be difficult and for certain so-called transition minerals this determination remains subject to interpretation.

For infinitely thick samples, XRF intensity emitted from element i (I_i) at the wavelength λ_i characteristic of element i can be obtained from the Sherman Equation [37]:

$$I_i(\lambda_i) = g_i C_i \int_{\lambda_0}^{\lambda_{edge\ i}} \frac{I_0(\lambda)\mu_i(\lambda)}{\mu_{s'}(\lambda) + \mu_{s''}(\lambda_i)} \left[1 + \sum_j C_j \delta_{ij}(\lambda)\right] d\lambda \qquad (9)$$

where I_0 is the intensity of excitation source at incident wavelength λ, C_i is weight fraction of analyte i, g_i a proportionality constant depending on the instrument used, δ_{ij} is the enhancement contribution of each matrix element j, λ_0 is minimum wavelength of the incident radiation, $\lambda_{edge\ i}$ is wavelength of the edge of considered line of analyte i, $\mu_i(\lambda)$ mass absorption coefficient of element i at wavelength λ, $\mu_{s'}(\lambda)$ and $\mu_{s''}(\lambda_i)$ mass absorption coefficient of the specimen for incident radiation and for emitted radiation. The two main elements of interest here, Ni and Fe, are subject to enhancement effect because of the proximity of the Ni Kα peak (7.48 keV) to the K absorption edge of Fe (7.11 keV). In such a case, the characteristic X-ray peak of Ni is absorbed by Fe and the characteristic peak of Fe is enhanced by the presence of Ni. Specimen mass absorption coefficient also depends on concentration of all elements. Concentrations values evaluated by devices differ from real concentrations so enhancement effect correction and specimen mass absorption coefficient cannot be accurately corrected. In our samples, Fe is the major element quantified by pXRF and its concentration is up to 55%. As XRF intensity of Ni is influenced by enhancement effect due to Fe, a new term proportional to Ni concentration multiplied by Fe concentration is added to Ni corrective law for the Niton and Xmet devices.

This correction law remains simple and could easily be used by operators on mine. This new correction law divides by 4 the RMSE value obtained with a classical linear correction law and has the major advantage to be usable for saprolitic or limonitic ores.

A challenge for in situ measurements, especially in tropical humid conditions, is to obtain the concentration of elements in dry soil, although the measurement is made on soil that may contain high water content. As usually observed, concentrations of Ni and Fe decrease when water content increases (Figure 9). The measured variations depend on the device used. The concentration range is larger for Niton than for Xmet and Titan.

Niton uses intensity of the Compton line to quantify the dark matrix. The X-ray absorption by the sample is thus better quantified. This method makes results less sensitive to matrix variations and in consequence increases accuracy of results. The addition of water increases the sample mass and therefore decreases the mass fraction of each of the elements to be quantified. However, concentrations in the dry sample can be obtained from the water content and concentrations measured in the wet sample using dilution law. As expected, dilution law corrects Niton results more effectively than those obtained with the other two devices. For Ni, RMSE is more than seven times smaller for Niton than for Xmet or Titan. Even for Niton, this correction does not fully correct water effects.

Use of Ge and coauthors [16] hypothesis corrects measured concentrations, which become closer to ones observed in dry sample. The three devices give direct access to concentrations but not to intensities, so concentrations were used rather than intensities as in Ge and coauthors initial study. Because of matrix effects, concentration of the analyte is not necessarily proportional to XRF intensity characteristic of the element. This may therefore lead to an error in the correction. For the same reason, water content was estimated from sample weight and not from intensity of primary X-rays scattering. On average RMSE is 1.1% for Fe and 0.08% for Ni.

The attenuation coefficient was considered to be constant for all samples, but it appears that the correction could be improved if an attenuation coefficient specific to each sample is used (see Figure 15). This is consistent with the observations of Stockmann and coauthors [38], who could not find a single attenuation coefficient for three soil samples studied. This suggests that the attenuation coefficient depends on the nature of the soil and therefore in our case on the nature of the ore. For the three instruments, correlation coefficients between the attenuation coefficient and concentrations of the different elements or the density of the sample were calculated. There is no significant correlation. Therefore, a law modifying the attenuation coefficient according to properties of the sample cannot be established.

Laboratory measurements of representative samples of New-Caledonian ores show a linear relationship between the concentrations in dry sample and concentrations measured on samples saturated with water. For all instruments and for Ni and Fe, the coefficient of determination is greater than 0.98, which attests the quality of this adjustment. Sahraoui and Hachicha [35] measured concentrations for 60 soil samples from the North East of Tunisia. They obtained, as for our samples, a good correlation between the Fe concentration in saturated sample and Fe concentration in dry sample (R = 0.949). The coefficient of determination for Ni was 0.819 in [35], so a better correlation is observed for our samples.

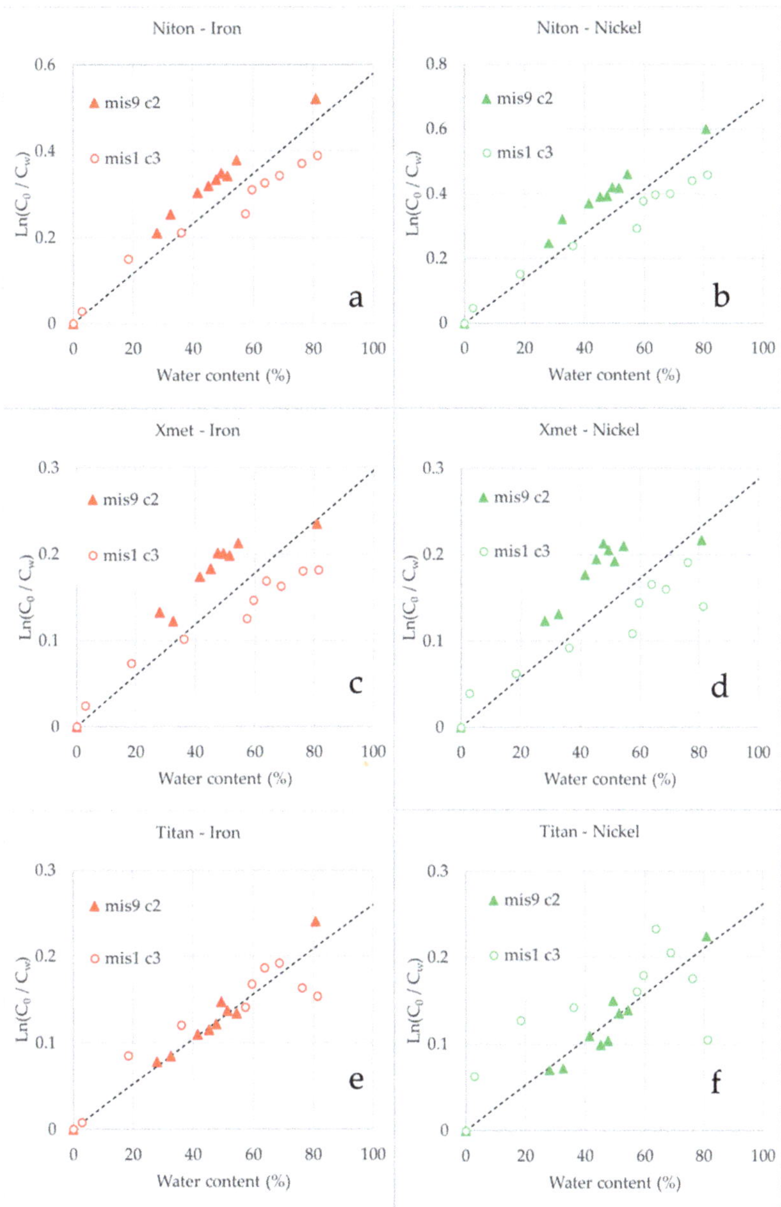

Figure 15. Ni and Fe concentrations in wet conditions divided by concentrations in dry conditions according to water content (expressed in wt%) for 2 samples. The dashed line corresponds to the fitted regression where attenuation coefficients are given in Table 5: (**a**,**b**) Niton, (**c**,**d**) Xmet and (**e**,**f**) Titan.

By using this linear correction, RMSE is approximately equal of those obtained by using the correction proposed by [16]. These two values of RMSE are not strictly comparable because the correction by [16] method was carried out for all humidities whereas for linear correction only the maximum humidity is considered. This indicates that a satisfactory estimation of the concentration of dry sample can be obtained by linear regression. The RMSE value is almost identical for Titan and Niton and is slightly lower for Xmet (1.5 times lower).

In situ, the correction law obtained in laboratory is no longer valid. In particular, in situ type of protective film, particle size and compaction differ from laboratory samples, which has a significant impact on pXRF measurements and on water content of saturated sample. New linear regressions were established, but directly with reference values. Probably due to dilution, Fe concentration has only a negligible influence on Ni correction. A simple linear regression is therefore sufficient for Ni and Fe. The determination coefficients are all higher than 0.97, which shows that a linear correction law is quite appropriate and also shows the validity of this method. RMSE value is approximately 2.3% for Fe and 0.15% for Ni regardless of the considered apparatus. These results are very satisfactory for Ni Root-mean-square error values are almost identical to those calculated from the linear regression linking pXRF measurements on dry samples and reference concentrations. For Fe, these values are two times greater than those obtained by saturation method applied in laboratory. They are also four times larger than those calculated when converting pXRF measurements on dry samples into reference values. In situ, these two corrections are performed in a single step but this does not justify RMSE increase. This difference is certainly explained by the greater dispersity in particle size and in compaction of soils in the natural environment. For some elements, sample grinding may be necessary to improve the quality of the quantification.

5. Conclusions

Although imperfect, correction found by [16] improves the accuracy of the determination of the Ni and Fe contents of the dry sample. Since pXRF equipment only gives concentrations, it is necessary to determine the water content by another method which limits its use.

One way to overcome the difficulty in determining the water content is to fix its value or at least limit its variation between samples, for example by saturating the sample with water. In situ, the Ni and Fe concentrations measured on water saturated samples are a linear function of reference concentrations measured by conventional methods. The coefficient of determination is high ($R^2 > 0.97$) which shows the relevance of this method. For some elements, an initial preparation of the sample before adding water may further improve these results. The pXRF measurements are carried out on wet samples, and XRF intensities will therefore be lower, which will reduce the accuracy of the measurement and increase minimum detectable concentration. Application to a larger number of samples is now necessary to estimate more precisely precision and accuracy of this method.

The underlying phenomenon and mechanisms that imply water content, sample nature and XRF are not yet well understood. Some corrections have been proposed in the literature but in Ni ore mining context, none are able to finely correct XRF measurement on wet samples. Our saturation method is an empirical method that may be calibrated for each environment and probably for each operator. Nerveless, in a mining context, this method has two advantages: (1) no other measuring device is necessary and (2) calibration coefficients can be entered directly in the pXRF device in the same way it is done for dry samples. Hence, the corrected value can be read directly on the device screen, allowing the operator to make decisions in the field.

This work shows that some fundamental questions are still pending. Indeed, our experiments with the Ge method reveal that the physical model is not adapted to our context or that it is not complete, and at least one parameter is missing to take into account the sample nature. Moreover, it is generally admit that moisture is not a major source of errors when moisture content is less than 20% but this aspect has not been properly addressed when a higher content is considered, in particular near or well above saturation. Some additional works should be undertaken to address those two questions.

Author Contributions: Conceptualization, T.Q., C.M. and V.L.; methodology, T.Q., C.M., J.G., P.W. and V.L.; validation, T.Q., C.M. and V.L.; formal analysis, T.Q., C.M., J.G., P.W. and V.L.; investigation, T.Q., C.M., J.G., P.W. and V.L.; writing—original draft preparation, T.Q., C.M. and V.L.; writing—review and editing, T.Q., C.M. and V.L.; visualization, T.Q., C.M. and V.L.; supervision, V.L.; project administration, V.L.; funding acquisition, V.L. and T.Q. All authors have read and agreed to the published version of the manuscript.

Funding: The authors gratelly acknowledge financial support by the CNRT (Centre National de Recherche Technologique) "Nickel et son Environnement". The authors would like to thank the anonymous reviewers for their help in improving the manuscript.

Acknowledgments: During this project, three portable devices of different brands were used. ThermoFisher ScientificTM Niton™ GOLDD+ (Fondis-Bioritech and Intermed companies), Oxford™ XMET (Xpertis company) and Bruker™ Titan (Cipac company) were made available free of charge.

Conflicts of Interest: The funders had no role in the design of the study; in the collection, analyses, or interpretation of data; in the writing of the manuscript, or in the decision to publish the results. The authors declare no conflict of interest.

References

1. Ross, P.-S.; Bourke, A.; Fresia, B. Improving lithological discrimination in exploration drill-cores using portable X-ray fluorescence measurements: (1) testing three Olympus Innov-X analysers on unprepared cores. *Geochem. Explor. Environ. Anal.* **2014**, *14*, 171–185. [CrossRef]
2. Fisher, L.; Gazley, M.F.; Baensch, A.; Barnes, S.J.; Cleverley, J.; Duclaux, G. Resolution of geochemical and lithostratigraphic complexity: A workflow for application of portable X-ray fluorescence to mineral exploration. *Geochem. Explor. Environ. Anal.* **2014**, *14*, 149–159. [CrossRef]
3. Arne, D.C.; Mackie, R.A.; Jones, S.A. The use of property-scale portable X-ray fluorescence data in gold exploration: Advantages and limitations. *Geochem. Explor. Environ. Anal.* **2014**, *14*, 233–244. [CrossRef]
4. Gazley, M.F.; Tutt, C.M.; Fisher, L.A.; Latham, A.R.; Duclaux, G.; Taylor, M.D.; de Beer, S.J. Objective geological logging using portable XRF geochemical multi-element data at Plutonic Gold Mine, Marymia Inlier, Western Australia. *J. Geochem. Explor.* **2014**, *143*, 74–83. [CrossRef]
5. Sarala, P.; Taivalkoski, A.; Valkama, J. Portable XRF: An advanced onsite analysis method in till geochemical exploration. *Geol. Surv. Finl.* **2015**, *57*, 63–86.
6. Bourke, A.; Ross, P.-S. Portable X-ray fluorescence measurements on exploration drill-cores: Comparing performance on unprepared cores and powders for 'whole-rock' analysis. *Geochem. Explor. Environ. Anal.* **2015**, *16*, 147–157. [CrossRef]
7. Andrew, B.S.; Barker, S.L.L. Determination of carbonate vein chemistry using portable X-ray fluorescence and its application to mineral exploration. *Geochem. Explor. Environ. Anal.* **2018**, *18*, 85–93. [CrossRef]
8. Sitko, R.; Zawisza, B. Quantification in X-ray Fluorescence Spectrometry. In *X-ray Spectrosc*; InTech Open: London, UK, 2012; Chapter 8; pp. 137–162. [CrossRef]
9. Gazley, M.F.; Fisher, L.A. A review of the reliability and validity of portable X-ray fluorescence spectrometry (pXRF) data. In *Mineral Resource and Ore Reserve Estimation—The AusIMM Guide to Good Practice*, 2nd ed.; The Australasian Institute of Mining and Metallurgy: Melbourne, Australia, 2014; pp. 1–13.
10. Le Vaillant, M.; Barnes, S.J.; Fisher, L.; Fiorentini, M.L.; Caruso, S. Use and calibration of portable X-ray fluorescence analysers: Application to lithogeochemical exploration for komatiite-hosted nickel sulphide deposits. *Geochem. Explor. Environ. Anal.* **2014**, *14*, 199–209. [CrossRef]
11. Quiniou, T.; Laperche, V. An assessment of field-portable X-ray fluorescence analysis for nickel and iron in laterite ore (New Caledonia). *Geochem. Explor. Environ. Anal.* **2014**, *14*, 245–255. [CrossRef]
12. Sarala, P. Comparison of different portable XRF methods for determining till geochemistry. *Geochem. Explor. Environ. Anal.* **2016**, *16*, 181–192. [CrossRef]
13. Gazley, M.F.; Bonnett, L.C.; Fisher, L.A.; Salama, W.; Price, J.H. A workflow for exploration sampling in regolith-dominated terranes using portable X-ray fluorescence: Comparison with laboratory data and a case study. *Aust. J. Earth Sci.* **2017**, *64*, 903–917. [CrossRef]
14. Hughes, R.; Barker, S.L.L. Using portable XRF to infer adularia halos within the Waihi Au-Ag system, New Zealand. *Geochem. Explor. Environ. Anal.* **2018**, *18*, 97–108. [CrossRef]
15. Knight, R.D.; Kjarsgaard, B.A.; Russel, H.A. An analytical protocol for determining the elemental chemistry of Quaternary sediments using a portable X-ray fluorescence spectrometer. *Appl. Geochem.* **2021**, *131*, 105026. [CrossRef]
16. Ge, L.; Lai, W.; Lin, Y. Influence of and correction for moisture in rocks, soils and sediments on in situ XRF analysis. *X-ray Spectrom* **2005**, *34*, 28–34. [CrossRef]
17. Weindorf, D.C.; Bakr, N.; Zhu, Y.; Mcwhirt, A.; Ping, C.L.; Michaelson, G.; Nelson, C.; Shook, K.; Nuss, S. Influence of ice on soil elemental characterization via portable X-ray fluorescence spectrometry. *Pedosphere* **2014**, *24*, 1–12. [CrossRef]

18. Kido, Y.; Koshikawa, T.; Tada, R. Rapid and quantitative major element analysis method for wet fine-grained sediments using an XRF microscanner. *Mar. Geol.* **2006**, *9*, 209–225. [CrossRef]
19. US Environmental Protection Agency. Method 6200: Field portable X-ray fluorescence spectrometry for the determination of elemental concentrations in soil and sediment. In *Test Methods for Evaluating Solid Waste*; US Environmental Protection Agency: Washington, DC, USA, 2007.
20. Kalnicky, D.J.; Singhvi, R. Field portable XRF analysis of environmental samples. *J. Hazard. Mater.* **2001**, *83*, 93–122. [CrossRef]
21. Laiho, J.V.-P.; Perämäki, P. Evaluation of portable X-ray fluorescence (PXRF) sample preparation methods. *Geol. Surv. Finland Spec. Pap.* **2005**, *38*, 73–82.
22. Bastos, R.O.; Melquiades, F.L.; Biasi, G.E.V. Correction for the effect of soil moisture on in situ XRF analysis using low-energy background. *X-ray Spectrom* **2012**, *41*, 304–307. [CrossRef]
23. Schneider, A.R.; Cancès, B.; Breton, C.; Ponthieu, M.; Morvan, X.; Conreux, A.; Marin, B. Comparison of field XRF and aqua regia/ICPAES soil analysis and evaluation of soil moisture influence on FPXRF results. *J. Soils Sediments* **2016**, *16*, 438–448. [CrossRef]
24. Phedorin, M.A.; Goldberg, E.L. Prediction of absolute concentrations of elements from SR XRF scan measurements of natural wet sediments. *Nucl. Instrum. Methods Phys. Res. Sect. A Accel. Spectrometers Detect. Assoc. Equip.* **2005**, *543*, 274–279. [CrossRef]
25. Ribeiro, B.T.; Weindorf, D.C.; Silva, B.M.; Tassinari, D.; Amarante, L.C.; Curi, N.; Guimarães Guilherme, L.R. The Influence of Soil Moisture on Oxide Determination in Tropical Soils via Portable X-ray Fluorescence. *Soil Sci. Soc. Am. J.* **2018**, *82*, 632–644. [CrossRef]
26. Rousseau, R. Corrections for matrix effects in X-ray fluorescence analysis—A tutorial. *Spectromchim. Acta Part B* **2006**, *61*, 759–777. [CrossRef]
27. Ravansari, R.; Wilson, S.C.; Tighe, M. Portable X-ray fluorescence for environmental assessment of soils: Not just a point and shoot method. *Environ. Int.* **2020**, *134*, 105250. [CrossRef]
28. Paris, J.P. Géologie de la Nouvelle-Calédonie. *Mémoire Bur. Rech. Géologique Minière* **1981**, *113*, 1–279.
29. Cluzel, D.; Aitchison, J.C.; Picard, C. Tectonic accretion and underplating of mafic terranes in the Late Eocene intraoceanic fore-arc of New Caledonia (southwest Pacific): Geodynamic implications. *Tectonophysic* **2001**, *340*, 23–59. [CrossRef]
30. Ulrich, M.; Picard, C.; Guillot, S.; Chauvel, C.; Cluzel, D.; Meffre, S. Multiple melting stages and refertilization as indicators for ridge to subduction formation: The New Caledonia ophiolite. *Lithos* **2010**, *115*, 223–236. [CrossRef]
31. Maurizot, P.; Robineau, B.; Vendé-Leclerc, M.; Cluzel, D. Introduction to New Caledonia: Geology, geodynamic evolution and mineral resources. In *New Caledonia: Geology, Geodynamic Evolution and Mineral Resources*; Maurizot, P., Mortimer, N., Eds.; Memoirs, 51; Geological Society: London, UK, 2020; Chapter 1; pp. 1–12.
32. Maurizot, P.; Sevin, B.; Lesimple, S.; Bailly, L.; Iseppi, M.; Robineau, B. Mineral resources and prospectivity of the ultramafic rocks of New Caledonia. In *New Caledonia: Geology, Geodynamic Evolution and Mineral Resources*; Maurizot, P., Mortimer, N., Eds.; Memoirs, 51; Geological Society: London, UK, 2020; Chapter 10; pp. 247–277.
33. Trescases, J.J. L'évolution géochimique supergène des roches ultrabasiques en zone tropicale: Formations des gisements nickélifères de Nouvelle-Calédonie. *Mémoires Orstom* **1975**, *78*, 259.
34. Shuttleworth, E.L.; Evans, M.G.; Hutchinson, S.M. Assessment of Lead Contamination in Peatlands Using Field Portable XRF. *Water Air Soil. Pollut.* **2014**, *225*, 1844. [CrossRef]
35. Sahraoui, H.; Hachicha, M. Effect of soil moisture on trace elements concentrations using portable x-ray fluorescence spectrometer. *J. Fundam. Appl. Sci.* **2017**, *9*, 468. [CrossRef]
36. Santana, M.L.T.; Carvalho, G.S.; Guilherme, L.R.G.; Curi, N.; Ribeiro, B.T. Elemental concentration via portable X-ray fluorescence spectrometry: Assessing the impact of water content. *Ciência Agrotecnologia* **2019**, *43*. [CrossRef]
37. Rousseau, R.M.; Boivin, J.A. The fundamental algorithm: A natural extension of Sherman equation. *Rigaku J.* **1998**, *15*, 13–28.
38. Stockmann, U.; Jang, H.J.; Minasny, B.; McBratney, A.B. The Effect of Soil Moisture and Texture on Fe Concentration Using Portable X-ray Fluorescence Spectrometers. In *Digital Soil Morphometrics*; Hartemink, A., Minasny, B., Eds.; Progress in Soil Science Book Series; Springer: Cham, Switzerland; Berlin/Heidelberg, Germany, 2016. [CrossRef]

Article

Objective Domain Boundaries Detection in New Caledonian Nickel Laterite from Spectra Using Quadrant Scan

Ayham Zaitouny [1,2,*], Erick Ramanaidou [3], June Hill [3], David M. Walker [1] and Michael Small [1,2,3]

1 Department of Mathematics and Statistics, University of Western Australia, Crawley, WA 6009, Australia; david.walker@uwa.edu.au (D.M.W.); michael.small@uwa.edu.au (M.S.)
2 ARC Training Centre for Transforming Maintenance through Data Science, University of Western Australia, Crawley, WA 6009, Australia
3 Mineral Resources, CSIRO, Kensington, WA 6151, Australia; erick.ramanaidou@csiro.au (E.R.); june.hill@csiro.au (J.H.)
* Correspondence: ayham.zaitouny@uwa.edu.au

Abstract: Modelling of 3D domain boundaries using information from drill holes is a standard procedure in mineral exploration and mining. Manual logging of drill holes can be difficult to exploit as the results may not be comparable between holes due to the subjective nature of geological logging. Exploration and mining companies commonly collect geochemical or mineralogical data from diamond drill core or drill chips; however, manual interpretation of multivariate data can be slow and challenging; therefore, automation of any of the steps in the interpretation process would be valuable. Hyperspectral analysis of drill chips provides a relatively inexpensive method of collecting very detailed information rapidly and consistently. However, the challenge of such data is the high dimensionality of the data's variables in comparison to the number of samples. Hyperspectral data is usually processed to produce mineral abundances generally involving a range of assumptions. This paper presents the results of testing a new fast and objective methodology to identify the lithological boundaries from high dimensional hyperspectral data. This method applies a quadrant scan analysis to recurrence plots. The results, applied to nickel laterite deposits from New Caledonia, demonstrate that this method can identify transitions in the downhole data. These are interpreted as reflecting mineralogical changes that can be used as an aid in geological logging to improve boundary detection.

Keywords: spectral data; mineralogical data; lithological boundaries; nickel laterite; New Caledonia; quadrant scan

Citation: Zaitouny, A.; Ramanaidou, E.; Hill, J.; Walker, D.M.; Small, M. Objective Domain Boundaries Detection in New Caledonian Nickel Laterite from Spectra Using Quadrant Scan. *Minerals* **2022**, *12*, 49. https://doi.org/10.3390/min12010049

Academic Editors: Cristina Domènech and Cristina Villanova-de-Benavent

Received: 7 October 2021
Accepted: 24 December 2021
Published: 29 December 2021

Publisher's Note: MDPI stays neutral with regard to jurisdictional claims in published maps and institutional affiliations.

1. Introduction

A fundamental task for any exploration geologist is to log core or samples from drilling programs. However, the traditional method of manually logging drill holes has proved problematic because the logging may be inconsistent between geologists. Not only do different geologists provide different labels for the same rock type, but they may also divide the hole into rock types (or domains) of different sizes depending on their perceived task (i.e., splitters vs. lumpers). To overcome this problem, exploration companies are increasingly collecting numerical data using sensors, such as hyperspectral scanners, which can provide a consistent output. However, numerical data still requires interpretation to provide geological meaning. The same issues of consistent interpretation arise for numerical data as for core or chip logging, but the problem is exacerbated by the high dimensionality of much of these data. For example, hyperspectral scanners may provide measurements for ~500 bands. Hence, it is becoming increasingly popular to use machine learning to overcome the task of providing consistent labelling for samples with the same properties [1].

Machine learning provides labelling for each sample independently, which means that any drill hole logged using machine learning will provide sample-scale results; however, the actual rock unit boundaries are not known. The location of boundaries as indicated by changes in classification of sequential samples can be deceptive due to misclassification problems; for example, due to mixed samples or noisy data where the sample composition is close to a classification decision boundary. Multiscale methods utilising the continuous wavelet transform (CWT) have been proposed as a solution to this problem of detecting the rock type boundaries [2,3]. CWT methods are based on well-established multiscale boundary detection methods from image analysis [4–10]. The fundamental shortcoming of wavelet transforms is that they are univariate. While boundaries can be combined for several variables to provide a multivariate result [2], this is not a practical solution for very high dimensional data, such as hyperspectral data. Due to computational load, dimensionality reduction is essential to the use of efficient boundary detection techniques.

We propose an alternative method of boundary detection, based on non-linear time series analysis [11–14], that provides simultaneous dimensionality reduction and boundary detection. This method has two parts, first a recurrence plot is used to simplify the similarity structure of the data and reduce the data dimensionality to two dimensions [12,15]. Second, the quadrant scan method is applied to the recurrence plot to further reduce the dimensionality to one dimension and to produce a profile line from which transition points can be extracted [16–19]. It is proposed that the transition points indicate the depths of geological boundaries. The type of geological boundary depends on the features that are measured by the input data.

The method does not provide labelling of rock types and so cannot indicate what rock types the boundary is separating. It is expected that this method would be used in conjunction with a classification scheme, which may be derived using machine learning or using more traditional user-selected threshold-based techniques. The boundaries detected by these methods indicate a significant change in the input data, which may be attributable to any compositional change including primary rock type, alteration or weathering. There are several parameters that may be adjusted. We demonstrate the effect of adjusting the parameters and how these may be used to detect the domaining boundaries with different scales and significance.

In this paper, we demonstrate the efficacy of applying the quadrant scan method to recurrence plots to identify compositional boundaries resulting from mineralogical change detected in hyperspectral scanning data of nickel laterite deposits in New Caledonia [20]. For these deposits, logging of drill core is conducted by visual examination of the cores by geologists. This results in subjective logging of domaining boundaries, which has proved to be inconsistent. Unreliable logging is very difficult to use when comparing one drill hole to another, for example, to construct a geological cross-section. In an attempt to remediate this problem, the Corescan Hyperspectral Core Imager was used to collect reflectance hyperspectral data from the nickel laterite deposits [21]. Hyperspectral data have been shown to be useful in identifying key mineral groups from drill hole samples [22,23]. For the New Caledonian deposit, the hyperspectral data were used to successfully identify all the minerals present in the lateritic profile [21]. While the identification of mineral groups can be used to facilitate logging of rock types, alteration styles and weathering zones intersected by the drill hole, the process of interpreting hyperspectral data and producing consistent mineralogical logs is time-consuming and requires high-level expertise. The ability to automate any part of the workflow would be beneficial, including the use of the data to detect important geological boundaries. Hyperspectral core scanning provides continuous data down hole, which can be used to derive very accurate compositional boundary locations.

The New Caledonian case study is used here to show that the quadrant scan method can handle high dimensional hyperspectral data efficiently and has the additional advantage in that it can process irregular sampling intervals. We apply a naive approach of using all the bands in the hyperspectral data set. However, with the inclusion of expert knowledge on hyperspectral mineralogy and an understanding of which minerals are important for distinguishing the rock types in the deposit, it would be possible to remove the hyperspectral bands that are unlikely to provide useful information, making the results easier to interpret. In addition, although the reflectance spectra were used in this study, it is possible that removing the continuum could have an influence on the results of the Quadrant Scan method.

2. Geological Background and Data Overview

Nickel laterite ores represent over 60% of total nickel reserves; the remaining 40% is associated with nickel sulphide. New Caledonia is endowed with very large lateritic nickel deposits (Figure 1) that correspond to 11% of the nickel reserve worldwide with 6.7 Mt reserve and a production in 2019 of 280,000 t [24]. Nickel laterite ores are the product of intensive deep weathering of ultramafic rocks under humid tropical conditions. The resulting thick lateritic mantle extends from the bedrock to the surface and includes distinctive layers of altered and weathered rock as shown in Figure 2. Lithological units may include: (1) fresh or, totally or partially serpentinised dunite and/or harzburgite that consist mainly of olivine, pyroxene and serpentine minerals such as lizardite, antigorite and chrysotile; (2) saprock where weathering begins; (3) coarse saprolite with Ni-rich garnierite; (4) yellow laterite where Ni-goethite is dominant; (5) red laterite with a mixture of Ni-goethite and Ni-hematite and (6) lateritic duricrust [25]. In New Caledonia, the nickel mineralisation occurs as "oxide type" mainly in iron and manganese oxides or as "garnierite" type in Mg-Ni-silicates (Figure 2).

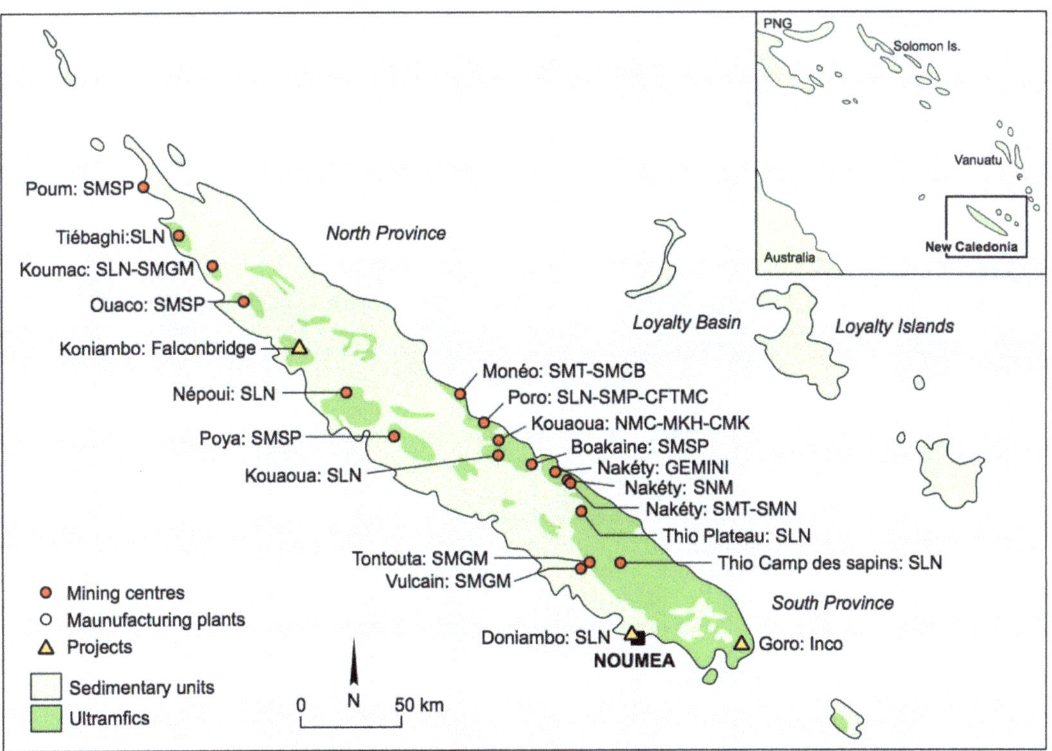

Figure 1. Location of the peridotites and the nickel deposits of New Caledonia [20].

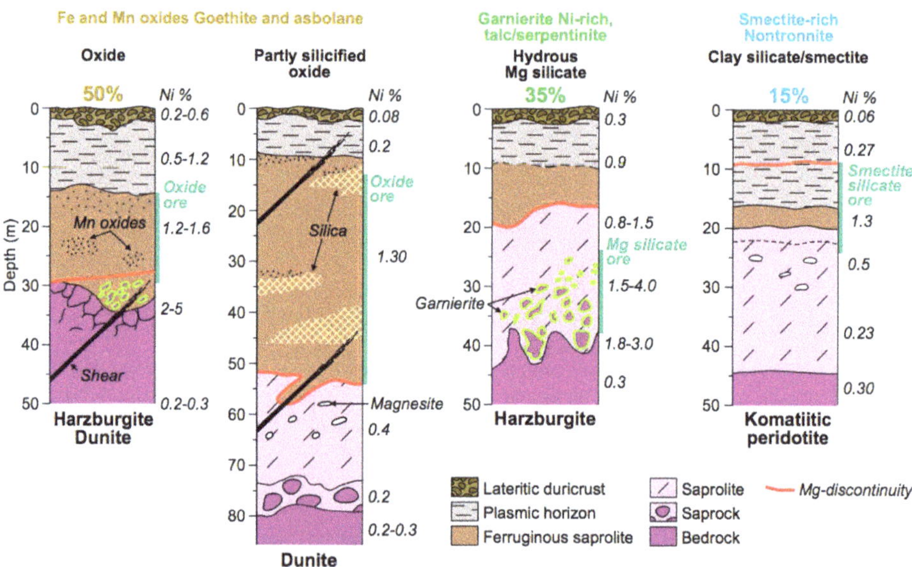

Figure 2. Typical weathering profiles developed on ultramafic rocks with absolute and residual accumulation of Ni and Co (Modified from [25]).

At present, logging of drill core is conducted by visual examination of the core by geologists, which results in subjective logging of lithological boundaries. The nickel laterite companies of New Caledonia have developed an ore type classification based on lithology, degree of serpentinisation and intensity of weathering (Table 1). Classification is dependent on mineral percentages estimated by the logging geologist. The lithology classification is based on the proportion of olivine and pyroxene as follows: if the proportion of olivine is more than 90%, it is dunite; otherwise, the label depends on the proportion of pyroxene. Harzburgite is 10 to 60% pyroxene, and pyroxenite is >60% pyroxene. Five serpentinization classes have been characterised (1) Superior (S) with 1–15% serpentine; (2) Intermediary (I) with 15–45% serpentine; (3) Normal (N) with 45–70% serpentine; (4) Basal (B) with 70–100% serpentine and (5) Green (Vert–V) with 45–70% serpentine (in this facies, the serpentinisation of the peridotites precedes the obduction). The degree of weathering is categorised from 0 to 5 based on the rock hardness and the presence of iron oxides (0 indicating no weathering and 5 indicating complete weathering).

The ability to develop a method for the collection of objective mineralogical data coupled with consistent lithological interpretation (including the identification of different ore types) would be of great value for the mining companies. Given that measurements of lateritic nickel diamond cores and drill chips demonstrated that the Corescan Hyperspectral Core Imager Mark III (HCI-3) could identify all the minerals present in the lateritic profile [21], it is proposed that the reflectance spectra can provide sufficient data for precise, objective, primary and alteration lithology and weathering boundary detection when used as input to an automated boundary detection algorithm.

Table 1. Interpretation of geological domaining samples.

Laterite/Primary Lithology		Serpentinization		Weathering	
Logging Code	Interpretation	Logging Code	Interpretation	Logging Code	Interpretation
LR	Latérite Rouge (Red Laterite)	NA	Not Assigned	NA	Not Assigned
LJ	Latérite Jaune (Yellow Laterite)	I	Intermediary	1	Weak Weathering
LT	Latérite de Transition (Transition Laterite)	V	Vert (green)	2	
BS	Basal Serpentine	N	Normal	3	
H	Harzburgite	B	Basal	4	
D	Dunite			5	Strong weathering
HD	Harzburgite/Dunite				
DH	Dunite/Harzburgite				

2.1. Samples

All the samples used for this study were provided by the Nickel Mining Company (NMC). The samples were collected from the deposits of Boulinda Monique, N'Go and Ouaco Mousquetaire. NMC has generally a sampling density of 1 m for drill chip with some exceptions with denser sampling. A representative sub-sample of each 1 m drill chips sample was transferred in a 20 cm core tray compartment (Figure 3) before been measured with the imaging spectrometer.

Figure 3. Drill chips from OUACO deposit prepared in a core tray divided into 20 cm compartments.

2.2. Spectral Collection

Corescan's HCI-3 combines reflectance spectroscopy, visual imagery and 3D laser profiling to map the mineralogy and geochemistry of geological samples. HCI-3 covers the VNIR (450 to 1300 nm) and SWIR (1300 to 2500 nm) range of the electromagnetic spectrum from 450 nm to 2500 nm at a spectral resolution (FWHM) of 3.5 nm. High quality optics focus the spectral measurement to a 0.5 mm point on the core. A RGB camera provides a high-resolution visual record of the core at 60 μm pixel size. Measurement of core surface features, texture and shape is complemented using a 3D laser profiler with a surface profile resolution of 20 μm. The system comprises a scan unit housing the optics, spectrometers, cameras and 3D profiling sensors; a translation table with conveyor driven core tray loading system and a high-speed data acquisition, processing and control computer. For each compartment sample, an average of 20,000 0.5 mm × 0.5 mm pixels were collected (depending on sample size) with an individual spectrum in each pixel. For this study, all the spectra from all pixels were averaged into 1 spectrum representing each sample.

2.3. Spectral Analysis

The HCI-3 can spectrally image drill core and cuttings at 0.5 mm resolution, hence providing distinct advantages by capturing pixels with less phases. Mineral mapping is undertaken using in-house proprietary expert system software. The reflectance spectrum in each pixel is either a pure mineral or a combination of two or more minerals; it is called a mineral class. For each mineral class, unique spectral indices were developed and consist of three parameters; (1) feature tracking that focuses on regions where absorptions exist; (2) matching regions that focuses on specific spectral regions and compare with the in-house database using Pearson correlation coefficient PCC and (3) spectral ratio which is a ratio of two reflectance values at specific wavelengths. Using the expert system, each pixel was mineralogically quantified and an average mineralogy for each sample was calculated.

3. Methods

The mathematical method described in this paper consists of three steps (Figure 4): (1) generating a 2D similarity matrix for the series of measurements in each drill hole; (2) converting the similarity matrix to a recurrence plot by applying a threshold, and (3) applying the quadrant scan method to the recurrence plot to generate a one-dimensional drill hole profile. Each of these three steps is described in detail in this section.

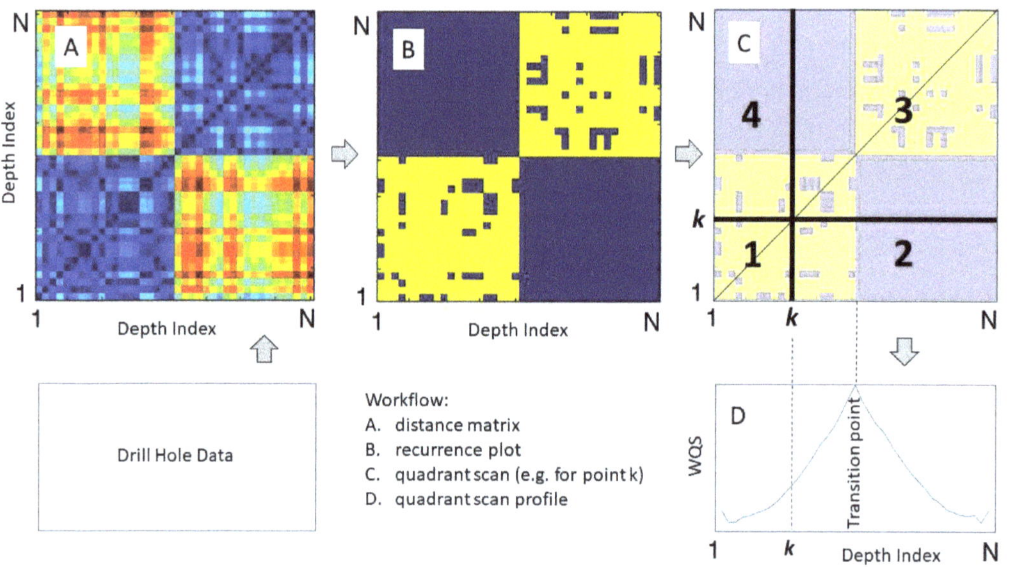

Figure 4. A flowchart demonstrating the steps of the proposed mathematical method. In subplot (**A**), the colours represent the difference between each sample and all other samples in the drill hole using the multivariate data. For the binary recurrence matrix, in subplot (**B**), blue represents 0 (above the threshold) and yellow represents 1 (below the threshold). At each depth index, the recurrence matrix is divided into four quadrants, shown in subplot (**C**), the black vertical and horizontal lines demonstrate the division for depth index k. The ratio of quadrant values transforms the 2D plot into a simple function, subplot (**D**), where maxima indicate transition points in the input data.

In the field of nonlinear time series analysis, recurrence is a fundamental characteristic of many dynamical systems. Recurrence is defined when two states of the system pass close to each other in phase space at different times. For example, in geology, recurrence occurs when geological processes produce similar (but not necessarily exactly the same) rock types because the geological processes are repeated over time. The recurrence plot was developed to reconstruct a time series as a two-dimensional plot to facilitate visual investigation of the recurrence structure in a system. This means that it can reduce any number of variables to a 2D representation (based on a similarity matrix), providing a very powerful dimensionality reduction tool. For example, a recurrence plot for drill hole data will indicate when similar rock types occur at different levels in the drill hole and where changes in rock type occur. Several measures have been developed to quantify features in recurrence plots, these are called recurrence quantification analysis measures (RQA). RQA measures can be used for quantitative investigation of the system's properties and used to assess its dynamics. The recurrence plot and RQA measures are widely used as a time series analysis tool in many applications, including engineering [26], physics [27], chemistry [28], finance [29] and geology [15,30], and [31] provide an overview of the method and its applications to different systems.

Recently, a new RQA method was developed to identify transition points in time series data from dynamical systems, namely, the quadrant scan method [16,18,19]. It has been used in medical applications for the interpretation of Electroencephalography (EEG) and Electrocardiography (ECG) signals [18]. More recently, it has been demonstrated that the quadrant scan method can be applied to spatial data by replacing the time index with a spatial index when used to detect geological boundaries in geochemical and petrophysical data from drill holes [19]. These applications have demonstrated that the quadrant scan is capable of detecting points of change in a data series for both univariate and multivariate data. For application to drill hole data, an index that provides the order of sampling down hole is used. If sampling occurs at regular intervals, then this index can be replaced by the actual depth measurements. For irregularly sampled data, the depth index can be used to avoid having to resample the data at regular intervals, as is the case for boundary detection methods involving wavelet transforms. The computational efficiency and low computational cost of the quadrant scan method [19] makes it an attractive tool to identify domaining boundaries from high-dimensional hyperspectral data.

The quadrant scan method has two versions, standard (density) or weighted quadrant scan. As elaborated in [18,19], the steps of the standard version are outlined in the flowchart (Figure 4) and summarised in the following paragraphs.

(1) First, a distance matrix is constructed (Figure 4A). Generally, the Euclidian norm is used to measure the pairwise distance between two states of the system. However, because of the unique structure and high dimensionality of the spectral data, the Euclidian norm was found to be an inappropriate and misleading measure due to correlations in the hyperspectral data. The Mahalanobis distance was found to be more appropriate, as it can account for correlations in the data [32]. However, due to the significant difference in the dimensionality between the number of the variables and the number of samples, finding the inverse of the covariance matrix is problematic. Alternatively, we can find the distance using the covariance matrix of the normalised samples. Using this method has the advantage that it is not necessary to have a separate dimensionality reduction step, before constructing the distance matrix. Equivalent to the Mahalanobis distance, the distance matrix is defined as follows:

$$A_{ij} = ||\vec{x}(i), \vec{x}(j)|| = \sqrt{(\vec{x}(i) - \vec{x}(j))^T S (\vec{x}(i) - \vec{x}(j))} \tag{1}$$

where $\vec{x}(d) \in \mathbb{R}^{514}$ is a spectrum profile at depth with index d (e.g., i and j refer to different depth indices), while S is the covariance matrix of the normalised data. Each index in the matrix refers to the index of the sample, i.e., the depth index. Hence, the matrix width and height are determined by the total number of samples.

(2) The recurrence plot matrix (Figure 4B) is constructed from the distance matrix A, by applying a threshold. If an entry in the distance matrix is less than the threshold, then the corresponding entry in the recurrence plot matrix is assigned to 1, otherwise it is 0. The threshold is controlled by a parameter α, which allows the user to control the scale of boundary detection [18,19]; that is, if α is small then the distance threshold will be small, which means that samples have to be very similar in composition to

be considered below the threshold. The recurrence plot matrix and the threshold are defined as follows:

$$R_{ij} = \begin{cases} 1, & if\ a_{ij} < \epsilon \\ 0, & otherwise \end{cases} \tag{2}$$

where

$$\epsilon = \alpha\left(\mu\left(A_{ij}\right) + 3\,\sigma\left(A_{ij}\right)\right); 0 < \alpha < 1 \tag{3}$$

where $\mu\left(A_{ij}\right)$ and $\sigma\left(A_{ij}\right)$ are the mean and standard deviation of the distance matrix entries and ϵ is the threshold.

(3) The quadrant scan profile (Figure 4D) is derived from the ratio of points in the binary recurrence plot above and below each depth index, by using the depth index to divide the recurrence plot into quadrants, as illustrated in Figure 4C. If $D_{1,3}$ and $D_{2,4}$ denote the density of the recurrent points in the first and third quadrants and the second and fourth quadrants, respectively, then the standard quadrant scan at depth index d for $d = 1, \cdots, N$ is defined as follows:

$$QS(d) = \frac{D_{1,3}}{D_{1,3} + D_{2,4}} \tag{4}$$

where, $D_{1,3} = \frac{\sum_{i,j<d} R_{ij} + \sum_{i,j>d} R_{ij}}{(d-1)^2 + (N-d)^2}$ and $D_{2,4} = \frac{\sum_{i<d,\ j>d} R_{ij} + \sum_{i>d,\ j<d} R_{ij}}{2(d-1)(N-d)}$.

In [19] a weighting scheme was applied in which higher weights were assigned to the states closer to the indexed point; this is called the weighted quadrant scan (WQS) method and showed considerable improvement over the standard quadrant scan method where all points are weighted equally. WQS requires two weighting parameters, m_1 and m_2, to determine the smoothness of the weighting scheme and selection of these is problem dependent [18]. The result of the WQS method is a profile in which the peaks can be used to identify the transitions (Figure 4D). The WQS provides a reduction of the 514 values in the original spectrum to a single value for each sample. When applied to hyperspectral data, the peaks in the WQS profile are used to identify the locations of lithological boundaries. Sharp peaks indicate rapid transitions and rounded peaks represent more gradual transitions [18].

As explained above, the implementation of the WQS method requires the setting of three parameters α, m_1 and m_2. Generally, as demonstrated in previous works, setting these parameters is problem dependent; however, guidance on how to choose these parameters and their effects on the WQS performance is discussed in [18]. Briefly, α adjusts and controls the recurrence threshold based on the distribution of the norm matrix and allows for multi-scale detection. The weighting scheme parameters m_1 and m_2 identify the gradient of the weighting function as well as the sizes of the neighbourhoods of the index point. Accordingly, the points within these neighbourhoods are assigned with maximum, transitional, or minimum weight. Finally, to avoid any artifact peaks in the WQS at the top or the bottom of any dataset which do not represent changes in the geological composition, the datasets are extended by copying the first and last samples four times.

4. Results

To test the algorithm, the WQS detection algorithm was applied to data from three drill-holes, namely OUACO, NGO_PB4 and BOULINDA. It is proposed that by using the reflectance spectra as input to the automated boundary detection algorithm, we can precisely and objectively detect boundaries for primary lithology, alteration or weathering where these are marked by changes in the spectra. For all the analyses, the weighting parameters were set at $m_1 = 10$ and $m_2 = 2$; this selection is based on the small number of the data samples, i.e., less than 50 samples per hole. The recurrence plot threshold was set at $\alpha = 0.03, 0.02$ and 0.01. The selection of these values is dependent on the distributions of the norm matrices derived by the Mahalanobis-like norms (Equation (3)). For spectral data, the largest variation of the WQS occurs when the value of α is within this range. For $\alpha < 0.01$ and $\alpha > 0.03$ there is less variation in the results. Therefore, we choose these values to demonstrate the robustness of our method. The variation of α allows detection of boundaries at different scales. In order to demonstrate the effectiveness of the WQS analysis for finding geological boundaries, we compare the results using a common unsupervised clustering technique, k-means [33]. K-means was chosen as it is very computationally efficient and one of the few clustering algorithms capable of handling such high dimensional data. To determine the optimal number of clusters for the k-means method, the Bayesian Information Criterion (BIC) [34] and the Elbow method are used [35] (Figure 5). The Elbow method plots the Residual Sum of Squares error (RSS) as a function of different values of k (the number of clusters), the optimal value of k is the point with most curvature. The BIC test penalizes for the number of clusters in addition to the RSS error; the minimal value of BIC identifies the optimal number of clusters. Figure 5 suggests that k = 8 is the optimal number of clusters to be used.

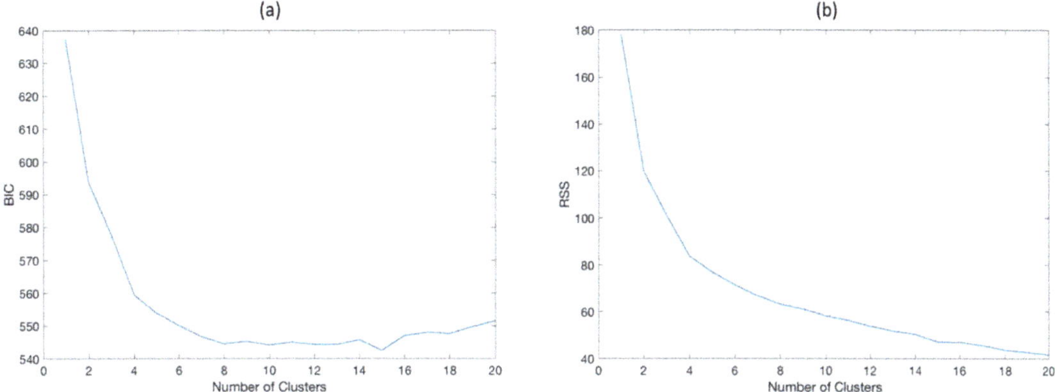

Figure 5. Tests to determine the optimal number of clusters for the k-means method. (**a**) The Bayesian Information Criterion (BIC) test shows that 8 is the optimal number of clusters at which the testing criterion is minimal, and; (**b**) The Elbow method test: it shows the Residual Sum of Squares error (RSS) as a function of the number of clusters.

The results for the three drill-holes are illustrated in Figures 6–8. Producing these results is quite fast; for example, generating a WQS profile from a 514-dimensional spectrum with 49 depth samples took approximately 6 s, including plotting three figures (the distance matrix, the recurrence plot matrix and the WQS profile), using MATLAB on a MacBook Pro. Figure 6a is a heat-map of the hyperspectral profiles at each depth interval in the OUACO hole (a total of 49 samples down hole). The heat-map is a graphical representation of spectral intensity down hole, wherein blue indicates low values and red indicates high values. Figure 6b shows the WQS profile derived from the spectral data using three different values of the parameter α (0.03, 0.02 and 0.01). The peaks in this curve identify the transition points in the data and, therefore, represent the predicted depth location of the lithological boundaries. Figure 6c demonstrates the results of the manual logging by domain experts; logging codes are described in Table 1. Figure 6d shows the results of the unsupervised k-means clustering technique. Figures 7 and 8 show the same information as Figure 5, but for the other two drill holes, NGO_PB4 and BOULINDA.

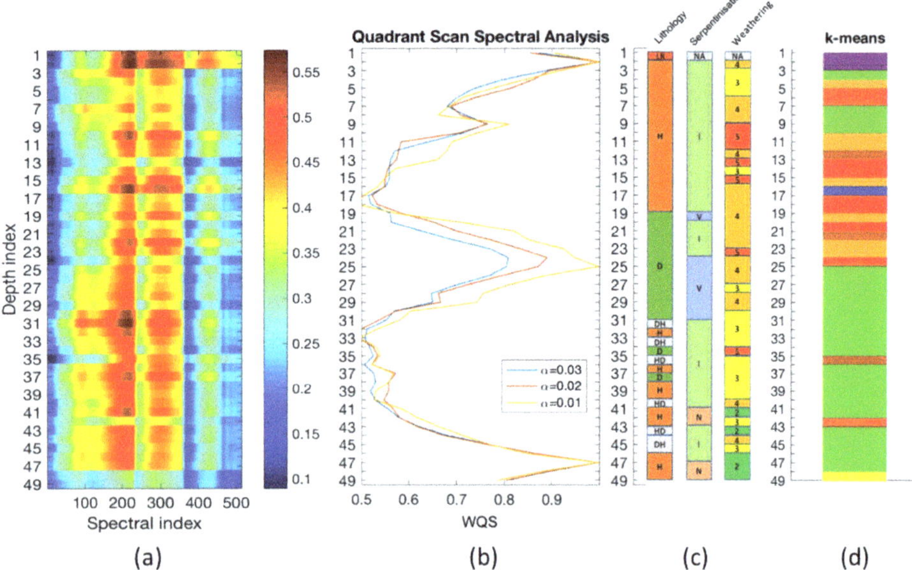

Figure 6. OUACO drill-hole. (**a**) Heatmap plot of depth index against spectral index (blue colours reflect low values and red colours reflect high values); (**b**) Weighted quadrant scan results from the spectral data using three different values of α, peaks indicate lithological boundaries; (**c**) The boundaries identified manually by domain experts (see Table 1 for geology logging codes), and; (**d**) Results of k-means clustering with k = 8 (the different colours represent different clusters).

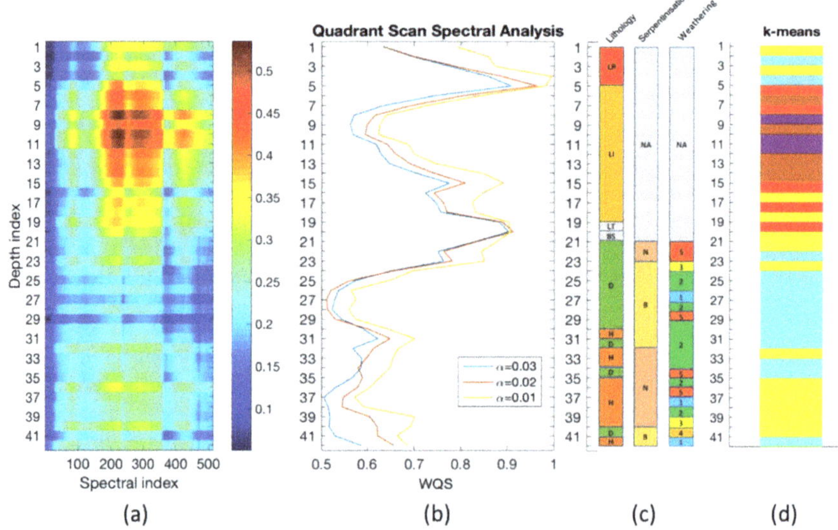

Figure 7. NGO_PB4 drill-hole. (**a**) Heatmap plot of depth index against spectral index (blue colours reflect low values and red colours reflect high values); (**b**) Weighted quadrant scan results from the spectral data using three different values of α, peaks indicate lithological boundaries; (**c**) The boundaries identified manually by domain experts (see Table 1 for geology logging codes), and; (**d**) Results of k-means clustering with k = 8 (the different colours represent different clusters).

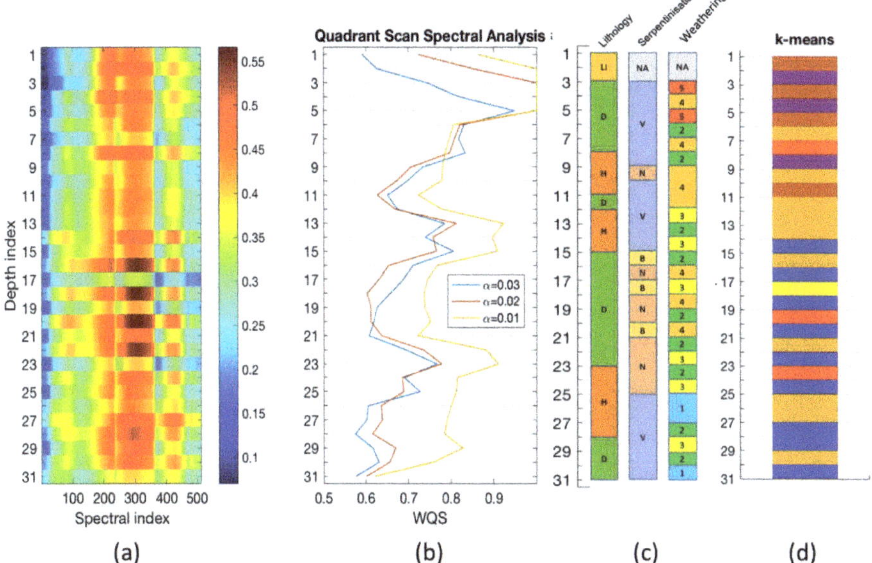

Figure 8. BOULINDA drill-hole. (**a**) Heatmap plot of depth index against spectral index (blue colours reflect low values and red colours reflect high values); (**b**) Weighted quadrant scan results from the spectral data using three different values of α, peaks indicate lithological boundaries; (**c**) The boundaries identified manually by domain experts (see Table 1 for geology logging codes), and; (**d**) results of k-means clustering with k = 8 (the different colours represent different clusters).

The depth sampling intervals are irregular; therefore, the figures in this manuscript show the results in terms of depth indices. The corresponding actual depths are shown for each hole in Table 2.

Table 2. The depth indices and corresponding actual depths in metres for each hole.

DEPTH INDEX	1	2	3	4	5	6	7	8	9	10	11	12	13	14	15	16	17	18	19	20	21	22	23	24	25
OUACO	1	2	3	4	5	5.5	6	7	7.3	8	9	10	10.6	11.5	12	12.5	13	14	15	16	17	18	19	20	21
NGO_PB4	0.3	1	2	3	4	5	6	7	8	9	10	11	12	13	14	15	16	17	18	19	19.2	20	21	21.5	22
BOULINDA	1	1.3	2	3	4	5	5.2	6	6.75	7	8	8.3	8.6	8.8	9.1	9.4	9.6	10.2	10.8	11.3	12	12.4	13.1	13.3	13.8

DEPTH INDEX	26	27	28	29	30	31	32	33	34	35	36	37	38	39	40	41	42	43	44	45	46	47	48	49	50
OUACO	22	22.65	23	24	25	26	27	28	29	30	31	32	33	34	35	36	37	38	38.6	39.6	40.3	41	41.6	42	_
NGO_PB4	22.5	23	23.75	24.75	25.3	26	27	28	28.5	29	30	30.3	32	32	32.5	32.9	33.5	_	_	_	_	_	_	_	_
BOULINDA	14	15	15.2	15.5	16	17	_	_	_	_	_	_	_	_	_	_	_	_	_	_	_	_	_	_	_

5. Discussion

5.1. Comparison with Geologist Logging

Geological logs are plotted alongside the spectral analysis in Figures 6–8 (panels c). There are three types of geological parameters in the logs: lithology, serpentinisation and weathering. Logging codes are decoded in Table 1 and, for example, a sample at depth index 4 will be coded as HI3 in Ouaco (Figure 6), whereas a sample at depth index 27 in NGO-PB4 will be coded as DB1 (Figure 7) and, lastly, a sample at depth index 26 will be coded HV1 at Boulinda (Figure 8). Caution must be exercised when comparing geological logs to logs derived from sensor data, such as hyperspectral imaging, because the input data to the two methods are not the same. For example, it would be unrealistic to expect hyperspectral data to recognise textural change that is not accompanied by mineralogical change. In addition, hyperspectral data may be able to distinguish changes in mineralogy that show no obvious visual change, and so will not be logged by the geologist.

Under lateritic conditions, the primary ultramafic rocks (dunite and harzburgite) are intensely serpentinised and weathered. The serpentinization (lizardite, nepouite) occurs first, affecting both olivine and pyroxene followed by the occurrence of weathering silicate minerals such as smectite and garnierite. The last and more intense weathering phase produces iron oxides such as goethite and hematite. Therefore, logging of lithology is largely based on the geologist's interpretation of textures, such as grain pseudomorphs, to indicate the primary rock lithology. Hence, there is a low expectation that WQS analysis of the hyperspectral data will indicate primary lithological boundaries within the ultramafic rocks. Additionally, when the rock is intensely weathered (e.g., with degree 5), the goethite/hematite-goethite will dominate, as all other mineralogy have disappeared, and the WQS analysis of the hyperspectral data will distinguish boundaries between the laterite and less weathered rocks such as Ouaco at depth index 2 and 9 (Figure 6), NGO_PB4 at depth index 20 and 23 (Figure 7) and Boulinda at depth index 5 (Figure 8).

WQS also detect boundaries between serpentinisation I and V at depth index 24 as well as serpentinisation I and N at depth index 47 for Ouaco (Figure 6). At NGO_PB4 (Figure 7), significant peaks mark the boundaries between two types of laterite, for example, between LR and LI at depth index 5, between LI and LT at depth index 19, and at depth index 20 between LT and BS. Additionally, boundaries were detected between serpentine N and B (depth index 23) and lithology H and D (depth index 31) and weathering 2 and 3 (depth index 39). At Boulinda, boundaries were identified between dunite (D) and harzburgite (H) and weathering 2 and 4 (depth index 8), between lithology H and D, serpentine V and B and weathering 2 and 3 (depth index 15), between lithology D and H and weathering 2 and 3 (depth index 23) and between weathering 2 and 3 (depth index 29). However, it is difficult to confidently assign any profile peaks to the different ultramafic rocks, and this is likely due to their altered and weathered state. It is often the case that boundaries for alteration and weathering are gradational and the distinction between the different groups (strong to weak) is highly subjective when logged by geologists. It is for these cases that the use of non-subjective methods can be most useful. The WQS technique can be used to identify significant changes in the cores that are difficult to pinpoint using manual identification. A more detailed investigation of the mineralogy and the core would be required to distinguish between these options. The value in the WQS technique is that it highlights these changes in the mineralogy, which might otherwise be missed by the logging geologist. It is also the case that some of these changes are occurring in parts of the hyperspectral bandwidth that do not provide useful information for geological logging. As previously mentioned, the detection of olivine and pyroxene can only be detected if the weathering is not intense, as iron oxides will mask their diagnostic features in the visible and near infrared part of the electromagnetic spectrum.

5.2. Comparison with k-Means Clustering

Panels (b) and (d) of Figures 6–8 compare the boundary identification results derived by the implementation of WQS with the samples classes derived from k-means clustering (for k = 8 clusters as suggested by the BIC and the Elbow method tests in Figure 5), respectively, on the high dimensional spectral data. K-means failed to distinguish important lithological boundaries, instead it provides many lithological boundaries based on changes in sample composition, which may be small or large. K-means attempts to divide a data set into approximately spherical clusters of approximately equal size. If the data structure is not compatible, then the class boundaries may be fairly arbitrary. Therefore, the change in class label may be due to arbitrary subdivision of the data into clusters, and may not signify an important change in composition. Because the WQS method takes the spatial correlation of the data into account, it is able to indicate where the more and less significant changes are using the peak height. Using the WQS method, we can be confident that the large peaks indicate significant change in the rock composition has occurred.

5.3. Comparison with the Spectrally Derived Mineralogy

Using the Corescan expert system, each pixel was mineralogically quantified and an average mineralogy for each sample was calculated. The results of this calculation are shown in Figures 9–11 as abundance percentages of the minerals. More detailed information on the analysis of the spectral mineralogy will be available in a forthcoming paper and, therefore will not be addressed further here. To understand how the boundaries detected by the WQS using the spectral data correspond to mineralogical changes the abundance ratio of the minerals is plotted alongside the spectral analysis in Figures 9–11.

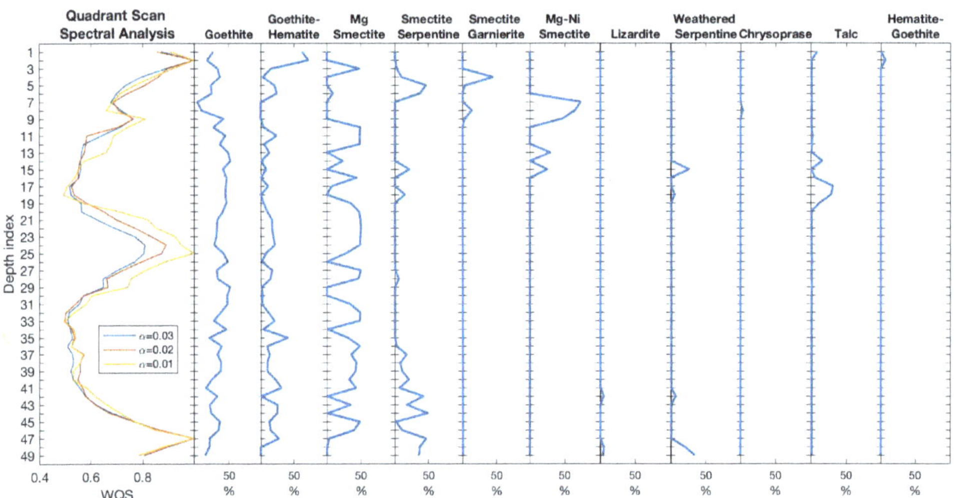

Figure 9. OUACO drill-hole: The WQS analysis in comparison with the abundance percentage of different minerals obtained with Corescan. The left panel is the WQS results from the spectral data using three different values of α, peaks indicate lithological, alteration and weathering boundaries. The rest of the panels show the abundance percentage of different minerals along the drill-hole.

For Ouaco (Figure 9), the WQS peaks indicate that the most significant changes in the mineralogy in the upper part of the profile are mainly depend on goethite, goethite-hematite and all the types of the smectite percentages. Toward the base of the profile, the main mineralogical variations are based on smectite derived from serpentine, lizardite and weathered serpentine percentages. These changes occur, respectively, (a) at depth index 2, with an increase in goethite-hematite (goethite dominant) and hematite-goethite (hematite dominant) percentages; (b) at depth index 9, with an increase in goethite and Mg-smectite percentages and a decrease in Mg-Ni smectite; (c) at depth index 25 with a decrease in Mg smectite percentage; (d) at depth index 29 with a decrease in Mg smectite percentage; (e) at depth index 37 with an increase in smectite serpentine percentage, and (f) at depth index 47 with an increase in smectite serpentine, lizardite and weathered serpentine percentages and a decrease in Mg smectite percentage.

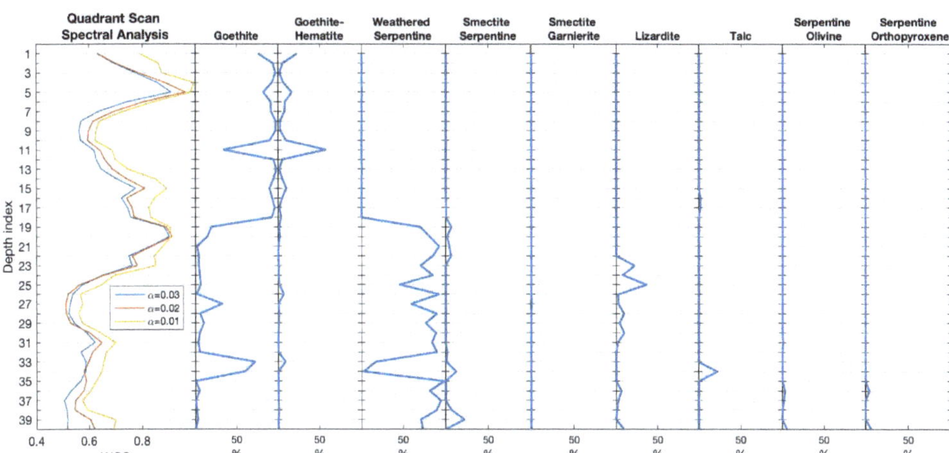

Figure 10. NGO_PB4 drill-hole: The WQS analysis in comparison with the abundance percentage of different minerals obtained with Corescan. The left panel is the WQS results from the spectral data using three different values of α, peaks indicate lithological, alteration and weathering boundaries. The rest of the panels show the abundance percentage of different minerals along the drill-hole.

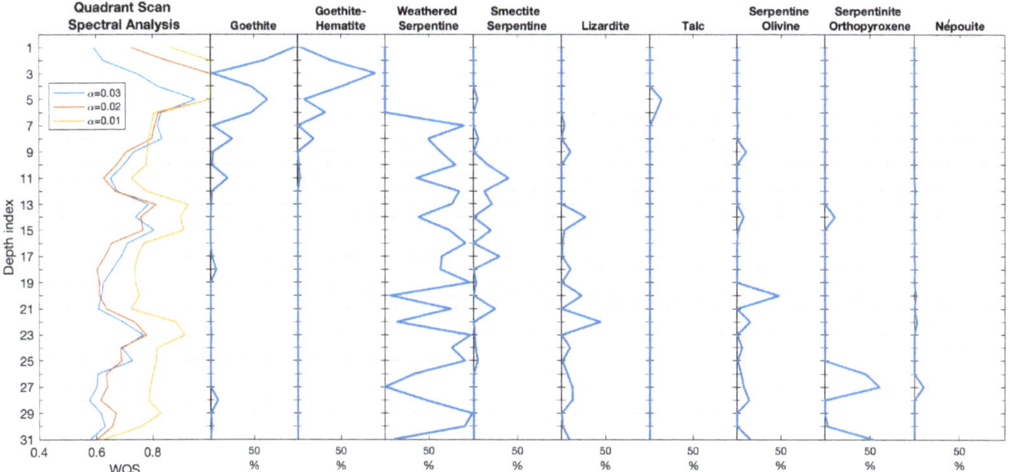

Figure 11. BOULINDA drill-hole: The WQS analysis in comparison with the abundance percentage of different minerals obtained with Corescan. The left panel is the WQS results from the spectral data using three different values of α, peaks indicate lithological, alteration and weathering boundaries. The rest of the panels show the abundance percentage of different minerals along the drill-hole.

For NGO_PB4 (Figure 10), the boundaries at depth index, 5, 11 and 15 correspond to a relative change between goethite and goethite-hematite abundance. The boundaries identified at depth index 20 correspond to a decrease in goethite percentage and an increase in weathered serpentine percentage whereas depth index 23 is linked to an increase in lizardite. The small change at depth index 27 correspond to a small goethite increase, while variation at depth index 31 indicates a large goethite increase with an equivalent weathered serpentine percentage decrease. The last change at depth index 39 is due the presence of a small amount of smectite derived from serpentine.

Similar observations can be made for boundaries detected in the Boulinda drill-hole (Figure 11). The different detection scales (α values) provide complementary results. The boundaries at depth indices at 3, 5, 8, 10, 13, 15, 20, 23, 25, 27, 29 and 30 all correspond to observable changes in mineral abundances for the iron oxides in the upper part of the profile, for weathered serpentine, smectite serpentine, lizardite and népouite, serpentine derived from olivine and serpentine derived from orthopyroxene in the lower part of the profile.

5.4. Precision of Boundary Detection and Dependance on Sample Size

For automated methods, the precision of boundary detection depends on the sampling interval. For boundary detection methods based on the continuous wavelet transform, a consistent size sampling interval is required and, therefore, if the data are provided at varying scales, all data must be re-scaled to the same scale resulting in a loss of information. This is not the case for WQS. For this study, most of the sampling intervals were approximately one metre (Table 2). However, for a large section of the BOULINDA hole, the sampling interval was much smaller. Using WQS automatically allows a more precise identification of boundary location in this section of the drill hole.

The user needs to take the sampling interval into account when interpreting the results, especially in the case where the sample size is not consistent down hole, as this affects the precision of the boundary location. Additionally, the spatial correlation depends on the number of samples not the length they occupy in the drill hole.

6. Conclusions

The WQS, based on the recurrence plot, allows identification of boundaries between geological domains from very high-dimensional data. The method has been tested on spectral data from three different drill-holes. The advantages of this algorithm are: (1) it is automated and so provides consistent data analysis and repeatable results; (2) it is computationally efficient and can be calculated quickly using a standard desktop or laptop computer; (3) the algorithm is unsupervised and does not require any geological knowledge; (4) it reduces very high dimensional spectral data ($D = 514$) into a single dimension, i.e., the WQS profile, in which the peaks identify the lithological boundaries; (5) the significance of the boundary can be inferred from the peak height; (6) sample intervals do not have to be of consistent size and localised finer scale sampling produces more precise results. This is beneficial for geologists and geological applications as it provides rapid results and indicates to the geologist the locations of significant changes in the geology based on the data.

Author Contributions: Conceptualization, A.Z., E.R. and J.H.; methodology, A.Z., D.M.W. and M.S.; software, A.Z.; validation, A.Z., E.R., J.H., D.M.W. and M.S.; formal analysis, A.Z.; investigation, A.Z., E.R., J.H., D.M.W. and M.S.; resources, E.R. and J.H.; data curation, E.R.; writing—original draft preparation, A.Z., E.R. and J.H.; writing—review and editing, A.Z., E.R., J.H., D.M.W. and M.S. All authors have read and agreed to the published version of the manuscript.

Funding: A.Z. is supported by the Australian Research Council through the Centre for Transforming Maintenance through Data Science (grant number IC180100030), funded by the Australian Government. A.Z. and M.S. acknowledge the support of the Australian Research Council through the Discovery Grant DP 180100718. D.M.W. and M.S. are partially supported by the Australian Research Council (DP200102961).

Conflicts of Interest: The authors declare no conflict of interest.

Computer Code and Software

The MATLAB code for the WQS method is available in the following public GitHub repository https://github.com/AyhamZaitouny/Boundaries-Detection-Weight-Quadrant-Scan- (accessed on 7 October 2021). For more information, contact the corresponding author.

References

1. Singh, H.; Seol, Y.; Myshakin, E.M. Automated Well-Log Processing and Lithology Classification by Identifying Optimal Features Through Unsupervised and Supervised Machine-Learning Algorithms. *SPE J.* **2020**, *25*, 2778–2800. [CrossRef]
2. Hill, E.J.; Pearce, M.A.; Stromberg, J.M. Improving Automated Geological Logging of Drill Holes by Incorporating Multiscale Spatial Methods. *Math. Geol.* **2021**, *53*, 21–53. [CrossRef]
3. Hill, E.; Robertson, J.; Uvarova, Y. Multiscale hierarchical domaining and compression of drill hole data. *Comput. Geosci.* **2015**, *79*, 47–57. [CrossRef]
4. Marr, D.; Hildreth, E. Theory of edge detection. *Proc. R. Soc. Lond. Ser. B Biol. Sci.* **1980**, *207*, 187–217.
5. Canny, J. A computational approach to edge detection. *IEEE Trans. Pattern Anal. Mach. Intell.* **1986**, *6*, 679–698. [CrossRef]
6. Mallat, S. Zero-crossings of a wavelet transform. *IEEE Trans. Inf. Theory* **1991**, *37*, 1019–1033. [CrossRef]
7. Perez-Muñoz, T.; Velasco-Hernandez, J.; Hernandez-Martinez, E. Wavelet transform analysis for lithological characteristics identification in siliciclastic oil fields. *J. Appl. Geophys.* **2013**, *98*, 298–308. [CrossRef]
8. Cooper, G.R.J.; Cowan, D.R. Blocking geophysical borehole log data using the continuous wavelet transform. *Explor. Geophys.* **2009**, *40*, 233–236. [CrossRef]
9. Davis, A.C.; Christensen, N.B. Derivative analysis for layer selection of geophysical borehole logs. *Comput. Geosci.* **2013**, *60*, 34–40. [CrossRef]
10. Arabjamaloei, R.; Edalatkha, S.; Jamshidi, E.; Nabaei, M.; Beidokhti, M.; Azad, M. Exact lithologic boundary detection based on wavelet transform analysis and real-time investigation of facies discontinuities using drilling data. *Pet. Sci. Technol.* **2011**, *29*, 569–578. [CrossRef]
11. Walker, D.M.; Zaitouny, A.; Corrêa, D.C. On using the modularity of recurrence network communities to detect change-point behaviour. *Expert Syst. Appl.* **2021**, *176*, 114837. [CrossRef]
12. Goswami, B. A Brief Introduction to Nonlinear Time Series Analysis and Recurrence Plots. *Vibration* **2019**, *2*, 332–368. [CrossRef]
13. Goswami, B.; Boers, N.; Rheinwalt, A.; Marwan, N.; Heitzig, J.; Breitenbach, S.F.M.; Kurths, J. Abrupt transitions in time series with uncertainties. *Nat. Commun.* **2018**, *9*, 48. [CrossRef]
14. Aminikhanghahi, S.; Cook, D.J. A survey of methods for time series change point detection. *Knowl. Inf. Syst.* **2017**, *51*, 339–367. [CrossRef]
15. Marwan, N.; Carmenromano, M.; Thiel, M.; Kurths, J. Recurrence plots for the analysis of complex systems. *Phys. Rep.* **2007**, *438*, 237–329. [CrossRef]
16. Rapp, P.E.; Darmon, D.M.; Cellucci, C.J. Hierarchical Transition Chronometries in the Human Central Nervous System. *IEICE Proc. Ser.* **2013**, *2*, 286–289. [CrossRef]
17. Walker, D.M.; Tordesillas, A.; Ren, J.; Dijksman, J.A.; Behringer, R.P. Uncovering temporal transitions and self-organization during slow aging of dense granular media in the absence of shear bands. *EPL Europhys. Lett.* **2014**, *107*, 18005. [CrossRef]
18. Zaitouny, A.; Walker, D.M.; Small, M. Quadrant scan for multi-scale transition detection. *Chaos Interdiscip. J. Nonlinear Sci.* **2019**, *29*, 103117. [CrossRef]
19. Zaitouny, A.; Small, M.; Hill, J.; Emelyanova, I.; Ben Clennell, M. Fast automatic detection of geological boundaries from multivariate log data using recurrence. *Comput. Geosci.* **2019**, *135*, 104362. [CrossRef]
20. Wells, M.A.; Ramanaidou, E.R.; Verrall, M.; Tessarolo, C. Mineralogy and crystal chemistry of "garnierites" in the Goro lateritic nickel deposit, New Caledonia. *Eur. J. Mineral.* **2009**, *21*, 467–483. [CrossRef]
21. Ramanaidou, E.; Fonteneau, L.; Sevin, B.; Foucher, W. Characterisation of New Caledonian nickel laterite using hyperspectral imaging. In Proceedings of the Australia Geological Council Convention, Adelaide, Australia, 14–18 October 2018.
22. Cracknell, M.J.; Jansen, N.H. National Virtual Core Library HyLogging data and Ni-Co laterites: A mineralogical model for resource exploration, extraction and remediation. *Aust. J. Earth Sci.* **2016**, *63*, 1053–1067. [CrossRef]
23. Cardoso-Fernandes, J.; Silva, J.; Perrotta, M.M.; Lima, A.; Teodoro, A.C.; Ribeiro, M.A.; Dias, F.; Barrès, O.; Cauzid, J.; Roda-Robles, E. Interpretation of the Reflectance Spectra of Lithium (Li) Minerals and Pegmatites: A Case Study for Mineralogical and Lithological Identification in the Fregeneda-Almendra Area. *Remote Sens.* **2021**, *13*, 3688. [CrossRef]
24. USGS. Science for a Changing World. National Minerals Information Center. 2021. Available online: https://www.usgs.gov/centers/national-minerals-information-center/nickel-statistics-and-information (accessed on 7 October 2021).
25. Butt, C.R.; Cluzel, D. Nickel laterite ore deposits: Weathered serpentinites. *Elements* **2013**, *9*, 123–128. [CrossRef]
26. Elwakil, A.; Soliman, A. Mathematical Models of the Twin-T, Wien-bridgeand Family of Minimum Component Electronic ChaosGenerators with Demonstrative Recurrence Plots. *Chaos Solitons Fractals* **1999**, *10*, 1399–1412. [CrossRef]
27. Vretenar, D.; Paar, N.; Ring, P.; Lalazissis, G.A. Nonlinear dynamics of giant resonances in atomic nuclei. *Phys. Rev. E* **1999**, *60*, 308–319. [CrossRef] [PubMed]
28. Rustici, M.; Caravati, C.; Petretto, E.; Branca, M.; Marchettini, N. Transition Scenarios during the Evolution of the Belousov−Zhabotinsky Reaction in an Unstirred Batch Reactor. *J. Phys. Chem. A* **1999**, *103*, 6564–6570. [CrossRef]

29. Gilmore, C.G. Detecting Linear and Nonlinear Dependence in Stock Returns: New Methods Derived from Chaos Theory. *J. Bus. Financ. Account.* **1996**, *23*, 1357–1377. [CrossRef]
30. Marwan, N.; Thiel, M.; Nowaczyk, N.R. Cross recurrence plot based synchronization of time series. *Nonlinear Process. Geophys.* **2002**, *9*, 325–331. [CrossRef]
31. Marwan, N. A historical review of recurrence plots. *Eur. Phys. J. Spec. Top.* **2008**, *164*, 3–12. [CrossRef]
32. De Maesschalck, R.; Jouan-Rimbaud, D.; Massart, D.L. The mahalanobis distance. *Chemom. Intell. Lab. Syst.* **2000**, *50*, 1–18. [CrossRef]
33. Jain, A.K. Data clustering: 50 years beyond K-means. *Pattern Recognit. Lett.* **2010**, *31*, 651–666. [CrossRef]
34. Schwarz, G. Estimating the Dimension of a Model. *Ann. Stat.* **1978**, *6*, 461–464. [CrossRef]
35. Ketchen, D.J.; Shook, C.L. The application of cluster analysis in strategic management research: An analysis and critique. *Strateg. Manag. J.* **1996**, *17*, 441–458. [CrossRef]

Article

Pseudo-Karst Silicification Related to Late Ni Reworking in New Caledonia

Michel Cathelineau [1,*], Marie-Christine Boiron [1], Jean-Louis Grimaud [2], Sylvain Favier [1], Yoram Teitler [1] and Fabrice Golfier [1]

1 Université de Lorraine, CNRS, GeoRessources, F-54000 Nancy, France
2 PSL University/MINES Paris/Centre de Géosciences, 35 rue St Honoré, 77305 Fontainebleau Cedex, France
* Correspondence: michel.cathelineau@univ-lorraine.fr

Abstract: Silicification in New Caledonian pseudo-karsts developed on peridotite was assessed using $\delta^{18}O$ and $\delta^{30}Si$ pairs on quartz cements. The objective was to document the chronology of pseudo-karst development and cementation relative to geomorphic evolution. The latter began at the end of the Eocene with the supergene alteration of peridotites and the subsequent formation of extended lateritic weathering profiles. Neogene uplift favoured the dismantling of these early lateritic profiles and valley deepening. The river incision resulted in (i) the stepping of a series of lateritic paleo-landforms and (ii) the development of a pseudo-karst system with subvertical dissolution pipes preferentially along pre-existing serpentine faults. The local collapse of the pipes formed breccias, which were then cemented by white quartz and Ni-rich talc-like (pimelite). The $\delta^{30}Si$ of quartz, ranging between $-5‰$ and $-7‰$, are typical of silcretes and close to the minimum values recorded worldwide. The estimated $\delta^{18}O$ of -6 to $-12‰$ for the fluids are lower than those of tropical rainfall typical of present-day and Eocene–Oligocene climates. Evaporation during drier climatic episodes is the main driving force for quartz and pimelite precipitation. The silicification presents similarities with silcretes from Australia, which are considered predominantly middle Miocene in age.

Keywords: silcrete; pseudo-karst; laterite; nickel; oxygen and silicon isotopes; weathering

Citation: Cathelineau, M.; Boiron, M.-C.; Grimaud, J.-L.; Favier, S.; Teitler, Y.; Golfier, F. Pseudo-Karst Silicification Related to Late Ni Reworking in New Caledonia. *Minerals* **2023**, *13*, 518. https://doi.org/10.3390/min13040518

Academic Editors: Cristina Domènech and Cristina Villanova-de-Benavent

Received: 28 January 2023
Revised: 2 April 2023
Accepted: 4 April 2023
Published: 6 April 2023

1. Introduction

Karst is a geomorphological structure resulting from the hydro-chemical and hydraulic erosion of all soluble rocks. The main karsts are those developed in limestones, which are particularly soluble when subjected to dissolved CO_2-rich water. Other rock types can be affected by dissolution leading to geomorphological structures similar to those developed in carbonate rocks; in particular, other sedimentary formations such as evaporites (gypsum or salt layers).

A lesser-known type is dissolution conduits that form in ultramafic rocks such as peridotites, which are soluble when subjected to a warm, rainy climate for a prolonged time. They have been described as "karsts" by several authors, including [1–3] for New Caledonia, but also in Greece [4,5], Cuba [6], Papua New Guinea and Indonesia [7,8] or "pseudo-karsts" [9,10]. In this paper, the term "pseudo-karst" will be used.

When peridotites are submitted to a hot and humid climate, the complete olivine dissolution yields a residual soil, the laterite. The latter is formed essentially of goethite and hematite due to the very low solubility of Fe^{3+} iron oxides and hydroxides under oxidising conditions [11–14]. In the lower part of the lateritic profile, Mg–Ni-silicates form at the boundary with the bedrock in the so-called saprolite horizon [12,14]. The homogeneous lowering of the weathering front and the generation of a weathering horizon of several tens of meters thick implies a duration of at least 10 Ma [15,16].

In New Caledonia, supergene alteration of peridotites started shortly after ophiolite obduction, which occurred between Late Eocene and Early Oligocene [17,18]. This led to vast expanses of residual laterite soils. Paleomagnetic ages on ferruginous duricrusts suggest that the paleo-weathering is at least as old as 25 Ma [19]. The uplift phases that followed the development of lateritic soils carried the laterites to higher elevations, which triggered erosion after the Oligocene [20]. The relief erosion led to significant loss of the ophiolite, more than two-thirds of the ophiolite surface, and the deepening of the valleys.

Geomorphological changes linked to periods of uplift and erosion can profoundly affect the hydrological regimes and modify the geometry of dissolution patterns. The formation of the pseudo-karst systems was favoured by denudation and valley incision [3,9,21]. This consisted of the complete olivine dissolution along connected fractures and faults, yielding to forming dissolution pipes [9], similar to those defined in other lithological contexts [22]. Locally, the gravity collapse of the pipes caused the formation of dolines at the surface [3,9] and breccia pipes, which were partially cemented locally by white quartz and then by Ni-silicates of a dark green colour. The breccia cement appears, therefore, to be an essential witness of syn- to post-uplift phases, and its analysis may provide keys to the genesis of the preferential Ni enrichments.

Several types of silicification, sometimes so-called silcretes by previous works [2,20], were already identified and linked to the landscape's evolution or the well-known chronology of Ni–ore formation [23–25]. Despite their common occurrence, the mechanisms and timing of silicification in surficial processes remain challenging to constrain [26–28].

Several silicified pipes have been observed in the Koniambo massif, New Caledonia (Figure 1a), mainly from fresh cuts made during open-pit mining operations and digging access tracks to the top of the plateau. These cuts provided a good vertical cross section from bedrock to saprock, with the opportunity to observe and sample breccias preserved from more recent supergene alteration and oxidation. This study aimed to combine the petrological information on the chronology of mineral assemblages with the geomorphic evolution. The conditions of silicification and precipitation of Ni-silicates were derived from a textural study by SEM and cathodoluminescence on breccia quartz cement, combined with a geochemical characterisation (trace elements, $\delta\,^{30}$Si, $\delta\,^{18}$O). Field observations were used to constrain the chronology of erosion surfaces with valley incision and to interpret the pseudo-karst system and its cementation in the geomorphic evolution. The resulting conceptual model aimed to unravel the relative timing and geometry of Ni transfer episodes, which is still an open question for future mining prospection.

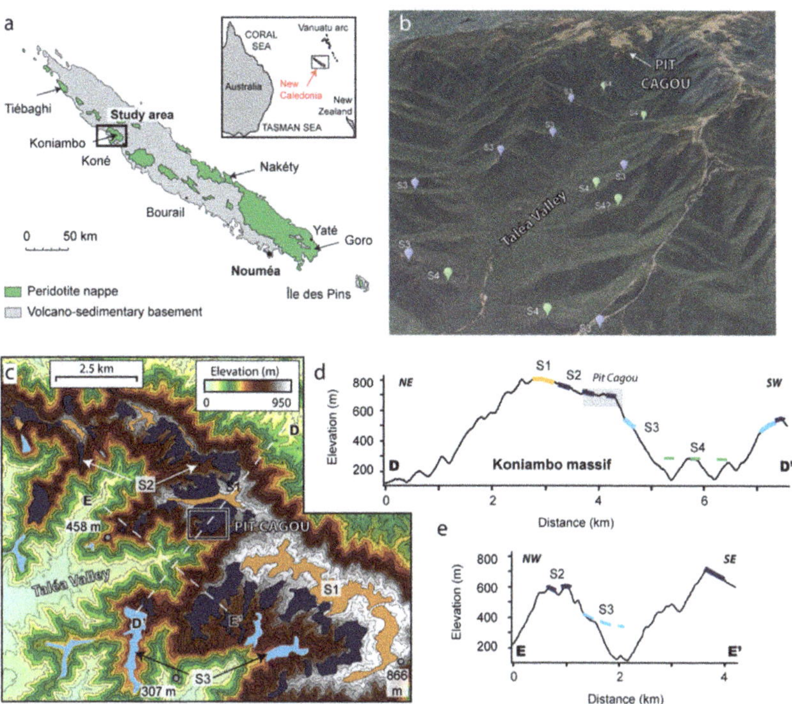

Figure 1. (**a**) Location of New Caledonia and Koniambo. (**b**) Aerial view with the indication of paleosurfaces (S1 to S4). (**c**) Location of the study site (open pit Cagou from KNS mining company) in the Koniambo massif. (**d**,**e**) Cross sections are in (**c**) (D-D$'$ and E-E$'$), with the indication of S1, S2, S3 paleosurface (in yellow, dark blue and light blue respectively) and the potential location of S4 (green dashed line).

2. Geological Setting of the Koniambo Massif

In the Koniambo massif, the Peridotite Nappe is exposed from its base, near sea level, up to ~800 m elevation (Figure 1b,c). The massif essentially consists of harzburgites with interlayers of dunites. In the higher part of the massif, peridotites are moderately serpentinised, contrary to the sole of the nappe, which is highly deformed and serpentinised [29]. At the Koniambo site, numerous serpentine fractures and faults related to syn- to slightly post-obduction stages were reactivated during compressional and extensional phases [29]. These same fractures play an essential role in the drainage of meteoric fluids responsible for the mineralisation of nickel or the current degradation of the old alteration profiles and the formation of the still-active pseudo-karst system [30,31].

Above ~400–600 m in the massif, the nappe is capped by a highly dissected and partly reworked lateritic profile [32]. This is composed of two lateritic remnant surfaces (referred to as S1 and S2), similar to the Tiébaghi plateau further north, as already suggested by Chevillotte (2005) [33].

A succession of ores has been described by Cathelineau et al [23,24] at the Koniambo mine site: (i) Type 1 ore [24] consists of crack-sealed veins with a succession of fillings comprising Mg–Ni talc, followed by red microcrystalline quartz full of microinclusions of iron oxide. These fillings are related to tectonic events that allowed low temperature (50–70 °C) reduced fluids to circulate and mix with oxidising waters, probably at a greater depth than today, (ii) Type 2 ore [23] formed later, closer to the surface, by evaporation in joints developed in metre-sized boulders and consists of films with concentric zones of pimelite (Ni-rich talc) at the periphery and Mg-talc-like in the centre. The ores mined today (ore Type 3) result from the deepening of the lateritic dissolution front, which affects all previous mineral assemblages and redistributes the nickel into a complex fine-grained talc, nontronite and goethite mineral assemblage. This saprolitic horizon forms a metric to decametric fine-grained layer enclosing preserved boulders and lies between the bedrock and the yellow lateritic horizon where goethite predominates.

3. Materials and Methods

Most observations and sampling were made in the Koniambo massif, located along the coast in the northwest part of New Caledonia, during several field campaigns from 2011 to 2021. Samples were taken in the Cagou pit, a location studied in detail for ore Types 1 and 2 and their structural context [29].

3.1. Petrographic, Cathodoluminescence and Micro-XRF Images

A Scanning Electron Microscopy (SEM) JEOL J7600F field-effect coupled with SDD type electron dispersive spectrometer, wave wavelength dispersive spectrometer (Oxford Instruments, Abingdon-on-Thames, Oxfordshire, UK) and cathodoluminescence (CITL Cold Cathodoluminescence device Model MK5-1) were used to document mineral assemblages.

Micro-XRF mapping was carried out using the Bruker-Nano M4 Tornado instrument. This system has an Rh X-ray tube with a Be side window and polycapillary optics giving an X-ray beam with a diameter of 25–30 µm on the sample. The X-ray tube was operated at 50 kV and 200 µA and a 2 kPa vacuum. X-rays were detected by a 30 mm^2 xflash®SDD with an energy resolution of <135 eV at 250,000 cps. Main elements such as Mg, Mn, Fe, Ni, Co, Cr, and Si were mapped and composite chemical images were generated.

High-Resolution Transmission Electron Microscopy (HRTEM), energy dispersive spectra and electron diffraction patterns were performed on representative samples of pimelite to observe, at a nano-metric scale, the Ni-talc-like texture and particles and obtain an elemental composition. A CM20-Philips instrument with a Si-Li detector operating at 200 kV and 10 eV was used at SCMEM (GeoRessources, Vandœuvre-lès-Nancy, France).

3.2. LA-ICPMS Analyses of Trace Elements in Quartz

Trace element abundances in quartz were analysed by a laser ablation inductively-coupled plasma mass spectrometry (LA-ICPMS) at GeoRessources laboratory, University of Lorraine (Vandœuvre-lès-Nancy, France). Analyses were performed using an Agilent 7500c quadrupole spectrometer interfaced with a GeoLas Pro 193 nm ArF excimer laser ablation system (Lambda Physik, Göttingen, Germany). Operating conditions are a 5 Hz repetition rate, a ~10 J/cm^2 fluence, and a beam size ranging from 32 to 60 µm. Helium was used as the carrier gas (0.8 L/min) and mixed with Ar before introducing it in the plasma (1.5 L/min). Analysed elements were the following: ^{27}Al, ^{29}Si, ^{45}Sc, ^{49}Ti, ^{51}V, ^{53}Cr, ^{55}Mn, ^{57}Fe, ^{59}Co, ^{60}Ni, ^{63}Cu, ^{66}Zn, ^{69}Ga, ^{74}Ge. Acquisition times for background and ablation signals were 40 and 50 s, respectively, allowing the measurement of duplicate spots per analysis. The NIST 610 reference material was used for external and ^{29}Si for internal standardisation. The NIST SRM 612 glass was employed as a secondary standard and yielded trace element abundances in agreement with the reference values. Data reduction was carried out following the standard methods from Longerich et al. (1996) [34]. The accuracy was around 10% depending on the element.

3.3. Oxygen and Silicon Isotopes

The oxygen and silicon isotopic composition of the quartz samples was determined using the CAMECA IMS 1270 ion microprobe at CRPG (Nancy, France) using classical procedures previously described by Rollion-Bard et al. (2007), Robert and Chaussidon (2006) and Marin et al. (2010) [35–37]. Oxygen and silicon isotopic compositions are reported here as per mil deviations from SMOW standard (Standard Mean Ocean Water) and NBS 28, respectively, using the conventional notation:

$$\delta^{18}O = [(^{18}O/^{16}Osample)/(^{18}O/^{16}O\ SMOW) - 1] \times 1000.$$

$$\delta^{30}Si = [(^{30}Si/^{28}Sisample)/(^{30}Si/^{28}Si\ NBS28) - 1] \times 1000.$$

The sample was analysed with a primary ion beam diameter of 20 μm on standard polished sections coated with gold for oxygen and silicon isotope measurements. The sample was sputtered with a 20 to 30 μm diameter Cs^+ primary beam of ~8–10 nA for oxygen measurements and ~20–25 nA for silicon measurements and 10 kV acceleration voltage. Secondary ions were accelerated at 10 kV and detected in multicollection mode. The mass resolving power was set at ≈4000, and the H_2O- interference on ^{18}O-being resolved around 1600. The total analytical time was 4 min, including a presputtering (60 sec), the centring of the magnetic field and analyses (500 sec) for oxygen and silicon measurements. The external reproducibility of the $\delta^{18}O$ measurement was determined using the quartz so-called "Bresil" ($\delta^{18}O$ SMOW= 9.6‰) and was ± 0.20‰ (2σ). The external reproducibility of $\delta^{30}Si$ measurements, determined using the quartz standards NBS28 ($\delta^{30}SiNBS28= -0.16 \pm 0.18$), was ± 0.3‰ (2σ).

3.4. Geochemical Modelling

The simulations were run on phreeqC code [38,39], with the llnl.dat database edited to account for the Ni-phyllosilicates [16], using data on rainwater from [2]. The Geochemists' Workbench software [40] made the activity diagrams using the thermodynamic data from the edited llnl.dat database.

4. Results

4.1. Field Observations

Field observations and considering available topographic maps allowed us to determine the relationships among paleosurfaces, S1 to S4, along two perpendicular profiles noted D-D′ and E-E′. S1 and S2 paleosurfaces form two distinctive levels on the plateau, observing the D-D′ topographic profile (Figure 1d). The S1 surface appears as a gently rolling convex hill, while the S2 surface is concave up. The plateau is gently dipping towards the southwest. S1 and S2 are thus in topographic continuity, suggesting a connection between their weathering profiles. Surface 3 (S3) is developed on the southwest foothills of the massif, particularly on the gently westerly dipping low-elevation (120–240 m) Kaféaté plateau along the coast. This observation confirms preliminary conclusions from [41,42]. In the Kaféaté plateau, new field observations indicate a slope break associated with a slight slope change with evidence of reworking of the S3 ferricrete, which is found in the form of boulders in the S4 ferricrete. Hence, this indicates that the S4 surface incised the S3 surface in this area. In the Koniambo massif, relics of the S3 are found at higher elevations (300–600 m), marking the headwaters of erosional glacis developed during the formation of S3 (Figure 1d,e). The silicified rocks and the white quartz from pseudo-karst have similar features to quartz from silcretes on a textural and isotopic basis, as shown later. The term "silcrete" was mentioned by Chardon and Chevillotte (2006) and Chevillotte et al. (2006) [20,42] as a typical weathering product associated with the S4 surface based on the Nepoui cross section on a textural basis. The silcretes are described as silicified conglomerates by Chevillotte et al. (2006) [42] However, it is rather hard to determine if they are similar to the silcrete described in this study, as petrographic descriptions, particularly of the cement, are lacking in their work.

Topographic analysis of the Taléa valley to the southwest of Pit Cagou shows relics of relatively flat surfaces of reduced extent (Figure 1e). Field reconnaissance allowed identifying partly eroded weathering profiles associated with these remnants at lower elevations than S1 and S2 surfaces (Figure 1).

4.2. Dissolution Pipes

Near the top of the plateau, in several places, large cavities and open channels are observed in the peridotite. They are best revealed along the main access road to the high plateau (Figure 2). They are located along fractures, mostly filled with serpentine, and follow the inherited fracture network. They are several decimetres to a few metres in aperture and extend over tens of meters in their outcropping part. The bulk vertical extension is inferred from a hydrological study and is about several hundred meters in Koniambo [43]. Only iron hydroxides (goethite) are observed along fracture walls. The protolith is similar on both edges of the fractures and dissolution features are identical to the boulders from the bedrock-saprolite interface.

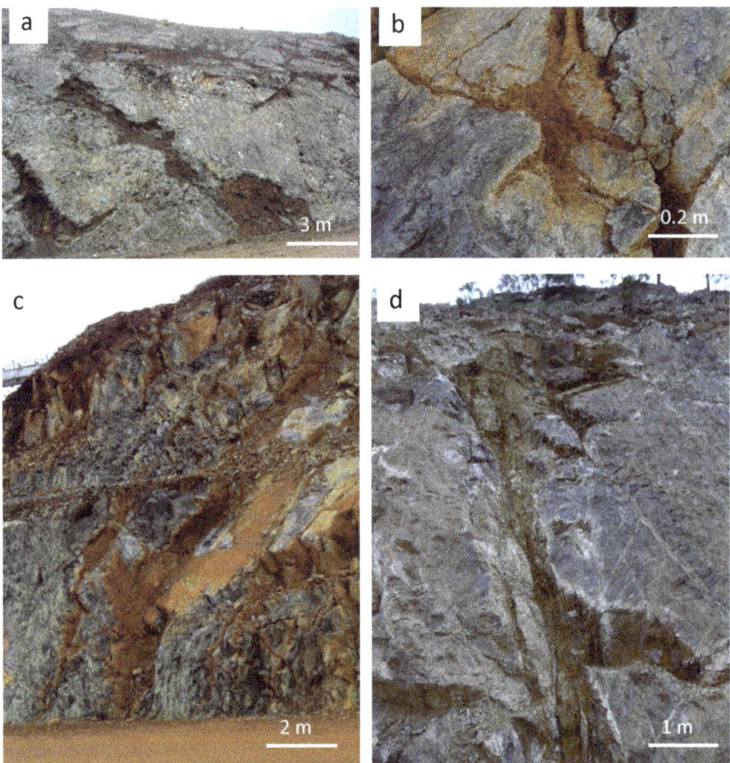

Figure 2. Dissolution pipes in the peridotite (main mining track at Koniambo). (**a**) Network of dissolution pipes near the high of the plateau. (**b**) Detail of the dissolution, which leaves only iron hydroxides as a coating on the edges of the pipe. (**c**) Vertical extension of the dissolution cavities (main track). (**d**) Dissolution along serpentinised fractures.

In the Cagou Pit, several dissolution pipes suffered complete collapse (Figure 3a,b). The cavities develop on the network of faults, the sides of which are usually serpentinised. They are filled with blocks of varied sizes. Blocks include weathered and silicified host rocks, serpentinised wall rocks, and fragments of kerolite crack-seals (Type I ores [24] which occur close to the serpentine fault. The breccias are cemented with white quartz and green pimelite (Ni-rich talc-like) (Figure 3c–e).

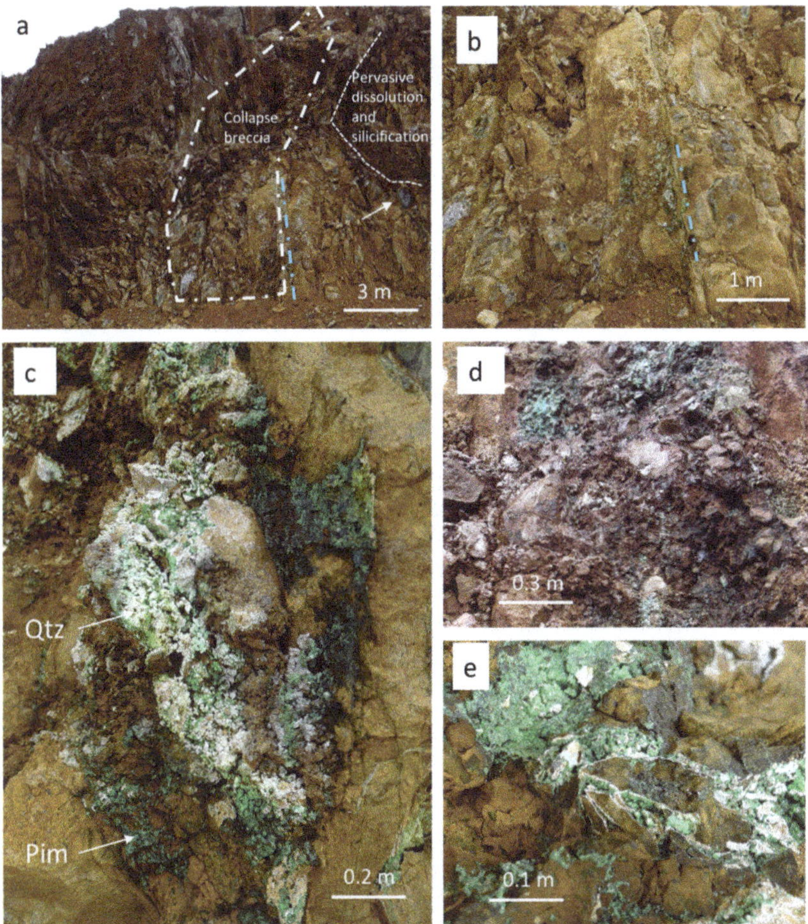

Figure 3. Collapse breccia (indicated by a white dashed line) in the open pit Cagou (former "207") from the Koniambo mining site. (**a,b**) Photographs were taken in the upper part of the open pit showing the penetration of the dissolution pipes, which are accompanied by pervasive silicification (dashed line) and quartz fracture infillings. The white arrow indicates a target like ore type II in (**a**). In (**a,b**), the blue dashed lines indicate fractures filled with blueish kerolite and red quartz (ore Type I), as described in [24]. (**c**) Cemented breccia by Ni-silicate (pimelite) and quartz. (**d,e**) Details of the cement. The arrows indicate the location of target-like ores (without quartz), as described in [23]. Qtz: quartz, Pim: pimelite.

4.3. Petrography and Mineralogy of Breccias

Collapse breccias constituted of blocks, including silicified host rocks and clasts of ore Type I, are cemented by quartz and pimelite (Figures 4 and 5).

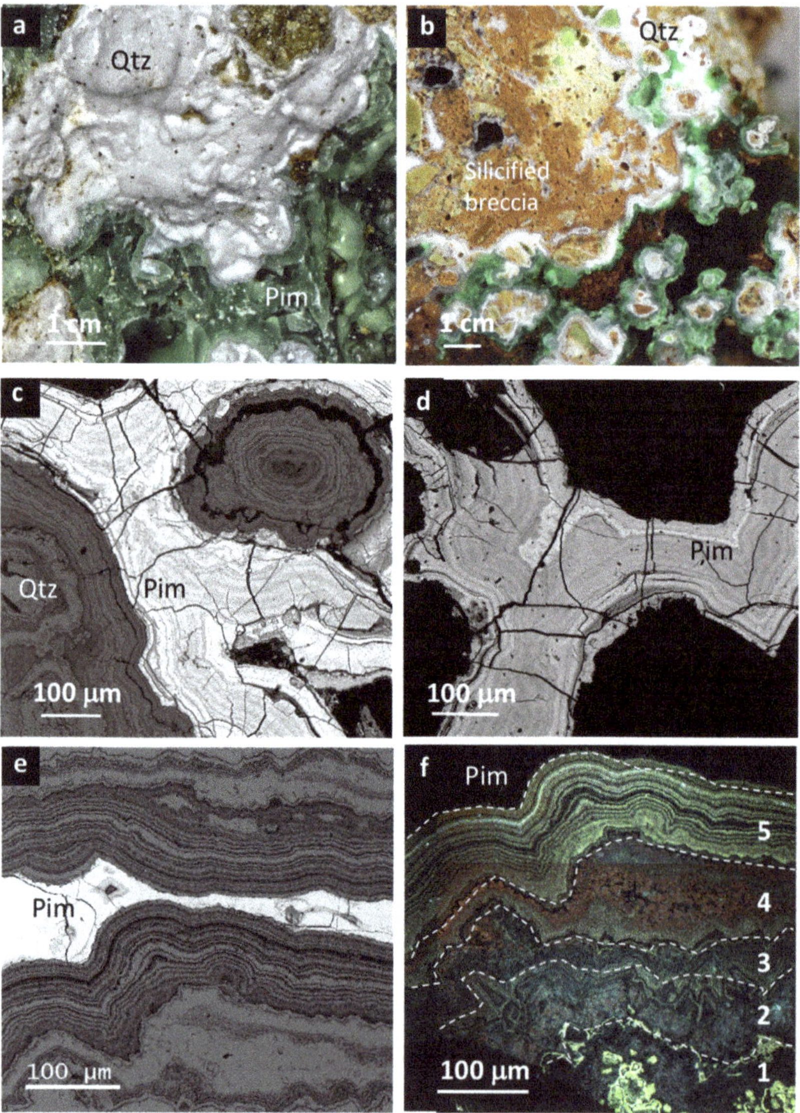

Figure 4. (**a**) Ni-silicate (pimelite, noted Pim) crystallised on white quartz (Qtz). (**b**) Pimelite on white quartz cementing silicified clasts. (**c**) BSE image showing the rims of quartz flowed by pimelite. (**d**) Detail of the chemical zoning in the pimelite colloform textures. (**e**) White quartz observed by BSE, showing a decreasing size of crystals; (**f**) Cathodoluminescence image showing white quartz rims from 1 (earliest rim) to 5 (colloform quartz).

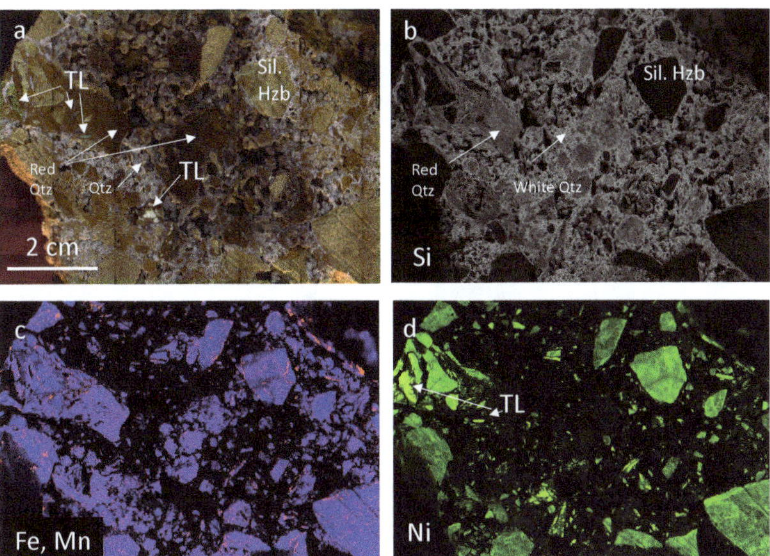

Figure 5. Micro-XRF images: (**a**) Silicified breccia sample from Cagou pit (Koniambo) with clasts of silicified harzburgite (Sil. Hzb), red-brown microcrystalline quartz (Red Qtz) and Ni-rich talc-like ore (TL). (**b**) Si map. (**c**) Superimposed Fe and Mn maps: Mn as Mn oxyhydroxides is located in microfractures and microvugs; Fe is present mainly in silicified harzburgite and red-brown microcrystalline quartz. (**d**) Ni map showing the location of mineralised clasts of ore Type 1 and clast of silicified harzburgite particularly enriched in Ni (TL: Ni-rich talc-like). All chemical images (**b–d**) are at the same scale than the macrophotograph shown in (**a**).

Breccia clasts include predominantly silicified rock fragments (noted silicified clasts) and a few clasts of ore Type I crack-seal infillings. The latter are brownish iron-rich microcrystalline quartz with relics of early blueish-green talc-like (Figure 4a,b). They are enriched in Ni and have micro-fissures and microvugs filled with Mn-oxides. Silicified clasts of host rocks are enriched in iron but are less rich in metals than red-brownish quartz. Micro-XRF images of a breccia sample (Figure 5a) and the corresponding chemical maps (Figure 5b–d) show that white quartz rims are almost free of metals, which are, on the contrary, relatively abundant in the other clasts. The clasts of microcrystalline red quartz are Fe-rich and contain spots and micro-fissures of Mn-oxides (Figure 5c). Most clasts, including silicified host-rock, contain Ni in their mass (Figure 5d), besides the visible ore clasts composed of Ni-Mg talc-like inclusions (noted TL for talc-like).

White quartz cement forms successive white, transparent quartz rims on the clasts and it is free from any other mineral inclusions. The late overgrowths are porous compared with the earlier quartz rims, as SEM back-scattered images show (Figure 4c,f). Cathodo-luminescence images also reveal thin layering with rims about 10 microns thick. Quartz luminescence intensity is relatively weak, with a 20 s exposure time required to obtain enough signals for imaging (Figure 4f). Quartz cementing micro-breccias close to the silicified clasts have light yellow luminescence. Brown-orange luminescence corresponds to an intermediate rim of subhedral quartz below the latest colloform rims. To the end of the quartz sequence, the most recurrent colour observed under cathodoluminescence is a greenish-blue colour for microcrystalline quartz alternating with quartz rims having a very low or absent luminescence appearing as black or dark grey.

Pimelite formed on the quartz rims and constituted the final filling of the breccia (Figure 3c,e and Figure 4d,e). This constantly develops over a succession of quartz cement increments described above, ending in a geode. Pimelite forms botryoidal layers of dark green colour, showing strong contrasts of chemical mass (Z) in the section, displaying the growth zones (Figure 4c,d and Figure 6a). As shown in Figure 6a, magnesium is always lower than 2 wt% and has limited fluctuation, with NiO contents remaining between 43 and 49 wt%. Such Ni concentrations correspond to the extreme pole of the Ni-Mg kerolite solid solution (NiO = 39% to 49%) (Figure 6b). This pole is characterized by a high-frequency Raman spectrum marked by a principal OH band at 3650 cm^{-1} and a shoulder at 3660 cm^{-1}, typical of pimelite [23]. As already observed in the case of target-like Ni-rich talc-like, pimelite does not show extended Ni-Mg substitution and is close to the end member of the solid solution.

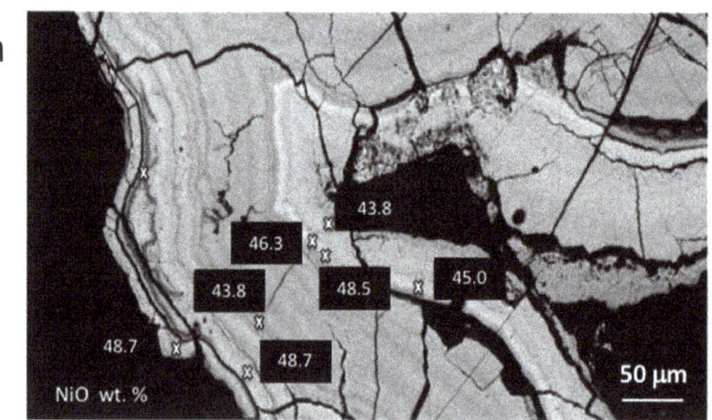

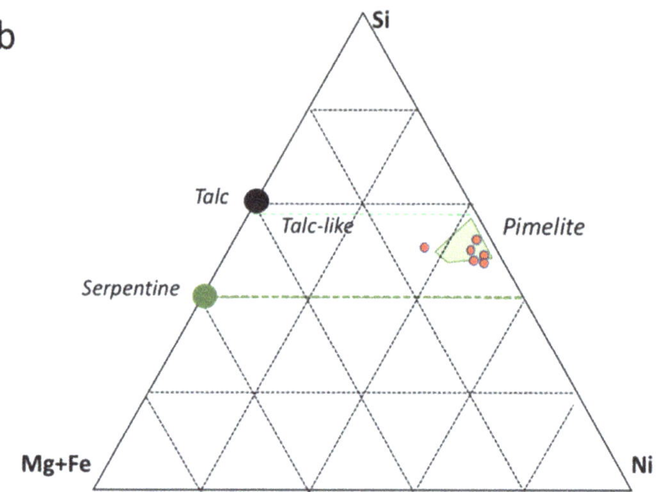

Figure 6. (**a**) Distribution of Ni (NiO%) in colloform pimelite. Each band has a high and relatively constant Ni concentration characterised by its specific mean Z (average atomic weight). (NiO concentrations from EMPA). (**b**) Si-Mg + Fe-Ni diagram applied to electron microprobe analyses of pimelites from Koniambo compared to pimelite from other New Caledonian sites (domain in green for other Caledonian deposits such as Thio and Poro, unpublished data). The reference line for talc-like solid-solution has already been proposed by [23,44].

When observed under TEM, the pimelite botryoidal texture consists of subparallel flexuous flakes (Figure 7a), which are sometimes associated with bi-pyramidal quartz (Figure 7b) but are primarily mono-mineral associations of tiny crystals of 50 to 150 nm in length (Figure 7c,d). The HRTEM images show that the flakes are made of the piling of no more than fifteen to twenty layers, generally less [23,44]. The spacing is close to 10 ± 0.5 Å among incertitude due to particle orientation and typical of talc-like [23] (Figure 7e,f). Although well crystallised, the coherent domain is small, explaining the relatively smooth XRD patterns and the difficulties in obtaining good diffraction patterns under TEM. The pimelite flakes are not associated with serpentine-like layers, which are entirely lacking.

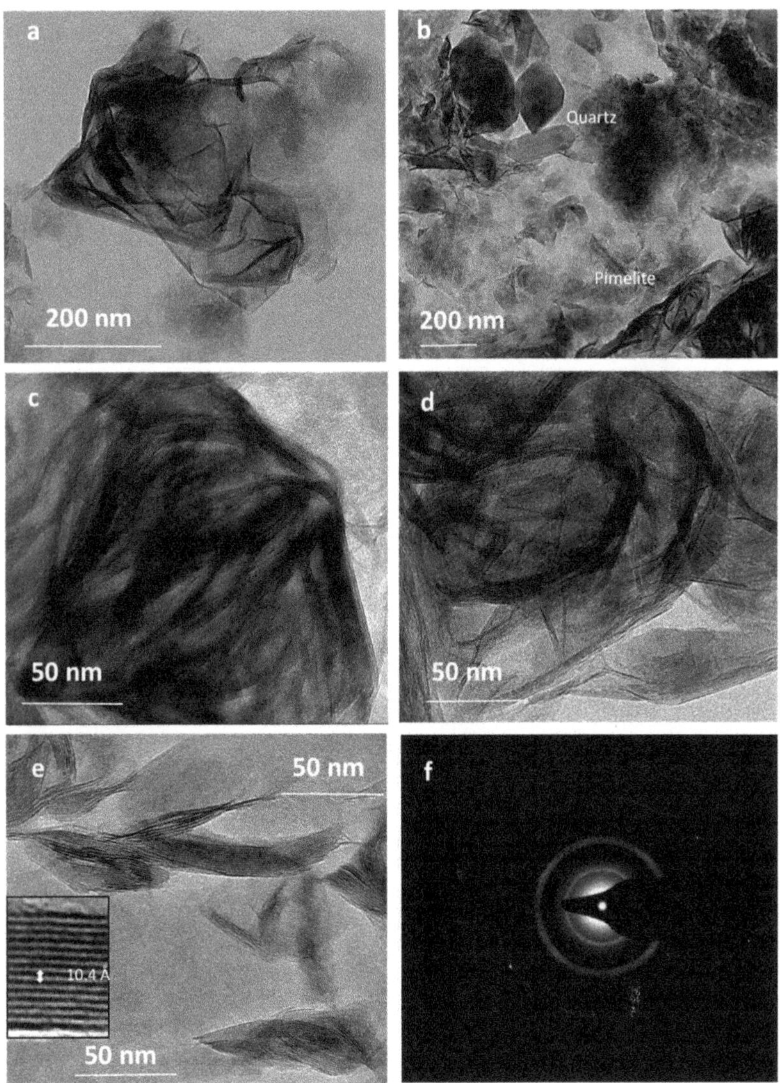

Figure 7. (**a**) TEM images of the Ni-silicate (pimelite). (**b**) Pimelite and euhedral quartz. (**c,d**) Flexuous assemblage of small size pimelite particles. (**e**) HRTEM image showing that the number of talc-like layers is generally low, about 5 to 15 in each particle, with no interstratification with other minerals. Indicative layer spacing is provided. (**f**) Diffraction pattern of one pimelite particle, with talc-like spacing.

4.4. Geochemical Data on Breccia Clast and Cement

Element concentrations in quartz types obtained by LA-ICP-MS were normalised to fresh surrounding rock (harzburgite) and presented in the box plots from Figure 8a. Red quartz has higher trace element concentrations than other quartz types, around one to two orders of magnitude higher. The richness in metals of the red microcrystalline quartz from the ore Type I is probably related to numerous metal-bearing micron-sized iron oxide particles. Al, Fe and Ga are two to five times higher in red microcrystalline quartz than in harzburgite. In white quartz, most elements are in low concentrations, around a few ppm, e.g., two to three orders of magnitude less concentrated than in the harzburgite, except Zn and Fe.

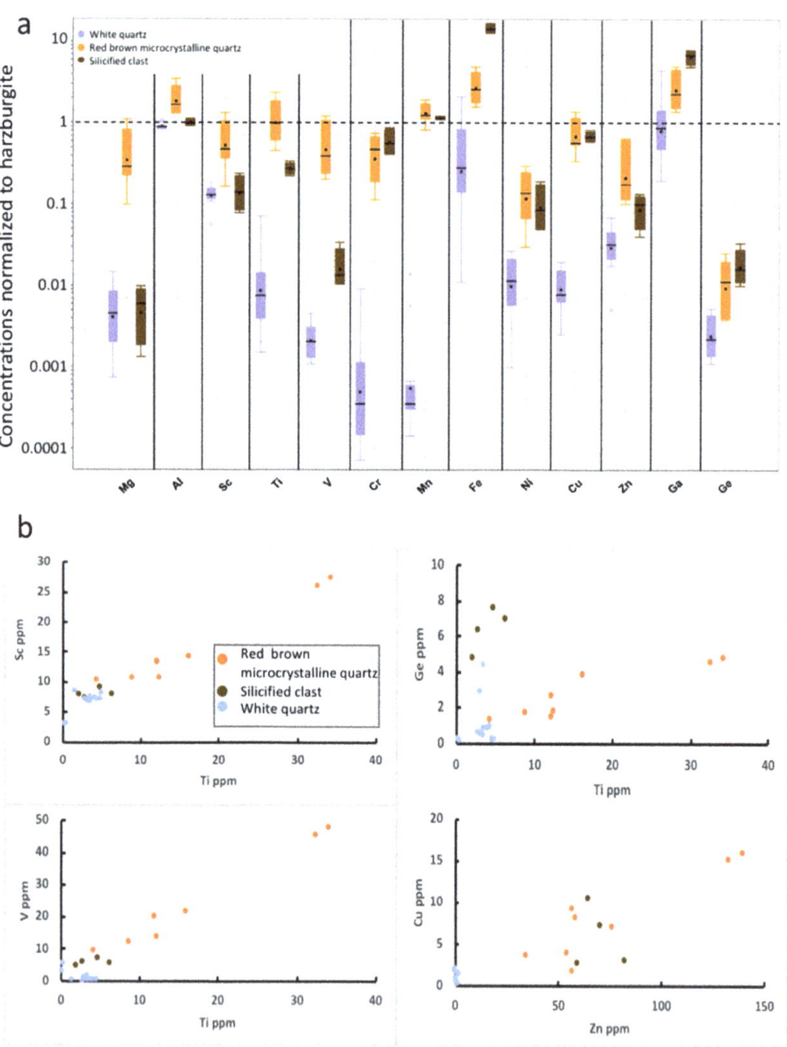

Figure 8. Trace elements in red-brown microcrystalline quartz, silicified clast and white quartz cement. (**a**) Normalised concentrations to harzburgite concentrations. (**b**) Binary plots, Sc-Ti, Ge-Ti, V-Ti and Cu-Zn for the three types of quartz.

The binary graphs, Sc-Ti, Ge-Ti, V-Ti and Cu-Zn established for the three types of quartz show that the red microcrystalline quartz has Ti, V, Cu and Sc concentrations of about two to three (up to ten) times those of the white quartz and are significantly enriched in Cu and Zn (Figure 8b). Ge concentrations are unusually high in silicified rocks, two to three times the concentrations determined in the other two types of quartz.

4.5. Quartz Oxygen Isotope Composition

Two $\delta^{18}O$ measurement profiles (twenty-five analyses) were made in the white quartz cement from the breccia. The $\delta^{18}O$ values of the white quartz range from +16.7‰ to +25.7‰ in the profile with twelve analyses shown in Figure 9. The second profile (13 analyses) was located near the same geode with $\delta^{18}O$ values ranging from +15.2‰ to +23.5‰. The $\delta^{18}O$ values evolve from +15.2‰ to +17‰ in the silicified breccia towards +24‰ to +25‰ in the centre of the geode, as shown by Figure 10a.

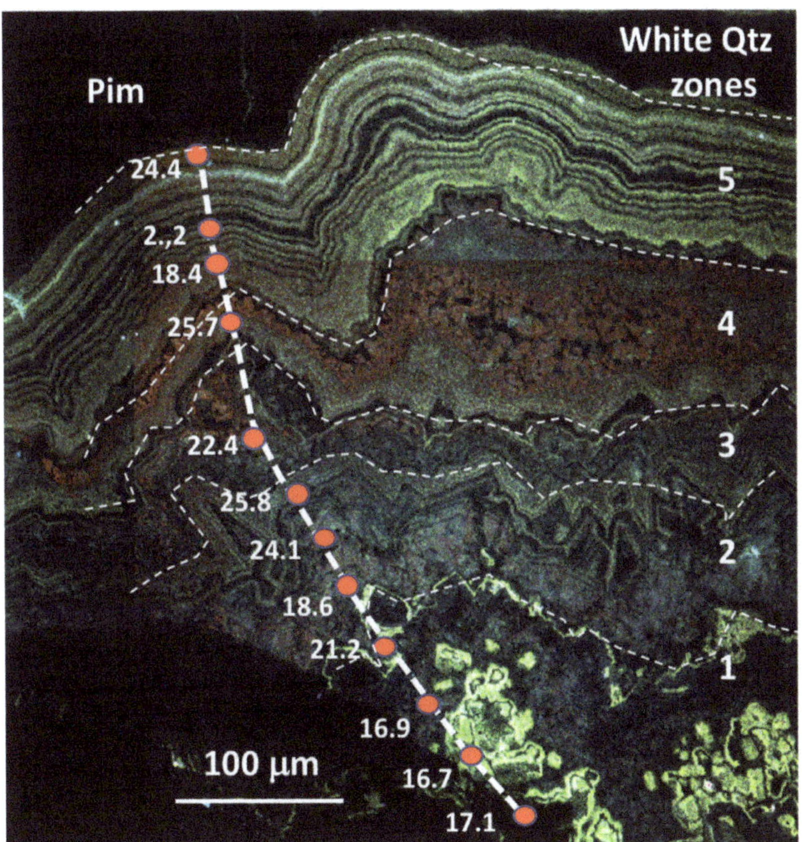

Figure 9. $\delta^{18}O$ (‰) values along a profile crosscutting white quartz from silicified breccia to the centre of the cavity (Zones 1 to 5). The colloform pimelite appears in black in the central vug. $\delta^{18}O$ values are reported on the cathodoluminescence image.

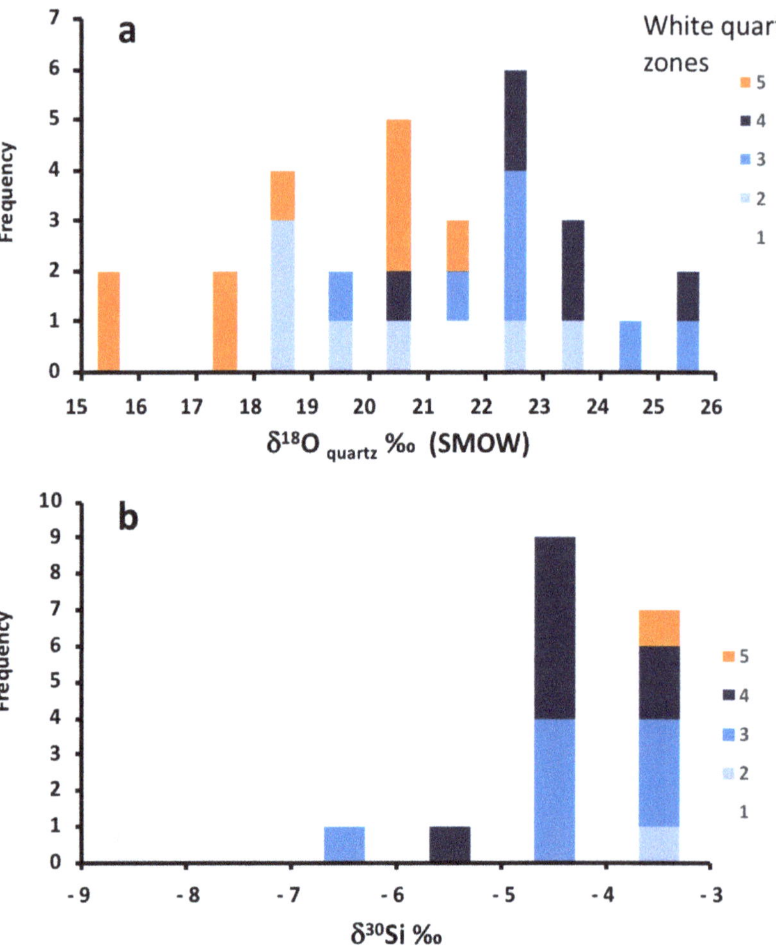

Figure 10. (**a**) $\delta^{18}O$ values along two profiles across the white quartz from the edge to the centre of the vug (five zones described in Figures 4f and 9). (**b**) $\delta^{30}Si$ quartz (‰) along the same profiles.

4.6. Silicon Isotope Composition of White Quartz

Analyses of $\delta^{30}Si$ were performed along the same profiles of white quartz (sixteen measurements, Figure 10b) as those used for the oxygen analysis spot (every two oxygen spots). The white quartz cement of the breccia sample shows a range of $\delta^{30}Si$ values from $-3.0‰$ to $-6.3‰$, with a mean value of $-4.8‰$ and a whole range of 3.3‰. A lower value of $-8.8‰$ characterises the earliest rim of white quartz.

5. Discussion

5.1. Dissolution of Primary Silicates and Formation of Dissolution Pipes

The dissolution pipes constitute an open network that may be considered a pseudo-karst. Such cavities in peridotite cannot result here from other processes than supergene dissolution, as shown by the mineralogy of the remaining iron (iron hydroxides) similar to that of laterite, and the dissolution features similar to subsurface boulders at the bedrock/saprolite boundary.

During intense rainfall, all primary silicates of the peridotites (olivine and pyroxene), the serpentines (early mesh and fracture infillings) and the secondary silicates related to the saprolite (talc-like, nontronite) are dissolved. Part of the rainwater gullies down and part of it feeds the laterite water table with a gravity-driven flow in the slope direction, most often close to the interface between the laterite and the saprolite [16,30,31,45]. However, some water flows along sub-vertical to relatively steep faults through fracture sets in the damaged areas of major faults. The consequence is the expulsion of significant masses of silica, magnesium and nickel, preferentially in the transmissive fracture network. In some zones, silicification of the saprolite occurs, preferentially at the top of the relief. Such silicification of the weathered harzburgite is described in the Cagou pit [24,25]. Following Butt's (2014) discussion [46], silicified saprolites have been found at several locations, such as Cawse (Australia) and Caldag (Turkey).

Dissolution pipes develop where drained waters initialise the dissolution of primary and secondary silicates. Continued dissolution along fractures then leads to the formation of channels (pseudo-karstification of the peridotite). The preferential water movements yield an empty structure where around 10% of the initial rock forms a brown coating along the fracture, constituted of goethite. Such dissolution processes leaving open cavities are distinct from those associated with the subparallel deepening of weathering front in lateritic profiles where pores resulting from dissolution collapse progressively, yielding dense sub-horizontal laterite horizons at the basis of the weathering profile.

Along these drains, the interaction between the renewed rainwater and the fracture walls leads to a total dissolution of primary and secondary silicates, leaving only the insoluble iron as oxyhydroxide (goethite) along the walls. In the absence of collapse of the fracture edges, the drain widens until it reaches a width of several tens of centimetres, up to one metre locally. The downward deepening of the alteration depends on the gravitational force, the pre-existing fracture networks and the differential between the upper part of the profiles and the outlet zones in the valley. The channels thus form relays and allow the formation of a "karst" or "pseudo-karst" within the peridotite massif.

5.2. Chronology of the Karstification Related to River Incision

The river incision in the Taléa valley was studied, complementing earlier findings in the Koniambo and Kaféaté plateaux [20,42]. There was difficulty in distinguishing whether the relics in the Taléa valley belong to the S3 surface, the S4 surface or both. Their presence, nevertheless, suggests a strong relief inversion since S3 surface formation. The inversion was then enhanced during the S4 surface stage. The attribution to S3 and S4 can, nevertheless, be attempted based on the relative elevation between relics; the higher relics being S3 and the lower relics being S4.

If the paleo landforms of the Koniambo plateau are synchronous with that of Thiébaghi, an age of ca. 25 Ma for the S1-S2 surfaces weathering phase could be proposed. This would contrast with an early suggestion (ca. 30 Ma) by Chevillotte et al. (2006) [42]. Sevin et al. (2014) [47] considered, based on the sedimentary record in the Nepoui area, that a significant uplift event started around 22 Ma. Depending on location in New Caledonia, formations were uplifted of ~200 m at minimum to more than 500 m, explaining that the S1-S2 paleosurfaces occur at Koniambo around 800 m of altitude (>500 m). As a consequence, erosion started, and the valley incision quickly reached hundreds of meters. The beginning of the pseudo-karst development could date from that period. Subsequent valley incisions likely continued during the Neogene, leaving remnants of the S3 and S4 surfaces in the Taléa valley and their lower elevation equivalent in the Kaféaté plateau.

The breccia formation mechanism implies relief and a gap between the valley table elevation and the highs. After the first stages of laterite formation (S1-S2, most likely around 25–20 Ma), probably under more moderate relief, the effects of rock uplift brought these laterite profiles up to their current elevation (over 800 m). In this context, a series of erosional cycles occurred, leading to several stepped surfaces, as described by [19].

Pre-existing fracturing related to ante- to syn-obduction deformation stages was reactivated at several periods, notably during extensional stages from Miocene to the late Miocene. This extension is at the origin of the reactivation of normal faults of serpentine planes, which are most often oriented in the island's axis, i.e., NNW–SSE, and generally a receptacle for fluids charged with Ni by the dissolution of both overlying rocks and pre-existing mineralisation. A tentative reconstruction of this dynamic evolution is proposed in Figure 11, with four corresponding stages: first, the late Paleogene period with two main plateau surfaces, followed during the Neogene period, by the formation of S3-S4 paleo-weathering surfaces synchronous of the karst formation and deepening.

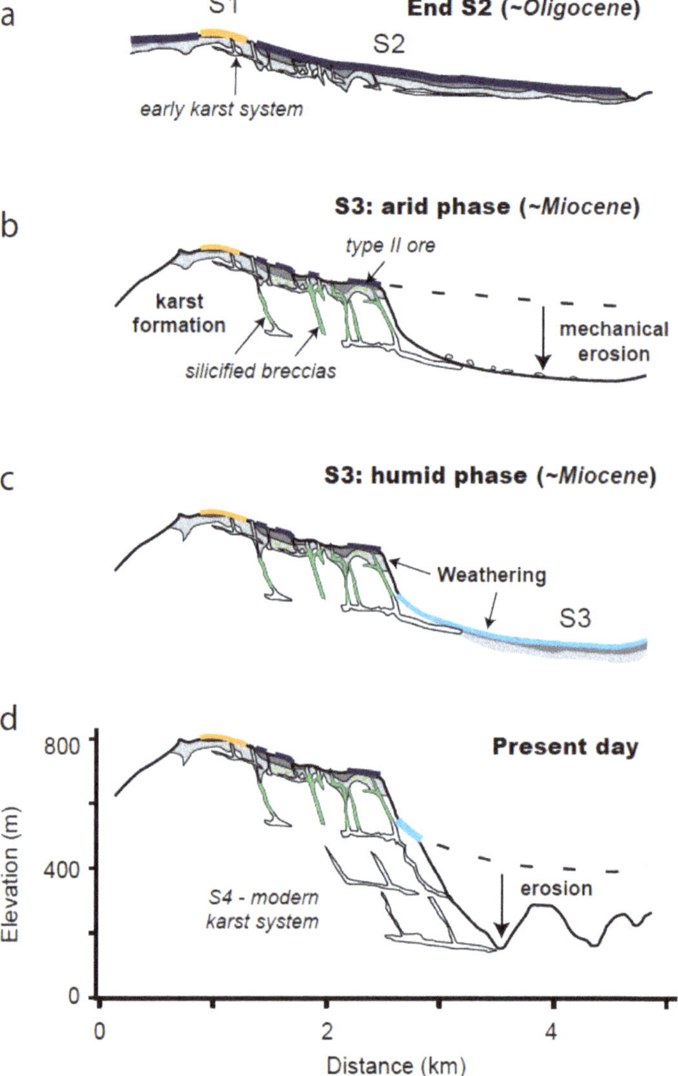

Figure 11. Conceptual model of the pseudo-karst development at Koniambo, with the paleosurface formation following the valley's deepening and progress in the erosion of the uplifted ophiolite.(**a**) Formation of the paleosurface S2 at the expense of S1, and initiation of the early karst system, (**b**) Deepening of the valley, and karst silicification during arid periods, (**c**) Formation of S3 related to humid periods, (**d**) Present day relief and karst.

The cavities result from the downward underdrawing, i.e., the gravity collapse, which is the principal driving force behind the brecciation. The resulting fill is a so-called collapse breccia, forming a "breccia pipe", as described in other geological contexts. The formation of these breccias is well constrained in the tectonic history, as they postdate the sealing of the kerolite crack-seals (Type I ores [24]) and are probably sub-synchronous with the formation of the target-like (Type II) ores [23]. They postdate the main laterite and are synchronous to the dissolution pipe formation, as they require a high dissolution rate along preferential meteoric water conduits. Thus, these breccias likely postdate S1-S2 surfaces, and are probably syn- to post-S3 at Koniambo.

5.3. Paleo-Climatic Conditions of the Formation of Quartz Cements

As quartz does not exhibit any studiable fluid inclusions but only tiny monophase fluid inclusions, generally considered metastable and formed below 60–80 °C, higher temperatures cannot be considered. The mineralogy of the pseudo-karst pipe is the same as around boulders issued from the supergene weathering of the peridotite in the saprolite, which produces only one run product in the laterite horizon: goethite. Goethite is not stable above 80 °C, as shown by thermodynamic data. If the activity of water is close to one, the boundary between goethite and hematite is around 35 °C, and the goethite-hematite conversion is favoured in the unsaturated zone explaining the formation of the hematite-rich duricrust on the laterite surface at the expense of goethite. Along the dissolution pipe, there is no evidence of hematite, but the only presence of goethite indicates that temperatures never reached high temperatures.

The relatively tight grouping of $\delta^{18}O$ values for quartz suggests that the quartz formed from a fluid that has a similar oxygen isotopic composition during a single event. This event is distinct from the previous quartz stages characterized by different chemical features, particularly red microcrystalline quartz with hematite micro-inclusions and high metal contents. $\delta^{18}O$ values between +18‰ and +24‰ would correspond to the lower values of the present-day waters around −4‰ within the 60–80 °C range. Such water values typical of tropical climate, with ambient temperatures of around 25 ± 10 °C, would yield $\delta^{18}O$ values for quartz higher than +30‰, which were never found in New Caledonia. Only some opals were thus found with such values of +32–33‰ and correspond to recent silica films and draperies [25]. Either the temperatures were higher or the water $\delta^{18}O$ was lower than the present-day ones. Temperatures around 70 °C were invoked for red microcrystalline quartz from crack-seals, supposed to be issued from low-temperature hydrothermalism, during tectonic compression events [24]. In the present case of very late cementing of quartz near the surface in pseudo-karst, a low temperature is the most realistic hypothesis. Ambient temperatures yield water oxygen isotopic values of −8 to −12‰, symptomatic of a cooler climate, as already invoked for Australian silcretes [48]. Adequate rainfall for the formation of the dissolution pipe is required. This could suggest a humid tropical or subtropical environment [27], but the alternation of seasons of rain and drying episodes could also happen under a less humid climate.

Considering that silicification was favoured by evaporation under ambient temperatures (mean temperature of 20 ± 5 °C), the $\delta^{18}O$ values of the fluid are estimated to range between −12‰ and −8‰ (Figure 12). By comparison, $\delta^{18}O$ values of chemical and possibly biogenic cherts range from +15‰ to +22‰, and $\delta^{18}O$ of silicified volcanoclastic cherts between +9‰ to +17‰ [49]. Features of New Caledonian silicifications display similarities with silcretes. For instance, the silcretes from Australia have values of +24‰ to +29‰ [48,50]. In the case of Lake Eyre, late Eocene to early Oligocene silcretes are considered as precipitated at temperatures of 15–20 °C from waters having $\delta^{18}O$ values of −6.9‰ to −12.2‰ [50]. Silcretes from Victoria and South Australia are thought to have precipitated from an undefined age between Eocene–Miocene to present at temperatures of 14–15 °C from rainwaters having $\delta^{18}O$ values of −5‰ to −1‰ [48]. In other examples of silcretes, a value of +29.3 ± 1‰ was found for quartz from silcretes in St. Peter sandstones (SW Wisconsin) which precipitated at 10–30 °C, from waters with $\delta^{18}O$ of −5‰ to

−10‰ [51]. Similarly, Bustillo et al. 2014 [52] found $\delta^{18}O$ values of +25–26‰ for the silcrete found in Miocene sediments from Torrijos (Madrid basin, Spain), which are considered post-Miocene and precipitated at 15–25 °C, from meteoric waters with $\delta^{18}O$ of −6‰ to −8‰.

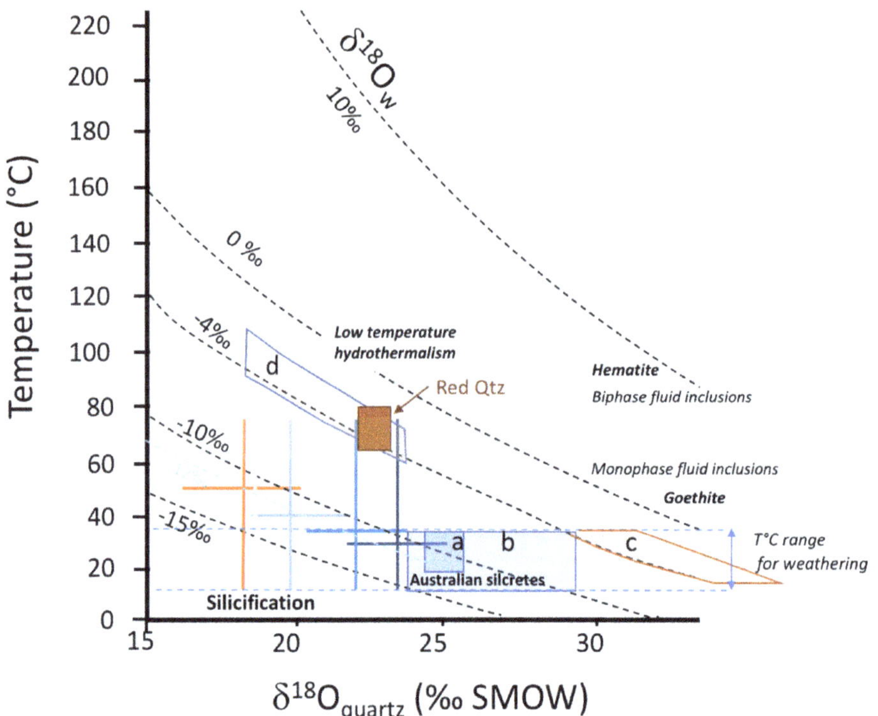

Figure 12. $\delta^{18}O$ quartz versus temperature diagram with reference lines for oxygen isotopic compositions of water ($\delta^{18}O$ w) reported from Matsuhisa et al. (1979) [53]. Data for the white quartz from Koniambo pseudo-karst are represented as crosses using mean values and ranges for the five zones described in Figures 4f and 9 with the same colours as in Figure 9. Temperatures are reported arbitrarily between 15 and 75 °C and crosses are organized to follow the water $\delta^{18}O$ value of −12‰ found in other silcretes. For comparison, $\delta^{18}O$ ranges and estimated temperatures for Lake Eyre silcretes are reported in the box a from [50] and other Australian silcretes in box b from [49]. Two domains calculated for present-day waters in tropical countries of −2‰ and −4‰ are reported for two temperature ranges (15–35 °C (domain c) and 60–110 °C (domain d). The estimated temperature range for red microcrystalline quartz is reported in the brown box. The boundary in grey indicates the limit of stability of goethite and hematite, as well the temperature fields for monophase and biphase fluid inclusions.

5.4. Analogies with Silcretes

The range of $\delta^{30}Si$ values for white quartz is quite similar to silcretes values, which range from −6‰ to +1‰ following [54]. This range is much lighter than most other Si reservoirs compiled in [55]. The number of available $\delta^{30}Si$-$\delta^{18}O$ is much more restricted, and most data concern archean cherts. The $\delta^{30}Si$ versus $\delta^{18}O$ diagram from Figure 13 confirms the specific features of silcretes and Koniambo pseudo-karst cements compared to most available silicon-oxygen isotope pairs, particularly those obtained on recent and old cherts and Icelandic hydrothermal systems [56]. A small number of silicon data, not plotted in Figure 13, overlap the range for silcrete and white quartz from Koniambo pseudo-karst, principally the dispersed data over an extensive range of values obtained on quartz amygdules from Isua basalts [57]. However, the reason for the scattering of $\delta^{30}Si$ is not fully understood, as the magmatic vesicles underwent hydrothermalism and metamorphism up to the amphibolite facies.

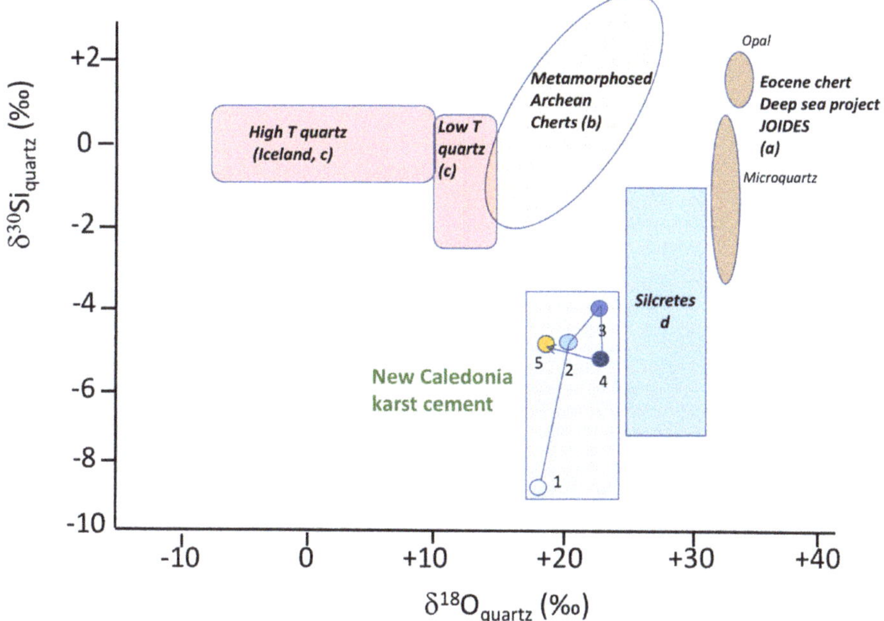

Figure 13. In situ $\delta^{30}Si$ versus $\delta^{18}O$ for white quartz from Koniambo silicified breccia (numbers refer to zoning from Figures 4f and 9), compared to the principal domains of values available for recent cherts (a—[49]) and metamorphosed Archean cherts (b—[58]), hydrothermal quartz of high and low temperature from Icelandic fields (c—[56]), and silcretes (d—composite box from several data sources [50,51,54]).

As the $\delta^{30}Si$ values are similar to silcretes, quartz is most probably formed through evaporation, as decreasing temperature, the most straightforward way to deposit quartz, is difficult to invoke. The evaporation was probably favoured by the heterogeneous permeability of the breccia, characterised by large clasts and blocks yielding a relatively high bulk permeability at the scale of the pipe. Quartz cementation of pseudo-karst breccia pipes occurred thus probably during a climatic period different from the present-day conditions, as suggested by the oxygen isotope data. Silicification also affects microfracture rocks and quartz filling sets of microfractures at all scales with no specific orientation. The host rocks at Koniambo, on both sides of the breccia pipe, are highly silicified. In the upper levels of the quarry, the complete dissolution of the remaining silicates yields quartz boxworks made of hundreds of microfractures filled by quartz, the remaining parts of the rocks being entirely dissolved.

Although amorphous silica is considered the most probable silica form to precipitate at low temperatures, only quartz was found in the fissures and cemented breccia. The necessary transformation of amorphous silica into quartz, invoked by Thiry and Milnes [59], is a conceptual model that is not based on systematic observation of the transition between these two phases at the micrometer scale. Suppose opal or other amorphous silica phases are thermodynamically unstable and must transform by ageing to quartz; this does not explain why opal is found in some silcrete layers and quartz in other layers from the same age in Australia. Precipitation of quartz instead of amorphous silica indicates mostly lower H_4SiO_4 activity in solution when it precipitates.

In addition, there are a lot of examples of quartz precipitation near the subsurface in the context of silicified caps in laterite and silicified peridotite, such as in United Arab Emirates [60], Turkey (Caldag laterite, [61]) and Australia (Mt Keith, [62]).

5.5. Comparison with Silcretes from Australia

In Australia, the timing of the silcrete formation is under debate. Some authors consider the main period of silicification as Eocene–Miocene, generally older than deep kaolinisation [26,63] and younger than the formation of ferruginous duricrust.

Van der Graaff (1983) [64] considers that Late Tertiary aridity cannot account for silcrete formation as silcretes predate Fe-duricrust. Considering North Queensland, Li and Vasconcelos [65] proposed an age of 14–10 Ma for a cooling period after the subtropical environment's 22 to 15 Ma period. More recently, Mathian et al. (2022) [66] obtained ages of 10 and 3.8 Ma to form kaolinite at Syerston. Such ages are close to the 10.8 Ma (and younger) data obtained by ESR on Queensland silcretes [67].

Seasonal climate or better alternate rain and drying episodes of very short durations may provide a favourable environment for silicate dissolution, followed by active silica (quartz) precipitation after lowering the water table level.

The data are, therefore, convergent about the formation of silcretes during the Miocene -likely after 15 Ma from waters having a $\delta^{18}O$ lighter than present ones under climatic conditions of moderate precipitation (<1 m/year) and average temperature. The climate was probably favourable both in Northern Australia and New Caledonia during this period to forming silcretes during the dry periods, alternating with rains under paleoclimatic conditions, which were different from the present.

5.6. A Late Silicification Stage into the Ni–Ore Chronology

The chronology of the development of the silicification found in the Koniambo massif includes several stages, summarized in Figure 14. During the first ones, dissolution affects the joint and fault network leaving boulders in the saprolite (Figure 14a) and then deepens along faults (Figure 14b,c). As pseudo-karst conduits widen, they can become progressively fragile due to the excessive size of the open pipes, and the collapse of the walls fills the conduits with blocks of any size and orientation (Figure 14d). The resulting collapse breccias are well timed in the tectonic history, as they postdate the sealing of the kerolite crack-seals (Type I) [24] and are probably sub-synchronous with the formation of the target-like ore type (Type II) [23].

The white quartz free of iron oxides attests to the total iron immobility during its crystallisation and the supergene feature of the process. The latter is thus very different from the formation mechanisms of the red quartz described in the earlier crack-seals, where iron was transported in the form of Fe^{2+} by low-temperature hydrothermal fluids [24]. The silicification occurs after the earliest lateritic stage (S1-S2) as brownish clasts of laterites are cemented by quartz precipitated from dissolved silica (Figure 14d,e). Karstification is enhanced by relief and favoured by incision starting at stage S3. This requires the release of silica by chemical weathering, by the dissolution of the remaining silicates in the weathered horizons and along the pipes. The silica release is not related to developing sub-horizontal residual soil but to water-rock interactions in the pseudo-karst along the drainage zone (faults) and their micro-fractured, damaged zone. The lack of laterite contamination of the breccias indicates that the collapse process affects and dissolves rocks along the drains but is sufficiently far from sub-surface conditions. The karst, therefore, develops at depth, and the increasing incision of the valley favours the downward deepening of the pseudo-karst.

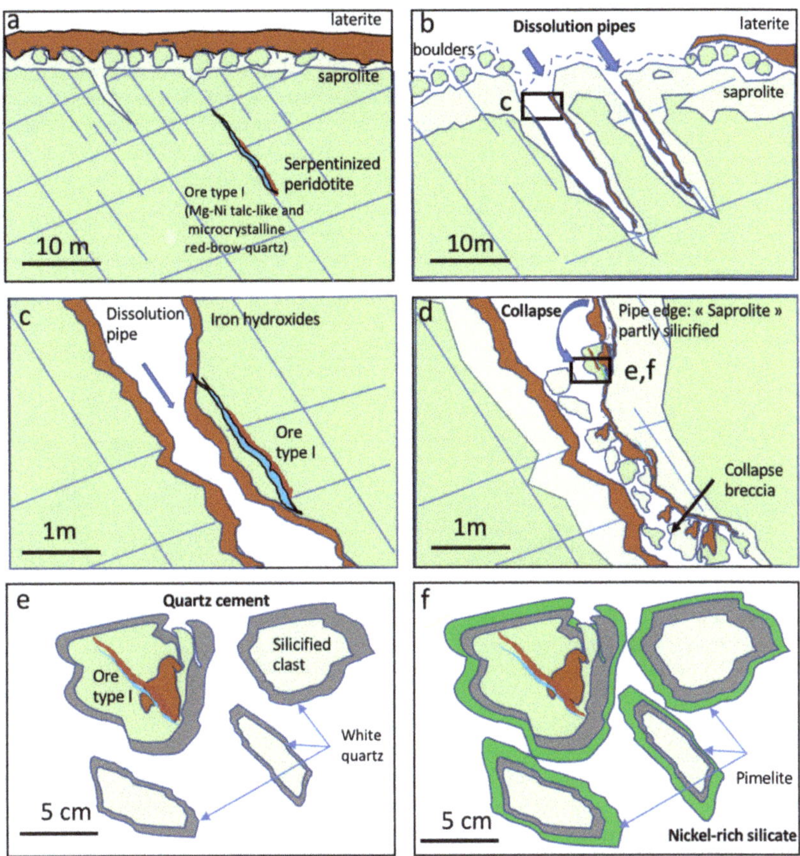

Figure 14. Main stages of silicified collapse breccia formation: (**a**) Formation of laterite and ore Type I as crack-seal fractures. (**b**) Development of dissolution pipes along previous fractures. (**c**) Detail of the dissolution pipes (inset c from b) with edges affected by saprolitisation and a layer of iron hydroxides. (**d**) Silicification and then the collapse of the dissolution pipes. The box refers to figures (**e**,**f**). (**e**) White quartz cement precipitation on clasts of silicified host rocks, and clasts of early ores. (**f**) Precipitation by evaporation of nickel-rich silicate (pimelite) on the white quartz rims.

White quartz formed onto the block surface is posterior to the quartz crack-seals found as brecciated clasts (Figure 14e). Correlations observed for red quartz are similar to those already determined in laterites, as iron hydroxides are enriched in all transition metals. The white quartz is almost free of metals. Variations in elements such as Ti, Ge, Al and V seem more dependent on the nature of the quartz and probably substituted to Si^{4+} in the quartz lattice. The contrasted chemical features of the microcrystalline red quartz associated with ore Type I crack-seals compared to the white quartz and silicified clasts of host rocks indicate a drastic change in the conditions of quartz precipitation and mobility of metals. The metals mobilized together with iron, probably under reducing conditions during ore stage I [24], are no more trapped during the late white quartz stage. Pimelite occurs after quartz (Figure 14f) and the lack of recurrence indicates that the process occurred only once.

5.7. Geochemical Modelling of the Cement Formation

The model consists of the progressive 1D dissolution of a fully saturated 4.5 mlong olivine column by rainwater, followed by a progressive evaporation step. Olivine composition was selected according to probe measurements on fresh olivine: $Mg_{1.865}Fe_{0.127}Ni_{0.008}SiO_4$

and its initial porosity of 1% was taken from [68]. The initial rainwater composition was taken from [2] (Table 1). The precipitation rate driving the advection was assumed to be 1500 mm/year, while the diffusivity was assumed to be spatially constant and equal to 4.16×10^{-10} m^2/s in agreement with [16]. The points represented in Figure 15 consist of the chemical composition of the solution in the middle of the column. During the dissolution step, mass transport occurs along the cells of the column. Water-rock interactions are governed by olivine dissolution kinetics and thermodynamic equilibrium for forming newly-formed minerals [16].

Table 1. Initial solution composition for the evaporation step from [2].

pH	Mg (mg/L)	Si (mg/L)	Ni (mg/L)	Fe (mg/L)
7.34	1.15	5.30×10^{-1}	2.33×10^{-3}	5.04×10^{-5}

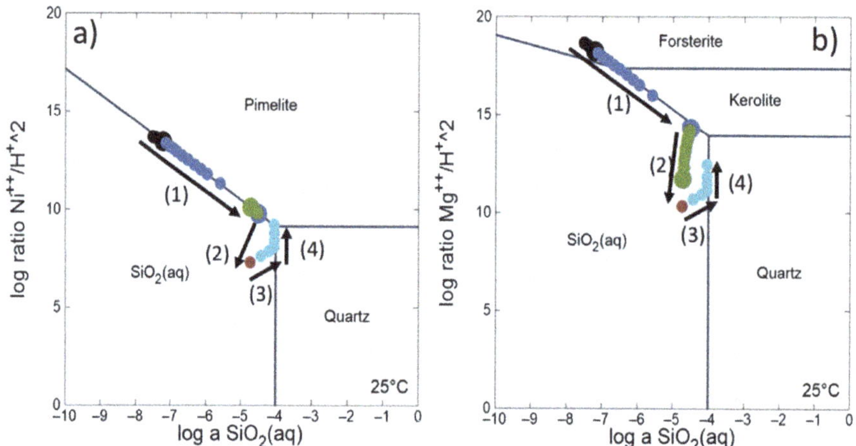

Figure 15. Chemical path of the mineralogical evolution in fractures in activity diagrams. (**a**) the Ni^{2+}/(H$^+$)^2 − SiO$_2$ and (**b**) Mg^{2+}/(H$^+$)^2 − SiO$_2$. In colours, the predominant species in the system: Black: Olivine; Blue: Mg-Kerolite; Green: Pimelite; Red: Goethite; Light Blue: Quartz. (1) Progressive silicate dissolution. (2) Goethite dissolution in the overlaying laterite. (3) Evaporation up to quartz saturation. (4) Evaporation up to pimelite saturation.

As the main driving force for silica precipitation (silcrete) at sub-constant ambient temperatures is evaporation [69], once the solution has reached equilibrium with goethite, another simulation is performed to model the progressive evaporation. For this step, the initial water composition is the solution in equilibrium with goethite. An arbitrary amount of water is removed from the previous solution at each stage. Then the saturation indexes of the considered minerals and the concentrations of Mg, Si, Ni and pH are calculated. These steps are repeated until only 0.2% of the initial water remains in the model to avoid changing the water activity from 1. The modelling has been carried out using either amorphous silica, in case of the formation of a precursor of quartz easier to precipitate at low temperature, or directly quartz. Results are not significantly different, except the inversion in the order of pimelite and silica-phase precipitation is not in agreement with field observation; therefore, justifying the choice of the quartz option in Figure 15.

After quartz precipitation, Ni silicate (pimelite) enrichment is explained by the redistribution of Ni from previous Ni silicates in fractures (ore stage 1) from the overlying levels and from harzburgite silicate dissolution. There is a need to increase the Ni^{2+}/(H$^+$)^2 ratio to reach oversaturation with respect to talc-like at sub-constant or slightly increasing silica activity in a solution (step (4) in Figure 15a). Notably, the solution reaches its saturation with respect to the pimelite (Figure 15a) and not to kerolite for a partial evaporation state (Figure 15b). Hence, total evaporation seems to be never reached, explaining the absence of Mg-talc-like (kerolite) after the quartz precipitation, while pimelite is precipitated.

6. Conclusions

Breccias are posterior to the main laterite formation stage and are closely related to pseudo-karst development as they are collapse breccias requiring a high rate of preferential conduit dissolution. The preferential water movements along the fracture differ from the downward migration of the bedrock-saprolite development sub-parallelly to the bedrock surface. This corresponds to the progressive deepening of the pseudo-karst as a function of the valley incision with the following sequence:

The preferential subvertical drainage along pre-existing serpentine faults favoured the pseudo-karst network initiation. This episode corresponds to the formation of the S3 paleosurface. The pipe collapse formed breccias, which were then partly silicified and cemented by white quartz and pimelite, the extreme Ni-containing pole of the talc-like solid solution. The formation of this assemblage was favoured by a progressive increase of $Ni^{2+}/(H^+)^2$ and $Mg^{2+}/(H^+)^2$ ratios during interstitial water evaporation in the upper part of the pseudo-karst pipes. As shown by modelling, the fluid chemistry, however, reached only pimelite and not Mg-kerolite, explaining the predominance of pimelite over Mg-kerolite.

Quartz precipitation occurred in conditions identical to those invoked for silcretes. The low negative $\delta^{30}Si$ values ranging between $-5‰$ and $-7‰$ are typical of silcretes and close to the minimum values recorded worldwide. This silicification may have occurred during drier climatic episodes than in tropical climates from Eocene–Oligocene. Meteoric waters with lower $\delta^{18}O$ values in the order of $-6‰$ to $-12‰$ favour this hypothesis, which is proposed for the first time in New Caledonia.

The isotopic characteristics of the silicified breccias are similar to those of silcretes described in Australia, which are predominantly from the Miocene age. Ni reworking in the form of pimelite could have occurred at this period as New Caledonia was in a similar geographical and paleoclimatic position to north-eastern Australia.

Author Contributions: Conceptualisation, M.C. and M.-C.B.; methodology, M.C., M.-C.B. and J.-L.G.; software, J.-L.G. and S.F.; analysis, M.C. and M.-C.B.; writing—original draft preparation, M.C., M.-C.B. and J.-L.G.; review and editing, all authors; supervision, M.C., M.-C.B. and J.-L.G.; project administration, M.C. and Y.T. funding acquisition, M.C. and Y.T. All authors have read and agreed to the published version of the manuscript.

Funding: This work has been funded and logistically supported by the CNRT research contract CSF N° 9PS2017-CNRT. GEORESSOURCE/TRANSNUM« Facteurs d'enrichissements et transferts de Ni, Co-Sc dans les saprolites de Nouvelle Calédonie: approche géométrique, minéralo-géochimique et numérique », and by the French National Research Agency (ANR) through the national program "Investissements d'avenir" of the Labex Ressources 21 with the reference ANR-10-LABX-21-01/LABEX RESSOURCES21.

Data Availability Statement: Data will be available in the CNRT final report of the contract mentioned above, which will be in open access on the CNRT's website when completed.

Acknowledgments: The authors would like to thank Andreï Lecomte for cathodoluminescence imaging, J. Gambaja for TEM imaging (SCMEM, GeoRessources, Vandœuvre-lès-Nancy, France) as well as C. Peiffert (LA-ICP-MS facilities, GeoRessources), and Platform LG-SIMS at CRPG-Nancy in particular, N. Bouden and A. Gurenko for technical support. Sampling benefited from the help of Koniambo geologists, C. Couteau and participants to the first samplings (B. Quesnel, Ph. Boulvais from Géosciences Rennes). Three anonymous reviewers and the editor are warmly acknowledged for constructive comments.

Conflicts of Interest: The authors declare no conflict of interest.

References

1. Avias, J. Note sur les facteurs contrôlant la genèse et la destruction des gîtes de nickel de la Nouvelle- Calédonie. Importance des facteurs hydrologiques et hydrogéologiques. *C. R. Acad. Sci.* **1969**, *268*, 244–246.
2. Trescases, J.J. *L'évolution Supergène des Roches Ultrabasiques en Zone Tropicale: Formation de Gisements Nickélifères de Nouvelle Calédonie*; Mémoire ORSTOM: Bondy, France, 1979; Volume 78, 259p.
3. Genna, A.; Maurizot, P.; Lafoy, Y.; Augé, T. Contrôle karstique de minéralisations nickélifères de Nouvelle-Calédonie. *C. R. Geosci.* **2005**, *337*, 367–374. [CrossRef]
4. Valeton, I.; Biermann, M.; Reche, R.; Rosenberg, F. Genesis of nickel laterites and bauxites in Greece during the Jurassic and Cretaceous, and their relation to ultrabasic parent rocks. *Ore Geol. Rev.* **1987**, *2*, 359–404. [CrossRef]
5. Riedl, H.; Papadopoulou-Vrinioti, K. Comparative investigations on Karst generations mainly in the Aegean Archipelago. *Mitt. Des Nat. Ver. Für Steiermark* **2001**, *131*, 23–39.
6. Nunez-Jimenez, A.; Korin, I.Z.; Finko, V.I.; Fomell-Cortina, F. Notas preliminarias acerca del caso en peridotita, Sierra de Moa, Oriente, Cuba. *Rev. De Geol.* **1967**, *1*, 5–28.
7. Löffler, E. *Geomorphology of Papua New Guinea*; Australian National University Press: Canberra, Australia, 1977; 195p.
8. Hope, G. Extended vegetation histories from ultramafic karst depressions. *Aust. J. Bot.* **2015**, *63*, 222–233. [CrossRef]
9. Jeanpert, J.; Genthon, P.; Maurizot, P.; Folio, J.L.; Vendé-Leclerc, M.; Serino, J.; Join, J.L.; Iseppi, M. Morphology and distribution of dolines on ultramafic rocks from airborne LIDAR data: The case of southern Grande Terre in New Caledonia (SW Pacific). *Earth Surf. Process. Landf.* **2016**, *41*, 1854–1868. [CrossRef]
10. Patino-Rojas, S.M.; Jaramillo, M.; Espinosa-Espinosa, C.; Arias-Lopez, M.F. Preferential groundwater flow directions in a pseudokarst system in Colombia, South America. *J. South Amer. Earth Sci.* **2021**, *112*, 103572. [CrossRef]
11. Byrne, R.H.; Kester, D.R. Solubility of hydrous ferric oxide and iron speciation in sea water. *Mar. Chem.* **1976**, *4*, 255–274.
12. Butt, C.W.; Cluzel, D. Nickel laterite ore deposits: Weathered serpentinites. *Elements* **2013**, *9*, 123–128. [CrossRef]
13. Teitler, Y.; Cathelineau, M.; Ulrich, M.; Ambrosi, J.P.; Minoz, M.; Sevin, B. Petrology and geochemistry of scandium in New Caledonian Ni-Co laterites. *J. Geochem. Explor.* **2019**, *196*, 131–155. [CrossRef]
14. Teitler, Y.; Favier, S.; Ambrosi, J.P.; Sevin, B.; Golfier, F.; Cathelineau, M. Evaluation of Sc concentrations in Ni Co laterites using Al as a geochemical proxy. *Minerals* **2022**, *12*, 615. [CrossRef]
15. Tardy, Y.; Roquin, C. *Dérive des Continents, Paléoclimats et Altérations Tropicales*; BRGM: Orléans, France, 1998.
16. Myagkiy, A.; Truche, L.; Cathelineau, M.; Golfier, F. Revealing the conditions of Ni mineralisation in the laterite profiles of New Caledonia: Insights from reactive geochemical transport modelling. *Chem. Geol.* **2017**, *466*, 274–284. [CrossRef]
17. Paquette, J.L.; Cluzel, D. U-Pb zircon dating of post-obduction volcanic-arc granitoids and a granulite-facies xenolith from New Caledonia. Inference on Southwest Pacific geodynamic models. *Int. J. Earth Sci.* **2007**, *96*, 613–622. [CrossRef]
18. Maurizot, P.; Cluzel, D. Pre-obduction records of Eocene foreland basins in Central New Caledonia: An appraisal from surface geology and Cadart-1 borehole data. New Zealand. *J. Geol. Geophys.* **2014**, *57*, 300–311. [CrossRef]
19. Sevin, B.; Ricordel-Prognon, C.; Quesnel, F.; Cluzel, D.; Lesimple, S.; Maurizot, P. First palaeomagnetic dating of ferricrete in New Caledonia: New insight on the morphogenesis and palaeoweathering of 'Grande Terre'. *Terra Nova* **2012**, *24*, 77–85. [CrossRef]
20. Chardon, D.; Chevillotte, V. Morphotectonic evolution of the New Caledonia ridge (Pacific Southwest) from post-obduction tectonosedimentary record. *Tectonophysics* **2006**, *420*, 473–491. [CrossRef]
21. Join, J.-L.; Robineau, B.; Ambrosi, J.-P.; Costis, C.; Colin, F. Système hydrogéologique d'un massif minier ultrabasique de Nouvelle-Calédonie. *C. R. Geosci.* **2005**, *337*, 9. [CrossRef]
22. Lundberg, J.; Taggart, B.E. Dissolution pipes in northern Puerto Rico: An exhumed paleokarst. *Carbonates Evaporites* **1995**, *10*, 171–183. [CrossRef]
23. Cathelineau, M.; Quesnel, B.; Gautier, P.; Boulvais, P.; Couteau, C.; Drouillet, M. Nickel dispersion and enrichment at the bottom of the regolith: Formation of pimelite target-like ores in rock block joints (Koniambo Ni deposit, New Caledonia). *Miner. Deposita* **2016**, *51*, 71–282. [CrossRef]
24. Cathelineau, M.; Myagkiy, A.; Quesnel, B.; Boiron, M.-C.; Gautier, P.; Boulvais, P.; Ulrich, M.; Truche, L.; Golfier, F.; Drouillet, M. Multistage crack seal vein and hydrothermal Ni enrichment in serpentinised ultramafic rocks (Koniambo massif, New Caledonia). *Miner. Depos.* **2017**, *52*, 945–960. [CrossRef]
25. Quesnel, B.; Boulvais, P.; Gautier, P.; Cathelineau, M.; John, C.M.; Dierick, M.; Agrinier, P.; Drouillet, M. Paired stable isotopes (O, C) and clumped isotope thermometry of magnesite and silica veins in the New Caledonia Peridotite Nappe. *Geochim. Cosmochim. Acta* **2016**, *183*, 234–249. [CrossRef]
26. Wopfner, H. *Silcretes of Northern South Australia and Adjacent Regions*; Langford-Smith, T., Ed.; Silcrete in Australia; University of New England Press: Armidale, Australia, 1978; pp. 93–141.
27. Summerfield, M.A. Silcrete as a palaeoclimatic indicator: Evidence from southern Africa. *Palaeogeogr. Palaeoclimatol. Palaeoecol.* **1983**, *41*, 65–79. [CrossRef]
28. Taylor, G.; Eggleton, R.A. Silcrete: An Australian perspective. *Aust. J. Earth Sci.* **2017**, *64*, 987–1016. [CrossRef]
29. Quesnel, B.; Gautier, P.; Cathelineau, M.; Boulvais, P.; Couteau, C.; Drouillet, M. The internal deformation of the Peridotite Nappe of New Caledonia: A structural study of serpentine-bearing faults and shear zones in the Koniambo Massif. *J. Struct. Geol.* **2016**, *85*, 51–67. [CrossRef]

30. Jeanpert, J.; Dewandel, B. *Analyse Préliminaire des Données Hydrogéologiques du Massif du Koniambo, Nouvelle-Calédonie*; SGNC/DIMENC. BRGM/RP-61765-FR; SGNC: Noumea, New Caledonia, 2013; 95p.

31. Jeanpert, J. Structure et Fonctionnement Hydrogéologique des Massifs de Péridotites de Nouvelle-Calédonie. Ph.D. Thesis, Université de la Réunion, Saint-Denis, France, 2017.

32. Maurizot, P.; Lafoy, Y.; Poupée, M. *Cartographie des Formations Superficielles et des Aléas Mouvements de Terrain en Nouvelle-Calédonie*; Zone du Koniambo. BRGM, Public Report RP51624-FR; BRGM: Orléans, France, 2002; 45p.

33. Chevillotte, V. Morphogenèse Tropicale en Contexte Epirogénique Modéré. Exemple de la Nouvelle-Calédonie (Pacifique Sud-Ouest). Ph.D. Thesis, Université de la Nouvelle-Calédonie, Nouméa, France, 2005.

34. Longerich, H.P.; Jackson, S.E.; Gunther, D. Laser ablation inductively coupled plasma mass spectrometric transient signal data acquisition and analyte concentration calculation. *J. Anal. At. Spectrom.* **1996**, *11*, 899–904. [CrossRef]

35. Rollion-Bard, C.; Mangin, D.; Champenois, M. Development and application of oxygen and carbon isotopic measurements of biogenic carbonates by ion microprobe. *Geostand. Geoanal. Res.* **2007**, *31*, 39–50. [CrossRef]

36. Robert, F.; Chaussidon, M. A palaeotemperature curve for the Precambrian oceans based on silicon isotopes in cherts. *Nature* **2006**, *443*, 969–972. [CrossRef]

37. Marin, J.; Robert, F.; Chaussidon, M. Microscale oxygen isotope variations in 1.9 Ga Gunflint cherts: Assessments of diagenesis effects and implications for oceanic paleo-temperature reconstructions. *Geochim. Cosmochim. Acta* **2010**, *74*, 116–130. [CrossRef]

38. Parkhurst, D.L.; Appelo, C.A.J. User's guide to PHREEQC (Version 2): A computer program for speciation, batch-reaction, one-dimensional transport, and inverse geochemical calculations. *Water Resour. Investig. Rep.* **1999**, *99*, 312. [CrossRef]

39. Parkhurst, D.L.; Appelo, C.A.J. Description of input and examples for PHREEQC version 3—A computer program for speciation, batch-reaction, one-dimensional transport, and inverse geochemical calculations. *US Geol. Surv. Tech. Methods* **2013**, *6*, 497.

40. Bethke, C.M. *Geochemical and Biogeochemical Reaction Modeling*, 2nd ed.; Cambridge University Press: Cambridge, UK, 2007.

41. Latham, M. On geomorphology of northern and western New-Caledonian massifs. In Proceedings of the International Symposium on Geodynamics in the Southwest Pacific, Noumea, New Caledonia, 27 August–2 September 1977; pp. 235–244.

42. Chevillotte, V.; Chardon, D.; Beauvais, A.; Maurizot, P.; Colin, F. Long-term tropical morphogenesis of New Caledonia (Southwest Pacific): Importance of positive epeirogeny and climate change. *Geomorphology* **2006**, *81*, 361–375. [CrossRef]

43. Jeanpert, J.; Iseppi, M.; Adler, P.M.; Genthon, P.; Sevin, B.; Thovert, J.F.; Dewandel, B.; Join, J.L. Fracture controlled permeability of ultramafic basement aquifers. Inferences from the Koniambo massif, New Caledonia. *Eng. Geol.* **2019**, *256*, 67–83. [CrossRef]

44. Villanova-De-Benavent, C.; Nieto, F.; Viti, C.; Proenza, J.A.; Galí, S.; Roqué-Rosell, J. Ni-phyllosilicates (garnierites) from the Falcondo Ni-laterite deposit (Dominican Republic): Mineralogy, nanotextures, and formation mechanisms by HRTEM and AEM. *Am. Mineral.* **2016**, *101*, 1460–1473. [CrossRef]

45. Myagkiy, A.; Golfier, F.; Truche, L.; Cathelineau, M. Reactive transport modeling applied to Ni laterite ore deposits in New Caledonia: Role of hydrodynamic factors and geological structures in Ni mineralisation. *Geochem. Geophys. Geosyst.* **2019**, *20*, 1425–1440. [CrossRef]

46. Butt, C.R.M. Discussion of 'Silicified serpentinite—A residuum of a Tertiary palaeo-weathering surface in the United Arab Emirates. *Geol. Mag.* **2014**, *151*, 1144–1146. [CrossRef]

47. Sevin, B.; Cluzel, D.; Maurizot, P.; Ricordel-Prognon, C.; Chaproniere, G.; Folcher, N.; Quesnel, F. A drastic lower Miocene regolith evolution triggered by post obduction slab break-off and uplift in New Caledonia. *Tectonics* **2014**, *33*, 1787–1801. [CrossRef]

48. Webb, J.A.; Golding, S.D. Geochemical mass-balance and oxygen-isotope constraints on silcrete formation and its palaeoclimatic implications in southern Australia. *J. Sediment. Res.* **1998**, *68*, 981–993. [CrossRef]

49. Marin-Carbonne, J.; Robert, F.; Chaussidon, M. The silicon and oxygen isotope compositions of Precambrian cherts: A record of oceanic paleo-temperatures? *Precambrian Res.* **2014**, *247*, 223–234. [CrossRef]

50. Alexandre, A.; Meunier, J.D.; Llorens, E.; Hill, S.M.; Savin, S.M. Methodological improvements for investigating silcrete formation: Petrography, FT-IR and oxygen isotope ratio of silcrete quartz cement, Lake Eyre Basin (Australia). *Chem. Geol.* **2004**, *211*, 261–274. [CrossRef]

51. Kelly, J.L.; Fu, B.; Kita, N.T.; Valley, J.W. Optically continuous silcrete quartz cements of the St. Peter Sandstone: High precision oxygen isotope analysis by ion microprobe. *Geochim. Cosmochim. Acta* **2007**, *71*, 3812–3832. [CrossRef]

52. Bustillo, M.Á.; Alonso-Zarza, A.M.; Plet, C. Isotopic composition (δ^{13}C, δ^{18}O and δD) of calcrete-silcrete intergrades: Paleoenvironmental significance (Mioceno, Torrijos area, Madrid Basin). *Geogaceta* **2014**, *56*, 67–70.

53. Matsuhisa, Y.; Goldsmith, J.R.; Clayton, R.N. Oxygen isotopic fractionation in the system quartz-albite-anorthite-water. *Geochim. Cosmochim. Acta* **1979**, *43*, 1131–1140. [CrossRef]

54. Basile-Doelsch, I.; Meunier, J.D.; Parron, C. Another continental pool in the terrestrial silicon cycle. *Nature* **2005**, *433*, 399–402. [CrossRef]

55. Opergelt, S.; Delmelle, P. Silicon isotopes and continental weathering processes: Assessing controls on Si transfer to the ocean. *C. R. Geosci.* **2012**, *344*, 723–738. [CrossRef]

56. Kleine, B.L.; Steffansson, A.; Halldorsson, S.A.; Whitehouse, M.J.; Jonasson, K. Silicon and oxygen isotopes unravel quartz formation processes in the Icelandic crust. *Geochem. Perspect. Lett.* **2018**, *7*, 5–11. [CrossRef]

57. Brengman, L.A.; Fedo, C.M.; Whitehouse, M.J. Micro-scale silicon isotope heterogeneity observed in hydrothermal quartz precipitates from the >3.7 Ga Isua Greenstone Belt, SW Greenland. *Terra Nova* **2016**, *28*, 70–75. [CrossRef]

58. Marin-Carbonne, J.; Chaussidon, M.; Robert, F. Micrometer-scale chemical and isotopic criteria (O and Si) on the origin and history of Precambrian cherts: Implications for paleo-temperature reconstructions. *Geochim. Cosmochim. Acta* **2012**, *92*, 129–147. [CrossRef]

59. Thiry, M.; Milnes, A.R. Silcretes: Insights into the occurrences and formation of materials sourced for stone tool making. *J. Archaeol. Sci. Rep.* **2016**, *15*, 500–513. [CrossRef]

60. Lacinska, A.M.; Styles, M.T. Silicified serpentinite—A residuum of a Tertiary palaeo-weathering surface in the United Arab Emirates. *Geol. Mag.* **2013**, *150*, 385–395. [CrossRef]

61. Thorne, R.; Herrington, R.; Robert, S. Composition and origin of the Çaldag oxide nickel laterite, W. Turkey. *Min. Depos.* **2009**, *44*, 581–595. [CrossRef]

62. Butt, C.R.M.; Nickel, E.H. Mineralogy and geochemistry of the weathering of the disseminated nickel sulfide deposit at Mt. Keith, Western Australia. *Econ. Geol.* **1981**, *76*, 1736–1751. [CrossRef]

63. Idnurm, M.; Senior, B.R. Palaeomagnetic ages of late Cretaceous and Tertiary weathered profiles in the Eromanga Basin, Queensland. *Palaeogeogr. Palaeoclimatol. Palaeoecol.* **1978**, *24*, 263–272. [CrossRef]

64. Van der Graaff, W.J.E. Silcrete in Western Australia: Geomorphological settings, textures, structures, and their possible genetic implications. In *Residual Deposits: Surface Related Weathering Processes and Materials*; Wilson, R.C.L., Ed.; Special Publications: London, UK, 1983; Volume 11, pp. 159–166.

65. Li, J.-W.; Vasconcelos, P. Cenozoic continental weathering and its implications for the palaeoclimate: Evidence from ^{40}Ar/^{39}Ar geochronology of supergene K-Mn oxides in Mt Tabor, Central Queensland, Australia. *Earth Planet. Sci. Lett.* **2002**, *200*, 223–239. [CrossRef]

66. Mathian, M.; Chassé, M.; Calas, G.; Griffin, W.L.; O'Reilly, S.Y.; Buisson, T.; Allard, T. Insights on the Cenozoic climatic history of Southeast Australia from kaolinite dating. *Palaeogeogr. Palaeoclimatol. Palaeoecol.* **2022**, *604*, 111212. [CrossRef]

67. Radtke, U.; Brückner, H. Investigation on age and genesis of silcretes in Queensland (Australia)-Preliminary results. *Earth Surf. Process. Landf.* **1991**, *16*, 547–554. [CrossRef]

68. Favier, S.; Teitler, Y.; Golfier, F.; Cathelineau, M. Multiscale physical–chemical analysis of the impact of fracture networks on weathering: Application to Nickel redistribution in the formation of Ni–laterite ores, New Caledonia. *Ore Geol. Rev.* **2022**, *147*, 104971. [CrossRef]

69. Butt, C.R.M. Granite weathering and silcrete formation on the Yilgarn Block, Western Australia. *Aust. J. Earth Sci.* **1985**, *32*, 415–432. [CrossRef]

minerals

MDPI

Article

Occurrence of SiC and Diamond Polytypes, Chromite and Uranophane in Breccia from Nickel Laterites (New Caledonia): Combined Analyses

Yassine El Mendili [1,2,*], Beate Orberger [3], Daniel Chateigner [1], Jean-François Bardeau [4], Stéphanie Gascoin [1], Sébastien Petit [1] and Olivier Perez [1]

check for updates

Citation: El Mendili, Y.; Orberger, B.; Chateigner, D.; Bardeau, J.-F.; Gascoin, S.; Petit, S.; Perez, O. Occurrence of SiC and Diamond Polytypes, Chromite and Uranophane in Breccia from Nickel Laterites (New Caledonia): Combined Analyses. *Minerals* **2022**, *12*, 196. https://doi.org/10.3390/min12020196

Academic Editors: Cristina Domènech and Cristina Villanova-de-Benavent

Received: 8 December 2021
Accepted: 28 January 2022
Published: 2 February 2022

Publisher's Note: MDPI stays neutral with regard to jurisdictional claims in published maps and institutional affiliations.

[1] CRISMAT-ENSICAEN, UMR CNRS 6508, Université de Caen Normandie, IUT Caen, Normandie Université, 6 boulevard Maréchal Juin, 14050 Caen, France; daniel.chateigner@ensicaen.fr (D.C.); stephanie.gascoin@ensicaen.fr (S.G.); sebastien.petit@ensicaen.fr (S.P.); olivier.perez@ensicaen.fr (O.P.)
[2] Laboratoire ESITC Caen-COMUE Normandie Université, 1 Rue Pierre et Marie Curie, 14610 Epron, France
[3] GEOPS, Université Paris Saclay-Paris Sud, UMR 8148 (CNRS-UPS), Bât 504, 91405 Orsay, France; beate.orberger@universite-paris-saclay.fr
[4] IMMM, Le Mans Université, UMR6283 CNRS, Avenue Olivier Messiaen, 72085 Le Mans, France; jean-francois.bardeau@univ-lemans.fr
* Correspondence: yassine.el-mendili@esitc-caen.fr; Tel.: +33-231-452-628

Abstract: Different techniques have been combined to identify the structure and the chemical composition of siliceous breccia from a drill core of nickel laterites in New Caledonia (Tiebaghi mine). XRD analyses show quartz as a major phase. Micro-Raman spectroscopy confirmed the presence of reddish microcrystalline quartz as a major phase with inclusion of microparticles of iron oxides and oxyhydroxide. Lithoclasts present in breccia are composed of lizardite, chrysotile, forsterite, hedenbergite and saponite. The veins cutting through the breccia are filled with Ni-bearing talc. Furthermore, for the first time, we discovered the presence of diamond microcrystals accompanied by moissanite polytypes (SiC), chromite ($FeCr_2O_4$) and uranophane crystals ($Ca(UO_2)_2(SiO_3OH)_2 \cdot 5(H_2O)$) and lonsdaleite (2H-[C-C]) in the porosities of the breccia. The origin of SiC and diamond polytypes are attributed to ultrahigh-pressure crystallization in the lower mantle. The SiC and diamond polytypes are inherited from serpentinized peridotites having experienced interaction with a boninitic melt. Serpentinization, then weathering of the peridotites into saprolite, did not affect the resistant SiC polytypes, diamond and lonsdaleite. During karstification and brecciation, silica rich aqueous solutions partly digested the saprolite. Again, the SiC polymorph represent stable relics from this dissolution process being deposited in breccia pores. Uranophane is a neoformed phase having crystallized from the silica rich aqueous solutions. Our study highlights the need of combining chemical and mineralogical analytical technologies to acquire the most comprehensive information on samples, as well as the value of Raman spectroscopy in characterizing structural properties of porous materials.

Keywords: Ni-laterites; mineralogy; breccia; uranium; SiC; combined analyses; Raman

1. Introduction

The occurrences of moissanite (SiC) associated with NiMnCo, FeSi and FeC alloys, are described from many peridotites and podiform chromitites belonging to ophiolites in Turkey, China, Myanmar, Russia, Mexico and Cuba [1–5]. Moissanite, natural SiC, is extremely rare. It was found in meteorite, kimberlites [6], metasomatic rocks [7], peridotites, serpentinites and podiform chromitites [8,9]. Moissanite intergrowth with other minerals have also been documented in eclogites and serpentinites from the Dabie-Sulu belt in China [9,10] and in volcanic rocks [7,11]. The more frequent natural moissanite are the 3C, 4H, 6H and 15R polytypes, where C, H and R polytypes are cubic, hexagonal, and rhombohedral, respectively. All these phases form at extremely low oxygen fugacity, 5 to 6 log units below the IW buffer [12]. The stability of these phases at surface conditions is explained by rapid tectonic uplift, while slow tectonic uplift would lead to diamond decomposition, as the observed graphitized diamonds from the Beni Bousera, Morocco [13] and the Ronda massifs, Spain [14]. Based on mineralogical studies, the super reduced association of moissanite, native Cu, Fe–Mn alloy in Al rich chromitite from the Mercedita ophiolite in Cuba was attributed to a low temperature serpentinization process in a very reduced microenvironment [4,15,16]. A similar model was proposed for micro diamonds found in chromite in podiform chromitite in serpentinite from the Tehuitzingo ophiolite in South Mexico [16].

None of these minerals and mineral association have been so far described from nickel laterite profiles. Recent studies by El Mendili et al. [17] give a detailed description of the 6O-SiC polytype in the siliceous breccia studied here. Its presence in an alteration profile is attributed to a three-stage process: formation in the lower mantle, transport into the upper mantle peridotite by second stage melts, weathering, and preservation in breccia pores.

Uranophane is a hydrous silicate ($Ca(UO_2)_2(SiO_3OH)_2.5(H_2O)$) occurring commonly as secondary minerals in oxidizing silica rich environment, fractures, mainly as result of uraninite (UO_2) alteration related to uranium and associated metal mineralization or nuclear fuel [18–20]. In laterites (Parana sedimentary basin, Brazil, hosting sedimentary and ultramafic-mafic intrusive rocks), uranophane was discovered in uraniferous nodules described from iron crusts [21]. Uraniferous minerals are not yet described from nickel laterites and New Caledonian rocks so far.

In this study, we report the first occurrence of diamond, lonsdaleite, moissanite among other reduced phases, and uranophane from silica rich rocks at the base of a nickel laterite profile, located in saprolite, at the Tiebaghi mine, New Caledonia. The silica rich rocks (sample ER-NC00-0001) was analyzed in the frame of the H2020 EU project SOLSA to evaluate the influence of surface roughness on close sensor drill core analyses [22,23].

2. Materials and Methods

2.1. Material

The siliceous dissolution breccia (SB; SOLSA label: ER-NC00-0001, Geographic coordinates: Latitude: 20°26′24.59″ S, Longitude: 164°13′1.20″ E) was sampled from a drill core performed in the nickel laterite profile at the Tiébaghi Mine in the northwestern part of New Caledonia.

2.2. Geological Setting

New Caledonia located in the Pacific Ocean 200 km off-shore NE Australia, represents a microcontinent composed of an autochtonous basement of metamorphic and metavolcanosedimentary rocks of Upper Carboniferous to Lower Cretaceous age, and Late Cretaceous to Lower Tertiary allochtonous terranes. Part of these terranes is the New Caledonia ophiolite [24–26]. The lower part of this ophiolite covers large terranes of the southern part of the island, while tectonic ophiolitic klippes occur in the northwestern part. The lithologies comprise serpentinized harzburgite (±lherzolite), and serpentinized pyroxenite, wherlite and dunite overlain by gabbro [25,26]. At Tiébaghi, the peridotites are much richer in chromite and show a higher diversity (dunites, harzburgite, lherzolite, pyroxenites), which may be related to magmatic differentiation and melt rock interactions [27]. Numerous dykes, sometimes pegmatitic, cut through the ultramafic rocks [28].

Lateritization of the ophiolite started in Oligocene (24 ± 5 Ma) [29] and resulted in the formation of world-class nickel laterite ore deposits in the lower part of the regolith. These nickel deposits represent ~9% of the world reserves [30]. Classically, the serpentinized peridotites decomposed into (a) sap-rock, representing serpentinized lithoclasts occurring in a soil matrix, (b) the nickel-rich saprolite, overlain successively by yellow and red laterite, with (c) a top layer of ferricrete. During lateritization, silicates (relict olivines and pyroxenes) are dissolved, and serpentines of several generations are gradually transformed into clay minerals and talc, while dissolved iron precipitates as oxy hydroxides and dissolved silica as quartz [31]. Nickel is often enriched in the latest serpentine generation (3 wt.%) and talc-like phases such as kerolite (about a factor 10 richer than serpentines: 30 wt.% [23,32–34]. Meteoric alteration starting at 24 ± 5 Ma, is favored by an intense fracture network related to orogenic and post-orogenic (relaxation) tectonics [35–39]. In this high hydraulic conductivity environment, aquifers percolate at different levels of the laterite profile reserves [40] and form pseudo-karstic structures [36,41,42]. At Tiébhagi, the lower aquifer drains the fine and coarse saprolite and the highly fractured serpentinized peridotites [40,41]. Dissolution, and resorption of serpentinized peridotite related to down drainage meteoric water are observed [40]. This alteration caused the emplacement of a collapse breccia (>50% clasts) [41] along vertical channels and egg-shape morphologies at the base of the laterite profile related in particularly, the egg-shape morphology at the base of the saprolites, to alteration at hydrothermal temperatures (70–90 °C) [40,43]. Based on fluid flow modelling, the authors suggest that meteoric waters heated-up through exothermic reactions associated with serpentine weathering.

2.3. Sample Description

At Tiébaghi, the lower part of the regolith hosts a white-beige-brownish, porous silica rich rock. It has a typical boxwork structure (Figure 1). This rock hosts a few fragments, relicts of saprolite. It is a dissolution breccia. We call it hereafter siliceous breccia. The overall grain size of the breccia is <2 mm. The breccia is mainly composed of quartz of millimetric size, red spots represent iron oxyhydroxides, and locally black oxides occur (<1 mm). Crosscutting veins are filled with talc and clay minerals [22]. The studied samples were cut out from the drill core surfaces as prisms of $\approx 20 \times 20 \times 6$ mm^3 size. From macroscopic observations, the formation of this breccia can be related to the epikarstic processes [40].

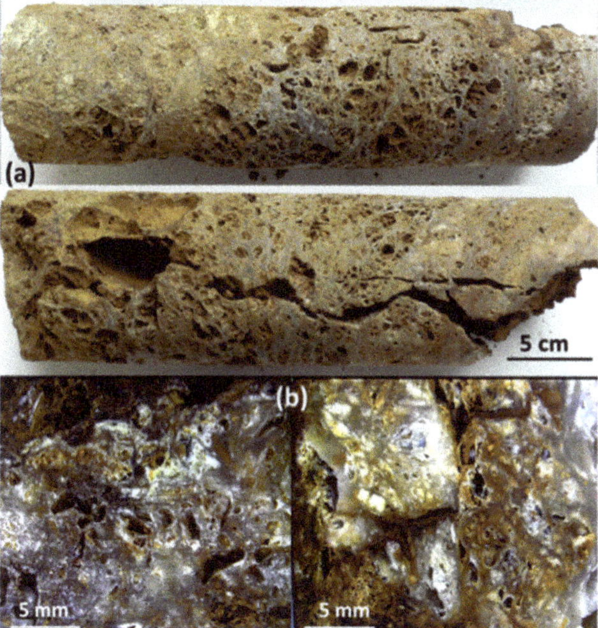

Figure 1. (**a**) Photographs of the breccia drill cores and (**b**) binocular images of breccia squares studied in this work.

2.4. Analytical Methods

Imaging, mapping, and compositional analysis of SiC polymorphs were performed using scanning electron microscope (SUPRA™ 55 SAPPHIRE; Carl Zeiss, Jena, Germany) equipped with energy dispersive spectrometer (EDS). The measurements were done at 15 kV.

X-ray fluorescence spectra were performed using an Inel Equinox 3500 spectrometer (Inel Equinox 3500, Thermo Fisher Scientific Inc., Waltham, MA, USA), equipped with a Cu microfocus source and an Amptek X-123SDD Silicon Drift Detector placed vertically 10 mm over the sample to ensure high sensitivity even with low-atomic number elements. XRF data were collected with an integration time of 600s.

High resolution X-ray diffraction of breccia sample was performed on a D8 Advance Vario 1 Bruker two-circles diffractometer (θ–2θ Bragg-Brentano scan) (D8 Advance Vario 1, Bruker, Karlsruhe, Germany) using a Cu Kα radiation (λ = 1.54059 Å) selected by an incident beam Ge (111) monochromator (Johansson type) and equipped with a LynxEye detector. XRD diagram is collected at room temperature for 2θ varying from 10° to 100° for 1 s per 0.0105° step (12 h/scan).

The Raman spectra were recorded at room temperature using a DXR Raman microscope (DXR, Thermo Fisher Scientific Inc., Waltham, MA, USA) equipped with a 900 lines/mm diffracting grating. The highest quality spectra (highest signal-to-noise) were obtained using green excitation (532 nm, Nd: YAG laser). Raman measurements were carried out at low laser power (1 mW) and with a 100× magnification long working distance. Raman spectra were collected over a range of 80–2000 cm^{-1} (systematically recorded twice with an integration time of 60 s). The spectral region 3300–4000 cm^{-1} was investigated for minerals species with expected OH stretching vibration modes. The laser spot diameter was estimated at 0.8 µm. The peak deconvolution was performed with the Origin software. The mineralogical composition was determined by comparing the collected Raman signals with those reported in the ROD database [44].

3. Results

3.1. Major Composition of the Siliceous Breccia

The chemical analyses by X-Ray Fluorescence (XRF) of the breccia surface indicates ~88 wt.% of silica and 1.9 wt.% Fe_2O_3 (as total iron). Traces of CaO and MgO (0.1 and 0.7 wt.%, respectively) are also present. In term of elemental composition, XRF indicate the presence of Si, Mg, Fe, and other trace elements such as Ni and Cr (Table 1).

Table 1. Elemental composition of breccia in weight, as obtained from XRF measures.

Element	Composition (wt.%)
O	42.7 ± 1.5
Si	25.3 ± 0.8
Mg	21.7 ± 0.8
Fe	4.9 ± 0.4
Ni	0.6 ± 0.2
Al	1.1 ± 0.1
Ca	1.0 ± 0.1
Cr	1.2 ± 0.1
Na	1.25 ± 0.2
K	0.20 ± 0.02

X-ray diffraction pattern (Figure 2) of breccia samples only exhibits the signature of highly crystalline quartz (100 wt.%). The Rietveld refinement performed with a trigonal space group P3$_2$21 converged to quartz lattice parameters of a = 4.916 Å and c = 5.408 Å with a mean crystallite size of 161 nm and the microstrain of 8×10^{-4} rms. These refined lattice parameters are very close to that reported in literature [45]. The general R-factors which indicate the overall goodness of fit between the model and experimental data are:

Rwp = 6.3% and Rb = 5.6%, giving goodness of fit of 1.3. It is important to note that no other phase was detected by XRD even though EDS and XRF analysis indicate the presence of 5–7% of iron and other elements. However, the mineral species potentially formed by such elements are not detected by XRD because they are present at very low concentrations (<3%).

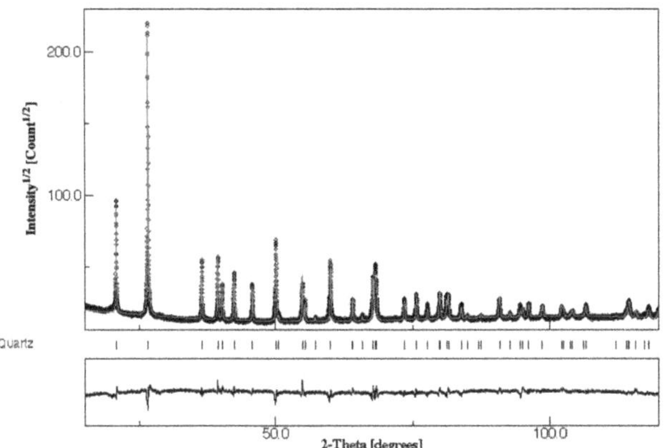

Figure 2. X-ray powder diffraction pattern of breccia sample and corresponding MAUD refinement (MAUD: Material Analysis Using Diffraction, refinement software, http://maud.radiographema.eu/ (accessed on 10 January 2021)). Calculated (red line) and observed patterns (coarse line) are shown. The difference curve ($I_{obs} - I_{calc}$) is shown at the bottom.

Iron is within 5–7 wt.% and is related to iron oxides and oxyhydroxide (hematite, maghemite, magnetite and goethite) present in the quartz matrix as micro to nano-inclusions (Figure 3b). Raman analyses of the breccia matrix confirm that quartz is predominant. The well-defined and mostly narrow Raman bands of quartz attested to its highly crystalline form (Figure 3a). Raman spectroscopy also evidences the presence of serpentine minerals (Figure 3c), lizardite and chrysotile, which crystallized in densely packed fibers. Forsterite, hedenbergite and yellow saponite occur as clots scattered throughout the fine-grained matrix (Figure 3d). Minor phases, e.g., anatase, pyrite and apatite were also observed (Figure 3e). The microfractures in the breccia (Figure 3f–h) are filled with calcite and kerolite 'Ni-rich talc' [33].

The kerolite is the only Ni-bearing phase. The vibrational modes observed for kerolite (Figure 3h) are slightly shifted toward lower wavenumbers compared to pure talc [46] suggesting high Ni substitution in Mg-talc [47]. However, kerolite and Ni-rich talc exhibit three main bands on similar positions (674, 188 and 108 cm^{-1}). In the high wavenumber range (3500–3750 cm^{-1}), the vibrations of the OH groups can be used to discriminate the talc-like phases: the positions and relative intensities of OH contributions are controlled by the types and number of crystallographically equivalent OH groups sites, and by the types of cation occupancies around OH sites, as well as the probabilities of these occupancies.

For Mg-talc, only a strong OH stretching band is expected around 3675 cm^{-1} and for Ni-talc-like (kerolite), the spectrum shows three Ni^{2+} additional vibrations at 3654, 3625, and 3610 cm^{-1} [23,47]. These bands correspond to OH bonded to 2Mg + 1Ni, 1Mg + 2Ni, and 3 Ni, respectively [23,47]. The intensity of the vibration is proportional to the amount of the cation configuration (in our case 3Mg (20.8%), 2Mg + 1Ni (62.4%), 1Mg + 2Ni (1.7%), and 3 Ni (15.1%)). This result thus confirmed the presence of high Ni substitution in Mg-talc.

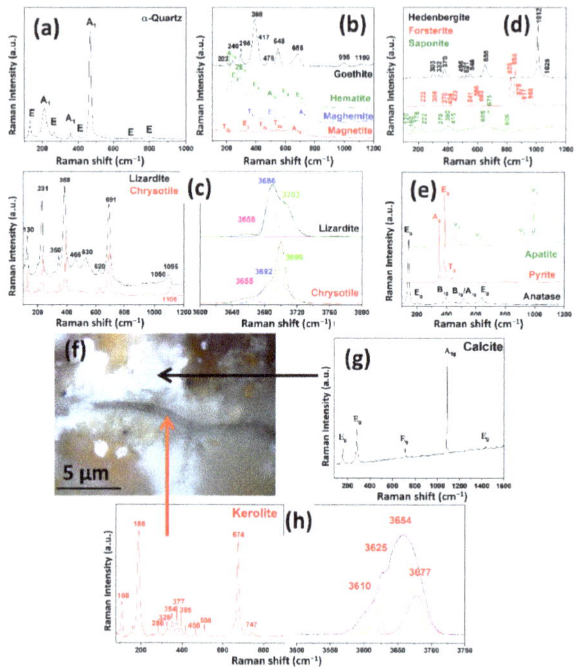

Figure 3. Raman spectra of phases present in the breccia core samples (**a**) α-Quartz (SiO$_2$), (**b**) Microparticles of iron oxides [hematite (α-Fe$_2$O$_3$), maghemite (γ-Fe$_2$O$_3$), magnetite (Fe$_3$O$_4$)] and goethite (α-FeOOH), (**c**) Lizardite and chrysotile (OH)$_3$Mg$_3$[Si$_2$O$_5$(OH)], (**d**) Hedenbergite (CaFeSi$_2$O$_6$), forsterite (Mg$_2$SiO$_4$) and saponite (Ca$_{0.25}$(Mg,Fe)$_3$((Si,Al)$_4$O$_{10}$)(OH)$_2$·n(H$_2$O)), (**e**) Pyrite (FeS$_2$), apatite (Ca$_5$(PO$_4$)$_3$OH) and anatase (TiO$_2$) crystals, (**f**) Optical images of the breccia vein with corresponding Raman spectra of: (**g**) Calcite (CaCO$_3$) and (**h**) Kerolite ((Mg,Ni)$_3$Si$_4$O$_{10}$(OH)$_2$·H$_2$O).

3.2. Pore Fillings in the Siliceous Breccia

Micrometric porosities observed by optical microscopy (Figure 4a,b) incorporate small crystals and clusters of crystals varying from blue to white and colorless-transparent (Figure 4a). Interestingly, micro-Raman measurements on these clusters reveal the presence of chromite and SiC (Figure 4g). SiC (moissanite) occurs either as single grains (approximated size: 5 × 8 × 10 μm^3), sometimes having light blue color to green and colorless irregular flakes or fragments of 10–25 μm (Figure 4c). EDS analyzes on this moissanite crystal indicate C, 48.7 ± 1.9 at. wt.% and Si, 51.3 ± 1.6 at. wt.% (Figure 4d). Although EDS is not accurate for low atomic number elements and not recommended as a technique for the quantification of elements lighter than Na, EDS analyses of moissanite showed that the crystals contained no elements heavier than carbon and Si.

The SiC grain in Figure 4b was selected for a full single crystal diffraction, SCXRD (Figure 5). This SiC grain crystallizes in the space group Cmc21 with 36 atoms per unit cell [a = 3.0778(6) Å, b = 5.335(2) Å, c = 15.1219(6) Å, α = 90°, β = 90°, γ = 120°]. Reflection's splitting can be evidenced in some experimental frames and could be attributed to twinning features. Data were then integrated. The different attempts of integration of reflections using the previous unit cell were not satisfactory: the internal reliability factor R$_{int}$ = 35%, quantifying the symmetry deviation from the intensity of the reflections expected to be equivalent, excludes unambiguously the hexagonal or trigonal symmetries. The reciprocal space was then interpreted in a different way by considering the orthorhombic unit cell and twin components related by a tri-fold axis parallel to c with an internal reliability R$_{int}$ of 7.4%. The structure has been determined by charge in the Cmc21 space group and then refined. The final agreement factor is Rf = 0.0329(2) for 194 reflections with 20 refinement parameters and I ≥ 3σ(I).

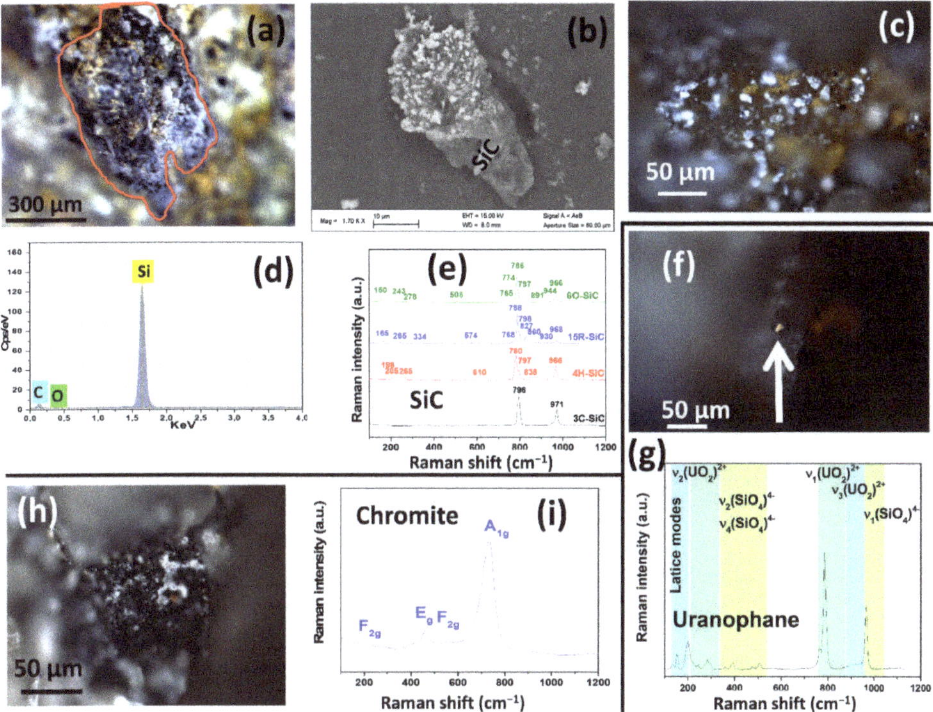

Figure 4. Porosities of siliceous breccia (ER-NC00-0001) (**a**) Optical images of porosities, (**b**) Scanning electron microscope (SEM) image showing a cluster composed of SiC, (**c**) Optical images of various SiC crystals with different colors, (**d**) EDS spectrum of SiC, (**e**) Raman spectra of various SiC crystals with different colors, (**f**) optical images of uranophane, (**g**) Raman spectrum of uranophane, (**h**) optical images of chromite matrix bearing SiC and diamond and (**i**) Raman spectrum of chromite.

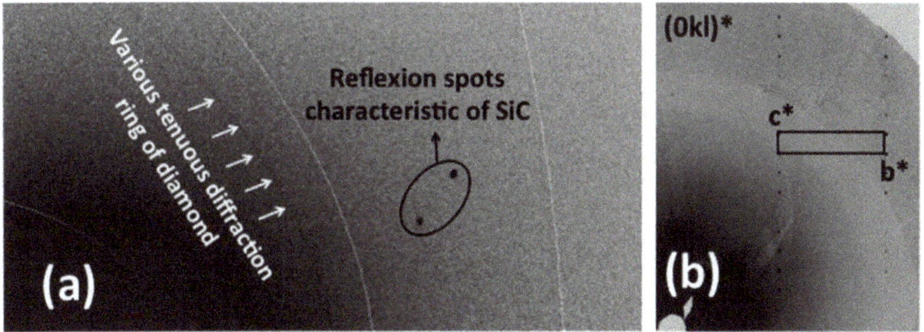

Figure 5. SCXRD of 6O-SiC grain (**a**) Experimental frame exhibiting SiC diffraction spot and (**b**) Part of the (0kl)* reciprocal plan of SiC assembled from the whole experimental frames.

A view of the structure is proposed in Figure 6a. The main characteristics of the 6H-SiC polytype can be observed. The structure can be described from SiC4 tetrahedra all sharing their neighbour edges. El Mendili et al. [17] showed that 6O-SiC results from the 6H-SiC wurtzite to 3C-SiC rock-salt phase transformation as an intermediate state. The 6O-SiC formation requires at least 4 GPa of pressures and high temperatures 2027–2527 °C.

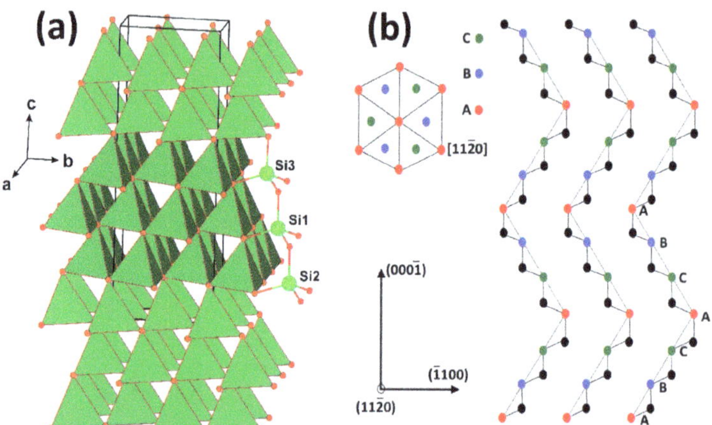

Figure 6. (**a**) Crystal structure refined for the orthorhombic 6O-SiC and (**b**) Stacking sequence of 6O-SiC in the [1120] plane.

Many SiC polytypes exist. They show differences in the stacking sequences of double atomic planes of Si-C along the c-direction. The most common polytypes are 3C, 2H, 4H, 6H, 8H, 9R, 10H, 14H, 15R, 19R, 20H, 21H, and 24R [48]. The three more frequent naturally occurring ones are the cubic 3C (β-type), hexagonal 4H and 6H polytypes. The Raman spectra of SiC polytypes (Figure 4e) show four of the SiC polytypes (3C, 4H, 6O and 15R) [49]. Vibrational bands attributions of SiC polytypes (3C, 4H and 15R) are summarized in the Table 2.

Table 2. Raman shifts of the SiC vibration bands shown in Figure 4e, with reference to [50].

Polytype	Phonon Modes–Frequency (cm^{-1})			
	Transversal Optic TO	Transversal Acoustic TA	Longitudinal Optic LO	Longitudinal Acoustic LA
3C	796	-	971	-
4H	797 780	196 205 256	838 966	610
15R	798 768 788	150 256	860 930 968	334 574

For all the SiC polytypes [50], the Raman modes consist of transverse acoustic and optical phonons (TA, TO), longitudinal acoustic and optical phonons (LA, LO) and longitudinal acoustic phonons (LA). In the case of 3C-SiC, only two strong Raman bands are observed, at 972 cm^{-1} attributed to the longitudinal optical phonon (LO) and at 796 cm^{-1} attributed to the transverse optical phonon mode (TO). In the case of 6H-SiC, the experimental characteristic phonon modes are assigned as follow [50]: E_2(TA) at 151 cm^{-1}, E_1(TA) at 240 cm^{-1}, A_1(LA) at 504 cm^{-1}, E_2(TO) at 762 and 783 cm^{-1}, E_1(TO) at 791 cm^{-1} and A_1(LO) at 887 and 963 cm^{-1}, respectively. For SiC in the siliceous breccia, the main Raman peaks were confirmed and matched with Raman-active vibrational modes of orthorhombic SiC structure with Space Group $Cmc2_1$ [17].

Concerning diamond, a series of polytypes are described and predicted in literature [51–54]. Two of them are observed in natural fabrics, the 3C (cubic diamond) and 2H (hexagonal lonsdaleite) diamond.

The Raman spectra of natural breccia diamonds show a sharp, first order peak of sp^3-bonded carbon centered between 1331.4 and 1318 cm^{-1} (Figure 7) with FWHM (Full Width at Half Maximum) from 4.9 to 7.3 cm^{-1}, respectively. Only the 1100–1500 region is illustrated since no other vibrational mode was observed in the region from 60 to 4000 cm^{-1}. Furthermore, no graphite was detected in our samples.

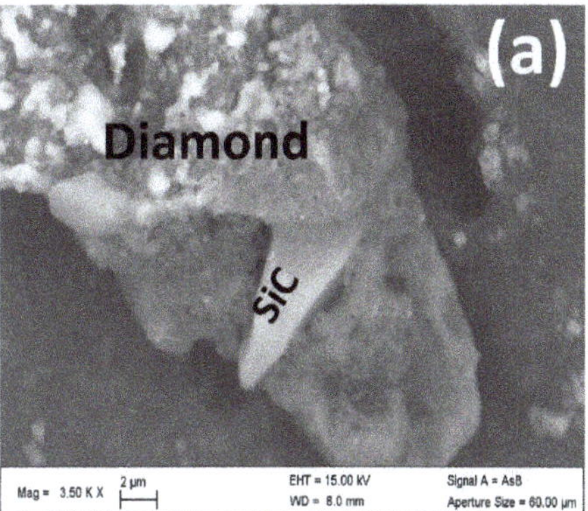

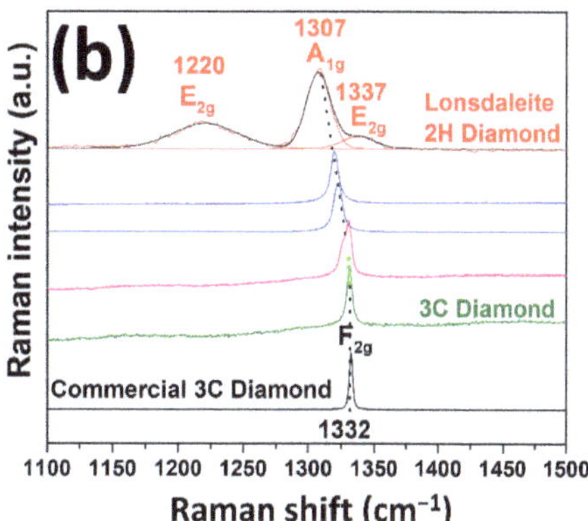

Figure 7. (**a**) Scanning electron microscope (SEM) image showing a cluster composed of SiC and diamond polytypes and (**b**) Raman spectra of five different diamonds from different zones and one commercial cubic diamond.

From literature, it is admitted that the well-crystallized cubic diamond peak is observed at 1332 cm^{-1} and that the sp^3 breathing vibration mode of lonsdaleite (hexagonal diamond) can vary between 1320 and 1327 cm^{-1} [54].

In comparison, a commercial well-crystallized cubic diamond exhibits sp^3 mode centered at 1332 cm^{-1} with a FWHM of 3 cm^{-1}. This value corresponds to one of the spectra observed in Figure 7 (1331.0 ± 0.5 cm^{-1} with a FWHM of 4.9 ± 0.1 cm^{-1}). The slightly larger FWHM is related to a slightly lower crystallinity of the natural cubic diamond.

Raman spectrum (Figure 7, pink) exhibiting two overlapped bands was recorded, the first at 1330.5 ± 0.2 cm^{-1} corresponding to cubic symmetry and the second at 1326.2 ± 0.2 cm^{-1} assigned to hexagonal diamond. Under the microscope, we could not distinguish individual crystals, and these two contributions can be due to the existence of distorted crystals or to the presence of some sub-micrometric diamond inclusion pockets.

For the sp3 in the 1320–1327 cm^{-1} region, the spectra can be attributed to hexagonal diamond polytypes. Indeed, Authors in [54] earlier interpreted Raman spectra of impact diamond-bearing rocks containing the most intense band within 1320–1327 cm^{-1} as the lonsdaleite contribution. The presence of lonsdaleite in their samples was also confirmed by X-ray diffraction.

In (Figure 7, blue), spectra with the sp3 mode at values below 1320 cm^{-1}, were observed. Authors in [55] suggested that this shift is due to stacking faults of diamond structure, leading to the cubic-hexagonal transition. This structural change from cubic to hexagonal can be explained as the change of stacking sequence of the (111) plane. The FWHM values ranging from 6.4 to 7.3 cm^{-1} are also larger, as expected, than those of cubic diamond. We can then hypothesize that the hexagonal structure is kept in these diamonds, however with some distortions explaining its slightly lower Raman shift and its larger FWHM. Finally, (Figure 7, brown) shows a Raman spectrum with three Raman-active vibrational modes around 1212, 1307 and 1328 cm^{-1}. This spectrum is similar to those reported for lonsdaleite [56].

In literature, although the theoretical calculations of the lonsdaleite vibration spectrum have been the subject of many investigations for years, several ambiguities and contradictions remain present for the vibrational attribution and the position of the bands [56–58].

For this reason, the Raman-active vibrational modes of cubic and hexagonal diamonds were calculated through Density Functional Theory (DFT). We have used the CRYSTAL software [59]. We performed these calculations applying harmonic approximation at the Γ point as it was done by [56]. for Raman identification of lonsdaleite. We have found for cubic diamond that the position of single Raman vibration band, corresponding to the first order scattering of F_{2g} symmetry, is 1331.99 cm^{-1}. This value is in good agreement with the experimental value. For hexagonal diamond (2H, lonsdaleite), we predict three fundamental vibrational modes: 1207 cm^{-1} (E_{2g}), 1307 cm^{-1} (A_{1g}), and 1330 cm^{-1} (E_{1g}). The theoretical intensity ratio of the Raman modes is: 0.5(E_{2g}):1(A_{1g}):0.3(E_{1g}). Hence, the A_{1g} mode is expected to be the most intense line in the Raman spectrum of lonsdaleite.

Based on the comparison between calculated and experimental results (Figure 8), the most intense band in the experimental Raman spectrum at 1307 cm^{-1} can be attributed unambiguously to the longitudinal optical vibrational mode A_{1g}.

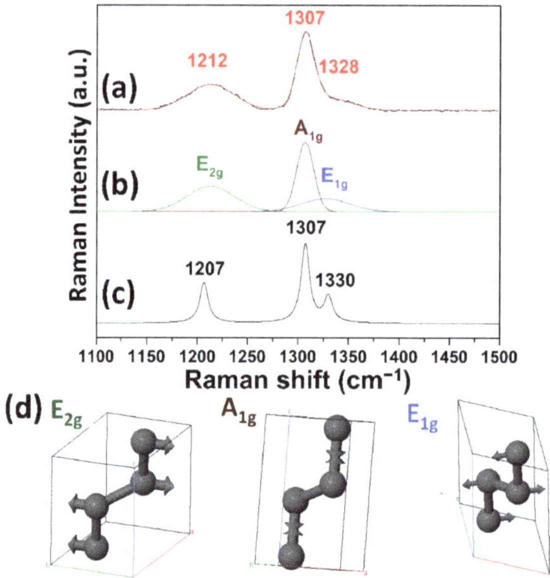

Figure 8. Raman spectrum of lonsdaleite (**a**) Experimental Raman spectrum of lonsdaleite zoomed from (Figure 2f), (**b**) The three Raman-active vibrational modes E_{2g}, A_{1g} and E_{1g} obtained via deconvolution of the experimental spectrum, (**c**) Theoretical Raman spectrum of lonsdaleite obtained through ab initio calculations with Lorentzian line shape and FWHM of 10 cm^{-1} and (**d**) The three Raman vibration modes of lonsdaleite.

The deconvolution analysis of the Raman spectrum let us identify the shoulder as a Raman contribution of the second intensity of the transverse optical vibrational mode (E_{2g}) observed at 1212 cm^{-1}. This agrees with our ab-initio calculations. DFT calculations also predict that the third Raman-active mode of lonsdaleite E_{1g} (transverse optical) should be observed at 1330 cm^{-1}, with its theoretically predicted intensity being close to that of the second intensity mode E_{2g}. The E_{2g} mode is observed in our spectrum around 1328 cm^{-1}.

Experimentally, the lonsdaleite bands are highly broadened (FWHM are in the range of 18–52 cm^{-1}), compared to the cubic diamond observed in our sample. This effect is due to lonsdaleite imperfection and to small dimensions of crystallites.

The originality of these results comes from the fact that lonsdaleite has been observed so far only in the meteorite from the Meteor Crater [51,56]. Probable contribution of other hexagonal diamond polytypes to Raman spectra can be masked by the presence of large and asymmetric bands [54].

In Figure 4g, the Raman spectrum of chromite shows four bands associated to the CrO bond-stretching region at 905, 730, 560 and 445 cm^{-1} [60]. The very intense and broad band at 730 cm^{-1} is assigned to symmetric stretching vibrational mode, $A_{1g}(v_1)$. The two peaks at 560 and 445 cm^{-1} are attributed to $F_{2g}(v_4)$ and $E_g(v_2)$ modes respectively, and the small band at 215 cm^{-1} to the $F_{2g}(v_3)$ mode.

Uranophane was unambiguously determined in the Raman spectrum shown in Figure 4f containing the characteristic bands reported in [61,62]. The identification of the different secondary phases is based on the analysis of the symmetrical stretching vibration of the uranyl group UO_2^{2+}, which allows the identification of individual uranyl phases and can be used as a fingerprint. For uranyl, two Raman bands close to 790 and 800 cm^{-1} are attributed to the doubly degenerate v_1 mode of UO_2^{2+} symmetric stretching modes, while a weak band close to 919 cm^{-1} is assigned to the $v_3(UO_2^{2+})$ anti-symmetric stretching vibrations, and the bands in the 200–300 cm^{-1} are assigned to $v_2(UO_2^{2+})$ symmetric bending mode and to UO ligand vibration [63]. Multiple bands in this region indicate the non-equivalence of the UO bonds and the lifting of the degeneracy of $v_2(UO_2^{2+})$ vibrations. In addition, the band at 148 cm^{-1} is attributed to lattice mode. Raman spectrum shows the expected $v_2(SiO_4^{4-})$ and $v_4(SiO_4^{4-})$ symmetric bending modes at 485 and 510 cm^{-1}, respectively. The anti-symmetric stretching modes $v_1(SiO_4^{4-})$ is present at 970 cm^{-1}. Figure 9 shows the SEM image and the chemical composition (EDS) of the uranium rich zone. The observation of uranium, silicon and calcium in the EDS spectrum is clear evidence of the presence of U-Ca-silicate.

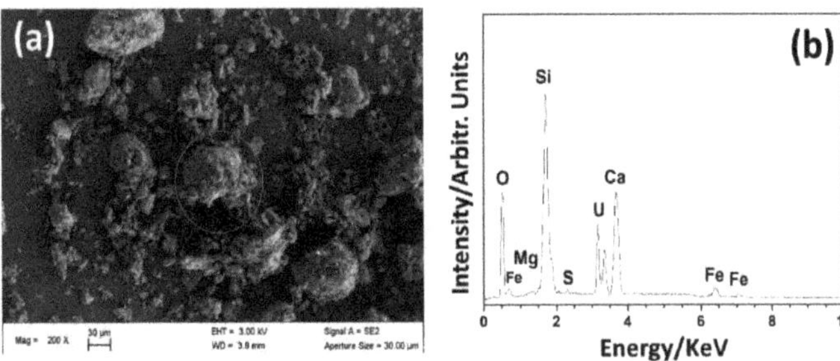

Figure 9. Ca-Uranyl silicate. (**a**) SEM image and (**b**) EDX spectrum.

4. Discussion

Siliceous breccia, or vuggy silica is described from laterite profiles world-wide, such as in the Nickel Mountain, Sierra Nevada, Douglas County, Oregon, or the Klamath Mountains [64,65]. At the Tiébaghi Mine (New Caledonia), the siliceous breccia is widespread along the saprolite, and was formed after saprolitization as it hosts saprolite fragments [20]. High hydraulic conductivity in the saprolite (8.10^{-7}m/s) [41] and meteoric water circulation is the origin of this siliceous breccia. The presence of trace amounts of olivine and the Ca-bearing pyroxene (hedenbergite) suggests additional serpentinization reactions. Saponite is an important compound in this saprolite. It was detected in the silica matrix in this study. Dissolution of saponite generates high amounts of silica: 18 saponite (Mg) + 23 H$_2$O = 19 serpentine + 6 diaspore + 28 SiO$_2$ (aq). This hydration reaction is exothermic [66,67], and may have just let to temperature increase as suggested by [43]. Saprolite is composed of a mixture of serpentines, clay minerals, and oxides. Saponite dissolution is pH dependent. Laboratory experiments have shown that at acidic pH = 3, silica is massively produced. For the Tiébaghi siliceous breccia formation, we suggest that most likely reaction (2) and (3) may be responsible for the silica generation. The dissolution of saprolite leads to the formation of cavities. These cavities are then filled by the porous breccia at Tiébaghi. For the sudden precipitation of massive porous micro-quartz in the saprolite horizon, one possible model is the precipitation due to a sudden change of pH of the aqueous fluids. The water resurgences at the base of the aquifers are basic with a pH up to 11 [68], whereas pH in coarse saprolite level is around ~8 [33].

Before discussing the origin of diamond and the associated strongly reducing phases appearing in the pores of the breccia, we argue against that these phases are artefacts, generated during sample preparation through diamond sawing: (i) the sample preparation was carried out identically for the three other rocks (sandstone, harzburgite, granite), all sawn with the same diamond blade [22]. (ii) all the samples were studied by micro-Raman spectroscopy [22,23], but only the siliceous breccia hosts diamond polymorphs. (iii) the intergrowth of diamond and lonsdaleite is unlikely being a feature for diamond saw derived particles. (iv) clusters of chromite, diamond and SiC polymorph are observed in natural environment [2,5,28,31]. Therefore, the idea of contamination was ruled out.

Diamonds and exotic minerals such as moissanite (SiC) and metal alloys were described from different ophiolite environments (e.g., Luobusa ophiolite (Tibet), in the early Paleozoic Ray-Iz ophiolite (polar Ural, Russia); Pozanti-Karsanti ophiolite, Eastern Tauride belt, southern Turkey, Mao Baracoa Ophiolite, Cuba and the Tehuitzingo ophiolite, Mexico) [2,4,15,16,28,31]. In all these cases, UHP and diamond are associated with peridotites and chromitite. The Tiébaghi nickel laterites were formed on ophiolitic upper mantle serpentinzed peridotites. Serpentinites formed under reducing conditions frequently host metal alloys such as $FeNi_3$ (awaruite), the Fe-rich variety, taenite, or even native Fe [26,69]. Moreover, magnetite and chromite are stable phases under high pH and reducing conditions (fO_2 < FMQ (Fayalite–Magnetite–Quartz–buffer)) [12,65]. Saprolitization, as incipient weathering of serpentinite, still occurs under reducing conditions and alkaline pH (around 8) as serpentine (formed during weathering) and smectite are stable. Both phases are major Fe and Ni carriers. An organic matter and nickel-sulfide rich horizon (so called gley) even occurs at the interface between yellow laterite and saprolite [65].

At Tiébaghi, the largest chromium mine in the world operated in 1941 produced about 54,000 tons per year with an overall production of 3.3 million tons of Cr-rich lumpy chromite within 88 years (1902–1990). Cr-spinel is detected in the siliceous breccia and can be interpreted as relicts having survived serpentinization and saprolitization of the peridotites under alkaline conditions. Based on point analyses, the Cr-spinel detected in the siliceous breccia show high Cr# typical for chromite from boninitic melts. Thi finding is in agreement with studies by [68] showing that lherzolite from at Tiébaghi and Poum presents a refertilized harzburgite having experienced an interaction with a boninitic melt [68].

Associated with the Cr-rich spinels in the siliceous breccia, diamond including lonsdaleite and moissanite as 6O-SiC occur. Different models were proposed to explain the diamond and moissanite occurrence in ophiolites, based on C-isotopic compositions, trace element studies on diamonds and diamond inclusion mineralogy [4,15,16]. Moissanite xenocrysts in chromitite are suggested being originated from a deep mantle source in the Luobusa ophiolite, as chromite crystals represent coesite lamellae [70]. Organic carbon derived from the subduction slab was proposed as C source as the involvement of high temperature fluids, such as second stage melts [71]. New results, based on mineralogical studies on microdiamonds from ophiolites in Cuba and Mexico [4,15,16], and thermodynamic calculation by McCollom and Bach [72] supports a low- pressure origin of microdiamonds.

In nature, lonsdaleite is mainly associated with cubic diamond. It can form under ultra-high pressures (>130 kbar), at depths > 150 km, at temperatures above 950 °C at PO_2 close to IW (Iron-wüstite) [73,74]. Since it is described from meteorites and rocks having experienced shock metamorphism [75], intensive studies have been performed [76]. Lonsdaleite is difficult to analyze as is always tightly intergrown with diamond, sometimes graphite, and is of submicrometric size [76]. These phases often present dislocations, stacking faults, twins, and grain boundary disordering. Therefore, microanalysis and structural interpretations of XRD and TEM-SAED (transmission electron microscopy (TEM), selected area electron diffraction (SAED)) may be erroneous due to fuzzy spectra [77].

In this study, analyzes of spectral data from micro-Raman spectroscopy were compared and supplemented by results from DFT. This methodology allowed unambiguously to identify lonsdaleite and cubic diamond. The mineral association (lonsdaleite, moissanite, chromite, native chromium) observed in the unaltered chromitite and peridotite from the above-mentioned localities is like that observed in the siliceous breccia from the Tiébaghi mine [4]. The moissanite described from peridotites and chromitites in ophiolites are mainly subhedral to anhedral, similarly to those observed in the siliceous breccia pores at Tiébaghi [4].

At Tiébaghi, not only microdiamonds were detected but also lonsdaleite and moissanite Our data set is not sufficient to conclude on the ultimate origin of these reducing phases. However, we suggest that these phases are inherited from the serpentinized peridotites. Further studies must be carried out to support a deep or shallow mantle origin. This is beyond the scope of this study.

The uranyl-silicates are the most abundant group of uranyl minerals because of the ubiquity of dissolved Si in most groundwaters. Uranophane is the most common uranyl mineral, precipitating from near neutral to alkaline groundwaters that contain dissolved Si and Ca. Uranophane is a common neoformed mineral derived from the alteration of uraninite (UO_2) [78] in large surroundings of uranium deposits such as in the *Koongarra* deposit (Australia) [79]. At this location uranium mineralization occurs in breccia and associated faults surrounding the breccia. This is related to the high mobility of uranium in oxidizing environments due to U(VI) formation.

Uranyl silicates are formed in the surface of uraninite under oxidizing conditions [80–83]. However, the direct precipitation of uranophane, as observed in the siliceous breccia in this study, requires a high concentration of Si in solution [80,82]. The stabilization of uranophane in high silica, high pH and Ca^{2+} bearing environments is also documented [80,82].

The origin of the uranium in the siliceous breccia, can be only roughly discussed, as no uranium mineralization is known in New Caledonia. Uranium bearing minerals, such as apatite was detected by micro-Raman analyses in this study. Apatite commonly contains several tens to hundreds of ppm of uranium [83]. However, under the above-mentioned conditions, apatite tends to be stable [84], which makes such an origin unlikely. At Tiébaghi, pegmatitic dykes crosscutting the laterite profile are described [28]. However, no petrographical study on these rocks is published so far. These rocks need to be further investigated as pegmatites may host uranium and uranium bearing accessory minerals.

5. Conclusions

In this paper, we reveal for the first time the association of SiC and diamondpolytypes, chromite and uranophane in a nickel laterite profile at Tiébaghi (New Caledonia), unambiguously defined by Micro Raman spectroscopy and XRD. The diamond and moissanite polytypes are inherited from serpentinized peridotites.

Based on these studies, the breccia at Tiébaghi needs to be further investigated and may present exploration potential, as it acts as a trap for weathering-resistant valuable minerals, and elements migrating in low temperature silica rich environments, such as U, Ni, Cr.

This study also highlights the potential and importance of micro-Raman spectroscopy to achieve the structural properties of porous materials. This technique is thus crucially important for mining companies to rapidly access detailed mineralogical compositions without any sample preparation.

Author Contributions: Conceptualization, Y.E.M., B.O., D.C. and S.G.; methodology, Y.E.M., B.O., D.C., J.-F.B., S.G., S.P. and O.P.; software, Y.E.M., D.C., J.-F.B., S.G., S.P. and O.P.; validation, Y.E.M., B.O. and D.C.; formal analysis, Y.E.M., B.O., D.C., J.-F.B. and S.G.; investigation, Y.E.M., B.O., D.C., J.-F.B., S.G., S.P. and O.P.; resources, Y.E.M., B.O. and D.C.; writing—original draft preparation, Y.E.M.,

D.C., J.-F.B. and S.G.; writing—review and editing, Y.E.M., D.C., J.-F.B. and S.G.; visualization, Y.E.M., D.C., J.-F.B. and S.G.; All authors have read and agreed to the published version of the manuscript.

Funding: This research was funded by the European Commission in the frame of the SOLSA project (H2020 program): SC5-11d-689868.

Data Availability Statement: The experimental and computational data presented in the present paper are available from the corresponding author upon request.

Acknowledgments: We thank the SLN for providing the sample material, and the BRGM staff for preparing the samples.

Conflicts of Interest: The authors declare no conflict of interest.

References

1. Gevorkyan, R.G.; Kaminsky, F.V.; Lunev, V.C.; Osovetsky, V.M.; Nachatryan, N.D. A new occurrence of diamonds in ultramafic rocks in Armenia. *Dokl. AN Armen. SSR* **1976**, *63*, 176–181.
2. Lian, D.; Yang, J.; Dilek, Y.; Wu, W.; Zhang, Z.; Xiong, F.; Liu, F.; Zhou, W. Deep mantle origin and ultra-reducing conditions in podiform chromitite: Diamond, moissanite, and other unusual minerals in podiform chromitites from the Pozanti-Karsanti ophiolite, southern Turkey. *Am. Mineral.* **2017**, *102*, 1101–1113.
3. Moe, K.S.; Yang, J.S.; Johnson, P.; Xu, X.; Wang, W. Spectroscopic analysis of microdiamonds in ophiolitic chromitite and peridotite. *Lithosphere* **2018**, *10*, 133–140. [CrossRef]
4. Pujol-Solà, N.; Proenza, J.A.; Garcia-Casco, A.; González-Jiménez, J.M.; Andreazini, A.; Melgarejo, J.C.; Gervilla, F. An Alternative Scenario on the Origin of Ultra-High Pressure (UHP) and Super-Reduced (SuR) Minerals in Ophiolitic Chromitites: A Case Study from the Mercedita Deposit (Eastern Cuba). *Minerals* **2018**, *8*, 433. [CrossRef]
5. Trumbull, R.B.; Yang, J.S.; Robinson, P.T.; Di Pierro, S.; Vennemann, T.; Wiedenbeck, M. The carbon isotope composition of natural SiC (moissanite) from the Earth's mantle: New discoveries from ophiolites. *Lithos* **2009**, *113*, 612–620. [CrossRef]
6. Gorshkov, A.I.; Titkov, S.V.; Bao, Y.N.; Ryabchikov, I.D.; Magazina, L.O. Microinclusions in diamonds of octahedral habit from kimberlites of Shandong Province, Eastern China. *Geol. Ore Deposit.* **2006**, *48*, 326–334. [CrossRef]
7. Di Pierro, S.; Gnos, E.; Grobety, B.H.; Armbruster, T.; Bernasconi, S.M.; Ulmer, P.J.S. Rock-forming moissanite (natural alpha-silicon carbide). *Am. Mineral.* **2003**, *88*, 1817–1821. [CrossRef]
8. Bai, W.; Robinson, P.T.; Fang, Q.; Yang, J.; Yan, B.; Zhang, Z.; Hu, X.-F.; Zhou, M.-F.; Malpas, J. The PGE and base-metal alloys in the podiform chromitites of the Luobusa ophiolite, southern Tibet. *Can. Mineral.* **2000**, *38*, 585–598. [CrossRef]
9. Xu, S.; Wu, W.; Xiao, W.; Yang, J.; Chen, J.; Ji, S.; Liu, Y. Moissanite in serpentinites from the Dabie Shan Mountains in China. *Mineral. Mag.* **2008**, *72*, 899–908. [CrossRef]
10. Qi, D.; DeYoung, B.J.; Innes, R.W. Structure-Function Analysis of the Coiled-Coil and Leucine-Rich Repeat Domains of the RPS5 Disease Resistance Protein. *Plant Physiol.* **2012**, *158*, 1819–1832. [CrossRef]
11. Bauer, J.; Fiala, J.; Hrichova, R. Natural a-silicon carbide. *Am. Mineral.* **1963**, *48*, 620–634.
12. Schmidt, M.W.; Gao, C.; Golubkova, A.; Rohrbach, A.; Connolly, J.A. Natural moissanite (SiC)—A low temperature mineral formed from highly fractionated ultra-reducing COH-fluids. *Prog. Earth Planet. Sci.* **2014**, *1*, 27–30. [CrossRef]
13. Pearson, D.G.; Davies, G.R.; Nixon, P.H. Geochemical Constraints on the Petrogenesis of Diamond Facies Pyroxenites from the Beni Bousera Peridotite Massif, North Morocco. *J. Petrol.* **1993**, *34*, 125–172. [CrossRef]
14. Davies, G.R.; Nixon, P.H.; Pearson, D.G.; Obata, M. Tectonic implications of graphitized diamonds from the Ronda peridotite massif, southern Spain. *Geology* **1993**, *21*, 471–474. [CrossRef]
15. Pujol-Solà, N.; Garcia-Casco, A.; Proenza, J.A.; González-Jiménez, J.M.; del Campo, A.; Colás, V.; Canals, À.; Sánchez-Navas, A.; Roqué-Rosell, J. Diamond forms during low pressure serpentinisation of oceanic lithosphere. *Geochem. Persp. Let.* **2020**, *15*, 19–24. [CrossRef]
16. Farré-de-Pablo, J.; Proenza, J.A.; González-Jiménez, J.M.; Garcia-Casco, A.; Colás, V.; Roqué-Rossell, J.; Camprubí, A.; Sánchez-Navas, A. A shallow origin for diamonds in ophiolitic chromitites. *Geology* **2019**, *47*, 75–78. [CrossRef]
17. El Mendili, Y.; Orberger, B.; Chateigner, D.; Bardeau, J.-F.; Gascoin, S.; Petit, S.; Perez, O.; Khadraoui, F. Insight into the structural, elastic and electronic properties of a new orthorhombic 6O-SiC polytype. *Sci. Rep.* **2019**, *10*, 7562. [CrossRef] [PubMed]
18. Lovering, T.G. *Radioactive Deposits in New Mexico*; Geological Survey Bulletin: Reston, VA, USA, 1956; pp. 315–390.
19. Min, M.; Fang, C.; Fayek, M. Petrography and genetic history of coffinite and uraninite from the Liueryiqi granite-hosted uranium deposit, SE China. *Ore Geol. Rev.* **2005**, *26*, 187–197. [CrossRef]
20. Shukla, M.K.; Sharma, A. A brief review on breccia: It's contrasting origin and diagnostic signatures. *Solid. Earth Sci.* **2018**, *3*, 50–59. [CrossRef]
21. Leonardos, O.H.; Fernandes, S.M.; Fyfe, W.S.; Powell, M. The mciro-chemistry of uraniferous laterites from Brazil: A natural example of inorganic chromatography. *Chem. Geol.* **1987**, *60*, 111–119. [CrossRef]
22. Duée, C.; Orberger, B.; Maubec, N.; Laperche, V.; Capar, L.; Bourguignon, A.; Bourrat, X.; Mendili, Y.E.I.; Chateigner, D.; Gascoin, S.; et al. Impact of heterogeneities and surface roughness on pXRF, pIR, XRD and Raman analyses: Challenges for on-line, real-time combined mineralogical and chemical analyses on drill cores and implication for "high speed" Ni-laterite exploration. *J. Geochem. Explor.* **2019**, *198*, 1–17. [CrossRef]

23. El Mendili, Y.; Chateigner, D.; Orberger, B.; Gascoin, S.; Bardeau, J.-F.; Petit, S.; Duee, C.; Guen, M.L.; Pilliere, H. Combined XRF, XRD, SEM-EDS, and Raman Analyses on Serpentinized Harzburgite (Nickel Laterite Mine, New Caledonia): Implications for Exploration and Geometallurgy. *ACS Earth Space Chem.* **2019**, *3*, 2237–2249. [CrossRef]

24. Cassard, D.; Nicolas, A.; Rabinovitch, M.; Moutte, J.; Leblanc, M.; Prinzhofer, A. Structural classification of chromite pods in New Caledonia. *Econ. Geol.* **1981**, *76*, 805–881. [CrossRef]

25. Marchesi, C.; Garrido, C.J.; Godard, M.; Belley, F.; Ferré, E. Migration and accumulation of ultra-depleted subduction-related melts in the Massif du Sud ophiolite (New Caledonia). *Chem. Geol.* **2009**, *266*, 171–186. [CrossRef]

26. Prinzhofer, A.; Allègre, C. Residual peridotite and the mechanism of partial melting. *Earth Planet. Sci. Let.* **1985**, *74*, 251–265. [CrossRef]

27. Moutte, J. Le massif de Tiebaghi, Nouvelle-Calédonie, et ses gites de chromite. Ph.D. Thesis, Ecole Nationale Supérieure des Mines de Paris, Paris, France, 1979.

28. Bailly, L.; Ambrosi, J.P.; Barbarand, J.; Beauvais, A.; Cluzel, D. Nickel-Typologie des latérites de Nouvelle-Calédonie. Gisements de nickel latéritique, volume II. In *Rapport de Recherche: Tome Nickel et Technologie*; HAL: Paris, France, 2014; pp. 1–448.

29. Sevin, B.; Ricordel-Prognon, C.; Quesnel, F.; Cluzel, D.; Lesimple, S.; Maurizot, P. First palaeomagnetic dating of ferricrete in New Caledonia: New insight on the morphogenesis and palaeoweathering of 'Grande Terre'. *Terra Nova* **2012**, *24*, 77–85. [CrossRef]

30. Mc Rae, E. *Nickel Statistics and Information*; Annual Publication; USGS: Reston, VA, USA, 2018.

31. Trescases, J.J. The lateritic nickel-ore deposits. In *Soils and Sediments Mineralogy and Geochemistry*; Paquet, H., Clauer, N., Eds.; Springer: Berlin, Germany, 1997; pp. 125–138.

32. Cathelineau, M.; Myagkiy, A.; Quesnel, B.; Boiron, M.-C.; Gautier, P.; Boulvais, P.; Ulrich, M.; Truche, L.; Golfier, F.; Drouillet, M. Multistage crack seal vein and hydrothermal Ni enrichment inserpentinized ultramafic rocks (Koniambo massif, New Caledonia). *Miner. Depos.* **2017**, *52*, 945–960. [CrossRef]

33. Dublet, G.; Juillot, F.; Morin, G.; Fritsch, E.; Fandeur, D.; Ona-Nguema, G.; Brown, G.E. Ni speciation in a New Caledonian lateritic regolith: A quantitative. X-ray absorption spectroscopy investigation. *Geochim. Cosmochim. Acta* **2012**, *95*, 119–133. [CrossRef]

34. Manceau, A.; Calas, G.; Decarrau, A. Nickel-bearing clay minerals: I. Optical study of nickel crystal chemistry. *Clay Miner.* **1985**, *20*, 367–487. [CrossRef]

35. Cluzel, D.; Meffre, S.; Maurizot, P.; Crawford, A.J. Earliest Eocene (53 Ma) convergence in the Southwest Pacific: Evidence from pre-obduction dikes in the ophiolite of New Caledonia. *Terra Nova* **2006**, *18*, 395–402. [CrossRef]

36. Jeanpert, J.; Lesimple, S.; Sevin, B.; Maurizot, P.; Robineau, B.; Maréchal, J.-C.; Dewandel, B. Exploration des Galeries Chromical-Massif de Tiébaghi. In *Observations Géologiques et Hydrogéologiques à L'intérieur D'un Massif de Péridotites*; SGNC/DIMENC: New Caledonia, France, 2015; pp. 1–35.

37. Lagabrielle, Y.; Chauvet, A.; Ulrich, M.; Guillot, S. Passive obduction and gravity-driven emplacement of large ophiolitic sheets: The New Caledonia ophiolite (SW Pacific) as a case study? *Bull. Soc. Géol. Fr.* **2013**, *6*, 545–556. [CrossRef]

38. Maurizot, P. Formations miocènes de Népoui. In *Compilation des Connaissances*; BRGM-SGNC: Nouméa, France, 2011; pp. 1–96.

39. Maurizot, P.; Cluzel, D. Pre-obduction records of Eocene foreland basins in central New Caledonia: An appraisal from surface geology and Cadart-1 borehole data. *New Zealand J. Geol. Geophys.* **2014**, *57*, 300–311. [CrossRef]

40. Camus, H.; Leveneur, D.; Bart, F. Structuration karstique des aquifères dans les massifs ophiolitiques de Nouvelle-Calédonie. In *Aquifères de Socle: Le Point Sur Les Concepts et Les Applications Opérationnelles*; La Roche-sur-Yon, Pays de la Loire, France, 2015.

41. Jeanpert, J. Structure et fonctionnement hydrogéologiques des massifs de péridotites d Nouvelle-Calédonie. Ph.D. Thesis, Université de la Réunion, Saint-Denis, France, 2017.

42. Trittschack, R.; Grobéty, B.; Koch-Müller, M. In situ high-temperature Raman and FTIR spectroscopy of the phase transformation of lizardite. *Am. Mineral.* **2012**, *97*, 1965–1976. [CrossRef]

43. Guillou-Frottier, L.; Beauvais, A.; Bailly, L.; Wyns, R.; Augé, T.; Audion, A.S. Transient hydrothermal corrugations within mineralized ultramafic laterites. In Proceedings of the SGA Biannual conference, Nancy, France, 24–27 August 2015.

44. El Mendili, Y.; Vaitkus, A.; Merkys, A.; Gražulis, S.; Chateigner, D.; Mathevet, F.; Gascoin, S.; Petit, S.; Bardeau, J.-F.; Zanatta, M.; et al. Raman Open Database: First interconnected Raman-XRD open-access resource for material identification. *J. Appl. Cryst.* **2019**, *52*, 618–625. [CrossRef]

45. Le Page, Y.; Donnay, G. Refinement of the Crystal Structure of Low-Quartz. *Acta Crystallogr. Sect. B Struct. Crystallogr. Cryst. Chem.* **1976**, *32*, 2456–2459. [CrossRef]

46. Neeway, J.; Abdelouas, A.; Ribet, S.; David, K.; El Mendili, Y.; Grambow, B.; Schumacher, S. Effect of Callovo-Oxfordian clay rock on the dissolution rate of the SON68 simulated nuclear waste glass. *J. Nucl. Mater.* **2015**, *459*, 291–300. [CrossRef]

47. Čermáková, Z.; Hradil, D.; Bezdička, P.; Hradilová, J. New data on "kerolite–pimelite" series and the colouring agent of Szklary chrysoprase, Poland. *Phys. Chem. Miner.* **2017**, *44*, 193–202. [CrossRef]

48. Cheung, R. *Silicon Carbide Microelectromechanical Systems for Harsh Environments*; Imperial College Press: London, UK, 2006.

49. Lu, Y.P.; He, D.W.; Zhu, J.; Yang, X.D. First-principles study of pressure-induced phase transition in silicon carbide. *Phys. B.* **2008**, *403*, 3543–3546. [CrossRef]

50. Nakashima, S.; Harima, H. Raman Investigation of SiC Polytypes. *Phys. Status Solidi* **1997**, *162*, 39–64. [CrossRef]

51. Frondel, C.; Marvin, U.B. Lonsdaleite, a hexagonal polymorph of diamond. *Nature* **1967**, *214*, 587–589. [CrossRef]

52. Spear, K.E.; Phelps, A.W.; White, W.B. Diamond polytypes and their vibrational spectra. *J. Mater. Res.* **1990**, *5*, 2277–2285. [CrossRef]

53. Kraus, D.; Ravasio, A.; Gauthier, M.; Gericke, D.O.; Vorberger, J.; Frydrych, S.; Helfrich, J.; Fletcher, L.B.; Schaumann, G.; Nagler, B.; et al. Nanosecond formation of diamond and lonsdaleite by shock compression of graphite. *Nat. Commun.* **2016**, *7*, 10970. [CrossRef]

54. Smith, D.C.; Godard, G. UV and VIS Raman spectra of natural lonsdaleites: Towards a recognised standard. *Spectrochim. Acta Part A Mol. Biomol. Spectrosc.* **2009**, *73*, 428–435. [CrossRef]
55. Nishitani-Gamo, M.; Sakaguchi, I.; Loh, K.P.; Kanda, H.; Ando, T. Confocal Raman spectroscopic observation on hexagonal diamond formation from dissolved carbon in nickel under chemical vapor deposition conditions. *Appl. Phys. Lett.* **1998**, *73*, 765–767. [CrossRef]
56. Goryainov, S.V.; Likhacheva, A.Y.; Rashchenko, S.V.; Shubin, A.S.; Afanas'ev, V.P.; Pokhilenko, N.P. Raman identification of lonsdaleite in Popigai impactites. *J. Raman Spectrosc.* **2014**, *45*, 305–313. [CrossRef]
57. Wu, B.R. Structural and vibrational properties of the 6H diamond: First-principles study. *Diam. Relat. Mater.* **2007**, *16*, 21–28. [CrossRef]
58. Denisov, V.N.; Mavrin, B.N.; Serebryanaya, N.R.; Dubitsky, G.A.; Aksenenkov, V.V.; Kirichenko, A.N.; Kuzmin, N.V.; Kulnitskiy, B.A.; Perezhogin, I.A.; Blank, V.D. First-principles, UV Raman, X-ray diffraction and TEM study of the structure and lattice dynamics of the diamond–lonsdaleite system. *Diam. Relat. Mater.* **2011**, *20*, 951–953. [CrossRef]
59. Dovesi, R.; Orlando, R.; Erba, A.; Zicovich-Wilson, C.M.; Civalleri, B.; Casassa, S.; Maschio, L.; Ferrabone, M.; De La Pierre, M.; D'Arco, P.; et al. CRYSTAL14: A program for the ab initio investigation of crystalline solids. *Int. J. Quantum Chem.* **2014**, *114*, 1287–1317. [CrossRef]
60. Reddy, B.J.; Frost, R.L. Spectroscopic characterization of chromite from the Moa-Baracoa Ophiolitic Massif, Cuba. *Spectrochim. Acta Part A: Mol. Biomol. Spectrosc.* **2005**, *61*, 1721–1728. [CrossRef]
61. Driscoll, R.J.P.; Wolverson, D.; Mitchels, J.M.; Skelton, J.M.; Parker, S.C.; Molinari, M.; Khan, I.; Geeson, D.; Allen, G.C. Raman spectroscopic study of uranyl minerals from Cornwall, UK. *RSC. Adv.* **2014**, *4*, 59137–59149. [CrossRef]
62. Bonales, L.J.; Menor-Salván, C.; Cobos, J. Study of the alteration products of a natural uraninite by Raman spectroscopy. *J. Nucl. Mater.* **2015**, *462*, 296–303. [CrossRef]
63. Frost, R.L.; Cejka, J.; Weier, M.L.; Martens, W. Molecular structure of the uranyl silicates—A Raman spectroscopic study. *J. Raman Spectrosc.* **2006**, *37*, 538–551. [CrossRef]
64. Alexander, E.B.; Coleman, R.G.; Keeler-Wolfe, T.; Harrison, S.P. *Serpentine Geoecology of Western North America: Geology, Soils and Vegetation*; Oxford University Press: Oxford, UK, 2007.
65. Frost, B.R.; Beard, J. On silica activity and serpentinization. *J. Petrol.* **2007**, *48*, 1351–1368. [CrossRef]
66. Fyfe, W.S. Heats of chemical reactions and submarine heat production. *Geophys. J. Roy. Astron. Soc.* **1974**, *37*, 213–215. [CrossRef]
67. Monin, C.; Chavagnac, V.; Boulart, C.; Ménez, B.; Gérard, M.; Gérard, E.; Quemeneur, M.; Erauso, G.; Postec, A.; Guentas-Dombrowski, L.; et al. The low temperature hyperalkaline hydrothermal system of the Prony Bay (New Caledonia). *Biogeosci. Discuss.* **2014**, *11*, 6221–6267.
68. Ulrich, M.; Picard, C.; Guillot, S.; Chauvel, C.; Cluzel, D.; Meffre, S. Multiple melting stages and refertilization as indicators for ridge to subduction formation: The New Caledonia ophiolite. *Lithos* **2010**, *115*, 223–236. [CrossRef]
69. Chamberlain, J.A.; McLeod, C.R.; Traill, R.J.; Lachance, G.R. Native metals in the Muskox Intrusion. *Can. J. Earth Sci.* **1965**, *2*, 188–215. [CrossRef]
70. Yamamoto, S.; Komiya, T.; Hirose, K.; Maruyama, S. Coesite and clinopyroxene exsolution lamellae in chromitites: In-situ ultrahigh-pressure evidence from podiform chromitites in the Luobusa ophiolite, southern Tibet. *Lithos* **2009**, *109*, 314–322. [CrossRef]
71. Cartigny, P. Stable Isotopes and the Origin of Diamond. *Elements* **2005**, *1*, 79–84. [CrossRef]
72. Bouilhol, P.; Debret, B.; Inglis, E.C.; Warembourg, M.; Grocolas, T.; Rigaudier, T.; Villeneuve, J.; Burton, K.W. Decoupling of inorganic and organic carbon during slab mantle devolatilisation. *Nature. Comm.* **2022**, *13*, 308. [CrossRef]
73. McCollom, T.M.; Bach, W. Thermodynamic constraints on hydrogen generation during serpentinization of ultramafic rocks. *Geochim. Cosmochim. Acta* **2009**, *73*, 856–875. [CrossRef]
74. Stagno, V.; Frost, D.J. Carbon speciation in the asthenosphere: Experimental measurements of the redox conditions at which carbonate-bearing melts coexist with graphite or diamond in peridotite assemblages. *Earth Planet. Sci. Lett.* **2010**, *300*, 72–84. [CrossRef]
75. Bundy, T.P.; Kasper, J.S. Hexagonal Diamond-A New Form of Carbon. *Chem. Phys.* **1967**, *46*, 3437–3446. [CrossRef]
76. Daulton, T.L.; Amari, S.; Scott, A.C.; Hardiman, M.; Pinter, N.; Anderson, R.S. Comprehensive Analysis of Nanodiamond Evidence Relating to the Younger Drays Impact Hypothesis. *J. Quat. Sci.* **2017**, *32*, 7–34. [CrossRef]
77. Cayron, C.; Hertog, M.D.; Latu-Romain, L.; Mouchet, C.; Secouard, C.; Rouviere, J.-L.; Rouviere, E.; Simonato, J.-P. Odd electron diffraction patterns in silicon nanowires and silicon thin films explained by microtwins and nanotwins. *J. Appl. Cryst.* **2008**, *42*, 242–252. [CrossRef] [PubMed]
78. Moore, R.O.; Gurney, J.J. *Proceedings of the Fourth International Kimberlite Conference, Kimberlites and Related Rocks*; Their Mantle/Crust Setting, Diamonds and Diamond Exploration; Geological Society of Australia Special Publication 14; Blackwell Scientic: Cambridge, UK, 1989; pp. 1029–1041.
79. Mudd, G.M. *Compilation of Uranium Production History and Uranium Deposit Data Across Australia*; SEA-US Inc.: Melbourne, Australia, 2006; pp. 1–46.
80. Finch, R.; Ewing, R. The corrosion of uraninite under oxidizing conditions. *J. Nucl. Mater.* **1992**, *190*, 133–156. [CrossRef]
81. Pérez, I.; Casas, I.; Martín, M.; Bruno, J. The thermodynamics and kinetics of uranophane dissolution in bicarbonate test solutions. *Geochim. Cosmochim. Acta* **2000**, *64*, 603–608. [CrossRef]
82. Shvareva, T.Y.; Mazeina, L.; Gorman-Lewis, D.; Burns, P.C.; Szymanowski, J.E.; Fein, J.B.; Navrotsky, A. Thermodynamic characterization of boltwoodite and uranophane: Enthalpy of formation and aqueous solubility. *Geochim. Cosmochim. Acta* **2011**, *75*, 5269–5282. [CrossRef]

83. Skinner, H.C.W.; Jahren, A.H. Biomineralization. In *Treatise on Geochemistry*; Holland, H.D., Turekian, K.K., Eds.; Elsevier: Amsterdam, The Netherlands, 2003; Volume 8, pp. 1–69.
84. Ayers, J.C.; Watson, E.B. Solubility of Apatite, Monazite, Zircon, and Rutile in Supercritical Aqueous Fluids with Implications for Subduction Zone Geochemistry. *Philos. Trans. Phys. Sci. Eng.* **1991**, *335*, 365–375.

MDPI

St. Alban-Anlage 66

4052 Basel

Switzerland

www.mdpi.com

Minerals Editorial Office

E-mail: minerals@mdpi.com

www.mdpi.com/journal/minerals